AF389474

Advances in Theoretical and Computational Energy Optimization Processes

Advances in Theoretical and Computational Energy Optimization Processes

Volume 2

Editors

Ferdinando Salata
Iacopo Golasi

MDPI • Basel • Beijing • Wuhan • Barcelona • Belgrade • Manchester • Tokyo • Cluj • Tianjin

Editors
Ferdinando Salata
University of Rome "Sapienza"
Italy

Iacopo Golasi
Sapienza University of Rome I
taly

Editorial Office
MDPI
St. Alban-Anlage 66
4052 Basel, Switzerland

This is a reprint of articles from the Special Issue published online in the open access journal *Processes* (ISSN 2227-9717) (available at: https://www.mdpi.com/journal/processes/special_issues/energt_optimization).

For citation purposes, cite each article independently as indicated on the article page online and as indicated below:

LastName, A.A.; LastName, B.B.; LastName, C.C. Article Title. *Journal Name* **Year**, *Article Number*, Page Range.

Volume 2
ISBN 978-3-03936-682-8 (Hbk)
ISBN 978-3-03936-683-5 (PDF)

Volume 1-2
ISBN 978-3-03936-684-2 (Hbk)
ISBN 978-3-03936-685-9 (PDF)

Contents

About the Editors

Ferdinando Salata was born in Rome in 1977, and is currently a researcher at the Department of Astronautical Engineering, Electrical and Energy (DIAEE) for the Scientific Sector ING-IND/11, at the University of Rome "Sapienza" (Italy) (according to the Italian university). He earned his degree in mechanical engineering, specialising in energy, at the University of Rome "Sapienza" in 2003, and went on to complete his PhD in "Technical Physic" in 2007 at the same university. For his thesis, he studied the use of ultraviolet radiation, coupled with HEPA filtration, for the disinfection of biological airborne contaminants in air conditioning systems. After completing his PhD, Ferdinando accepted a research grant from the Department where he completed his thesis. During this period, Ferdinando was named "Expert in the field" for the Scientific Sector ING-IND/11, at the Faculties of Civil and Industrial Engineering of the University of Rome "Sapienza". In 2017, he completed his national academic qualification (required for becoming an associate professor) at the "Ministry of University and Research". That same year, he joined DIAEE as an assistant professor. He is a member of the National Association of Italian Technical Physics. In recent years, he has studied CHP systems; the energy demand optimization of buildings; the energy and reliability optimization of conditioning and lighting systems; natural ventilation in buildings; desalination through absorption machines; urban microclimate and outdoor thermal comfort; thermal conductivity in soils. He has the role of docent (at Faculty of Architecture of University of Rome "Sapienza") for the course "Environmental Applied Physics" and teaching assistant (at Faculty of Civil and Industrial Engineering of University of Rome "Sapienza") for the courses: "Applied Physics" for Electrical Engineering, "Environmental Applied Physics" for Building Engineering—Architecture. He taught a course on "Energy certification of buildings" on behalf of the Lazio Region and on behalf of the Kyoto Club Italia. He is, and has been, the tutor or co-tutor of several MSc thesis dissertations at his faculty. He is a member of the Academic Board of the Ph.D. in Energy and Environment at DIAEE and a member of the International Advisory Board for the *Thermal Science Journal, Journal of Daylighting, Atmosphere Journal, Sustainable Cities and Society Journal*.

Iacopo Golasi was born in Rome in 1987 and achieved, in March 2014, his MSc, with honors in mechanical engineering, from the University of "Roma Tre". He completed his doctorate in "Energy and Environment" in the Department of Astronautics, Electrical and Energetics Engineering (DIAEE) at the University of Roma "Sapienza". At present, his interests lie in the development of new empirical outdoor comfort indices able to predict the thermal perception of people in the analysis of the influence of materials' thermo-physical properties on outdoor thermal comfort and buildings' energy demands, in the evaluation of natural ventilation inside buildings (with particular focus on innovative solutions as the solar chimney), in the buildings' energy efficiency assessment, through building a dynamic simulation analysis, also considering the inter-building effect and in the analysis of lighting installations with LED-type light sources (performing an energy, economic and maintenance comparison with traditional installations). He provides consulting services to private and public companies, such as Aeroporti di Roma S.p.A., Lamaro Appalti S.p.A. and ANCI (National Association of Italian Municipalities). He has been the co-tutor of various thesis dissertations, covering different topics related to the subjects of Applied Physics, Thermotechnics, Lighting and Thermodynamics. He is the co-author of more than 40 papers published in international peer review journals or presented at conferences.

 *processes* 

Article

Off-Grid Solar PV Power Generation System in Sindh, Pakistan: A Techno-Economic Feasibility Analysis

Li Xu [1,2], Ying Wang [1], Yasir Ahmed Solangi [1,*], Hashim Zameer [1] and Syed Ahsan Ali Shah [1]

[1] College of Economics and Management, Nanjing University of Aeronautics and Astronautics, Nanjing 211106, China; shirley.xu@nuaa.edu.cn (L.X.); yingwang@nuaa.edu.cn (Y.W.); hashimzameer@nuaa.edu.cn (H.Z.); ahsan.shah1@hotmail.com (S.A.A.S.)
[2] College of Finance, Jiangsu Vocational Institute of Commerce, Nanjing 211168, China
* Correspondence: yasir.solangi86@hotmail.com; Tel.: +86-186-5185-2672

Received: 6 April 2019; Accepted: 17 May 2019; Published: 22 May 2019

Abstract: The off-grid solar photovoltaic (PV) system is a significant step towards electrification in the remote rural regions, and it is the most convenient and easy to install technology. However, the strategic problem is in identifying the potential of solar energy and the economic viability in particular regions. This study, therefore, addresses this problem by evaluating the solar energy potential and economic viability for the remote rural regions of the Sindh province, Pakistan. The results recommended that the rural regions of Sindh have suitable solar irradiance to generate electricity. An appropriate tilt angle has been computed for the selected rural regions, which significantly enhances the generation capacity of solar energy. Moreover, economic viability has been undertaken in this study and it was revealed that the off-grid solar PV power generation system provides electricity at the cost of Pakistani Rupees (PKR) 6.87/kWh and is regarded as much cheaper than conventional energy sources, i.e., around PKR 20.79/kWh. Besides, the off-grid solar PV power generation system could mitigate maximum CO_2 annually on the condition that all of the selected remote rural regions adopt the off-grid solar PV system. Therefore, this study shall help the government to utilize the off-grid solar PV power generation system in the remote rural regions of Pakistan.

Keywords: off-grid Solar PV power generation; remote rural regions; economic feasibility; CO_2 mitigation; Pakistan

---

1. Introduction

Electricity is the main source for economic, environmental, and social growth of any country. Electricity is considered to be an ideal invention of humankind and has brought a lot of changes in human lives and society. Nevertheless, approximately 1.1 billion people of the earth are suffering or living without electricity [1]. The majority of the population suffering from this situation are located in rural areas of South-Asia and Sub-Saharan Africa [2]. Similarly, a large proportion of Pakistan are living in rural regions, and the majority of them do not have access to electricity. Pakistan is a developing country facing economic, environmental, and social development challenges which have led the country to an increased power demand. The country's total power demand is 25,000 Megawatts (MW) and this is estimated to be boosted up 40,000 MW by 2030 [3]. Whereas, the electricity supply remains around 17,000 MW, causing an electricity shortage of 8000 MW in the country [4]. In the results, the electricity shortfalls in both urban and rural areas around 12 to 18 h a day [5]. Furthermore, the condition of the remote rural regions of Sindh is very bad, where electricity remains inaccessible for many days.

Pakistan is enriched with a vast potential of energy sources such as oil, gas, coal, and renewable energy (i.e., solar, wind, hydro, and biomass). The estimated potential to generate electricity from solar energy is 2900 Gigawatts (GW), wind energy (346 GW), hydropower (6 GW), and biomass energy (5 GW) [6]. The province of Sindh is also enriched with renewable energy (RE) sources and the government needs to tap RE to generate electricity [7]. However, most of the rural regions do not have an electricity facility. Forty-eight percent of the population of the Sindh province are living in rural regions, and approximately 13,451 villages are un-electrified [8]. These villages are scattered near and far from the on-grid station, thus connecting to the grid is uneconomic and expensive. The demand for electricity in the rural regions is low when compared to urban areas, from only 50 to 100 Watts (W) per household [9]. Only a small number of lights and one to two fans are required in rural houses because each house is very small and generally built with one room. Providing on-grid transmission to these villages for such a minimum load is expensive and therefore, there is a very minimum chance of grid-connected electricity in the near future. Likewise, electricity generated from diesel generators does not propose an economical option because it is difficult to transport oil to remote rural regions, as well as ineffective for the environment. Pakistan has a structured energy sector for both international and local stakeholders. Moreover, the stakeholders are unwilling to invest in and participate in RE technologies due to high-investment cost, high-discount rates, short-back period requirements, the lack of infrastructural conditions, remoteness regions, and unavailability of the specific region's potential [10]. Currently, due to the worsening economic condition in Pakistan, the government has also called to shut-down all of the on-going RE projects in the Sindh and the Khyber Pakhtunkhwa (KPK) provinces, and this decision negatively affects the development of RE sources and over three billion dollars in investments [11]. The province of Sindh suffers the most from this government decision as its 53 projects are in-progress.

On the above-stated factors, the off-grid solar energy is the best option to generate electricity for rural regions of the province. The regions of the Sindh province receive a high amount of solar radiance throughout the year [12]. The province has enormous potential for solar energy and receives high solar irradiation, with more than 300 sunlight days with about 1800–2200 kWh/m^2 annual global horizontal irradiation [13]. Furthermore, the Asian Development Bank recommended that the off-grid solar photovoltaic (PV) is the best option, as it is easy to install, low-cost, and increases the socio-economic conditions of the rural regions [9,14]. Various studies have proposed the off-grid solar PV system solution to provide electricity in the rural regions [15,16]. Moreover, solar PV evades extra costs, fuel transportation, and makes the project simple by installing on-site resources. In reference [17], it was presented that off-grid solar PV is an appropriate and sustainable choice for rural electrification due to its life-cycle cost, net energy, and local environmental benefits. In another study [18], the authors identified that the development of the solar PV system improves the living standard of the people and also increases the economic and social conditions in the region. The solar PV system is very favorable for the environment because it has no noise impact, mitigates CO_2 emissions, and does not harm human health [19]. Moreover, numerous other studies, such as [20–22], have presented that the off-grid solar PV system is a significant application for electrification and is an economically viable option for the rural regions. In the US, the residential sector has built energy consumption-related heating and air-conditioning, which make-up of a total of 42% of a buildings total energy use [23]. In another study, the authors have assessed the wind energy potential to generate renewable hydrogen energy in the Sindh province [24]. The planning is the most important aspect for energy management and sustainable development, such as social, environmental, and economic [25].

For achieving the target of providing electricity to rural regions, there should be the proper policies implemented for solar PV power generation system. Extensive research is required to evaluate the particular regions for identifying solar energy potential, as well as to assess the economic viability of the regions. To the best of the authors' knowledge, no such research has been conducted for the Sindh province. Thus, this study aims to fill this research gap. In the study, five rural regions of the Sindh province, i.e., Panoaqil, Badin, Nawabshah, Mirpurkhas, and Kambar, are undertaken to investigate the solar energy potential for electricity generation. The main objectives of the study are:

- To evaluate the techno-economic feasibility of an off-grid solar PV system of five regions of Sindh
- To electrify the above-mentioned rural regions by an off-grid solar PV system

Therefore, this study shall help policy and decision-makers to establish solar PV power generation system rural programs in Sindh and also support unwilling stakeholders to invest by providing comprehensive techno-economic analysis. This study is a way forward for developing off-grid solar PV system in the rural regions of Sindh, Pakistan.

2. Electricity Background in Sindh Province

Sindh is the third largest province by area, and the second largest in terms of the population in Pakistan [26]. The location of the Sindh province holds strategic importance due to its long coastal line, as presented in Figure 1. The Karachi port also provides the best, most economical, and shortest route to the neighboring countries for transferring cargo. The geographical location of the port is very significant, thus it has attracted foreign investment, development projects, and overall contributes to both business and economic growth [27]. Therefore, the on-going projects have rapidly increased the electricity demand in Sindh.

Figure 1. Sindh province map [28].

The energy demand is increasing day by day, which results in a huge electricity shortfall in the country [29]. The Sindh province is being the most affected by the increasing electricity deficit as they face a fresh series of load-shedding between 2 to 17 h a day [30]. This situation is even worse in remote rural regions of Sindh, where power is inaccessible for many days. Moreover, the electricity consumption in the rural regions is considered very low, and the transmission lines are a long distance from the rural areas. Thus, it is considered as cost-intensive.

Pakistan has an estimated 2900 GW solar energy potential, however, this renewable source is still waiting to be harnessed [31]. Figure 2 presented the major share in the electricity generation comprised of the gas of 33.6%, oil 32.1%, hydropower 26.1%, nuclear 5.7%, renewable energy 2.2%, and coal 0.2%, respectively [32].

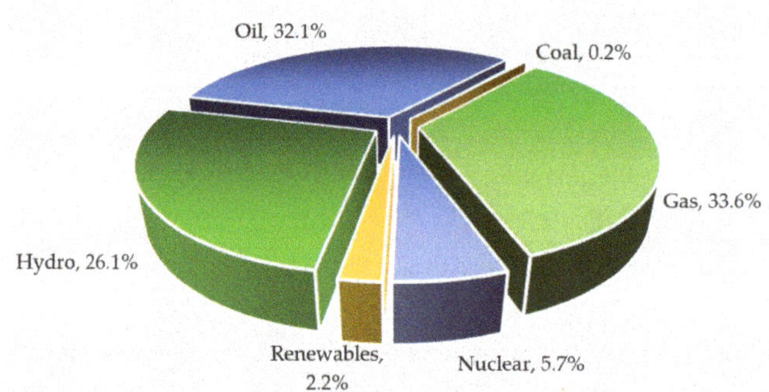

Figure 2. Energy mix of Pakistan.

The Sindh province is rich with renewable energy (RE) sources, such as wind, solar, mini-hydro, and biomass which could be easily utilized for electricity generation. However, the government of Sindh has not taken issues for the development of RE sources seriously, despite increasing demand for electricity. The government had planned a RE policy in 2006, but it is in the infancy stage due to the lack of interest of the government in exploiting these natural resources. Investors are worried and unwilling to invest in the remote rural regions of Sindh since a worse law and order situation, no infrastructure, and a low return on investment are the key factors behind obstructing private investment. However, recently the World Bank has announced that they will finance $100 million worth of loans for the installation of clean energy in Sindh, the target is to provide off-grid solar PV electricity to 200,000 households, equal to 1.2 million people [30].

Therefore, this study will help government, policymakers, and stakeholders in the implementation of solar PV projects in the rural regions of Sindh.

2.1. Solar PV Power Generation Progress in Remote Rural Regions

Pakistan has installed a small number of solar PV projects in the country, and the first solar PV project was installed in the 1980s. However, the project failed due to the lack of managerial and to technical mistakes [33]. Afterward, until 2005, the country did not develop and promote any RE-based project. Later in 2006, two organizations were established, the Alternative Energy Development Board (AEDB) and the Pakistan Commission of Renewable Energy Technologies (PCRET) to promote and develop RE resources for electricity production [34]. AEDB intends to install a solar PV system in 906 houses of rural regions [35]. Furthermore, the government has understood the advantage of a solar energy framework for enriching socio-economic development and saving the environment in rural regions.

2.2. Solar PV Power Generation Issues in Remote Rural Regions

Solar PV is the appropriate option for providing electricity to off-grid rural regions because of the low-cost technology, easy installation and being environmentally benign. Whereas the development of the solar PV system is substantially very low in rural villages of Pakistan, according to the National Electric Power Regulatory Authority (NEPRA), 40,000 villages in the country do not have access to electricity [36]. There must be robust coordination among organizations is required for a successful solar PV rural electrification programs [37]. Before 2006, no organization was established for developing and planning RE projects, PCRET and AEDB were established in 2006 to coordinate and develop plans for the installation of RE projects in Pakistan. Unfortunately, the progress of both organizations for the development of RE is very poor.

The government has failed to develop and plan innovate strategies and policies for the solar PV system in rural regions, and solar energy productions have failed to take-off, regardless of the electricity crisis in Pakistan [38]. The common users put themselves at risk by choosing a solar energy solution as an alternative energy. The high up-front cost is also an interruption in the development of solar PV technology. Moreover, the cost of solar PV is significantly higher compared to that in developed countries [31]. In the finance bill 2014–15, the government implemented a 32% tariff on the import of solar PV panels, which results in the low progress of solar PV. Therefore, the government took its decision back and reduced tariffs on solar panels. Despite tariffs on solar inverters, tariffs on batteries are still existing with around 50%. Additionally, the government failed to provide incentives to households on the installation of a solar PV system, which shows the lack of government policies for both investors and customers [39].

3. Research Framework

The research framework of the study has been divided into several sub-sections, which are briefly described as follows:

3.1. Determining the Solar Energy Potential

The average peak solar hours are used to identify and determine the solar irradiation in a particular region when the sunshine at its maximum value for a certain number of hours. The peak solar irradiation is 1 kW/m^2, the peak hours of sun are equal to the daily solar irradiation in kWh/m^2. For example, the daily solar array output can be projected to be 545 Wh, if we assume that a 100 Wp solar array is installed in the Panoaqil region with an average solar irradiation of 5.45 kWh/m^2/day. Therefore, the annual energy output can be computed for monitoring the PV system performance by using the Equation (1) [40]:

$$\text{Annual energy output } \left(\tfrac{\text{kWh}}{\text{kWp}}\right)$$
$$= \text{Global inplane irradition} \left(\left(\text{kWh/m}^2\right)/\text{year}\right) \times \text{Performance ratio} \tag{1}$$

3.2. Solar Irradiation and Determining the Optimal Tilt Angle

The solar irradiation is generally measured on a horizontal surface of the particular region. The direct solar irradiation received by a solar panel produces a high energy yield. Thus, usually solar panels are angle-tilted to enhance the efficiency of the solar irradiation, and it is necessary to maximize the solar energy yield to determine the optimal tilt angle [41]. The most effective way to increase and improve the solar energy yield is by using solar tracker, solar trackers help in providing maximum energy by changing the angle of solar panels. Nevertheless, solar trackers require high costs, and they utilize more energy for tracking [42]. Furthermore, these solar trackers are a multifaceted nature. Thus it is useless to install in remote rural regions. Consequently, it is more convenient and feasible to change the title angle of solar panels manually rather than installing solar trackers [43]. The various techniques have been employed to compute the ideal title angle of solar panels for exploiting the solar irradiance [44–46]. In this study, a titled horizontal surface obtains a direct beam, some irradiation is diffused, and some are absorbed, while some rays show off the ground, therefore the global horizontal irradiance on a tilted surface I_G^T is described as:

$$I_G^T = I_B^T + I_D^T + I_R^T \tag{2}$$

where I_B^T is a direct beam, I_D^T is diffuse irradiation, and I_R^T is reflected rays of solar energy on a tilted surface.

Let G_B be the ratio for the average daily direct beam on a horizontal surface and average daily direct beam on a tilted surface, then I_B^T can be altered as:

$$I_B^T = I_B G_B \tag{3}$$

where G_B is a geometric parameter, thus the value depends upon the declination angle, horizontal tilt, surface azimuth, and latitude, respectively. Here, the extensively employed Liu and Jordan model [47] is utilized for computing G_B,

$$G_B = \frac{\cos(L_1 - T_1)\cdot \cos Dsh\cdot \sin i_{ss} + i_{ss}\cdot \sin(L_1 - T_1)\cdot \sin Dsh}{\cos L_1\cdot \cos Dsh\cdot \sin i_{ss} + i_{ss}\cdot \sin L_1 Dsh} \tag{4}$$

where L_1 is the latitude, T_1 is the tilt angle, and i_{ss} and Dsh are declining angles and the sunshine hours.

For clarity, suppose an isotropic distribution of diffused irradiation. Therefore, the diffused region upon the diffused irradiation on the horizontal surface and the horizontal tilt angle λ:

$$I_D^T = I_D \frac{(\cos(\lambda) + 1)}{2}. \tag{5}$$

Here, a property which is famous as albedo factor ω. The range of albedo varies between 0.1 and 0.9 [48]. Thus the reflected beam can be computed as:

$$I_R^T = \omega(I_B + I_D)\frac{(-\cos(\lambda) + 1)}{2}. \tag{6}$$

3.3. The Economic Viability of Off-Grid Solar PV Power Generation System

The economic feasibility of the off-grid solar PV power generation system in rural regions can be described and identified in the following sub-sections:

3.3.1. Solar PV Power Generation System Size and Battery Storage

A normal solar PV system comprises a solar PV module, load or demand, battery storage, system controller, and DC-AC inverter. The solar PV panels receive solar energy and transfer it to the system controller, then transforming it to DC. Afterward, DC transmits the load to the DC and AC inverter. The electricity produced by a solar PV system relies on the solar irradiance obtained in a particular region. Whereas, several other criteria should be well-measured, such as optimal tilt, efficiency, and solar PV maintenance [49].

Moreover, it is essential to calculate the losses suffered during the DC-AC transformation. The various methods are available to forecast solar power yield on a tilted solar PV. The potential of solar PV to produce electricity and S_{pv} (kWh) is computed using Equation (7) [50].

$$S_{pv} = a_{pv}\cdot b_{pv}\cdot c_t\cdot PR \tag{7}$$

where, a_{pv} is the panel area, b_{pv} is the efficiency, c_t is the annual solar irradiation obtained on a tilted PV panel, and PR is the performance ratio used to determine the losses. Further, b_{pv} is computed as [51]:

$$b_{pv} = b_r[1 - \lambda_r[T_A - T_R + (T_N - T_{a.N})\frac{I_T}{I_N} \tag{8}$$

where b_r is the efficiency of solar panels, λ_r is the temperature of solar panels, T_A is the ambient temperature and T_R is the referenced temperature of solar panels, T_N is the nominal operating temperature of solar panel cell, $T_{a.N}$ is the ambient nominal operating temperature, and I_N is the solar radiation.

The designing of any solar PV is a very crucial task because it would have to approximate the load that the PV system supports. For any Solar PV, it is necessary to measure the demand of electricity per household, and it can be computed by multiple appliances, i.e., watt ratings, the number of operating hours, and summing up watt ratings. As presented in Table 1, the projected load is about 440 W per household in rural regions, comprising one pedestal fan, one ceiling fan, two charging slots, and three light-emitting diodes (LED) lights.

Table 1. Projected load requirement per household.

Appliance	No. in Use	Operational Hours	Watts Rating	Total Load (Watts-Hour)
Pedestal fan	1	8	12	96
Ceiling fan	1	12	12	144
Charging slot	2	2	5	20
LED light	3	5	12	180
Total Watts per day				**440**

The front end of the solar PV total electricity produced and demanded is presented here:

$$Electricity\ difference = \sum_{i=1}^{365}\left(S_{pv} - S_d\right) \tag{9}$$

where i is the day of the year, S_{pv} is the total electricity produced, and S_d is the total electricity demand.

The solar energy can be used in the sunshine hours, thus for the night hours, an energy storage technology is required for providing the electricity to the households. Most of the remote rural regions of Sindh are off-grid, thus, battery storage is required at an extra cost. The benefit of battery storage is that the electricity can be stored in the battery and can be utilized anytime, mostly in night hours or cloudy weather when sunshine is unavailable. If electricity produced is more than its demand, then there will be an electricity surplus, such as $S_{pv} > S_d$, and the additional energy will be kept in the battery. However, if the demand of the electricity is more than the electricity produced then $S_d > S_{pv}$ and the solar PV is supposed to be insufficient to meet the electricity demand and load at a particular period. The electricity required to be saved in a battery annually, K_b, is therefore:

$$K_b = \left(\sum SE - \sum FE\right)\cdot e_b \tag{10}$$

where SE is surplus electricity, FE is shortfall electricity, and e_b is the efficiency of the battery. Simultaneously, the daily storage capacity of a battery, S_b, is considered as:

$$S_b = \frac{K_b}{365} \tag{11}$$

3.3.2. Levelized Cost of Electricity (LCOE)

Levelized cost of electricity (LCOE) is an important metric employed to determine and compare the cost of electricity produced by several technologies and sources. It prioritizes numerous choices dependent on cost-effectiveness. This study compared the electricity generated by the off-grid solar PV system and a conventional on-grid system to determine the total cost of electricity in both systems. Therefore, the study has compared both alternative technologies through the estimated levelized cost of electricity in kWh unit and is computed by a simple LCOE formula [52]:

$$LCOE = \frac{\sum_{\alpha=1}^{n} \frac{I_\alpha + M_\alpha + F_\alpha}{(1+d)^\alpha}}{\sum_{\alpha=1}^{n} \frac{e_\alpha}{(1+d)^\alpha}} \tag{12}$$

where, I_α is the investment cost, M_i is the maintenance cost, F_i is the fuel cost, α is a year, e_i is the amount of electricity generated in kWh, d is the discounted rate, and n shows the working-life duration of the alternative technology.

3.4. CO$_2$ Emissions Mitigation from Solar PV Power Generation System

The clean energy is generated from the solar PV system through sunlight, which may help to support minimizing greenhouse gas (GHG) emissions. Therefore, the government should install a solar PV system in the rural regions, so it may also help to eliminate the use and the need of diesel generators which may possess high-carbon intensity and affect the environment and human health in a bad manner. The solar PV system generates very little or no CO_2 emissions during operation, but suffer emissions in the manufacturing period [53]. Environmental sustainability is a globally challenging issue since CO_2 emissions are increasing from the unwanted activities of humans, such as utilizing fossils fuels, which may directly affect the climate in a bad manner [54,55]. Thus, a solar PV system can significantly mitigate CO_2 emissions if it is replaced with a diesel generator. The amount of mitigating CO_2 emissions and diesel fuel kept or saved, F_k, is calculated by employing a solar PV system [56]:

$$F_k = S_{pv} \times F_R \tag{13}$$

where F_R is fuel required for a diesel generator for producing electricity of 1 kWh. For the solar PV system, the decrease in CO_2 is measured in kilograms (kg), the CO_2 emissions kept or saved is EM_k in the following Equation [56]:

$$EM_k = S_{pv} \times \left(C_d - C_{pv}\right) \tag{14}$$

where C_d is the emitted carbon in kg required for a diesel generator for producing 1 kWh of electricity, and C_{pv} is the emitted carbon in kg required for a solar PV system to produce electricity of 1 kWh.

4. Results and Discussion

The most important step before implementing and utilizing a solar PV system is the determination of the available solar energy in the considered region [57]. The daily solar irradiance values received in all of the five rural regions present the appropriate potential to generate electricity from solar PV energy. Data of solar irradiance is obtained from the NASA database [58]. The data of these five regions, i.e., Panoaqil, Badin, Nawabshah, Mirpurkhas, and Kambar regions, has been provided in Table 2. It is identified from Table 2 that all of the selected remote rural regions have enough daily solar irradiation throughout the year for electricity generation. Further, the daily solar irradiation received on a horizontal surface in each rural region is presented in Figures 3–7. Moreover, the average values of annual solar irradiation in the selected rural regions of Sindh is illustrated in Figure 8. The Nawabshah region receives the highest annual solar irradiation (5.49 kWh/m^2) followed by the Kambar region (5.48 kWh/m^2), the Panoaqil region (5.45 kWh/m^2), the Mirpurkhas region (5.41 kWh/m^2), and the Badin region (5.39 kWh/m^2), respectively.

Table 2. Solar data for five regions of Sindh, Pakistan [58].

Period	Panoaqil Region		Badin Region		Nawabshah Region		Mirpurkhas Region		Kambar Region	
	Daily Solar Irradiation (kWh/m^2/day)	Earth Temp (°C)	-	-	-	-	-	-	-	-
Jan	4.10	14.65	4.49	16.92	4.20	14.60	4.41	15.90	3.73	14.70
Feb	4.97	18.36	5.25	21.03	5.09	18.30	5.06	20.03	4.89	19.26
Mar	5.71	25.29	5.97	27.85	5.76	26.19	5.88	27.55	5.83	26.04
April	6.65	33.15	6.69	33.97	6.67	34.04	6.61	34.81	6.90	34.14
May	6.88	38.72	6.79	36.06	6.90	37.55	6.78	37.81	6.79	39.60
June	6.76	41.93	6.48	37.31	6.75	39.46	6.55	39.29	6.77	41.12
July	5.91	40.69	5.08	36.15	5.82	37.91	5.52	37.94	5.65	39.92
Aug	5.97	39.28	5.14	34.38	5.91	37.44	5.49	36.73	6.00	39.40
Sep	5.86	39.28	5.47	33.45	5.82	35.75	5.58	34.95	5.82	36.01
Oct	4.95	30.42	4.97	31.47	5.03	30.88	5.00	31.51	5.47	30.02
Nov	4.00	22.97	4.31	26.55	4.13	24.37	4.15	25.67	4.22	21.03
Dec	3.70	16.08	4.02	19.24	3.80	16.90	3.90	18.48	3.68	15.23
Avg. annual values	5.45	30.06	5.39	29.53	5.49	29.45	5.41	30.05	5.48	29.71

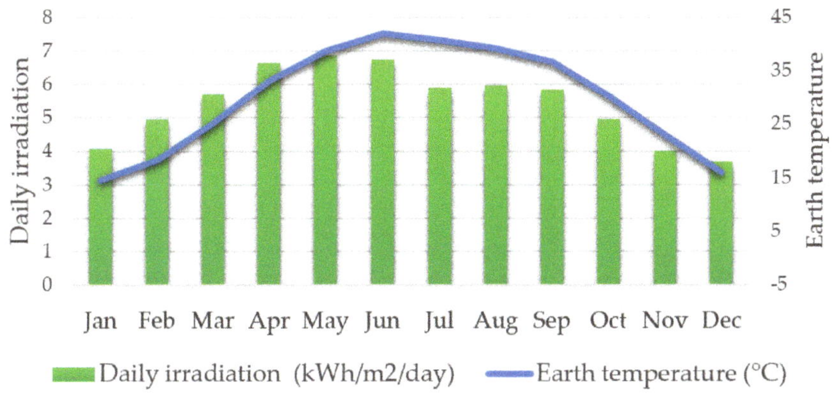

Figure 3. The Panoaqil region daily solar irradiance received.

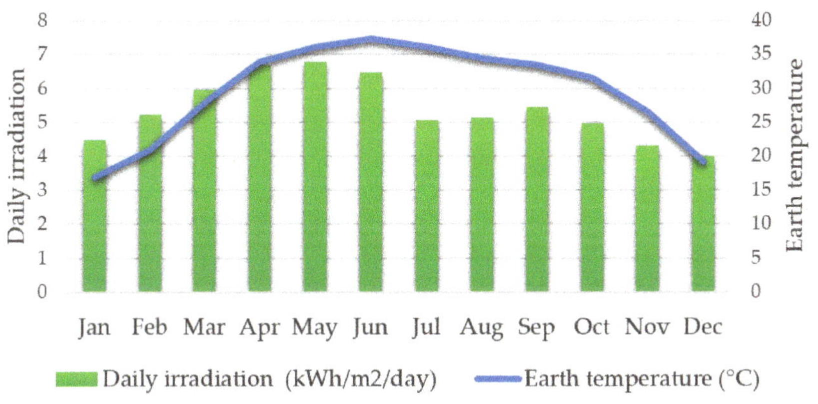

Figure 4. The Badin region daily solar irradiance received.

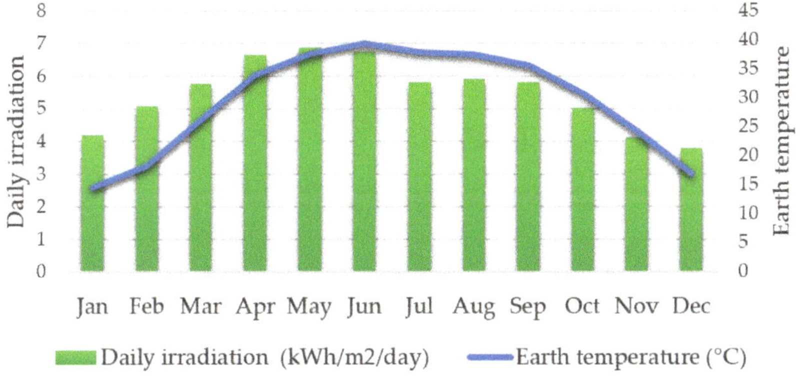

Figure 5. The Nawabshah region daily solar irradiance received.

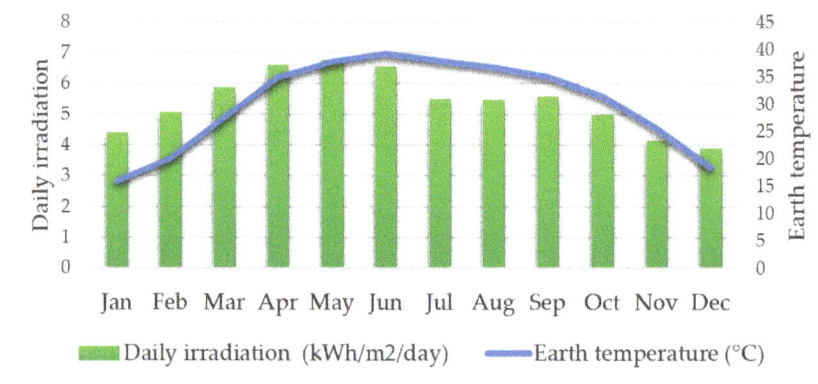

Figure 6. The Mirpurkhas region daily solar irradiation received.

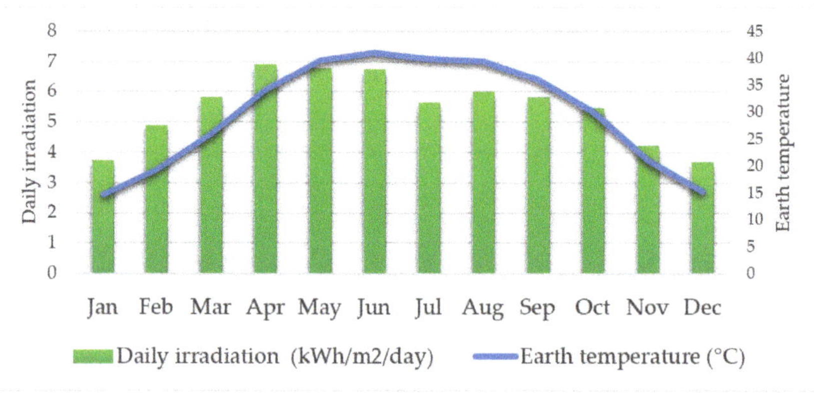

Figure 7. The Kambar region daily solar irradiation received.

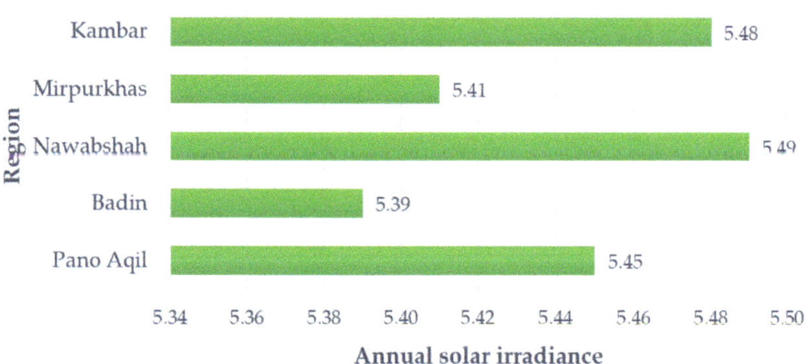

Figure 8. Annual solar irradiance received in five rural regions of Sindh.

4.1. Analyzing the Solar Energy Potential

The above-stated Figures present the average solar irradiation values for the selected regions of the Sindh province. The optimal average peak solar hours are also computed for the selected regions. Table 3 presents the total potential of using solar PV system in five regions of Sindh province, Pakistan, such as the Panoaqil, Badin, Nawabshah, Mirpurkhas, and Kambar regions. It presents that the potential for the solar PV power generation system is significant in these regions. For example, the Nawabshah and Kambar regions, with average solar irradiation of around 5.49 kWh/m^2/day and 5.48 kWh/m^2/day, have the probability of generating 1503 kWh/kWp and 1500 kWh/kWp annually. Furthermore, the daily energy produced from a solar PV panel is around more than 500 Wh in each region of the Sindh province, which can satisfy the need for a primary household energy consumption.

Table 3. Potential of solar photovoltaic (PV) power generation in selected regions of Sindh, Pakistan.

Region	Solar Irradiation (kWh/m^2/day)	Avg. Peak Solar Hours	Daily Energy Output (Wh)	Annual Energy Output (kWh/kWp)
Panoaqil	5.45	5.450	545	1492.21
Badin	5.39	5.390	539	1475.782
Nawabshah	5.49	5.490	549	1503.162
Mirpurkhas	5.41	5.410	541	1481.258
Kambar	5.48	5.480	548	1500.424

4.2. Solar Irradiation Obtained at the Optimal Tilt Angle

Using Equations (2)–(6), the mean values of daily solar irradiation received on a tilted angle are shown in Table 4. Similarly, [59] we conducted the techno-economic analysis to check the impact of title angle on the performance of a PV battery storage for a single household in Germany. The results indicate that a substantial percentage of solar energy yield can be annually elevated by changing the angle from 0 to 90 degrees of solar PV panels on the optimal tilt angle. Solar energy yield can be elevated by 9.87% in Panoaqil at the optimal tilt angle of 28.9°. Likewise, 9.98% in Badin on 29.1°, 10.11% in Nawabshah on 29.3°, 11.66% in Mirpurkhas on 30.6°, and 10.43% in Kambar at 29.5°.

Table 4. Daily solar irradiation (kWh/m^2/day) received on a tilted varying between 0 and 90 angle degrees.

Region	0 Tilt Angle Degree	15 Tilt Angle Degree	30 Tilt Angle Degree	45 Tilt Angle Degree	60 Tilt Angle Degree	75 Tilt Angle Degree	90 Tilt Angle Degree
Panoaqil	5.8	6.2	6.3	6.1	5.6	4.8	3.9
Badin	5.5	5.9	6.0	5.8	5.4	4.6	3.7
Nawabshah	5.7	6.2	6.3	6.2	5.7	4.9	4.0
Mirpurkhas	5.8	6.2	6.3	6.1	5.6	4.8	3.9
Kambar	5.8	6.2	6.3	6.1	5.6	4.8	3.9

4.3. The Economic Viability of Solar PV Power Generation System

An off-grid solar PV system is proposed in the study to provide electricity to rural households in the Sindh province. In Table 5, the simulation of the parameters has been employed and provided by [60]. Based on Equations (7)–(11), it is identified that the PV module area is 1.2 m^2, whereas maximum voltage and current are 26.3 V and 7.61 A. The solar PV system can be used for the electrification of one household, having a production capacity of 200 W or less [61]. Therefore, for a solar panel of 200 W with a 140 Ah/12 V battery is appropriate for the load of one household. Moreover, the conversion efficiency of a solar panel is 16%. Maximum electricity is produced in all of the selected rural regions from April to June, however, due to monsoon season, a slight decline is observed from July to October in the rural regions of Sindh. The cost specifications for the off-grid solar PV system are shown in Table 5. It was identified using Equation (12), that the levelized cost of energy (LCOE) of the proposed off-grid solar PV system is PKR 6.87/kWh, however, the electricity cost from the conventional energy sources is PKR 20.79/kWh [60]. Thus, electricity produced by the solar PV system saves each household approximately PKR 13.92 per kWh.

Table 5. Parameters employed in the study.

Parameter	Unit	Value
Panel area	m^2	1.2
Max: power current	A	7.61
Max: power voltage	V	26.3
PV panel power rating	WP	200
Ambient temperature	°C	20
Panel referenced temperature	°C	25
Panel referenced efficiency	%	16
Solar radiation for NOM	W/m^2	800
PV panel life	Year	25
Panel capital cost	PKR/WP	110
Battery efficiency	%	85
Battery cost	PKR/Ah	120
O/M cost	% of the total cost	4
Discount rate	%	9
Battery duration	Year	5

4.4. CO$_2$ Emissions Mitigation from the Solar PV Power Generation System

A 20 kW diesel generator produces the electricity of 4 kWh/liter. The annual demand of the electricity for the household mentioned above is 160 kWh, so the diesel generator would consume a total of 41.43 L/year. Therefore, based on Equation (14), as per household, the proposed solar PV system could mitigate CO_2 at about 97.50 kg/year. According to the sixth population and housing census of Pakistan, there are 4,185,828 households in rural regions of Sindh [62]. If the off-grid solar PV system electrified 100% of the rural areas, then the high level of CO_2 could be mitigated annually.

5. Conclusions and Policy Recommendations

5.1. Conclusions

An off-grid solar PV system is recognized as the optimal choice to provide electricity in remote rural regions. However, it is very necessary to assess the techno-economic feasibility of the particular region for installation of an off-grid solar PV power generation system. Therefore, this study has evaluated the techno-economic feasibility of five rural regions of the Sindh province, i.e., the Panoaqil, Badin, Nawabshah, Mirpurkhas, and Kambar regions. The study also suggests that by installing the off-grid solar PV system in the above-mentioned regions it would help to mitigate the CO_2 emissions from the particular regions. The solar energy potential, solar irradiance, and optimal tile angles of solar panels have been evaluated. For maximizing the solar energy, it was found that the output of energy could be significantly increased by varying the solar panels angle on the optimal tilt angle. The results also revealed that the off-grid solar PV system is a much cheaper option for electricity compared to other conventional electricity sources. This study identified that all five regions of the Sindh province have good potential for solar energy and are technically and economically feasible for producing electricity. Therefore, the government must plan and build a strong policy framework to install an off-grid solar system in the Sindh province. Meanwhile, this study provides key policy recommendations for the implementation of an off-grid solar PV system in the rural regions of Sindh.

5.2. Policy Recommendations

The following policy recommendations can be employed to develop an off-grid solar PV power generation system in remote rural regions of the Sindh province, Pakistan.

- The results of this study indicate that rural regions have a very good potential for solar-based electricity generation. Therefore, off-grid solar PV rural programs must be started with critical

- action plans and strategies and the concerned authorities must formulate planning and policies to implement such projects.
- According to the investigation, all of the selected regions have suitable solar energy potential to generate off-grid solar PV electricity. However, the Nawabshah is the most favorable due to its higher solar energy potential. Also, the geographical location of Nawabshah is technically and economically the best option for generating PV electricity.
- As the review of the existing scientific studies and government policies indicate, lack of supportive policies and political will are the main hurdles in the deployment of solar electricity generation systems in rural areas of the country [63,64].
- It is suggested that supportive policies must be planned for stakeholders so that stakeholders can easily invest in rural regions for developing a solar PV power generation system.
- The financial constraints due to poverty and a huge budget deficit are also recognized as another hurdle in the deployment of the solar PV system. Therefore, the government tries to ensure the availability of micro-financing projects which may aid rural communities to install an off-grid solar PV system.
- The policy framework should be formulated by giving high preference to a renewable-based solar power system instead of the conventional power generation system.
- The upfront cost of a solar PV system must be minimized, so that rural communities install and electrify with a solar PV system.
- The quality standards should be taken into account for the off-grid solar PV system.
- It is important that householders should be given loans and subsidies to utilize solar PV power generation systems at the domestic level.
- Educate people by organizing training and campaigns about the drawback associated with conventional energy and make them aware of the advantages related to the deployment of solar energy.
- Priority must be given to the local communities and train them to install, operate, and maintain the solar PV system.

5.3. Limitations of the Current Study and Future Research Direction

This study has considered only five particular regions of Sindh province, thus the results of the study are not feasible for other regions of Sindh and other provinces of Pakistan. Therefore, the techno-economic feasibility analysis must be carried out for other regions to identify solar power potential. In the future, the hybrid renewable energy system, solar and wind, can be developed in the potential regions of Pakistan because the hybrid renewable energy system is a more reliable and effective source of energy. The hybrid solar–wind energy system shall be employed in the remote rural regions of Pakistan to make them independent of grids. The government can play a pivotal role to end the energy crisis by facilitating rural regions with such a hybrid energy system.

Author Contributions: Conceptualization, Y.A.S., S.A.A.S.; methodology, L.X.; Y.W., Y.A.S. and S.A.A.S.; validation, S.A.A.S.; H.Z.; formal analysis, L.X.; Y.A.S.; and H.Z.; writing—original draft preparation, L.X.; Y.W.; and H.Z.; writing—review and editing, Y.W., H.Z.; and Y.A.S.; funding acquisition, Y.W.

Funding: 1. National Natural Science Foundation of China (Grant No. 71873064), Research on OFDI driving low-carbon upgrading of China's equipment manufacturing global value chain: theoretical mechanism, implementation path, and performance evaluation. 2. General Projects of Humanities and Social Sciences of the Ministry of Education (Planning Projects) (Grant No. 18YJA790085), Performance evaluation of OFDI driving low-carbon upgrading of China's equipment.

Acknowledgments: These data of Table 2 were obtained from the NASA Langley Research Center (LaRC) POWER Project funded through the NASA Earth Science/Applied Science Program.

References

1. Warner, K.J.; Jones, G.A. A Population-Induced Renewable Energy Timeline in Nine World Regions. *Energy Policy* **2017**, *101*, 65–76. [CrossRef]
2. International Energy Agency (IEA). *World Energy Outlook 2017*; International Energy Agency: Paris, France, 2017; pp. 1–15.
3. Rehman, S.A.U.; Cai, Y.; Fazal, R.; Das Walasai, G.; Mirjat, N.H. An Integrated Modeling Approach for Forecasting Long-Term Energy Demand in Pakistan. *Energies* **2017**, *10*, 1868. [CrossRef]
4. Raheem, A.; Abbasi, S.A.; Memon, A.; Samo, S.R.; Taufiq-Yap, Y.H.; Danquah, M.K.; Harun, R. Renewable Energy Deployment to Combat Energy Crisis in Pakistan. *Energy Sustain. Soc.* **2016**, *6*, 16. [CrossRef]
5. Mirjat, N.H.; Uqaili, M.A.; Harijan, K.; Das Valasai, G.; Shaikh, F.; Waris, M. A Review of Energy and Power Planning and Policies of Pakistan. *Renew. Sustain. Energy Rev.* **2017**, *79*, 110–127. [CrossRef]
6. Solangi, Y.A.; Tan, Q.; Mirjat, N.H.; Das Valasai, G.; Khan, M.W.A.; Ikram, M. An Integrated Delphi-AHP and Fuzzy TOPSIS Approach toward Ranking and Selection of Renewable Energy Resources in Pakistan. *Processes* **2019**, *7*, 118. [CrossRef]
7. Dailytimes. Sindh is Resource-Rich Province, Govt only Needs to Tap Them, Says CM. Available online: https://dailytimes.com.pk/317213/sindh-is-resource-rich-province-govt-only-needs-to-tap-them-says-cm/ (accessed on 25 April 2019).
8. Qayyum, S. Pakistan: A Renewable Energy State? Available online: http://www.technologyreview.pk/pakistan-a-renewable-energy-state/ (accessed on 22 October 2018).
9. Bhutto, A.W.; Bazmi, A.A.; Zahedi, G. Greener Energy: Issues and Challenges for Pakistan-Solar Energy Prospective. *Renew. Sustain. Energy Rev.* **2012**, *16*, 2762–2780. [CrossRef]
10. Mirza, U.K.; Ahmad, N.; Harijan, K.; Majeed, T. Identifying and Addressing Barriers to Renewable Energy Development in Pakistan. *Renew. Sustain. Energy Rev.* **2009**, *13*, 927–931. [CrossRef]
11. DAWN. Sindh, KP Ask PM to Reverse Decisions about Renewable Energy. Available online: https://www.dawn.com/news/1477631 (accessed on 25 April 2019).
12. Irfan, M.; Zhao, Z.-Y.; Ahmad, M.; Mukeshimana, M.C. Solar Energy Development in Pakistan: Barriers and Policy Recommendations. *Sustainability* **2019**, *11*, 1206. [CrossRef]
13. Wakeel, M.; Chen, B.; Jahangir, S. Overview of Energy Portfolio in Pakistan. *Energy Procedia* **2016**, *88*, 71–75. [CrossRef]
14. Hasnie, S. How the un-Electrified Poor can Leapfrog to Off-Grid Solar. Available online: https://blogs.adb.org/blog/how-unelectrified-poor-can-leapfrog-grid-solar (accessed on 6 April 2019).
15. Mamaghani, A.H.; Escandon, S.A.A.; Najafi, B.; Shirazi, A.; Rinaldi, F. Techno-Economic Feasibility of Photovoltaic, Wind, Diesel and Hybrid Electrification Systems for off-Grid Rural Electrification in Colombia. *Renew. Energy* **2016**, *97*, 293–305. [CrossRef]
16. Ghafoor, A.; Munir, A. Design and Economics Analysis of an Off-Grid PV System for Household Electrification. *Renew. Sustain. Energy Rev.* **2015**, *42*, 496–502. [CrossRef]
17. Sandwell, P.; Chan, N.L.A.; Foster, S.; Nagpal, D.; Emmott, C.J.M.; Candelise, C.; Buckle, S.J.; Ekins-Daukes, N.; Gambhir, A.; Nelson, J. Off-Grid Solar Photovoltaic Systems for Rural Electrification and Emissions Mitigation in India. *Sol. Energy Mater. Sol. Cells* **2016**, *156*, 147–156. [CrossRef]
18. Mishra, P.; Behera, B. Socio-Economic and Environmental Implications of Solar Electrification: Experience of Rural Odisha. *Renew. Sustain. Energy Rev.* **2016**, *56*, 953–964. [CrossRef]
19. Hosenuzzaman, M.; Rahim, N.A.; Selvaraj, J.; Hasanuzzaman, M.; Malek, A.B.M.A.; Nahar, A. Global Prospects, Progress, Policies, and Environmental Impact of Solar Photovoltaic Power Generation. *Renew. Sustain. Energy Rev.* **2015**, *41*, 284–297. [CrossRef]
20. Kerekes, T.; Koutroulis, E.; Séra, D.; Teodorescu, R.; Katsanevakis, M. An Optimization Method for Designing Large PV Plants. *IEEE J. Photovolt.* **2013**, *3*, 814–822. [CrossRef]
21. Islam, M.R.; Guo, Y.; Zhu, J. A Multilevel Medium-Voltage Inverter for Step-up-Transformer-Less Grid Connection of Photovoltaic Power Plants. *IEEE J. Photovolt.* **2014**, *4*, 881–889. [CrossRef]
22. Nasir, M.; Khan, H.A.; Hussain, A.; Mateen, L.; Zaffar, N.A. Solar PV Based Scalable DC Microgrid for Rural Electrification in Developing Regions. *IEEE Trans. Sustain. Energy* **2018**, *9*, 390–399. [CrossRef]
23. Wilson, M.; Luck, R.; Mago, P.; Cho, H. Building Energy Management Using Increased Thermal Capacitance and Thermal Storage Management. *Buildings* **2018**, *8*, 86. [CrossRef]

24. Iqbal, W.; Yumei, H.; Abbas, Q.; Hafeez, M.; Mohsin, M.; Fatima, A.; Jamali, M.A.; Jamali, M.; Siyal, A.; Sohail, N. Assessment of Wind Energy Potential for the Production of Renewable Hydrogen in Sindh Province of Pakistan. *Processes* **2019**, *7*, 196. [CrossRef]

25. Awan, U.; Imran, N.; Muneer, G. Sustainable Development through Energy Management: Issues and Priorities in Energy Savings. *Res. J. Appl. Sci. Eng. Technol.* **2014**, *7*, 424–429. [CrossRef]

26. Govt. of Sindh. Population Welfare Department Sindh Population Welfare Department Sindh. Available online: http://pwdsindh.gov.pk/ (accessed on 8 January 2018).

27. Kazmi, S.A.Z.; Naaranoja, M.; Kantola, J. Reviewing Pakistan's Investment Potential as a Foreign Investor. *Procedia Soc. Behav. Sci.* **2016**, *235*, 611–617. [CrossRef]

28. Map, G. Sindh Map, Pakistan. Available online: https://www.google.com/maps/place/Sindh,+Pakistan/ (accessed on 25 April 2019).

29. Khalil, H.B.; Zaidi, S.J.H. Energy Crisis and Potential of Solar Energy in Pakistan. *Renew. Sustain. Energy Rev.* **2014**, *31*, 194–201. [CrossRef]

30. Bhutta, Z. Power Shortfall Hits Record Peak. Available online: https://tribune.com.pk/story/1742548/1-electricity-deficit-power-shortfall-hits-record-peak/ (accessed on 22 October 2018).

31. Harijan, K.; Uqaili, M.; Mirza, U. Assessment of Solar PV Power Generation Potential in Pakistan. *J. Clean Energy Technol.* **2015**, *3*, 54–56. [CrossRef]

32. Hydrocarbon Development Institute of Pakistan. *Pakistan Energy Yearbook*; Hydrocarbon Development Institute of Pakistan: Islamabad, Pakistan, 2017.

33. Shaikh, P.H.; Shaikh, F.; Mirani, M. Solar Energy: Topographical Asset for Pakistan. *Appl. Sol. Energy* **2013**, *49*, 49–53. [CrossRef]

34. Solangi, Y.A.; Tan, Q.; Khan, M.W.A.; Mirjat, N.H.; Ahmed, I. The Selection of Wind Power Project Location in the Southeastern Corridor of Pakistan: A Factor Analysis, AHP, and Fuzzy-TOPSIS Application. *Energies* **2018**, *11*, 1940. [CrossRef]

35. AEDB. Progress so Far Made in Solar Power Sector in Pakistan. Available online: https://www.aedb.org/ae-technologies/solar-power/solar-current-status (accessed on 6 April 2019).

36. NEPRA. State of Industry Report 2016. Available online: http://www.nepra.org.pk/ (accessed on 6 April 2019).

37. Urmee, T.; Harries, D.; Schlapfer, A. Issues Related to Rural Electrification Using Renewable Energy in Developing Countries of Asia and Pacific. *Renew. Energy* **2009**, *34*, 354–357. [CrossRef]

38. Dawn. Solar Energy Production Fails to Take off Despite Electricity Crisis. Available online: https://www.dawn.com/news/1194261 (accessed on 25 April 2019).

39. Bhutta, F.M. Solar PV Opportunities and Challenges in Pakistan. April 2015. Available online: http://engghorizons.com/solar-pv-opportunities-and-challenges-in-pakistan/ (accessed on 6 April 2019).

40. Borhanazad, H.; Mekhilef, S.; Saidur, R.; Boroumandjazi, G. Potential Application of Renewable Energy for Rural Electrification in Malaysia. *Renew. Energy* **2013**, *59*, 210–219. [CrossRef]

41. Benghanem, M. Optimization of Tilt Angle for Solar Panel: Case Study for Madinah, Saudi Arabia. *Appl. Energy* **2011**, *88*, 1427–1433. [CrossRef]

42. Barker, L.; Neber, M.; Lee, H. Design of a Low-Profile Two-Axis Solar Tracker. *Sol. Energy* **2013**, *97*, 569–576. [CrossRef]

43. Shah, S.A.A.; Das Valasai, G.; Memon, A.A.; Laghari, A.N.; Jalbani, N.B.; Strait, J.L. Techno-Economic Analysis of Solar PV Electricity Supply to Rural Areas of Balochistan, Pakistan. *Energies* **2018**, *11*, 1777. [CrossRef]

44. Skeiker, K. Optimum Tilt Angle and Orientation for Solar Collectors in Syria. *Energy Convers. Manag.* **2009**, *50*, 2439–2448. [CrossRef]

45. Hartner, M.; Ortner, A.; Hiesl, A.; Haas, R. East to West—The Optimal Tilt Angle and Orientation of Photovoltaic Panels from an Electricity System Perspective. *Appl. Energy* **2015**, *160*, 94–107. [CrossRef]

46. Rowlands, I.H.; Kemery, B.P.; Beausoleil-Morrison, I. Optimal Solar-PV Tilt Angle and Azimuth: An Ontario (Canada) Case-Study. *Energy Policy* **2011**, *39*, 1397–1409. [CrossRef]

47. Liu, B.; Jordan, R. Daily Insolation on Surfaces Tilted towards Equator. *ASHRAE Trans.* **1962**, *67*, 526–541.

48. Kotak, Y.; Gul, M.S.; Muneer, T. Investigating the Impact of Ground Albedo on the Performance of PV Systems. In Proceedings of the CIBSE Technical Symposium, London, UK, 16–17 April 2015.

49. Stojanovski, O.; Thurber, M.; Wolak, F. Rural Energy Access through Solar Home Systems: Use Patterns and Opportunities for Improvement. *Energy Sustain. Dev.* **2017**, *37*, 33–50. [CrossRef]

50. Okoye, C.O.; Oranekwu-Okoye, B.C. Economic Feasibility of Solar PV System for Rural Electrification in Sub-Sahara Africa. *Renew. Sustain. Energy Rev.* **2018**, *82*, 2537–2547. [CrossRef]
51. Okoye, C.O.; Taylan, O.; Baker, D.K. Solar Energy Potentials in Strategically Located Cities in Nigeria: Review, Resource Assessment and PV System Design. *Renew. Sustain. Energy Rev.* **2016**, *55*, 550–566. [CrossRef]
52. Baurzhan, S.; Jenkins, G.P. Off-Grid Solar PV: Is It an Affordable or Appropriate Solution for Rural Electrification in Sub-Saharan African Countries? *Renew. Sustain. Energy Rev.* **2016**, *60*, 1405–1418. [CrossRef]
53. Jabeen, M.; Umar, M.; Zahid, M.; Rehaman, M.U.; Batool, R.; Zaman, K. Socio-Economic Prospects of Solar Technology Utilization in Abbottabad, Pakistan. *Renew. Sustain. Energy Rev.* **2014**, *39*, 1164–1172. [CrossRef]
54. Ikram, M.; Mahmoudi, A.; Shah, S.Z.A.; Mohsin, M. Forecasting Number of ISO 14001 Certifications of Selected Countries: Application of Even GM (1,1), DGM, and NDGM Models. *Environ. Sci. Pollut. Res.* **2019**, *26*, 12505–12521. [CrossRef]
55. Ali, S.; Xu, H.; Ahmed, W.; Ahmad, N.; Solangi, Y.A. Metro Design and Heritage Sustainability: Conflict Analysis Using Attitude Based on Options in the Graph Model. *Environ. Dev. Sustain.* **2019**, 1–22. [CrossRef]
56. Baurzhan, S.; Jenkins, G.P. On-Grid Solar PV versus Diesel Electricity Generation in Sub-Saharan Africa: Economics and GHG Emissions. *Sustainability* **2017**, *9*, 372. [CrossRef]
57. Kharseh, M.; Wallbaum, H. How Adding a Battery to a Grid-Connected Photovoltaic System Can Increase Its Economic Performance: A Comparison of Different Scenarios. *Energies* **2019**, *12*, 30. [CrossRef]
58. NASA-POWER Project Data Sets. Available online: https://power.larc.nasa.gov/ (accessed on 23 October 2018).
59. Lahnaoui, A.; Stenzel, P.; Linssen, J. Tilt Angle and Orientation Impact on the Techno-Economic Performance of Photovoltaic Battery Systems. *Energy Procedia* **2017**, *105*, 4312–4320. [CrossRef]
60. Ullah, H.; Kamal, I.; Ali, A.; Arshad, N. Investor Focused Placement and Sizing of Photovoltaic Grid-Connected Systems in Pakistan. *Renew. Energy* **2018**, *121*, 460–473. [CrossRef]
61. Abdullah; Zhou, D.; Shah, T.; Jebran, K.; Ali, S.; Ali, A.; Ali, A. Acceptance and Willingness to Pay for Solar Home System: Survey Evidence from Northern Area of Pakistan. *Energy Rep.* **2017**, *3*, 54–60. [CrossRef]
62. Pakistan Bureau of Statistics. 6th Population and Housing Census. Available online: http://www.pbs.gov.pk/ (accessed on 6 April 2019).
63. Zafar, U.; Ur Rashid, T.; Khosa, A.A.; Khalil, M.S.; Rahid, M. An Overview of Implemented Renewable Energy Policy of Pakistan. *Renew. Sustain. Energy Rev.* **2018**, *82*, 654–665. [CrossRef]
64. Zameer, H.; Wang, Y. Energy Production System Optimization: Evidence from Pakistan. *Renew. Sustain. Energy Rev.* **2018**, *82*, 886–893. [CrossRef]

Article

Determination of the Acidity of Waste Cooking Oils by Near Infrared Spectroscopy

Juan Francisco García Martín [1,*], María del Carmen López Barrera [1], Miguel Torres García [2], Qing-An Zhang [3] and Paloma Álvarez Mateos [1]

[1] Departamento de Ingeniería Química, Facultad de Química, Universidad de Sevilla, C/Profesor García González, 1, 41012 Seville, Spain; mlopez91@us.es (M.d.C.L.B.); palvarez@us.es (P.Á.M.)
[2] Departamento de Ingeniería Energética. E.T.S. de Ingeniería, Universidad de Sevilla, Camino de los Descubrimientos, s/n, 41092 Seville, Spain; migueltorres@us.es
[3] School of Food Engineering and Nutrition Sciences, Shaanxi Normal University, Xi'an 710062, Shanxi, China; qinganzhang@snnu.edu.cn
* Correspondence: jfgarmar@us.es; Tel.: +34-954-55-71-83

Received: 29 April 2019; Accepted: 15 May 2019; Published: 21 May 2019

Abstract: Waste cooking oils (WCO) recycling companies usually have economic losses for buying WCO not suitable for biodiesel production, e.g., WCO with high free acidity (FA). For this reason, the determination of FA of WCO by near infrared (NIR) spectroscopy was studied in this work to assess its potential for in situ application. To do this, FA of 45 WCO was measured by the classical titration method, which ranged between 0.15 and 3.77%. Then, the NIR spectra from 800 to 2200 nm of these WCO were acquired, and a partial least squares model was built, relating the NIR spectra to FA values. The accuracy of the model was quite high, providing r^2 of 0.970 and a ratio of performance to deviation (RPD) of 4.05. Subsequently, a model using an NIR range similar to that provided by portable NIR spectrometers (950–1650 nm) was built. The performance was lower (r^2 = 0.905; RPD = 2.66), but even so, with good accuracy, which demonstrates the potential of NIR spectroscopy for the in situ determination of FA of WCO.

Keywords: free acidity; NIRS; partial least squares; waste cooking oil

---

1. Introduction

Fossil fuel combustion has negative effects on the environment. Therefore, current research is mainly focused on the search for economic, environmentally friendly biofuels that can replace petroleum fuels. Diesel combustion in engines leads to air pollution by greenhouse gas emissions, namely NO_x, CO, and CO_2, and to the destruction of the ozone layer, due to photochemical interactions of the hydrocarbon, CO, and NO_x emissions. As an alternative to petroleum diesel fuel, the currently produced biofuel is biodiesel, which is composed of fatty acid methyl esters (FAME). Biodiesel is obtained by transesterification of vegetable oils with short-chain alcohols, mainly methanol [1].

The main problems of biodiesel production from vegetable oils are the high cost of the raw materials, the threat to food security [2,3], and the oversupply of glycerin as a byproduct [4]. The current alternative to vegetable oils is the use of waste cooking oils (WCO) as raw material for biodiesel production. WCO are much cheaper than vegetable oils from crops or trees, are not in conflict with food security, and are available as waste products [3,5]. The HORECA (hotels, restaurants, and catering) sector produces roughly 400,000 tons of WCO per year in Spain [3]. As a profitable market, many enterprises dealing with the collection and recycling of waste cooking oils for subsequent biodiesel production have been set up [6].

Frying consists of introducing food to an oil bath at temperatures between 160 and 200 °C, during a certain period in the presence of air. Because of this high temperature exposure, the oil undergoes numerous physical and chemical changes that make the characteristics of this oil different. Among these changes, the most notorious one is the increasing of acidity, mainly provoked by the release of free fatty acids from the partial hydrolysis of triglycerides.

WCO recycling companies usually collect these oils from clean points or special containers, or buy entire trucks from smaller companies. After filtering in their facilities to remove leftover food, flour, etc., WCO are left to decant to obtain three phases: Clean oil, water, and sludge [6]. This oil is then analysed. The main requirement is that the free acidity (FA) is lower than 2.5%. If not, in order to be suitable for biodiesel production, this oil must be subjected to a chemical process (esterification) to reduce FA. Most WCO recycling companies usually do not have either reactors nor qualified staff to carried out the esterification reaction, so these companies have to sell this oil to bigger companies that have the required equipment (mainly biodiesel producers) at a lower price than that of purchase, or to pay to an environmental manager to dispose it. This problem could be overcome if FA was measured in situ. The determination of oil FA is carried out by acid–base titration, and cannot be performed in situ because it needs reagents, sample preparation, and laboratory glassware, and generates chemical wastes.

Near-infrared spectroscopy (NIRS) is a low-cost, safe, and non-destructive technique, which generally does not require sample preparation or chemicals [7,8]. Nowadays, portable near-infrared spectrometers can be purchased for roughly 6000 € (e.g., Oceanoptics Flame-NIR Spectrometer). Therefore, NIRS can be suitable for in situ analysis. The potential of NIRS for olive and sunflower oils' free acidity determination has been demonstrated [7–9], which makes it feasible that WCO's FA can also be measured by NIRS.

The aim of this work is to assess the feasibility of determining the free acidity of waste cooking oils by near-infrared spectroscopy and verify whether this technique can be used for its in situ determination by WCO collection and recycling companies. To do this, FA determination in WCO was first assayed using the whole NIR spectrum. Subsequently, it was assayed in a reduced NIR interval, similar to that provided by portable NIR spectrometers, thus proving the applicability of the technique.

2. Materials and Methods

2.1. Waste Cooking Oils (WCO)

Fifty WCO were collected from the university canteen of the Reina Mercedes Campus (University of Seville) and different private households. These WCO were olive, sunflower and pomace oils, and mixtures of them. This ensured a wide variety of oil types and degradation degrees. Once received at the laboratory, WCO were filtered to remove solids such as leftover food, flour, etc.

2.2. Free Acidity (FA) Determination by Acid–Base Titration

The percentage of free fatty acids that WCO contain, expressed as oleic acid percentage, was measured according to the Official Methods of Analysis of the EC [10,11]. Briefly, 20 g WCO were placed into 250-cm^3 wide-mouth Erlenmeyer flasks along with 50 cm^3 ethanol/ethyl ether solution (1:1 v/v) and a few drops of phenolphthalein, and then neutralized with 0.1 N KOH, previously standardized with benzoic acid. The titration ends when a reddish-brown color change is observed. Determinations were performed in duplicate.

The percentage of acidity of the oil was calculated according to the following equation:

$$FA\ (\%) = \frac{V \times 0.1\ N \times 0.282}{m} \times 100$$

where V is the spent volume of KOH in mL, 0.1 N stands for the KOH normality, 0.282 is the equivalent weight of oleic acid in mequiv, and m is the mass of WCO sample in grams. The FA contents ranged between 0.15% and 3.77%, being the mean value and standard deviation, 0.94% and 0.79%, respectively.

2.3. Spectra Acquisition

Before spectra acquisition, WCO samples were kept at 32 °C for 30 min in a water bath because NIR radiation reflected and absorbed by a sample depends on its temperature and this temperature ensures that all oil compounds are completely dissolved. A Vis/NIR Labspec Pro model LSP 350-2500P (Analytical Spectral Devices Inc., Boulder, CO, USA) spectrophotometer with three detectors was used for spectral acquisition, as described in [7]. The spectrophotometer is equipped with internal shutters and automatic offset correction, the scanning speed time being 100 ms.

Samples were introduced in a 10-mm quartz cuvette, which was placed in a cuvette accessory joined by fibre optic connectors to the spectrophotometer light source on one side, and to the spectrophotometer detector on the opposite side. This optical path length was selected because it showed higher absorption intensity than 1 mm, 2 mm, and 5 mm path-length quartz cuvettes when acquiring olive oil NIR spectra [7]. NIR spectra from 800 to 2200 nm were then acquired in transmittance mode and recorded using the Indico Pro software (Analytical Spectral Devices Inc. Boulder, CO, USA), each spectral variable matching to a 1-nm interval. The spectrum of each sample was acquired in duplicate.

2.4. Calibration Procedure and Model Evaluation

Reflectance data were first transformed to absorbance and then maximum normalised. Partial least squares (PLS) models coupled to full-cross internal validation were built with The Unscrambler software (CAMO Software AS, Oslo, Norway) for the full NIR spectrum (800–2200 nm) and for the wavelength interval specified for the Ocean Optics Flame-NIR spectrometer (950–1650 nm).

The standard error of calibration (SEC) and the multiple correlation coefficient of calibration (r^2_c) were used to assess models' fitness. The prediction performance of the models was evaluated based on the standard error of prediction (SEP), which corresponds to the standard error of full-cross validation, the multiple correlation coefficient of full-cross validation (r^2_{cv}), and the ratio of performance to deviation (RPD). The RPD was defined as the ratio between the standard deviation from the FA reference data and SEP. Among them, the most important parameters to assess the performance of a PLS model are r^2_c (calibration) and RPD (validation): The higher these parameters are, the higher the accuracy of the model. To be specific, models with $r^2_c \geq 0.90$ are considered to have excellent precision, while models with $r^2_c = 0.70$–0.89 are regarded as good precision models [12]. As for RPD, this parameter must be higher than 3 for a PLS model to be considered of excellent precision [13], although another author has pointed out that predictive models suitable for routine analysis should have RPD values between 2 and 10 [14].

3. Results

3.1. Features of the NIR Spectra of Waste Cooking Oils

Absorbance in the NIR region is linear with the concentration of organic compounds. The NIR spectrum of a sample consists of first and second overtones (800–1800 nm) and combination bands (1800–2700 nm) of fundamental, largely hydrogenic vibrations that occur in the MIR region. The acquired WCO absorbance spectra (Figure 1) were practically identical to those previously obtained for olive oils (Figure 2), so their main features are described elsewhere [7]. Briefly, from left to right, a broad absorbance band is observed at 1210 nm, which is related to C–H second overtones and CH=CH– stretching vibrations. Next, a wide absorption band, due to the water first overtone, is observed in the 1350–1450 nm region. The intense absorption found at 1720 nm is related to the first overtone of the C–H vibration of several chemical groups (=CH–, –CH$_3$, –CH$_2$–) and is characteristic of triglycerides and fatty acids of vegetable oils [7]. Another broad water combination band is located at 1880–2100 nm. The two water bands (1350–1450 nm and 1880–2100 nm) show multiple overlapping bands. Finally, the absorption band of the C–H vibration of cis-unsaturation occurs at 2140 nm.

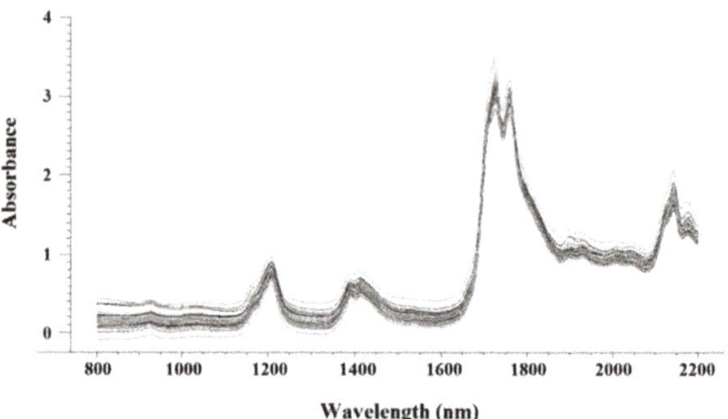

Figure 1. Near-infrared spectra of the waste cooking oils (WCO) used in this research.

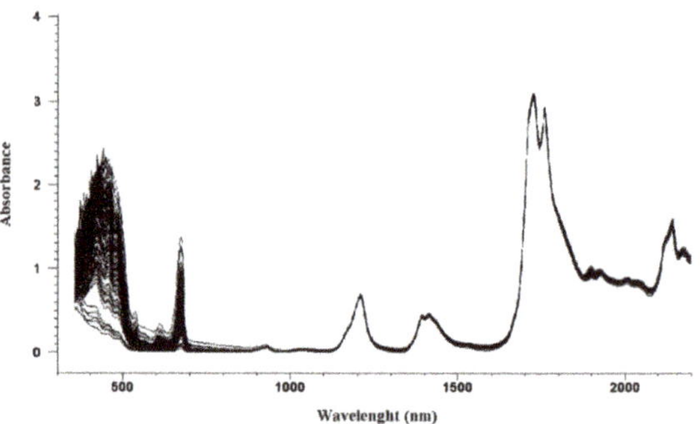

Figure 2. Visible/near-infrared spectra of olive oils [7].

3.2. NIR Partial Least Squares Model for the 800–2200 nm Spectrum

In light of the similarity between the NIR spectra of vegetable and waste cooking oils, it is reasonable to think that free acidity of WCO can be obtained from their NIR spectra, as it has been demonstrated for olive oil [7,8]. Indeed, the PLS calibration model for FA determination, built using the 800–2200 NIR absorbance spectra of the 45 WCO samples, achieved high r^2_c and RPD (0.970 and 4.05, respectively), which accounts for the excellent precision of the model taking into account the aforementioned criteria [12–14]. Figures 3 and 4 illustrate the fit of the data to the proposed model, along with the model statistics. The number of optimal principal components to build this PLS model was nine, explaining 95.1% of the variation in the FA data. Models with large total explained variance (close to 100%) explain most of the variation in the data. Ideally, in order to have simple models, the residual variance has to go down to zero with as few principal components as possible. If this were not the case, it would mean that there might be a large amount of noise in the data.

3.3. NIR Partial Least Squares Model for the 950–1650 nm Spectrum

Once it was proved the NIRS was suitable to determine FA in WCO, the second objective of this work was to try to quantify FA in WCO using a narrower NIR range, similar to the typical range provided by portable NIR spectrometers. In this case, the selected range was 950–1650 nm. The PLS model was built using nine principal components, which explained 90.5% of the variation in the FA data. According to the criteria established [12–14], this PSL model can be regarded as a good precision model. As shown in Figure 5, the statistics of the calibration model were satisfactory, achieving an acceptable r^2_c (0.90). However, SEP increased to 0.30% (Figure 6), thus decreasing the RPD value to 2.66. Compared to the model obtained for the 800–2200 nm NIR range, the precision of the model obtained for 950–1650 nm is markedly lower (Table 1).

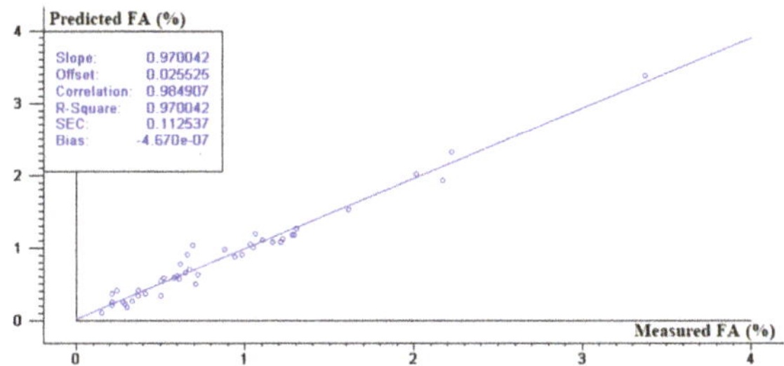

Figure 3. Performance of the 800–2200 partial least squares (PLS) model for free acidity (FA) determination.

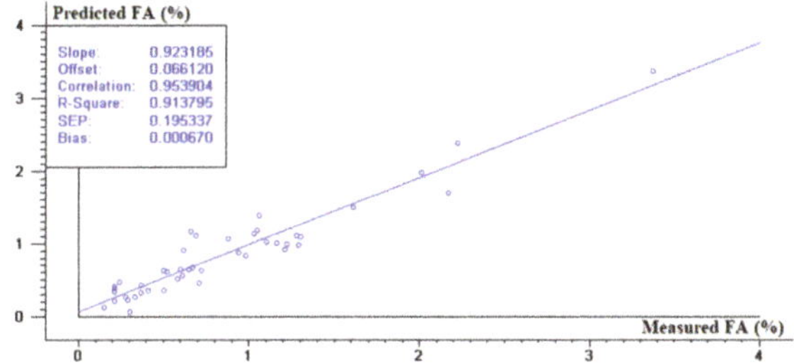

Figure 4. Validation of the 800–2200 PLS model for FA determination.

Table 1. Statistics of the PLS models obtained for the two assayed near-infrared spectroscopy (NIRS) intervals.

NIR Interval	n	r^2_c	r^2_{cv}	SEC	SEP	RPD
800–2200	1401	0.970	0.914	0.113	0.195	4.05
950–1650	701	0.905	0.800	0.201	0.297	2.66

n: Spectral variable number; r^2_c: Correlation coefficient of calibration; SEC: Standard error of calibration; SEP: Standard error of prediction; RPD: Ratio of performance to deviation.

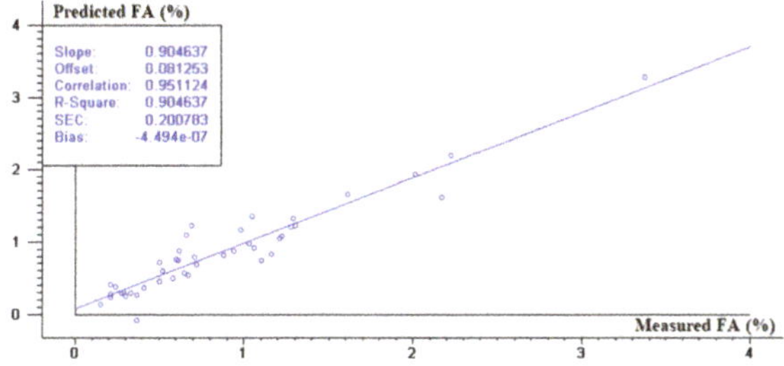

Figure 5. Performance of the 950–1650 PLS model for FA determination.

This could be due to the missing information related to the first overtone of the C–H vibration of several chemical groups (=CH–, –CH$_3$, –CH$_2$–), characteristic of fatty acids of vegetable oils, which was found to be 1720 nm. That aside, WCO contain not only the vegetable oil components, but also degradation products, due to thermal, oxidative, and hydrolytic reactions that occur during frying and food rests (flour, etc.). These components differ among samples and can decrease the PLS model performance. Their negative effect can be balanced, increasing the number of contributing spectral variables to the model, which accounts for the higher accuracy of the PLS model built for the NIR range 800–2200 nm. The precision of both PLS models could be improved, eliminating spectral variables without information related to the measured parameter, such as noise and background, or removing outliers in the calibration and validation sets [7]. However, the accuracy of the 950–1650 nm PLS model is enough for the proposed purpose: The in situ determination of FA of WCO. In fact, the achieved SEP is not high when compared to other authors' results. In this sense, the SEP using the whole NIR spectrum (from 750 to 2500 nm) was 0.35% for virgin olive oils [8], while in this work, the SEP for waste cooking oils was 0.30%. Furthermore, the use of techniques for uninformative spectral variable removing requires knowledge in programs such as MATLAB (The MathWorks, Inc., Natick, MA, USA), and the purchase of these programs and powerful computers, which will obviously decrease the feasibility and practical application of NIRS to in situ FA determination.

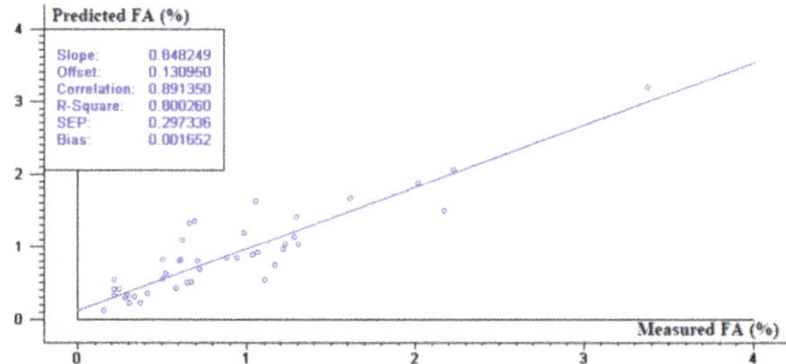

Figure 6. Validation of the 950–1650 PLS model for FA determination.

4. Conclusions

In the absence of tests at companies' facilities with different portable NIR spectrometers, the potential of near-infrared spectroscopy for in situ waste cooking oils' free acidity determination has been demonstrated. High accuracy was achieved when using the whole NIR range provided by a benchtop NIR spectrometer (r^2 = 0.970; RPD = 4.05). Although not perfect, the accuracy of the partial least squares model, built using a wavelength range similar to that incorporated in portable NIR spectrometers (950–1650 nm), was good (r^2 = 0.905; RPD = 2.66). Furthermore, the standard error of prediction achieved with this narrow NIR interval (0.30%) was in the range of that obtained by other authors for vegetable oils using the full NIR spectrum, which accounts for the ability of NIR spectroscopy to determine free acidity of waste cooking oils.

Author Contributions: Conceptualization, J.F.G.-M.; methodology, J.F.G.-M. and P.Á.-M.; formal analysis, J.F.G.-M. and Q.-A.Z.; investigation, M.d.C.L.-B.; data curation, M.d.C.L.-B. and P.Á.-M.; software, M.d.C.L.-B. and J.F.G.-M.; writing—original draft preparation, J.F.G.-M.; writing—review and editing, J.F.G.-M.; supervision, J.F.G.-M and P.Á.-M.; project administration, P.Á.-M.; funding acquisition, P.Á.-M. and M.T.-G.

Funding: This research was funded by European Union under grant LIFE 13-Bioseville ENV/1113.

Conflicts of Interest: The authors declare no conflict of interest.

References

1. Pereda Marín, J.; Barriga Mateos, F.; Álvarez Mateos, P. Aprovechamiento de las oleinas residuales procedentes del proceso de refinado de los aceites vegetales comestibles, para la fabricación de biodiesel. *Grasas y Aceites* **2003**, *54*, 130–137.

2. Sánchez, S.; Cuevas, M.; García-Martín, J.F. Bioethanol production: Corn v/s lignocellulose biomass from olive oil industry and the potential role in ensuring food security. *J. Fundam. Renew. Energy Appl.* **2015**, *5*, 18.

3. García-Martín, J.F.; Barrios, C.C.; Alés-Álvarez, F.J.; Dominguez-Sáez, A.; Alvarez-Mateos, P. Biodiesel production from waste cooking oil in an oscillatory flow reactor. Performance as a fuel on a TDI diesel engine. *Renew. Energy* **2018**, *125*, 546–556. [CrossRef]

4. García-Martín, J.F.; Alés-Álvarez, F.J.; Torres-García, M.; Feng, C.-H.; Álvarez-Mateos, P. Production of Oxygenated Fuel Additives from Residual Glycerine Using Biocatalysts Obtained from Heavy-Metal-Contaminated Jatropha curcas L. Roots. *Energies* **2019**, *12*, 740. [CrossRef]

5. García-Martín, J.F.; Alés-Álvarez, F.J.; del Carmen López-Barrera, M.; Martín-Domínguez, I.; Álvarez-Mateos, P. Cetane number prediction of waste cooking oil-derived biodiesel prior to transesterification reaction using near infrared spectroscopy. *Fuel* **2019**, *240*, 10–15. [CrossRef]

6. Álvarez-Mateos, P.; García-Martín, J.F.; Guerrero-Vacas, F.J.; Naranjo-Calderón, C.; Barrios, C.C.; Pérez-Camino, M.C. Valorization of a high-acidity residual oil generated in the waste cooking oils recycling industries. *Grasas y Aceites* **2019**, *40*, e335.

7. García Martín, J.F. Optical path length and wavelength selection using Vis/NIR spectroscopy for olive oil's free acidity determination. *Int. J. Food Sci. Technol.* **2015**, *50*, 1461–1467. [CrossRef]

8. Marquez, A.J.; Díaz, A.M.; Reguera, M.I.P. Using optical NIR sensor for on-line virgin olive oils characterization. *Sens. Actuators B Chem.* **2005**, *107*, 64–68. [CrossRef]

9. Lagardere, L.; Lechat, H.; Lacoste, F. Détermination de l'acidité et de l'indice de peroxyde dans les huiles d'olive vierges et dans les huiles raffinées par spectrométrie proche infrarouge à transformée de Fourier. *OCL Ol Corps Gras Lipides* **2004**, *11*, 70–75. [CrossRef]

10. EU Standard Methods. Commission Regulation (EEC) No 2568/91 on the characteristics of olive oil and olive-residue oil and on the relevant methods of analysis. *Off. J. Eur. Communities* **1991**, *24*, 1–15.

11. European Union. Commission implementing regulation (EU) No 1348/2013 amending Regulation (EEC) No 2568/91. *Off. J. Eur. Union* **2013**, *2013*, 31–67.

12. Shenk, J.S.; Westerhaus, M. Calibration the ISI way. In *Near Infrared Spectrosc Futur Waves*; Davies, A.M.C., Williams, P.C., Eds.; NIR Publications: Chichester, UK, 1996; pp. 198–202.

13. Shenk, J.S.; Workman, J.J.J.; Westerhaus, M.O. Application of NIR Spectroscopy to Agricultural Products. In *Handbook of Near-Infrared Analysis*; Burns, D.A., Ciurczak, E.W., Eds.; Dekker Inc.: New York, NY, USA, 2001; pp. 383–431.

14. Fearn, T. Assessing calibrations: SEP, RPD, RER and R2. *NIR News* **2002**, *13*, 12–13. [CrossRef]

Article

Multi-Objective Optimal Scheduling Method for a Grid-Connected Redundant Residential Microgrid

Weiliang Liu, Changliang Liu, Yongjun Lin, Kang Bai *, Liangyu Ma and Wenying Chen

State Key Laboratory of Alternate Electrical Power System with Renewable Energy Sources, North China Electric Power University, Baoding 071003, China; lwlfengzhiying@163.com (W.L.); 13603123513@163.com (C.L.); lin3172@126.com (Y.L.); maliangyu@ncepu.edu.cn (L.M.); 51651854@ncepu.edu.cn (W.C.)

* Correspondence: 51651798@ncepu.edu.cn; Tel.: +86-159-3357-9831

Received: 1 April 2019; Accepted: 15 May 2019; Published: 19 May 2019

Abstract: Optimal scheduling of a redundant residential microgrid (RR-microgrid) could yield economical savings and reduce the emission of pollutants while ensuring the comfort level of users. This paper proposes a novel multi-objective optimal scheduling method for a grid-connected RR-microgrid in which the heating/cooling system of the RR-microgrid is treated as a virtual energy storage system (VESS). An optimization model for grid-connected RR-microgrid scheduling is established based on mixed-integer nonlinear programming (MINLP), which takes the operating cost (OC), thermal comfort level (TCL), and pollution emission (PE) as the optimization objectives. The non-dominate sorting genetic algorithm II (NSGA-II) is employed to search the Pareto front and the best scheduling scheme is determined by the analytic hierarchy process (AHP) method. In a case study, two kinds of heating/cooling systems, the radiant floor heating/cooling system (RFHCS) and the convection heating/cooling system (CHCS) are investigated for the RR-microgrid. respectively, and the feasibility and validity of the scheduling method are ascertained.

Keywords: redundant residential microgrid (RR-microgrid); optimal scheduling; virtual energy storage system (VESS); non-dominate sorting genetic algorithm II (NSGA-II); analytic hierarchy process (AHP)

---

1. Introduction

In recent years, technologies for the utilization of clean energy generation and the improvement of energy efficiency have been attracting more and more attention for the growing concerns about energy exhaustion and environmental pollution all over the world. The European Union put forward targets for 2030, which will achieve a 40% reduction, at least, in emissions of greenhouse gases compared with 1990 levels, while increasing the renewable energy utilization to 27% of total energy consumption [1]. Similarly, the United States proposed its greenhouse gas emission target for 2025, which will attain a 26–28% reduction as compared to 2005 levels. With respect to the Chinese government, it has been committed that the reduction of greenhouse gas emissions per unit of GDP will be 40–45% at 2020 [2,3]. According to the International Energy Agency's report, buildings bring about 32% of the total energy expenditure while being responsible for approximately 30% of CO_2 emissions [4]. In China, buildings currently consume 27.6% of the total exhausted energy and it is predicted to be 35% by 2020 [5,6]. Therefore, it is of great importance to encourage the high penetration of clean energy generation and the reduction of energy consumption for buildings.

The application of microgrids has become increasingly popular which provides a desirable architecture able to improve the energy utilization efficiency. There are different definitions of microgrid in the literature, and a broadly cited definition provided by U.S. Department of Energy (DOE) is as follows: *"A microgrid is a group of interconnected loads and distributed energy resources within clearly defined electrical boundaries that acts as a single controllable entity with respect to the grid. A microgrid can connect and disconnect from the grid to enable it to operate in both grid-connected or island mode. A remote microgrid is a variation of a microgrid that operates in islanded conditions [7]".*

According to the tracker report from Navigant Research, at least 405 microgrid projects are currently proposed, planned, under development, or fully operating [8]. Feng et al. [9] pointed out that the world's microgrid projects are mainly located in North America and the Asia Pacific region, and present a review of microgrid development on policies, demonstrations, controls, and software tools. For research purposes focusing on topics like operation, control, and protection, many experimental projects have been built as the test beds for microgrids [10].

Building microgrids are generally comprised of combined cooling, heat and power (CCHP) systems, distributed generators (DGs), energy storage systems (ESS), electric loads, and heating/cooling demand. In order to provide economical, comfortable, and low-emission energy service to users, the microgrid operation should be scheduled reasonably, however, there are still many great challenges to face. For instance, the operating state of different kinds of energy supplies need to be reasonably coordinated. Meanwhile, the energy balance and operating constraints of energy supplies must be met simultaneously. Consequently, the intelligent scheduling method for building microgrids has been a current research hotspot.

The optimal scheduling problems of building microgrids have been treated as a linear programming (LP) problem [6,11,12], a non-linear programming (NLP) problem [13–15], and a multi-objective programming problem (MOP) [16–18]. Guan et al. [6] established an economic scheduling model of a building microgrid to minimize the total consumption of natural gas as well as electricity. Jaramillo et al. [11] presented a multi-objective mixed-integer linear programming (MILP) model for a hybrid energy microgrid to reduce its daily operating cost as well as its total emission. Wu et al. [12] presented a MILP model for a microgrid to realize its economic scheduling. Jiang et al. [13] proposed a double-layer coordination control method of microgrid based on NLP. Lu et al. [14] presented a mixed-integer nonlinear programming (MINLP) model for a building microgrid to realize the economic scheduling. Zhao et al. [15] presented a predictive control model for a building microgrid scheduling under dynamic electricity prices. Javidsharifi et al. [16] presented an intelligent evolutionary modified multi-objective bird mating optimizer (MMOBMO) algorithm for a renewable-based microgrid to realize its short-term optimal scheduling. Carpinelli et al. [17] presented a multi-objective scheduling approach for microgrid including different distributed resources. Lin et al. [18] proposed a two-stage multi-objective dispatching method for an integrated community energy system.

Generally, ESS play an important part in the scheduling of microgrids. The commonly used ESS includes electric ESS and thermal ESS. Electric ESS, like super capacitors or storage batteries, have the strong point of rapid response speed and high energy density, however, large-capacity configuration of them is quite expensive. Thermal ESS, like heat/cool storage tanks, have the virtue of low construction cost, nevertheless, they are usually unavailable in applications due to the distinct weakness of higher space requirements. Lately, it is a novel way to improve the performance of microgrid through scheduling the controllable load on the demand side—such as water heaters [19], air conditioners [20], heat pumps [21], refrigerators [22], electric vehicles (EVs) [23,24], etc.—of which the patterns of power consumption could be changed.

Considering the insulation characteristics of buildings and the heat capacity of the indoor air, Jin et al. [25,26] constructed a virtual energy storage system (VESS) and presented a scheduling method for a building microgrid to minimize the daily operating costs. Similarly, considering that the radiant floor heating/cooling system (RFHCS) has considerable thermal storage capacity and has been widely used in residential buildings, Liu et al. [27] treated it as a VESS, and proposed a scheduling method for two kinds of typical residential microgrids to lower the operating cost while ensuring the thermal comfort level (TCL). In these VESS-related papers, the optimal scheduling problem is treated as a single-objective problem which mainly focuses on the operation economy. However, economy, comfort, and low-emission are expected to be achieved simultaneously for the operation of a building microgrid in application.

The motivation of this paper is to propose a multi-objective optimal scheduling method for building microgrids, which could be considered as the extension and improvement of [27]. The main contributions are as follows:

(1) A new kind of building microgrid—a redundant residential microgrid (RR-microgrid)—is chosen as the investigated subject for optimal scheduling problem.
(2) ETP models are established for different heating/cooling systems, the RFHCS and the convection heating/cooling system (CHCS).
(3) Three optimization objectives—operating cost (OC), thermal comfort level (TCL), and pollution emission (PE)—are considered for the optimization model.
(4) The non-dominate sorting genetic algorithm II (NSGA-II) is applied to search the Pareto front of the presented multi-objective optimization model and the best scheduling scheme is determined by the analytic hierarchy process (AHP) method.

Accordingly, the contents of this paper includes six parts: an introduction of the RR-microgrid (Section 2); the equivalent thermodynamic parameters (ETP) model of the heating/cooling system in the RR-microgrid (Section 3); a multi-objective optimization model for the scheduling of the RR-microgrid (Section 4); a solution to the optimization model using NSGA-II and AHP (Section 5); a case study (Section 6); and the conclusion (Section 7).

2. Introduction of the Redundant RR-Microgrid

2.1. Structure of the RR-Microgrid

The structure of the RR-microgrid studied in this paper is shown in Figure 1, and it is known that the RR-microgrid consists of renewable DGs, such as photovoltaic generation (PV) and wind generation (WT), a battery energy storage system (BESS), a CCHP unit which is composed of micro-gas turbines (MTs), a waste heat recovery system (WHRS), absorption chillers (ACs), and other devices, like electric heaters (EHs) as well as electric chillers (ECs). In addition, the RR-microgrid is connected to an external grid so that the exchange of electric power is allowed. "Redundant" means that the heating/cooling demand of residential buildings could be satisfied by the CCHP unit as well as EHs/ECs, while the electric load of the residential buildings could be satisfied by clean energy generation, the external grid, and the CCHP unit.

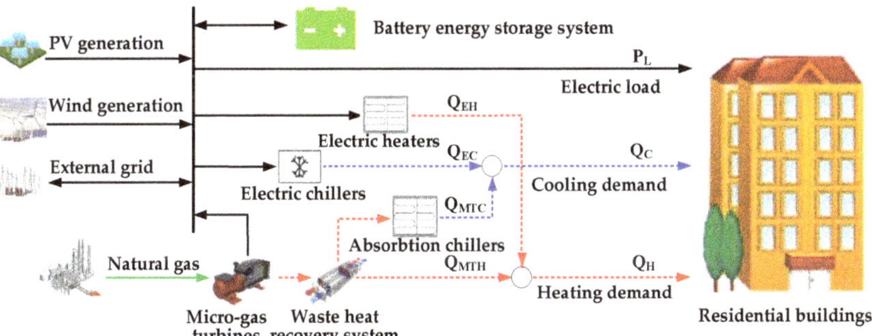

Figure 1. Structure of the studied RR-microgrid.

2.2. Models of Energy Supplies

(1)   CCHP Unit

MTs generate electricity through consuming natural gas, and the output electric power P_{MT} can be expressed as:

$$P_{MT} = F_{gas} \times L_{HVNG} \times \eta_{MTE} \tag{1}$$

meanwhile, the output thermal power Q_{MT} is:

$$Q_{MT} = F_{gas} \times L_{HVNG} \times \eta_{MTH} \tag{2}$$

where F_{gas} is natural gas MTs consumed per unit time, L_{HVNG} is the low calorific value for natural gas, and η_{MTE} and η_{MTH} are the electric power efficiency and thermal power efficiency for MTs, respectively.

Output thermal power Q_{MT} could be turned into the heating power Q_{MTH} by the WHRS:

$$Q_{MTH} = Q_{MT} \times \eta_{HE} \tag{3}$$

where η_{HE} is the conversion efficiency of WHRS, and could be further turned into the cooling power Q_{MTC} by the ACs:

$$Q_{MTC} = Q_{MTH} \times COP_{AC} \tag{4}$$

where COP_{AC} is coefficient of performance (COP) for ACs.

(2)   Electric Heaters/Chillers

The EHs consume electric energy to generate the heating power Q_{EH} expressed as:

$$Q_{EH} = P_{EH} \times COP_{EH} \tag{5}$$

where P_{EH} is the consumed electric power, COP_{EH} is the COP for EHs. The ECs consume electric energy to generate the cooling power Q_{EC} expressed as:

$$Q_{EC} = P_{EC} \times COP_{EC} \tag{6}$$

where P_{EC} is the consumed electric power, COP_{EC} is the COP for ECs.

(3)   Battery Energy Storage System

In practice, the charging and discharging processes of the BESS are usually not constant. However, for the sake of simplicity, the charging and discharging of the battery is regarded as a constant power load or supply during every scheduling period. The state of charge (SOC) for the BESS varies during charging/discharging process, which can be expressed as:

$$E^t = E^{t-1} + \Delta T \times U_{Si+}^t \times P_{Si+}^t \times \eta_c - \Delta T \times U_{Si-}^t \times \frac{P_{Si-}^t}{\eta_{disc}} \tag{7}$$

where E^t is the SOC at the end of scheduling period t, E^{t-1} is the SOC at the end of scheduling period $t-1$, P_{Si+}^t / P_{Si-}^t are the charging power and discharging power for scheduling period t, respectively, U_{Si+}^t / U_{Si-}^t are the charging status and discharging status for scheduling period t, respectively, η_c / η_{disc} are the charging efficiency and discharging efficiency, respectively, and ΔT is the time length of the scheduling period.

3. Equivalent Thermodynamic Parameters Model of the Heating/Cooling System in the RR-Microgrid

In this paper, two kinds of heating/cooling systems for residential buildings are investigated: the RFHCS as well as the CHCS. The RFHCS is different from CHCS in the way of transferring heat/cool to the human body, as the former does so mainly through thermal radiation of the floor and envelope structure, while the latter does so mainly through indoor air convection. Generally, the operative temperature is suitable to evaluate the TCL. Correspondingly, the mean value of the air temperature and indoor average radiation temperature could be regarded as the operative temperature for the RFHCS, while the indoor air temperature could be regarded as the operative temperature for the CHCS. The operative temperature is influenced mainly by the solar radiation load, heating/cooling demand, and the thermal runaway resulting from the difference between the indoor temperature and outdoor temperature. Accordingly, in this paper, equivalent thermodynamic parameter (ETP) models for the RFHCS and CHCS are established, respectively, to describe their mathematical relationships based on [16], as shown in Figure 2.

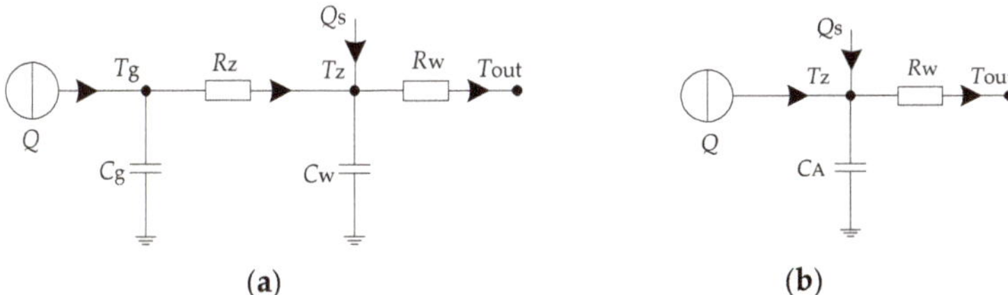

Figure 2. ETP models for different heating/cooling systems. (**a**) RFHCS; and (**b**) CHCS.

In Figure 2, Q and Q_s represent the heating/cooling demand (W) and solar radiation load (W), respectively; T_g, T_z, and T_{out} represent the radiant floor surface temperature (°C), the operative temperature (°C), and the outdoor temperature (°C), respectively, C_g, C_w, and C_A represent the equivalent heat capacities (J/°C) of the radiant floor, the envelope structure, and the indoor air, respectively; R_W represents the equivalent heat resistance (°C/W) for the envelope structure, while R_Z represents the equivalent heat resistance (°C/W) of convection and radiation from the radiant floor surface to the indoor air and envelope structure.

For RFHCS, the differential equations for corresponding ETP model are:

$$C_g \frac{dT_g}{dt} = Q - \frac{T_g - T_z}{R_z} \tag{8}$$

$$C_w \frac{dT_z}{dt} = \frac{T_g - T_z}{R_z} + Q_s - \frac{T_z - T_{out}}{R_w} \tag{9}$$

while for the CHCS, the differential equation for corresponding ETP model is:

$$C_A \frac{dT_z}{dt} = Q + Q_s - \frac{T_z - T_{out}}{R_w} \tag{10}$$

Taking into account the specific structure and material properties of residential buildings, Equations (8) and (9) can be converted to:

$$A_g \times C_{g1} \times \frac{dT_g}{dt} = Q - A_g \times h_z(T_g - T_z) \tag{11}$$

$$(A_{wi} \times C_{wi} + A_{wa} \times C_{wa}) \frac{dT_z}{dt} = A_g h_z(T_g - T_z) + A_{wi} \times I \times \alpha - (A_{wi}k_{wi} + A_{wa}k_{wa})(T_z - T_{out}) \tag{12}$$

while Equation (10) can be converted to:

$$\rho \times C \times V \times \frac{dT_z}{dt} = Q + A_{wi} \times I \times \alpha - (A_{wi}k_{wi} + A_{wa}k_{wa})(T_z - T_{out}) \tag{13}$$

where A_g, A_{wa}, and A_{wi} represent the total area (m^2) of the radiant floor, external walls, and external windows, respectively, in the residential building; C_{gl}, C_{wa}, and C_{wi} represent the equivalent heat capacity (kJ/(m^2·°C)) of the radiant floor, external walls, and external windows, respectively; ρ, C, and V represent, respectively, the density (kg/m^3), heat capacity (kJ/(kg·°C)), and volume (m^3) of the indoor air; h_z represents the comprehensive heat transfer coefficient (W/m^2·°C) from the radiant floor surface to the indoor air as well as the envelope structure; k_{wi} and k_{wa} represent the heat transfer coefficient (W/m^2·°C) for the external walls and external windows of the envelope structure, respectively; I represents the total solar radiation intensity (W/m^2); and α represents the shading coefficient of the residential building.

From Equations (11)–(13), it is known that owing to the heat capacity for the radiant floor, external windows, external walls, and indoor air, the heating/cooling demand could be adjusted to a certain extent while ensuring the operative temperature T_z changes within a reasonable range. Therefore, both the RFHCS and CHCS have charging/discharging characteristics, like the energy storage system, which could be considered as a virtual energy storage system (VESS).

4. Optimization Model for RR-Microgrid Scheduling

4.1. Optimized Variables

A day-ahead optimization model for scheduling of the RR-microgrid is presented based on MINLP, of which the time length of the scheduling period ΔT is 1 h and the number of the scheduling periods θ_T is 24. For the scheduling period $t \in \theta_T$, the variables that need to be optimized could be divided into control variables and state variables, as shown in Tables 1 and 2, respectively.

Table 1. Control variables.

Variables	Description
P^t_{MT}	Output electric power of MTs
P^t_{Si+}	Charging power for the BESS
P^t_{Si-}	Discharging power for the BESS
P^t_{grid+}	Electric power purchasing from the grid
P^t_{grid-}	Electric power selling to the grid
P^t_{EH}	Electric power consumed by the EHs
P^t_{EC}	Electric power consumed by the ECs

Table 2. State variables.

Variables	Description
Q^t_{MT}	Output thermal power for MTs
η^t_{MTE}	Electric power efficiency for MTs
E^t	SOC for the BESS
T^t_g	Surface temperature of the radiant floor
T^t_z	Operative temperature of the residential building

4.2. Objective Function

In this paper, there are three optimization objectives considered for RR-microgrid scheduling: operating cost (OC), thermal comfort level (TCL), and pollution emission (PE).

4.2.1. Objective Function for Operating Cost

The OC of the RR-microgrid consists of four parts: the consuming cost of natural gas, the charging/discharging cost of the BESS, the cost of exchanging electric power with the external grid, and the maintenance cost of clean energy generation and devices, of which the objective function is expressed as:

$$\min f_2(x_s) = f_G(x_s) + f_S(x_s) + f_{Grid}(x_s) + f_{RMC}(x_s) \tag{14}$$

In Equation (15), f_G is consuming cost of natural gas:

$$f_G(x_s) = \sum_{t \in \theta_T} c_{gas} F_{gas}^t \tag{15}$$

where c_{gas} is the price for natural gas, F_{gas}^t is the natural gas MTs consumed at scheduling period t.
f_S is charging/discharging cost of the BESS:

$$f_S(x_s) = \sum_{t \in \theta_T} (c_{Si+} P_{Si+}^t + c_{Si-} P_{Si-}^t) \Delta T \tag{16}$$

where c_{Si+} and c_{Si-} are the unit costs for charging/discharging.
f_{Grid} is cost of exchanging electric power with external grid:

$$f_{Grid}(x_s) = \sum_{t \in \theta_T} (c_{grid+}^t P_{grid+}^t - c_{grid-}^t P_{grid-}^t) \Delta T \tag{17}$$

where c_{grid+}^t and c_{grid-}^t are the prices for purchasing/selling electricity from the external grid at scheduling period t.
f_{RMC} is maintenance cost of clean energy generation and devices:

$$f_{RMC}(x_s) = \sum_{t \in \theta_T} (c_{WT} P_{WT}^t + c_{PV} P_{PV}^t + c_{MT} P_{MT}^t + c_{EH} P_{EH}^t + c_{EC} P_{EC}^t + c_{AC} Q_{MTC}^t) \Delta T \tag{18}$$

where P_{WT}^t and P_{PV}^t are, respectively, the output power of WT and PV at scheduling period t, c_{WT}, c_{PV}, c_{MT}, c_{EH}, c_{EC} and c_{AC} are, respectively, the unit maintenance cost for WT, PV, MTs, EHs, ECs, and ACs.

4.2.2. Objective Function for the Thermal Comfort Level

Ref. [28] presented the predicted mean vote (PMV) as well as the predicted percentage of dissatisfied (PPD) to describe peoples' subjective perception to the thermal environment. According to the national standards of the PRC (GB/T 18049-2000), the reasonable scopes of PMV and PPD are: $PPD \leq 27\%$, $-1 \leq PMV \leq +1$. It is calculated in [27] that, in winter, the optimum operative temperature T_{zopt} is about 22 °C, while in summer it is about 25 °C, corresponding to $PMV = 0$. In winter the variation range of T_z is 17~27 °C, while in summer is 21.5~29 °C, corresponding to $PPD \leq 27\%$, $-1 \leq PMV \leq +1$, which proves it is feasible to adjust the heating/cooling demand for economic benefits.

In the presented optimization model, the permitted adjustable scope of T_z during the scheduling is ±2.5 °C to guarantee the TCL, and the objective function for the TCL is expressed as quadratic sum of deviations between the optimum value and the actual value of the operative temperature during all scheduling periods:

$$\min f_1(x_s) = \sum_{t \in \theta_T} \left| T_z^t - T_{zopt} \right|^2 \tag{19}$$

4.2.3. Objective Function for Pollution Emission

The PE mainly includes the emission from the consumption of electricity purchased from the external grid and the emission from consumption of natural gas. In this paper, the electricity purchasing from the external grid is assumed generated by coal-fired power stations. Three types of polluting gas are taken into account for consumption of coal and natural gas, i.e., CO_2, SO_2, and NO_x, as shown in Table 3. Consequently, the objective function for PE is depicted in Equation (20):

$$\min f_3(x_s) = \sum_{t\in\theta_T} (P_{\text{grid}+}^t \lambda_e + F_{\text{gas}}^t \lambda_g)\Delta T \tag{20}$$

where λ_e and λ_g are the total emission coefficients for coal consumption and natural gas consumption, respectively.

Table 3. Emission coefficients of natural gas and coal [29].

Pollution Gas	CO_2	SO_2	NO_x	Total
Coal (kg/(MWh))	326.37	3.14	1.134	330.644
Natural Gas (kg/(MWh))	203.74	0.011	0.202	203.953

4.3. Constraints

(1)   Balance of electric power:

$$P_{\text{load}}^t + P_{\text{EH}}^t + P_{\text{EC}}^t + P_{\text{Si}+}^t - P_{\text{Si}-}^t = P_{\text{PV}}^t + P_{\text{WT}}^t + P_{\text{MT}}^t + P_{\text{grid}+}^t - P_{\text{grid}-}^t, \ \forall t \in \theta_T \tag{21}$$

where P_{load}^t is the forecast value of electric load at scheduling period t (electric power consumed by EHs/ECs is not taken into account).

(2)   Balance of thermal power

Considering the process of thermal runaway and operative temperature fluctuation in residential building are quite slow, for the convenience of solving the presented optimization model, Equations (11)–(13) are transformed to be difference equations expressing the constraint of the thermal power balance for the RR-microgrid:

$$T_g^{t+1} = T_g^t + \frac{\Delta T}{A_g C_{g1}} \times [Q^t - A_g h_z(T_g^t - T_z^t)], \ \forall t \in \theta_T \tag{22}$$

$$T_z^{t+1} = T_z^t + \frac{\Delta T}{(A_{\text{wi}}C_{\text{wi}}+A_{\text{wa}}C_{\text{wa}})} \times [A_g h_z(T_g^t - T_z^t) + A_{\text{wi}} \times I^t \times \alpha - (A_{\text{wi}}k_{\text{wi}} + A_{\text{wa}}k_{\text{wa}})(T_z^t - T_{\text{out}}^t)], \ \forall t \in \theta_T \tag{23}$$

$$T_z^{t+1} = T_z^t + \frac{\Delta T}{\rho \times C \times V} \times [Q^t + A_{\text{wi}} \times I^t \times \alpha - (A_{\text{wi}}k_{\text{wi}} + A_{\text{wa}}k_{\text{wa}})(T_z^t - T_{\text{out}}^t)], \ \forall t \in \theta_T \tag{24}$$

where for heating in winter, there is:

$$Q^t = Q_{\text{MT}}^t \eta_{\text{HE}} + P_{\text{EH}}^t \text{COP}_{\text{EH}}, \ \forall t \in \theta_T \tag{25}$$

while for cooling in summer, there is:

$$Q^t = -Q_{\text{MT}}^t \eta_{\text{HE}} \text{COP}_{\text{AC}} - P_{\text{EC}}^t \text{COP}_{\text{EC}}, \ \forall t \in \theta_T \tag{26}$$

(3) Battery energy storage system

Uniqueness constraints for charging/discharging states of the BESS are expressed as:

$$P_{\text{Si}+}^t P_{\text{Si}-}^t = 0, \; P_{\text{Si}+}^t \geq 0, \; P_{\text{Si}-}^t \geq 0, \; \forall t \in \theta_T \tag{27}$$

Meanwhile, the charging/discharging power should satisfy the upper limits as:

$$P_{\text{Si}+}^t \leq \overline{P_{\text{Si}+}}, \; \forall t \in \theta_T \tag{28}$$

$$P_{\text{Si}-}^t \leq \overline{P_{\text{Si}-}}, \; \forall t \in \theta_T \tag{29}$$

and the SOC should satisfy upper/lower limits as:

$$\underline{E} \leq E^t \leq \overline{E}, \; \forall t \in \theta_T \tag{30}$$

where $\overline{P_{\text{Si}+}}$ and $\overline{P_{\text{Si}-}}$ are, respectively, the upper limit for charging power and discharging power, while $\overline{E}$ and $\underline{E}$ are, respectively, the upper limit and lower limit for the SOC.

At the end of the last scheduling period, the SOC of the BESS should be equal with the initial value to ensure the balance of energy:

$$E^0 = E^N \tag{31}$$

(4) Power exchanged with the external grid

The uniqueness constraints for power exchanged with external grid are expressed as:

$$P_{\text{grid}+}^t P_{\text{grid}-}^t = 0, \; P_{\text{grid}+}^t \geq 0, \; P_{\text{grid}-}^t \geq 0, \; \forall t \in \theta_T \tag{32}$$

Meanwhile, the power exchanged should satisfy the upper limits:

$$P_{\text{grid}+}^t \leq \overline{P_{\text{grid}+}}, \; \forall t \in \theta_T \tag{33}$$

$$P_{\text{grid}-}^t \leq \overline{P_{\text{grid}-}}, \; \forall t \in \theta_T \tag{34}$$

where $\overline{P_{\text{grid}+}}$ and $\overline{P_{\text{grid}-}}$ are the upper limit, respectively, for power purchasing from, and selling to, the external grid.

(5) Micro-gas turbines

Output electric power for MTs should satisfy the upper limits:

$$P_{\text{MT}}^t \leq \overline{P_{\text{MT}}}, \; \forall t \in \theta_T \tag{35}$$

where $\overline{P_{\text{MT}}}$ is the upper limit for the output electric power for MTs.

In practice, the relationship between η_{MT} and P_{MT} is nonlinear. In the presented optimization model, the fourth-order polynomial is chosen to fit the relationship in order to facilitate subsequent calculation. Thus, the obtained polynomial equation could be expressed as:

$$\eta_{\text{MT}}^t = \alpha_1 \left(\frac{P_{\text{MT}}^t}{P_{\text{MT}}^{\text{max}}}\right)^4 + \alpha_2 \left(\frac{P_{\text{MT}}^t}{P_{\text{MT}}^{\text{max}}}\right)^3 + \alpha_3 \left(\frac{P_{\text{MT}}^t}{P_{\text{MT}}^{\text{max}}}\right)^2 + \alpha_4 \left(\frac{P_{\text{MT}}^t}{P_{\text{MT}}^{\text{max}}}\right) + \alpha_5, \; \forall t \in \theta_T \tag{36}$$

where $\alpha_1, \alpha_2, \alpha_3, \alpha_4,$ and α_5 are the fitting coefficients and $P_{\text{MT}}^{\text{max}}$ is the rated power.

(6) Electric heaters and electric chillers

Electric power EHs/ECs consumed should satisfy the upper limits:

$$P_{\mathrm{EH}}^t \le \overline{P_{\mathrm{EH}}}, \ \forall t \in \theta_{\mathrm{T}} \tag{37}$$

$$P_{\mathrm{EC}}^t \le \overline{P_{\mathrm{EC}}}, \ \forall t \in \theta_{\mathrm{T}} \tag{38}$$

where $\overline{P_{\mathrm{EH}}}$ and $\overline{P_{\mathrm{EC}}}$ are the upper limits for electric power consumed by EHs/ECs, respectively.

(7) Operative temperature

$$\underline{T_z} \le T_z^t \le \overline{T_z}, \ \forall t \in \theta_{\mathrm{T}} \tag{39}$$

where $\overline{T_z}$ and $\underline{T_z}$ are, respectively, the upper and lower limits of the operative temperature.

At the end of the last scheduling period, the operative temperature must be equal with the initial value for balance of thermal energy stored in residential building:

$$T_z^0 = T_z^N \tag{40}$$

In addition, in order to prevent the appearance of condensation for cooling in summer, the surface temperature of the radiant floor is required to be higher than the dewpoint temperature:

$$T_g^t > \underline{T_g}, \ \forall t \in \theta_{\mathrm{T}} \tag{41}$$

where $\underline{T_g}$ is the dewpoint temperature.

5. Solve the Optimization Model Using NSGA-II and AHP

A genetic algorithm (GA) is a kind of population-based search algorithm which is quite suitable for solving multi-objective optimization problems. The NSGA-II algorithm is one of the most effective and efficient multi-objective optimization algorithms [30]. Compared with the NSGA algorithm, the NSGA-II algorithm has a better sorting method and incorporates elitism, while no sharing parameters need to be chosen. According to the non-domination concept, the populations are combined and sorted at each generation. The N least crowded solutions are chosen based on the crowding distance and abandons the rest of the non-dominated solutions. Owing to the above improvements, both spreading and convergence are ensured for the solution front without requiring any external population [31]. The flowchart of the NSGA-II algorithm is shown in Figure 3.

The steps of the NSGA-II algorithm are as follows:

(1) Start, input basic system data.
(2) Initialize parameters of the NSGA-II algorithm which consist of the number of individuals in the population, N_{p}, the maximum number of generations, $g_{\max}$, the crossover probability, p_{c}, the mutation probability p_{m}, and generate N_{p} individuals randomly as the parent population, P_t.
(3) Calculate the objective functions, and generate the offspring population Q_t from P_t using selection, crossover, and mutation operators.
(4) Create the intermediate population $R_t = P_t \cup Q_t$.
(5) Perform a non-dominated sorting to R_t based on the calculation of the crowding distance and check constraints.
(6) Select the first N_{p} individuals as new parent population P_{t+1}.
(7) Check whether the result is equal with the maximum number of generations. If not, return to Step (3), otherwise continue to Step (8).
(8) Output the Pareto-optimal front.

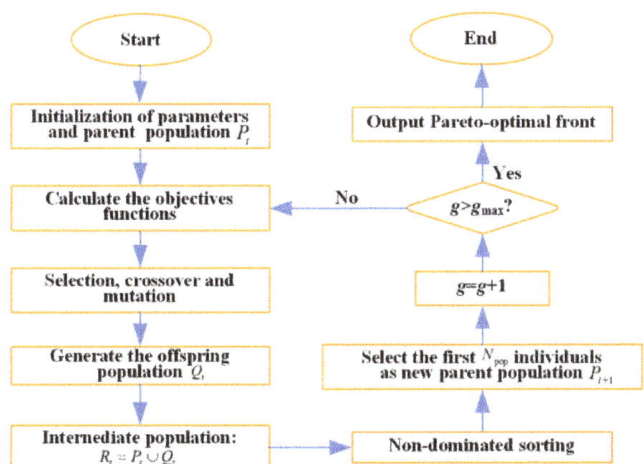

Figure 3. Flowchart of the NSGA-II algorithm.

The final output of above NSGA-II algorithm is the Pareto front which represents a set of non-dominated solutions, and the last step is to determine a solution representing the best scheduling, which is called multi-objective decision making (MODM). There are different methods that can be adopted for MODM, and the AHP method is utilized in this paper. The main steps of the AHP method are as follows: Above all, the relative importance of each objective is judged in accordance with the fundamental scale defined in [32], a pairwise comparison matrix B is constructed, like in Equation (42), and the largest eigenvalue λ_{max} and corresponding normalized eigenvectors set ω are calculated. Then, the consistency check of B is performed to ensure the process of the AHP method is effective. Finally, ω is taken as the set of weight of each objective, as shown in Equation (43):

$$B = \begin{bmatrix} 1 & 3 & 5 \\ 1/3 & 1 & 3 \\ 1/5 & 1/3 & 1 \end{bmatrix} \tag{42}$$

$$\lambda_{max} = 3.0385, \ \omega = \{0.6090, 0.2793, 0.1116\} \tag{43}$$

6. Case Study

In order to validate the feasibility and validity of the proposed scheduling method for the RR-microgrid, a case study is performed respectively considering scenes of heating in winter as well as cooling in summer.

6.1. Case Introduction

Take a residential building (30 m long, 20 m wide, and 70 m high) consisting of 100 households as an example, like in [27], the total areas of the envelope structure and radiant floor are 7100 m^2 and 10,600 m^2, respectively, while the shading coefficient α is 0.2 and the window-to-wall ratio is 0.3. Material properties of the RFHCS and the envelope structure of the residential building are shown in Tables 4 and 5, respectively.

Table 4. Structure and material properties of RFHCS [33].

Type	Structure and Materials	Comprehensive Heat Transfer Coefficient (W/(m^2·°C))	Equivalent Heat Capacity (kJ/(m^2·°C))
Heavy floor	25 mm cement mortar +25 mm marble + 70 mm concrete (embedded diameter 20 mm pipe, spacing 150 mm)	11	148.1

Table 5. Structure and material properties of the envelope structure [25].

Type	Structure and Materials	Heat Transfer Coefficient (W/(m^2·°C))	Equivalent Heat Capacity (kJ/(m^2·°C))
External window	ordinary hollow glass + PV plastic window	2.80	6.0
External wall	25 mm cement mortar + 190 mm single row hole block + 25 mm adiabatic mortar	1.50	62

The specifications of the MTs, BESS, EHs, ECs, WT, and PV in the RR-microgrid are shown in Tables 6–10. The price of natural gas c_{gs} is 2.4 CNY/m^3, and the calorific value L_{HVNG} is 34.92 MJ/m^3. The upper limits of power purchasing from and selling to the external grid are $\overline{P}_{grid+} = \overline{P}_{grid-} = 1000$ kW.

Table 6. Specification of MTs.

Type	Number of Units	$\overline{P_{MT}}$ (kW)	c_{MT} (CNY/MWh)	η_{MTH}	η_{HE}
C200	3	600	30	0.53	0.95

Table 7. Specification of the BESS.

η_c/η_{disc}	$\overline{P_{Si+}}$ (kW)	$\overline{E}$ (kWh)	$\underline{E}$ (kWh)	E^0 (kWh)	c_{Si+}/c_{Si-}
0.9	80	550	50	150	0.01

Table 8. Specification of EHs.

Type	$\overline{P_{EH}}$ (kW)	COP$_{EH}$	c_{EH} (CNY/MWh)
CWDZ1080-85/70	1080	0.99	10

Table 9. Specification of ECs.

$\overline{P_{EC}}$ (kW)	COP$_{EC}$	c_{EC} (CNY/MWh)
1000	4	10

Table 10. Specification of WT and PV.

Type	Rated Power (kW)	Maintenance Cost (CNY/MWh)
WT	400	110
PV	300	80

Typical days in summer and in winter are chosen to carry out the scheduling experiment in Hebei Province of China, while the corresponding forecasted curves of solar radiation intensity and outdoor temperature are shown in Figure 4, forecasted curves of WT output, PV output, electric load, and price curves of purchasing electricity from the external grid are shown in Figure 5. The peak-valley purchasing electricity price curves released by Hebei Southern Grid is utilized for the scheduling experiment, while the selling electricity price is set to be 80% of the purchasing electricity price.

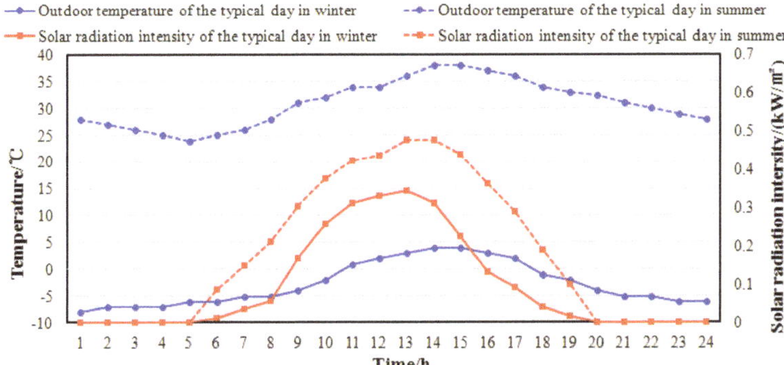

Figure 4. Forecasted curves of solar radiation intensity and outdoor temperature.

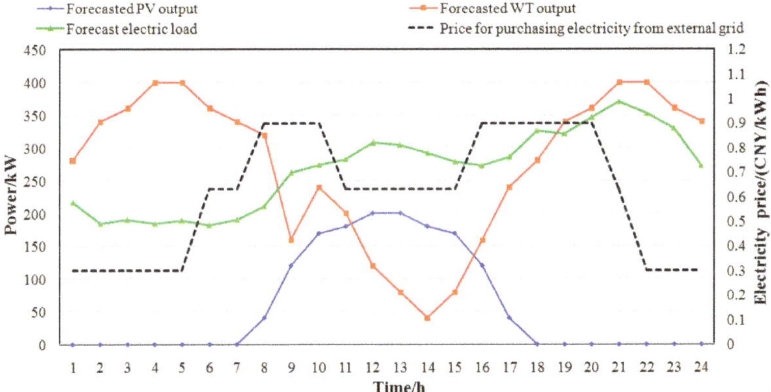

Figure 5. Forecasted curves of WT output, PV output, electric load, and electricity price curve.

The solving process of the presented optimization model for the RR-microgrid is implemented using MATLAB software from MathWorks Company of America. Set the parameters of the NSGA-II algorithm as: individual number in the population N_p = 200, maximum number of generation g_{max} = 60,000, crossover probability p_c = 0.9, and mutation probability p_m = 0.5.

6.2. Analysis of Scheduling Results

Optimal scheduling results are analyzed, respectively, for the RR-microgrid with RFHCS and the RR-microgrid with CHCS in this paper.

6.2.1. Scheduling Results of RR-Microgrid with RFHCS

The Pareto-optimal front of the presented multi-objective optimization model obtained by NSGA-II algorithm is shown in Figure 6. It is known that the Nhe edges SGA-II algorithm could gain enough optimal scheduling solutions, and the solution, which is at t of the Pareto-optimal front represent optimal scheduling schemes for minimized OC, TCL, and PE, respectively. The normalized objectives of the Pareto-optimal front sorted by TCL are shown in Figure 7 and, obviously, the OC and PE are two opposite objectives where increasing one of them decreases the other one when TCL is invariable.

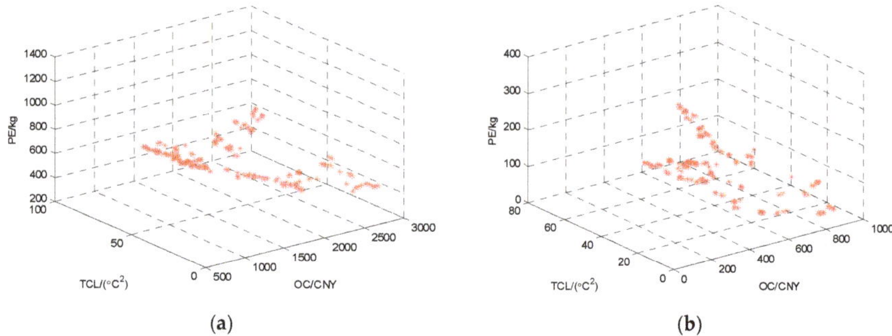

(**a**)

(**b**)

Figure 6. Pareto front of optimal scheduling for RR-microgrid with RFHCS. (**a**) A typical day in winter; and (**b**) a typical day in summer.

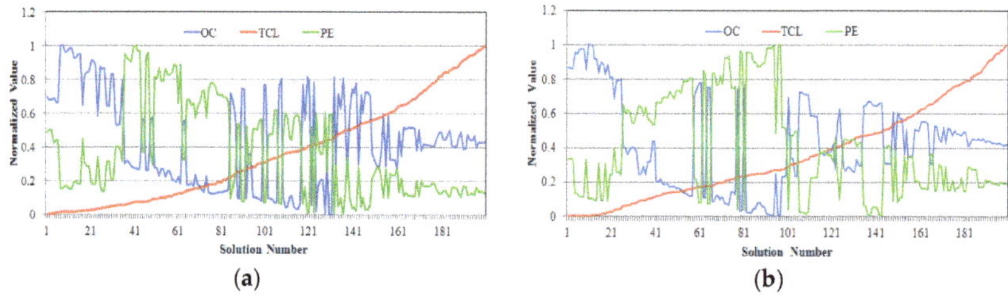

(**a**)

(**b**)

Figure 7. The normalized objectives of Pareto front of optimal scheduling for RR-microgrid with RFHCS. (**a**) A typical day in winter; and (**b**) a typical day in summer.

The optimal scheduling schemes for minimized TCL, OC, PE, as well as the best scheduling scheme determined by AHP method are shown in Figure 8, in which the discharging power of the BESS as well as the electric power selling to the external grid are taken as negative for the convenience of representation, and the corresponding value of the objectives are shown in Tables 11 and 12.

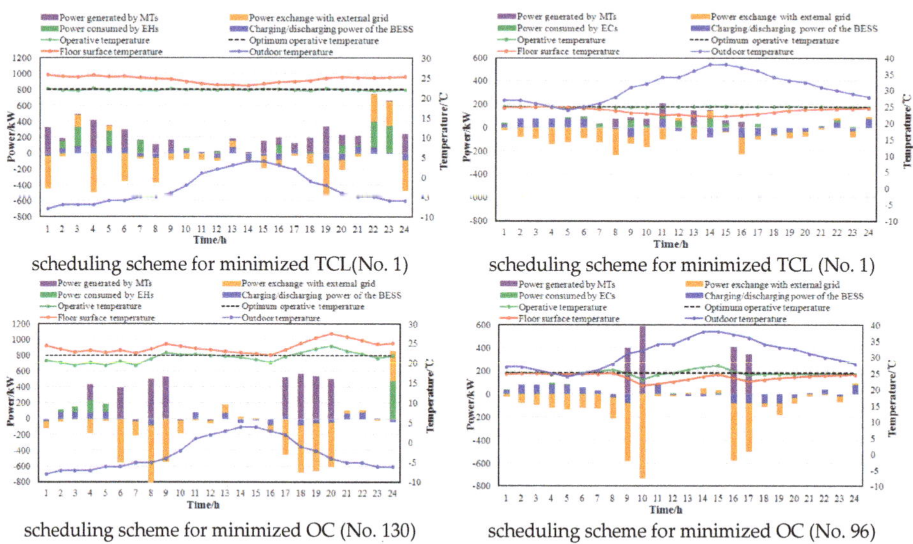

scheduling scheme for minimized TCL(No. 1)

scheduling scheme for minimized TCL (No. 1)

scheduling scheme for minimized OC (No. 130)

scheduling scheme for minimized OC (No. 96)

Figure 8. *Cont.*

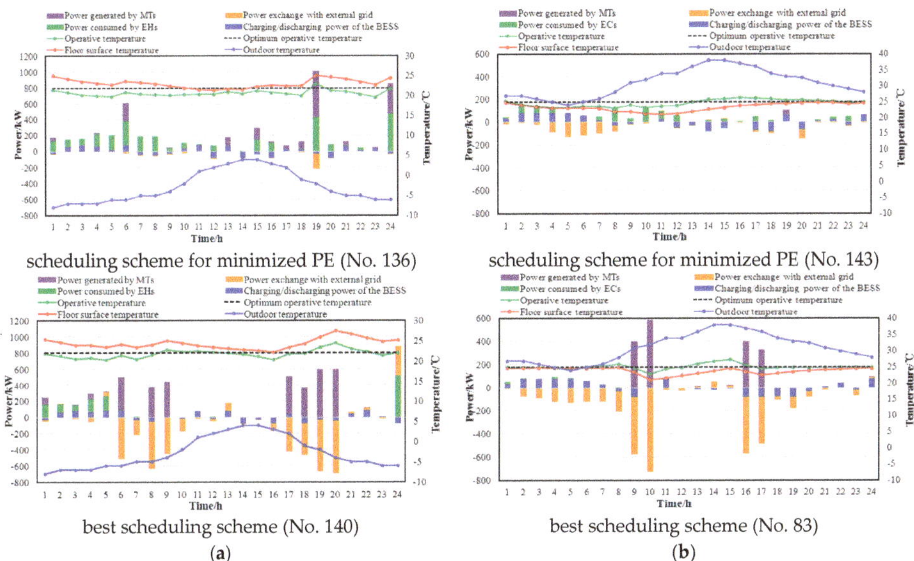

scheduling scheme for minimized PE (No. 136)

scheduling scheme for minimized PE (No. 143)

best scheduling scheme (No. 140)

best scheduling scheme (No. 83)

(**a**)

(**b**)

Figure 8. The optimal scheduling results for RR-microgrid with RFHCS. (**a**) A typical day in winter; and (**b**) a typical day in summer.

Table 11. Value of objective functions of different scheduling schemes for the typical day in winter.

Objective Functions	OC (CNY)	TCL ($^{\circ}$C^2)	PE (kg)
No. 1	2103.56	0.69	840.71
No. 130	736.49	44.65	949.51
No. 136	2326.27	48.89	348.62
No. 140	927.67	25.26	888.73

Table 12. Value of objective functions of different scheduling schemes for the typical day in summer.

Objective Functions	OC (CNY)	TCL ($^{\circ}$C^2)	PE (kg)
No. 1	759.75	0.09	135.97
No. 96	184.77	17.38	387.96
No. 143	623.80	31.87	9.46
No. 83	197.6	15.62	372.19

For all above scheduling schemes, the charging/discharging behaviors of the BESS are mainly effected by electricity price, that is, the charging behavior prefers to happen during lower price periods (0:00–5:00, 21:00–24:00), while the discharging behavior prefers to happen during higher price periods (7:00–10:00, 15:00–20:00), which is a benefit to the economy of the RR-microgrid, obviously.

As for the scheduling scheme for minimized TCL, it is observed that the operative temperature is quite close to the optimum operative temperature for all scheduling periods, which indicates that the thermal power generated by MTs and EHs/ECs could satisfy the heating/cooling demand quite well. According to the time-varying characteristic of the outdoor temperature, as well as solar radiation intensity, it is easily deduced theoretically that the cooling demand in summer concentrates on the daytime while the heating demand in winter concentrates on the nighttime. In addition, the heating demand in winter is obviously greater than the cooling demand in summer mainly because the difference between indoor temperature and outdoor temperature in winter is greater than in summer.

The distribution characteristic of electric power generated by MTs and power consumed by EHs/ECs on typical days in winter/summer agree quite well with the above-mentioned conclusion.

As for the scheduling scheme for minimized OC, on the typical day in winter, the EHs only consume power during lower price periods, while the MTs mainly work during higher price periods for at these time they could achieve better economy, respectively; on the typical day in summer, since to the cooling demand of the nighttime is quite small, the ECs almost stop work for all periods, while the MTs only work during higher price periods in daytime. On the whole, during lower price periods, less thermal energy are generated by MTs and EHs/ECs, therefore operative temperature falls slowly or maintains at a low level on the typical day in winter while it rises slowly or maintains at a high level on the typical day in summer; on the contrary, during higher price periods, more thermal power is generated by MTs and EHs/ECs, therefore, the operative temperature rises slowly on the typical day in winter while it falls slowly on the typical day in summer. Obviously, to achieve the minimized OC, there is little electric power purchasing from the external grid on the whole, but quite a lot of electric power selling to the external grid during higher price periods. It should be realized that both the MTs and EHs/ECs nearly stop working during the middle price periods (10:00–15:00) for typical days in winter/summer, which indicates that the RFHCS is able to pre-store considerable heat/cool energy enough to maintain the operating temperature within a reasonable scope for the next several scheduling periods.

As for the scheduling scheme for minimized PE, for both typical days, no electric power is purchased from the grid due to the coal consumption having greater a total emission coefficient than natural gas. Meanwhile, the electric power selling to the grid is obviously less than that of the scheduling schemes for minimized OC and TCL.

As for the best scheduling scheme determined by the AHP method, the variation trend of the operative temperature is quite similar with that of the scheduling scheme for minimized OC, while smaller TCL and PE are achieved through the power redistribution between EHs/ECs and MTs. Obviously, it is essentially a compromise scheduling scheme with overall consideration of multi-objectives according to the set of weight of each objective in AHP method.

In this paper, heating/cooling demand of the RR-microgrid is the sum of the heating/cooling power generated by EHs/ECs and MTs. Treating the scheduling scheme for minimized TCL as the condition without VESS, while the best scheduling scheme and scheduling scheme for minimized OC as the condition with the VESS, respectively, then the curves of the heating/cooling demand with/without the VESS are shown in Figures 9 and 10. Considering the curve of the heating/cooling demand without the VESS as the reference curve, it is known that the curve of the heating/cooling demand with the VESS fluctuates around the reference curve. Therefore, the part above the reference curve could be considered as 'charging', and the part below the reference curve could be considered as 'discharging', then charging/discharging power for the VESS is obtained as the difference of the heating/cooling demand between the two cases. It is found that the performance of the VESS in best scheduling scheme is quite similar with that in scheduling scheme for minimized OC for both the typical days. In addition, compared with BESS, the VESS has the contrary charging/discharging response to the changing of electricity price. For both typical days in winter/summer, the charging process of the VESS mainly happens during higher price periods, while the discharging process of the VESS mainly happens at lower price or middle-price periods.

Owing to the VESS capacity of RFHCS being quite considerable, the OC of the best scheduling scheme and the scheduling scheme for minimized OC has a dramatic decline compared with the condition without the VESS (55.90% and 64.98% for the typical day in winter, 73.99% and 75.68% for the typical day in summer).

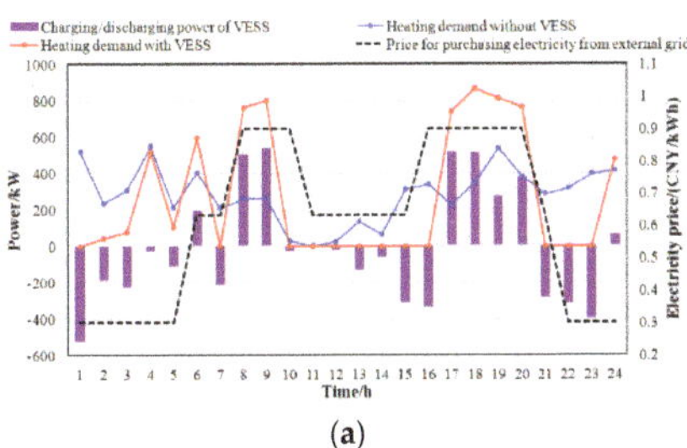
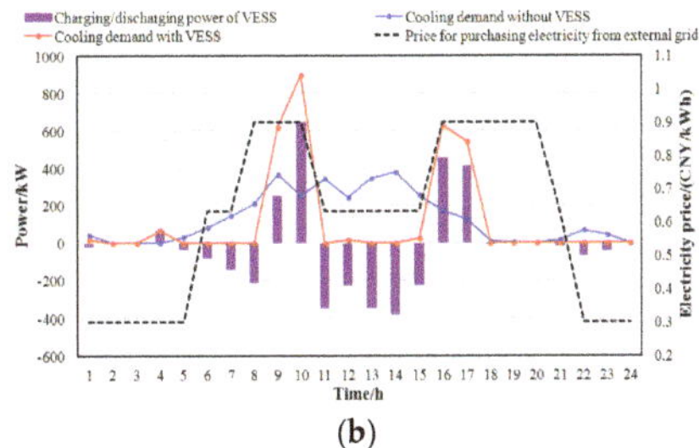

(a)   (b)

Figure 9. VESS charging/discharging power in minimized OC scheduling scheme for RR-microgrid with RFHCS. (**a**) A typical day in winter; and (**b**) a typical day in summer.

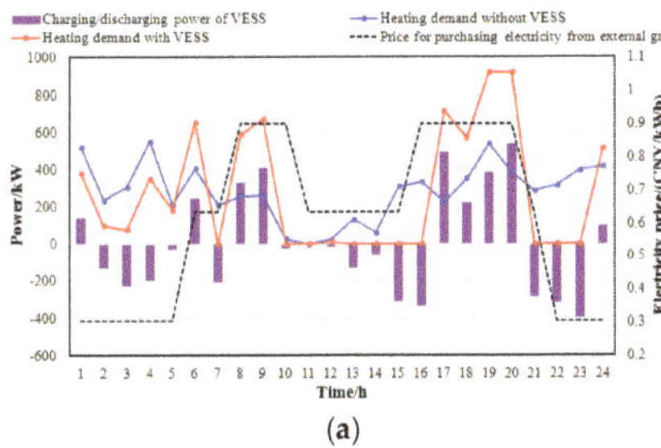
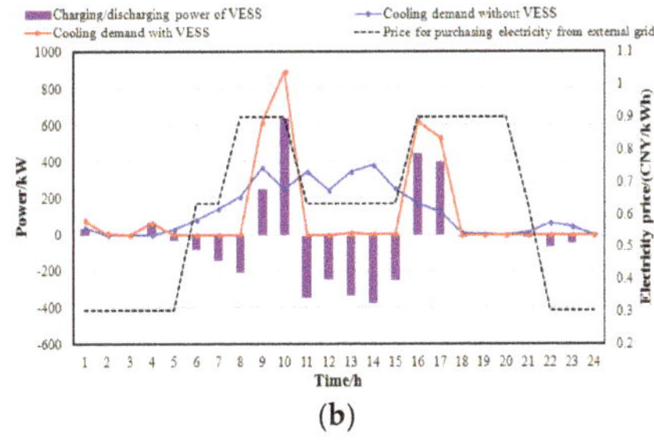

(a)   (b)

Figure 10. VESS charging/discharging power in best scheduling scheme for RR-microgrid with RFHCS. (**a**) A typical day in winter; and (**b**) a typical day in summer.

6.2.2. Scheduling Results of the RR-Microgrid with CHCS

The Pareto-optimal front of the presented multi-objective optimization model obtained by the NSGA-II algorithm is shown in Figure 11. Similarly, it can be known that the NSGA-II algorithm could also gain enough optimal scheduling solutions.

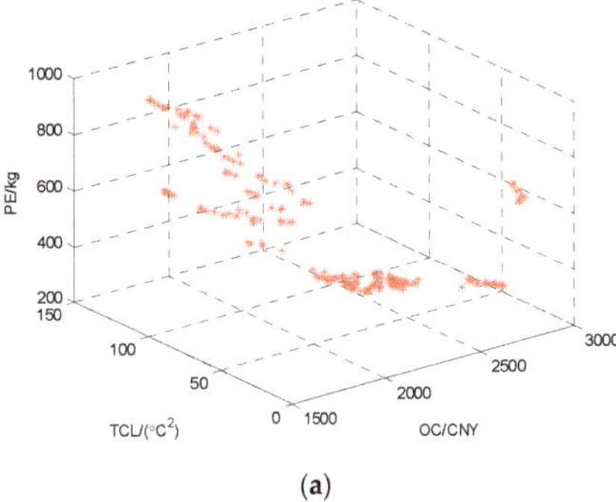
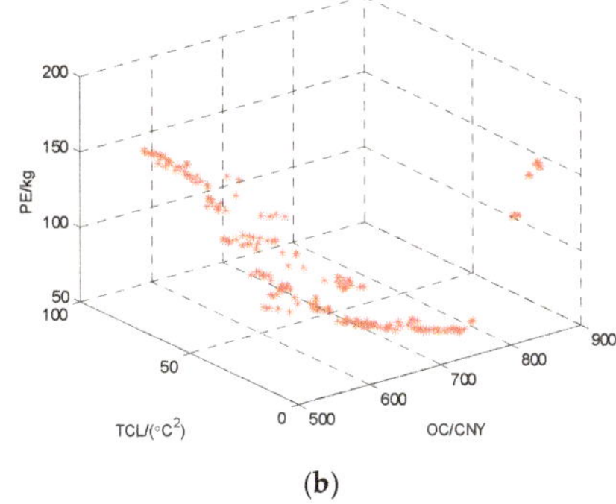

(a)   (b)

Figure 11. Pareto front of optimal scheduling for RR-microgrid with CHCS. (**a**) A typical day in winter; and (**b**) a typical day in summer.

The normalized objectives of Pareto-optimal front sorted by TCL are shown in Figure 12, obviously, and the OC and PE are also two opposite objectives where increasing one of them decreases the other one when TCL is invariable.

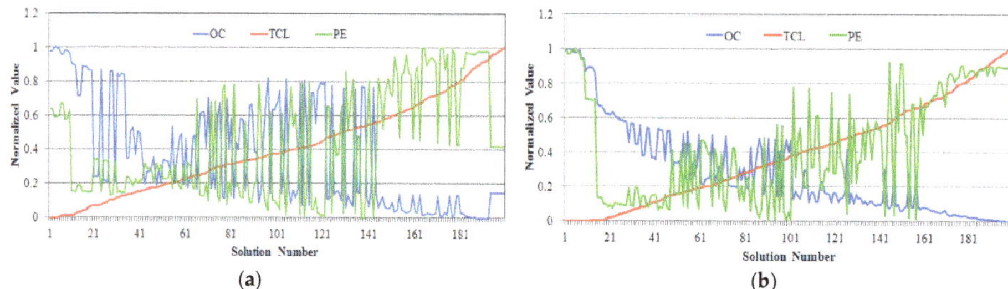

Figure 12. The normalized objectives of Pareto front of optimal scheduling for the RR-microgrid with CHCS. (**a**) A typical day in winter; and (**b**) a typical day in summer.

The optimal scheduling schemes for minimized TCL, OC, PE, as well as the best scheduling scheme determined by the AHP method are shown in Figure 13, and the corresponding objectives are shown in Tables 13 and 14. For the sake of convenience of representation, the discharging power of the BESS, as well as the electric power selling to the external grid, are also taken as negative in Figure 13.

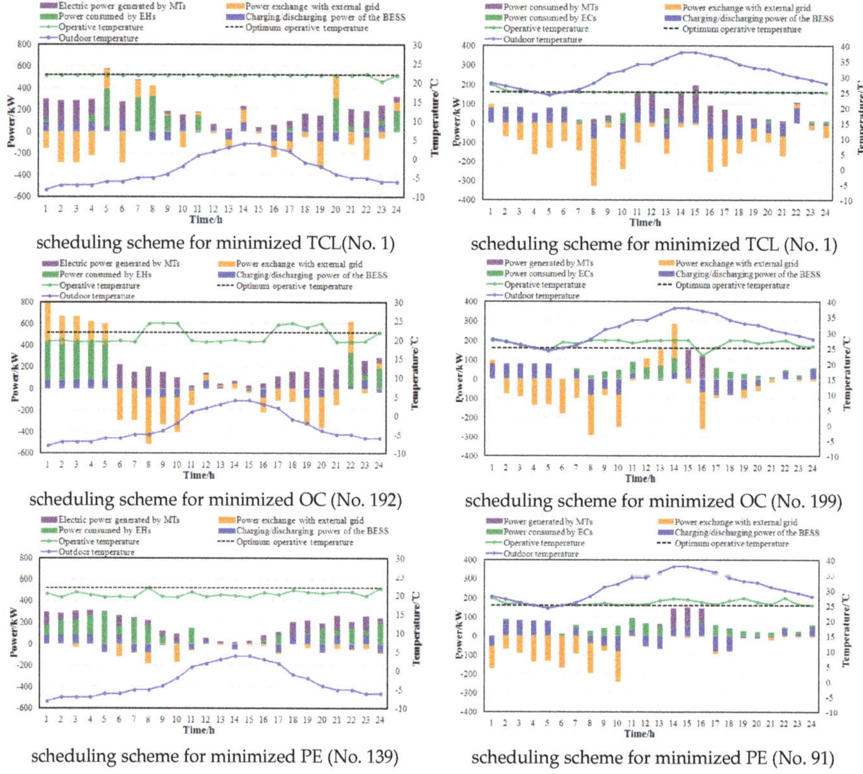

Figure 13. *Cont.*

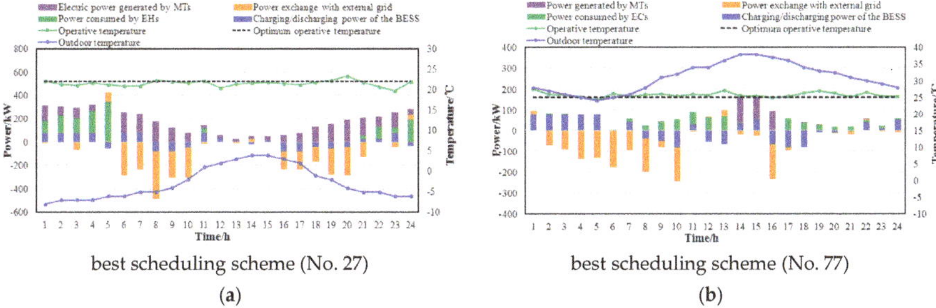

best scheduling scheme (No. 27)	best scheduling scheme (No. 77)
(**a**)	(**b**)

Figure 13. The optimal scheduling results for RR-microgrid with CHCS. (**a**) A typical day in winter; and (**b**) a typical day in summer.

Table 13. Value of objective functions of different scheduling schemes for a typical day in winter.

Objective Functions	OC (CNY)	TCL ($^\circ$C^2)	PE (kg)
No.1	2733.71	2.88	726.57
No.192	1686.66	123.22	957.37
No.139	2512.08	73.61	292.81
No.27	1911.02	15.24	518.05

Table 14. Value of objective functions of different scheduling schemes for a typical day in summer.

Objective Functions	OC (CNY)	TCL ($^\circ$C^2)	PE (kg)
No.1	840.75	0.73	166.17
No.199	545.72	85.99	154.49
No.91	689.18	28.64	57.64
No.77	604.29	22.72	84.35

For all of the above scheduling schemes, the charging/discharging behavior of the BESS are quite similar with that in the mentioned scheduling schemes for the RR-microgrid with RFHCS, that is to say, the charging behavior of the BESS prefers to happen during lower price periods, while the discharging behavior prefers to happen during higher price periods. However, the electric power consumed by EHs/ECs and generated by MTs are more balanced than the corresponding result in the RR-microgrid with RFHCS. The reason for this phenomenon is that the total heat capacity of the indoor air is quite smaller than that of the radiant floor, the heat/cool energy stored by the previous scheduling period is not enough to support the EHs/ECs and MTs to stop working for the next one or more scheduling periods.

As for the scheduling scheme for minimized TCL, the operative temperature is quite close to the optimum operative temperature at all of the scheduling periods, which also indicates that the thermal power generated by MTs and EHs/ECs could satisfy the heating/cooling demand quite well.

As for the scheduling scheme for minimized OC, on the typical day in winter, EHs only consume power during lower price periods, while the MTs mainly work during higher price periods; on the typical day in summer, MTs and ECs mainly work during the daytime for the cooling demand in the nighttime is quite small. The ECs consume power during most scheduling periods, while the MTs only work during the few higher outdoor temperature periods (14:00–16:00) when the cooling demand is greater to achieve better economy than ECs. On the typical day in winter, the operative temperature maintains at a low level during lower price periods while maintains at a high level during higher price periods. On the typical day in summer, the operative temperature maintains at a high level at most scheduling periods while maintaining at a low level during the few lower outdoor temperature periods (4:00–6:00) or the higher outdoor temperature periods (14:00–16:00). To achieve the minimized OC, on the typical day in winter, purchasing electric power from the external grid happens during lower-periods and selling electric power to the external grid happens during higher price periods;

on the typical day in summer, purchasing electric power from the external grid happens mainly at midday scheduling periods while selling electric power to the external grid happens during most scheduling periods.

As for the scheduling scheme for minimized PE, for both typical days, there is also no electric power purchasing from the external grid due to the coal consumption has bigger total emission coefficient than natural gas. On the typical day in winter, the electric power selling to the external grid is obviously less than that of the scheduling schemes for minimized OC and TCL. While on the typical day in summer, the electric power selling to the external grid concentrate on the lower outdoor temperature periods (0:00–10:00) due to the cooling demand at that time is small.

As for the best scheduling scheme determined by the AHP method, compared with the scheduling the scheme for minimized OC, the operative temperature has a similar variation trend and has an obviously smaller variation magnitude so that TCL is dramatically decreased. Meanwhile, PE is effectively decreased.

Similarly, we treat the scheduling scheme for minimized TCL as the condition without the VESS, while the best scheduling scheme and scheduling scheme for the minimized OC as the condition with the VESS, respectively; then the curves of heating/cooling demand with/without the VESS are shown in Figures 14 and 15. For both the best scheduling scheme and scheduling scheme for the minimized OC, on the typical day in winter, the charging process of the VESS mainly happens during the higher price periods, while discharging process of the VESS mainly happens during the lower price periods; on the typical day in summer, the charging process of the VESS rarely happens while the discharging process of the VESS happens during most scheduling periods. Compared with the result shown in Figures 9 and 10, it is found that the magnitude of the charging/discharging power of the VESS becomes significantly smaller. The reason for this phenomenon is that the VESS capacity of the CHCS is obviously smaller than that of the RFHCS. Consequently, compared with the condition without the VESS, the OC of the best scheduling scheme and scheduling scheme for the minimized OC has a relatively small decline (30.10% and 38.32% for a typical day in winter, 28.12% and 35.14% for a typical day in summer).

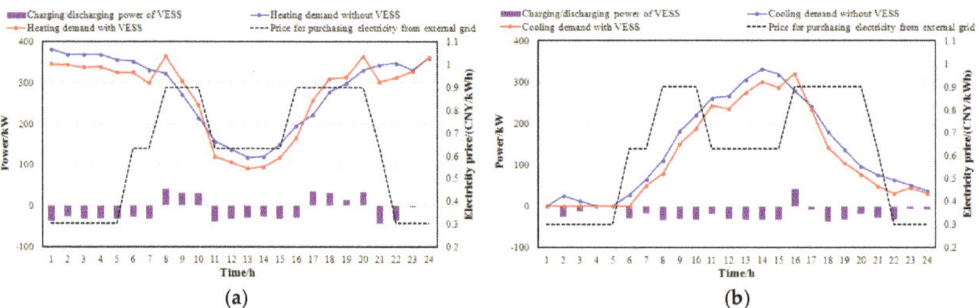

Figure 14. VESS charging/discharging power in minimized OC scheduling scheme for RR-microgrid with CHCS. (**a**) A typical day in winter; and (**b**) a typical day in summer.

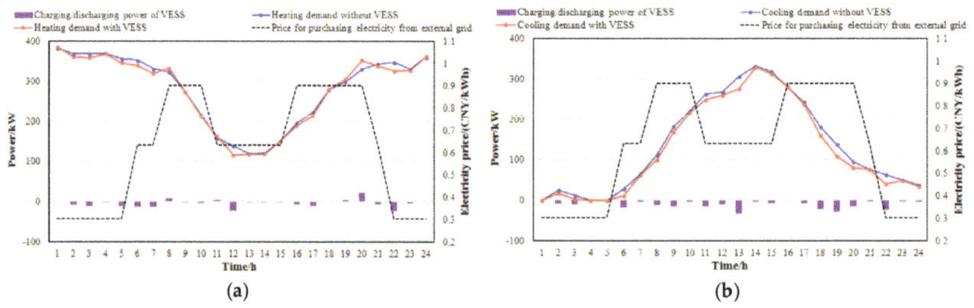

Figure 15. VESS charging/discharging power in best scheduling scheme for RR-microgrid with CHCS. (**a**) A typical day in winter; and (**b**) a typical day in summer.

The charge/discharge power characteristics of the two kinds of VESS in the minimized OC scheduling scheme and in the best scheduling scheme are calculated, as shown in Table 15, Table 16, and Figure 16. It can be known that, compared with the VESS of CHCS, the charge/discharge power characteristics of the VESS of the RFHCS in the minimized OC scheduling scheme are more similar with that in the best scheduling scheme. In addition, the mean charging/discharging power and the total charging/discharging amount are obviously greater, indicating that the heat capacity of the heavy radiant floor is much higher than that of the indoor air.

Table 15. Performance of the VESS of RFHCS for the typical days in winter/summer.

Scheduling Scheme		Maximum Charging Power	Maximum Discharging Power	Mean Charging/ Discharging Power	Total Charging Amount	Total Discharging Amount
Typical day in winter	No. 130	539.21	520.65	256.67	2989.17	3170.86
	No. 140	537.40	398.18	230.01	2851.14	2668.95
Typical day in summer	No. 96	647.90	382.67	166.20	1841.12	2158.66
	No. 83	638.87	379.26	166.66	1835.24	2153.42

Table 16. Performance of the VESS of CHCS for the typical days in winter/summer.

Scheduling Scheme		Maximum Charging Power	Maximum Discharging Power	Mean Charging/ Discharging Power	Total Charging Amount	Total Discharging Amount
Typical day in winter	No. 192	41.38	42.59	29.26	217.39	484.88
	No. 27	21.55	22.29	7.44	42.96	135.66
Typical day in summer	No. 199	41.70	36.93	22.10	41.69	488.79
	No. 77	0.58	31.58	9.54	0.58	228.34

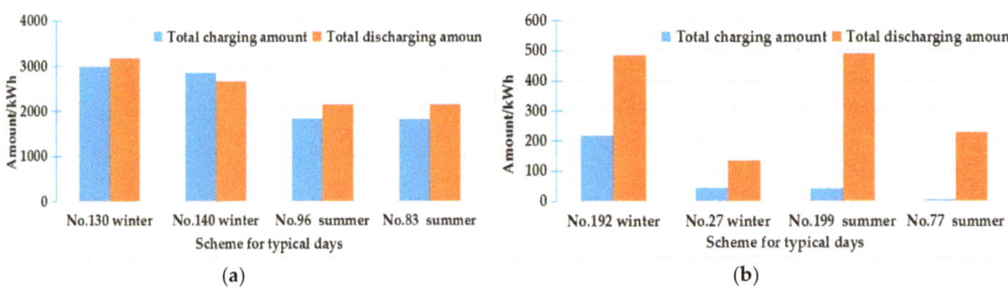

Figure 16. Comparison of total charging/discharging amount. (**a**) VESS of RFHCS; and (**b**) VESS of CHCS.

7. Conclusions

A novel multi-objective optimal scheduling method for a grid-connected RR-microgrid is presented in which the heating/cooling system of a residential building is considered as a virtual energy storage system. The following conclusions are drawn:

(1) The NSGA-II algorithm could obtain enough optimal scheduling schemes for the presented multi-objective optimization model of RR-microgrid, and the OC and PE are two opposite objectives where increasing one of them decreases the other one when TCL is invariable. The best scheduling scheme could be reasonably selected by the AHP method according to the set of weights of each objective.

(2)   As for the best scheduling scheme of RR-microgrid with RFHCS, the charging process of VESS mainly happens during higher price periods, while the discharging process of VESS mainly happens during lower price or middle price periods for both typical days in winter/summer.

(3)   As for the best scheduling scheme of the RR-microgrid with CHCS, on the typical day in the winter, the charging process of the VESS mainly happens during higher electricity price periods, and the discharging process of the VESS mainly happens during lower electricity price periods; on a typical day in the summer, the charging process of VESS rarely happens while the discharging process of VESS happens during most scheduling periods.

(4)   Due to the VESS capacity of the CHCS being obviously smaller than that of the RFHCS, as for the corresponding best scheduling scheme, the electric power consumed by EHs/ECs and generated by MTs in the RR-microgrid with CHCS are more balanced than that in the R-microgrid with RFHCS. Meanwhile, compared with the condition without VESS, the OC of the RR-microgrid with RFHCS has a greater decline than that of the RR-microgrid with CHCS.

Author Contributions: W.L. proposed the optimization model and translated the original manuscript. C.L. and Y.L. checked the results of the whole manuscript. K.B. and L.M. contributed to the case study. W.C. performed, in part, the research tasks.

Funding: This research was supported by Beijing Natural Science Foundation (4182061) and the Fundamental Research Funds for the Central Universities (2017MS134, 2018ZD05).

Acknowledgments: Great thanks for valuable comments and suggestions of the reviewers and the editors of Processes.

Conflicts of Interest: The authors declare no conflict of interest.

Nomenclature

F_{gas}	Natural gas MTs consumed per unit time	A_{wi}	Total area of external window
c_{gas}	Price of natural gas	C_{gl}	Equivalent heat capacity for radiant floor (kJ/(kg·°C))
P_{MT}	Output electric power of the MTs	C_{wa}	Equivalent heat capacity for external wall (kJ/(kg·°C))
Q_{MT}	Output thermal power of the MTs	C_{wi}	Equivalent heat capacity for external window (kJ/(kg·°C))
L_{HVNG}	Low calorific value for natural gas	ρ	Density of the indoor air
η_{MTE}	Electric power efficiency of the MTs	C	Heat capacity of the indoor air
η_{MTH}	Thermal power efficiency of the MTs	V	Volume of the indoor air
η_{HE}	Conversion efficiency of WHRS	h_z	Comprehensive heat transfer coefficient from radiant floor surface to indoor air and envelope structure
Q_{MTH}	Output heating power of the WHRS	k_{wa}	Heat transfer coefficient of the external wall
Q_{MTC}	Output cooling power of the ACs	K_{wi}	Heat transfer coefficient of the external window
COP_{AC}	Coefficient of performance of the ACs	c_{Si+}	Unit costs for charging of the BESS
COP_{EH}	Coefficient of performance of the EHs	c_{Si-}	Unit costs for discharging of the BESS
COP_{EC}	Coefficient of performance of the ECs	c^t_{grid+}	Price of purchasing electricity from grid at period t
Q_{EH}	Heating power generated by the EHs	c^t_{grid-}	Price for selling electricity to grid at period t
Q_{EC}	Cooling power generated by the ECs	P^t_{WT}	Output power of WT at period t
P_{EH}	Electric power consumed by the EHs	P^t_{PV}	Output power of PV at period t
P_{EC}	Electric power consumed by the ECs	c_{WT}	Unit maintenance cost of WT
E^t	SOC of the BESS at the end of period t	c_{PV}	Unit maintenance cost of PV
E^{t-1}	SOC of the BESS at the end of period $t-1$	c_{MT}	Unit maintenance cost of MTs
P^t_{Si+}	Charging power of the BESS at period t	c_{EH}	Unit maintenance cost of EHs
P^t_{Si-}	Discharging power of the BESS at period t	c_{EC}	Unit maintenance cost of ECs
U^t_{Si+}	Charging status of the BESS for period t	c_{AC}	Unit maintenance cost of ACs
U^t_{Si-}	Discharging status of the BESS for period t	I	Total solar radiation intensity
θ_T	Number of scheduling periods	$\overline{P_{Si+}}$	Upper limit for charging power of the BESS

η_c	Charging efficiency of the BESS	α	Shading coefficient of residential building
η_{disc}	Discharging efficiency of the BESS	λ_e	Total emission coefficient for coal consumption
ΔT	Time length of scheduling period	λ_g	Total emission coefficient for natural gas consumption
Q	Heating/cooling demand	$\overline{P_{Si-}}$	Upper limit for discharging power of the BESS
Q_s	Solar radiation load	$\overline{E}$	Upper limit for SOC
T_g	Surface temperature of radiant floor	$\underline{E}$	Lower limit for SOC
T_z	Operative temperature	$\overline{P_{grid+}}$	Upper limit for power purchasing from grid
T_{out}	Outdoor temperature	$\overline{P_{grid-}}$	Upper limit for power selling to grid
C_g	Equivalent heat capacity for radiant floor (J/°C)	$\overline{P_{MT}}$	Upper limit for output electric power of MTs
C_w	Equivalent heat capacity for envelope structure (J/°C)	$\overline{P_{EH}}$	Upper limit for electric power consumed by EHs
C_A	Equivalent heat capacities for indoor air (J/°C)	$\overline{P_{EC}}$	Upper limit for electric power consumed by ECs
R_W	Equivalent heat resistance for envelope structure	$\overline{T_Z}$	Upper limit of operative temperature
R_Z	Equivalent heat resistance for convection and radiation from the radiant floor surface to the indoor air and the envelope structure	$\underline{T_Z}$	Lower limit of operative temperature
A_g	Total area of radiant floor	$\underline{T_g}$	Dewpoint temperature
A_{wa}	Total area of external wall		

References

1. European Commission. A Policy Framework for Climate and Energy in the Period from 2020 up to 2030, Brussels. 2014. Available online: http://eur-lex.europa.eu/legal-content/EN/TXT/?uri=CELEX:52013DC0169 (accessed on 17 May 2019).
2. den Elzen, M.; Fekete, H.; Höhne, N.; Admiraal, A.; Forsell, N.; Hof, A.F.; Olivier, J.G.; van Soest, H. Greenhouse gas emissions from current and enhanced policies of China until 2030: Can emissions peak before 2030? *Energy Policy* **2016**, *89*, 224–236. [CrossRef]
3. Yang, L.; Wang, J.; Shi, J. Can China meet its 2020 economic growth and carbon emissions reduction targets? *J. Clean. Prod.* **2017**, *142*, 993–1001. [CrossRef]
4. Jin, X.; Wu, J.; Mu, Y.; Wang, M.; Xu, X.; Jia, H. Hierarchical microgrid energy management in an office building. *Appl. Energy* **2017**, *208*, 480–494. [CrossRef]
5. Yu, J.; Tian, L.; Xu, X.; Wang, J. Evaluation on energy and thermal performance for office building envelope in different climate zones of China. *Energy Build.* **2015**, *86*, 626–639. [CrossRef]
6. Guan, X.; Xu, Z.; Jia, Q. Energy-efficient buildings facilitated by microgrid. *IEEE Trans. Smart Grid* **2010**, *1*, 243–252.
7. Pesin Michael. U.S. Department of Energy Electricity Grid Research and Development. In Proceedings of the American Council of Engineering Companies, Environment and Energy Committee Winter Meeting, Washington, DC, USA, 9 February 2017.
8. Navigant Research. More Than 400 Microgrid Projects are Under Development Worldwide [EB/OL]. Available online: http://www.navigantresearch.com/newsroom/more-than-400-microgrid-projects-are-under-development-worldwide (accessed on 2 April 2013).
9. Feng, W.; Jin, M.; Liu, X.; Bao, Y.; Marnay, C.; Yao, C.; Yu, J. A review of microgrid development in the United States–A decade of progress on policies, demonstrations, controls, and software tools. *Appl. Energy* **2018**, *228*, 1656–1668. [CrossRef]
10. Cagnano, A.; De Tuglie, E.; Cicognani, L. Prince-Electrical Energy Systems Lab: A pilot project for smart microgrids. *Electr. Power Syst. Res.* **2017**, *148*, 10–17. [CrossRef]
11. Jaramillo, L.B.; Weidlich, A. Optimal microgrid scheduling with peak load reduction involving an electrolyzer and flexible loads. *Appl. Energy* **2016**, *169*, 857–865. [CrossRef]
12. Wu, X.; Wang, X.; Wang, J.; Bie, C. Economic generation scheduling of a microgrid using mixed integer programming. *Proc. CSEE* **2013**, *33*, 1–8. (In Chinese)
13. Jiang, Q.; Xue, M.; Geng, G. Energy management of microgrid in grid-connected and stand-alone modes. *IEEE Trans. Power Syst.* **2013**, *28*, 3380–3389. [CrossRef]

14. Lu, Y.; Wang, S.; Sun, Y.; Yan, C. Optimal scheduling of buildings with energy generation and thermal energy storage under dynamic electricity pricing using mixed-integer nonlinear programming. *Appl. Energy* **2015**, *147*, 49–58. [CrossRef]

15. Zhao, Y.; Lu, Y.; Yan, C.; Wang, S. MPC-based optimal scheduling of grid-connected low energy buildings with thermal energy storages. *Energy Build.* **2015**, *86*, 415–426. [CrossRef]

16. Javidsharifi, M.; Niknam, T.; Aghaei, J.; Mokryani, G. Multi-objective short-term scheduling of a renewable-based microgrid in the presence of tidal resources and storage devices. *Appl. Energy* **2018**, *216*, 367–381. [CrossRef]

17. Carpinelli, G.; Mottola, F.; Proto, D.; Russo, A. A multi-objective approach for microgrid scheduling. *IEEE Trans. Smart Grid* **2017**, *8*, 2109–2118. [CrossRef]

18. Lin, W.; Jin, X.; Mu, Y.; Jia, H.; Xu, X.; Yu, X.; Zhao, B. A two-stage multi-objective scheduling method for integrated community energy system. *Appl. Energy* **2018**, *216*, 428–441. [CrossRef]

19. Yin, Z.; Che, Y.; Li, D.; Liu, H.; Yu, D. Optimal scheduling strategy for domestic electric water heaters based on the temperature state priority list. *Energies* **2017**, *10*, 1425. [CrossRef]

20. Lu, N. An evaluation of the HVAC load potential for providing load balancing service. *IEEE Trans. Smart Grid* **2012**, *3*, 1263–1270. [CrossRef]

21. Wang, C.; Liu, M.; Lu, N. A tie-line power smoothing method for microgrid using residential thermostatically-controlled loads. *Proc. CSEE* **2012**, *2*, 36–43. (In Chinese)

22. Jia, H.; Qi, Y.; Mu, Y. Frequency response of autonomous microgrid based on family-friendly controllable loads. *Sci. China Technol. Sci.* **2013**, *43*, 247–256. (In Chinese) [CrossRef]

23. Van, R.J.; Leemput, N.; Geth, F.; Büscher, J.; Salenbien, R.; Driesen, J. Electric vehicle charging in an office building microgrid with distributed energy resources. *IEEE Trans. Sustain. Energy* **2014**, *5*, 1–8. [CrossRef]

24. Igualada, L.; Corchero, C.; Cruz-Zambrano, M.; Heredia, F.J. Optimal energy management for a residential microgrid including a vehicle-to-grid system. *IEEE Trans. Smart Grid* **2014**, *5*, 2163–2172. [CrossRef]

25. Jin, X.; Mu, Y.; Jia, H. Optimal scheduling method for a combined cooling, heating and power building microgrid considering virtual storage system at demand side. *Proc. CSEE* **2017**, *37*, 581–590. (In Chinese)

26. Jin, X.; Mu, Y.; Jia, H.; Wu, J.; Tao, J.; Yu, X. Dynamic economic dispatch of a hybrid energy microgrid considering building based virtual energy storage system. *Appl. Energy* **2016**, *194*, 386–398. [CrossRef]

27. Liu, W.; Liu, C.; Lin, Y.; Ma, L.; Bai, K.; Wu, Y. Optimal scheduling of residential microgrids considering virtual energy storage system. *Energies* **2018**, *11*, 942. [CrossRef]

28. Fanger, P.O. *Analysis and Applications in Environmental Engineering*; Thermal Comfort Analysis & Applications in Environmental Engineering; McGraw Hill: New York, NY, USA, 1970.

29. Wang, J. Optimal Design of Building Cooling Heating and Power System and Its Multi-Criteria Integrated Evaluation Method. Ph.D. Thesis, North China Electric Power University, Beijing, China, 2010.

30. Brownlee, A.E.I.; Wright, J.A. Constrained, mixed-integer and multi-objective optimisation of building designs by NSGA-II with fitness approximation. *Appl. Soft Comput.* **2015**, *33*, 114–126. [CrossRef]

31. Deb, K.; Pratap, A.; Agarwal, S.; Meyarivan, T. A fast and elitist multiobjective genetic algorithm: NSGA-II. *IEEE Trans. Evol. Comput.* **2002**, *6*, 182–197. [CrossRef]

32. Chen, W. Quantitative decision-making model for distribution system restoration. *IEEE Trans. Power Syst.* **2010**, *25*, 313–321. [CrossRef]

33. Zhao, K.; Liu, X.; Jiang, Y. Dynamic performance of water-based radiant floors during start-up and high-intensity solar radiation. *Sol. Energy* **2014**, *101*, 232–244. [CrossRef]

 *processes*

Article

Control Strategy of Electric Heating Loads for Reducing Power Shortage in Power Grid

Siyuan Xue [1], Yanbo Che [1,*], Wei He [2], Yuancheng Zhao [1] and Ruiping Zhang [3]

[1] Key Laboratory of Smart Grid of Ministry of Education, Tianjin University, Tianjin 300072, China; 1029348750@163.com (S.X.); cheng528zyc@163.com (Y.Z.)

[2] State Grid Jiangxi Electric Power Research Institute, Nanchang 330096, China; lanlyhw@163.com

[3] College of Automation and Electrical Engineering, Lanzhou Jiaotong University, Gansu 730070, China; zhrp74@mail.lzjtu.cn

* Correspondence: lab538@163.com; Tel.: +86-166-026-10538

Received: 17 March 2019; Accepted: 24 April 2019; Published: 9 May 2019

Abstract: With the development of demand response technology, it is possible to reduce power shortages caused by loads participating in power grid dispatching. Based on the equivalent thermal parameter model, and taking full account of the virtual energy storage characteristics presented during electro-thermal conversion, a virtual energy storage model suitable for electric heating loads with different electrical and thermal parameters is proposed in this paper. To avoid communication congestion and simplify calculations, the model is processed by discretization and linearization. To simplify the model, a control strategy for electric heating load, based on the virtual state ofcharge priority list, is proposed. This paper simulates and analyzes a control example, explores the relevant theoretical basis affecting the control effect, and puts forward an optimization scheme for the control strategy. The simulation example proved that the proposed method in this paper can reduce power storage in the grid over a long period of time and can realize a power response in the grid.

Keywords: power shortage; electric heating load; electric water heater; demand response; virtual energy storage (VES), virtual state of charge (VSOC)

---

1. Introduction

When a power shortage occurs in the generation side of the power system, the system frequency will be reduced, resulting in a series of power quality problems. Demand response [1,2] can solve the mismatch problem between supply and demand at a relatively lower cost, which is of great significance to absorb new energy sources, reduce power shortages, and reduce environmental pollution [3–5]. The popularization of advanced metering infrastructure [6] makes it possible to use the control strategy of thermostatically controlled load (TCL) and scale the application of trunked dispatching.

As an important part of the demand response, TCLs realize virtual power storage through indirect heat energy storage. The main principle of electric heating load is Joule's law, that is, the chamber temperature is maintained within a certain temperature range through the on-off state of resistance wire heating. On the premise of guaranteeing users' basic comfort [7], electrical heating equipment, such as electric water heaters [8,9], air conditioners, refrigerators, can be equivalent to virtual power storage devices, so electric heating loads can be regarded as good TCLs. An electrical heating device forced to shut down for several minutes can reduce power consumption by a small amount without dropping the temperature below the comfortable temperature range. When a large amount of electric heating loads are forced to shut down in order, they can release a large amount of electric power. The indirect energy storage capability of electric heating loads can reduce peak load and improve the reliability of power grid operation.

Several studies on the virtual energy storage (VES) model of electric heating loads have been undertaken. Reference [10] analyzed the mechanism of aggregated load oscillations caused by the traditional temperature adjusting method, on the basis of the equivalent thermal parameter (ETP) model, and modeled household electric heating load. However, the rationality for model linearization has not been quantitatively analyzed. Reference [11] presented a temperature state priority list method to suppress power flow and decreased power shortage demand for a given micro-grid; however, the modeling of the working state of the constant temperature state was not accurate enough. Reference [12] presented a load model suitable for terminal voltage control of electric water heaters, which could reduce the peak load of the power grid while ensuring the comfort of users, but the modeling process was not described in detail. In [13], using the characteristics of household electric heating loads such as electric water heaters, a high-precision model reflecting different working conditions of electric heating loads was proposed. However, due to the limitation of computational complexity, this model was suitable for small-scale regulation and control only, instead of for large power grid-trunked dispatching.

ETP modelling, as the theoretical basis of control, is widely used [14]. However, previous studies lack an analysis of power parameters, the complete VES index system, and the deep mining of the coupling relationship between variables, and do not accurately reflect the actual electro-thermal conversion relationship.

In terms of the control algorithm, reference [15] started from the macro-layer of the grid side and the micro-layer of the load aggregator, and presented a bi-level optimal dispatch and control model for air-conditioning loads based on direct load control. But the running states of loads before control right transferring need to be uniformly distributed. Reference [10] proposed a new temperature-adjusting method on the basis of the ETP model, to avoid load oscillations caused by the traditional temperature regulation method, but the parameters of the devices participating in the demand side response needed to be the same. Reference [16] proposed a demand-side decentralized control strategy with variable participation to provide directional control of the start-up and shutdown of TCLs, so as to improve the frequency regulation capability of isolated microgrid systems in collaboration with energy storage systems. But, the operation of the units in the cluster control was not analyzed. Reference [17] developed a weighting coefficient-queuing algorithm based on a modified coloredpower algorithm state-queuing model, which can be used to directly control the TCLs of electric heating equipment.

The daily load peak of a power grid generally lasts for several hours, thus the transfer of control rights can last from a few minutes to hours. Previous studies have paid less attention to the analysis of the control effect in the case of long-term (several hours) transfer of control rights and theoretical analysis of factors affecting the control effect.

In this paper, a load model and a trunked dispatching strategy for electric heating loads are analyzed deeply to solve the problem of electric heating loads with different parameters and demands participating in demand response at the same time. Based on a simplified first-order ETP model, a VES model, which can reflect the electro-thermal exchange, is proposed where the potential of demand response can be fully exploited. Based on this model, the trunked dispatching strategy based on virtual state of charge (VSOC) priority list, is proposed. The control effect under the condition of long-term control right transferring is analyzed, and the control strategy is optimized according to the analysis results. The validity and advancement of the optimized control strategy based on the VSOC priority list are proved by design and simulation examples.

2. Virtual Energy Storage (VES) Model of Electric Heating Load

2.1. Concepts of VES

When power consumption is increased (or reduced) by controlling the difference between the working states of the equipment before and after the control right transferring of the equipment, and this power is stored in other forms, the equipment can be equivalent to VES. When the electric power consumption of the equipment after control right transferring is greater than that before transferring, it can be considered to be virtual energy storage charging, whereas, when the electric power consumption of the equipment after control right transferring is less than that before transferring, it can be regarded as VES discharging. The control system of VES can offset the shortage of energy storage by guiding and intervening in energy demand, and can achieve the effect of reducing the energy storage capacity and cost.

VES is described using four indicators—charging/discharging power, switch state, charge/discharge time, and VSOC—which are defined as follows:

(1) Charging/discharging power: charging power is the difference in power consumption of the equipment after control right transferring minus that before transferring. A VES is in the charging state when the charging power is positive, and in the discharging state when the charging power is negative. The value of the discharging power is the opposite of the charging power;

(2) Switch state: refers to the switching state of electric heating equipment;

(3) Charge time: the length of time of the charging state; and discharge time: the length of time of the discharging state;

(4) VSOC: The United States Advanced Battery Consortium defines state of charge (SOC) as the ratio of the residual electricity to the rated capacity under the same conditions at a certain discharge rate. Similarly, virtual state of charge (VSOC) is defined as the ratio of the residual energy to the rated capacity under the same conditions at a certain charging and discharging power, which represents the responsiveness of VES at a given stage.

2.2. Equivalent Thermal Parameter (ETP) Model of Electric Heating Load

The main idea of the ETP modeling method is to equivalent the internal and external environment parameters of the room (chamber) and the refrigerating (heating) capacity of electric energy conversion, to circuit components, such as resistors, capacitors, and power supplies, then use circuit knowledge to analyze the relationship between temperature and energy conversion.

Considering the process of heat exchange between the medium and the mass in the room (chamber), and the exterior environment, the differential equation of the second-order ETP model is:

$$\dot{x} = Ax + Bu \qquad \dot{y} = Cy + Du$$

$$A = \begin{bmatrix} -\left(\frac{1}{R_2 C_a} + \frac{1}{R_1 C_a}\right) & \frac{1}{R_2 C_a} \\ \frac{1}{R_2 C_m} & -\frac{1}{R_2 C_m} \end{bmatrix} \qquad B = \begin{bmatrix} \frac{T_{out}}{R_1 C_a} + \frac{\eta P_{ele}}{C_a} \\ 0 \end{bmatrix} \tag{1}$$

$$\dot{x} = \begin{bmatrix} \dot{T}_{in\_g} \\ \dot{T}_{in\_m} \end{bmatrix} \qquad \dot{y} = \begin{bmatrix} T_{in\_g} \\ T_{in\_m} \end{bmatrix} \qquad u = 1 \quad C = \begin{bmatrix} 1 & 0 \\ 0 & 1 \end{bmatrix} \quad D = 0$$

The thermal energy storage process is described by heat capacity and heat transfer resistance. In Equation (1), P_{ele} represents electric power, η represents refrigeration or heating efficiency, and ηP_{ele} is refrigeration or heating power (kW). T_{in_g} represents the temperature of the medium in the room (chamber)(°C), T_{out} represents the ambient temperature (°C), T_{in_m} represents the temperature of the mass in the room (chamber) (°C), C_e represents the heat capacity of the medium (J/°C), C_m represents the heat capacity of the mass (J/°C), R_1 represents the heat transfer resistance of energy between the interior and the exterior environment of the room (chamber) (°C/W), and R_2 represents the heat transfer resistance of energy between the medium and the mass in the room (chamber) (°C/W).

The widely used second-order ETP model [18] is shown in Figure 1.

When the temperature change is relatively smooth, there is no obvious difference between the medium and the mass temperatures. In order to improve the practicability of the model, assuming $T_{in\_g} = T_{in\_m} = T_{in}$, the second-order ETP model can be reduced to the first-order ETP model:

$$\frac{T_{in} - T_{out}}{R_1} + C_e \frac{dT_{in}}{dt} = \eta P_{ele} \tag{2}$$

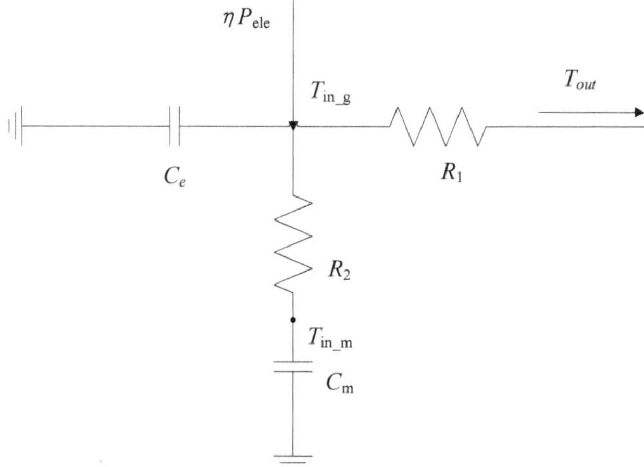

Figure 1. Second-order equivalent thermal parameter (ETP) model.

2.3. Thermal Parameters Part of VES Model for an Electric Heating Unit

On the basis of the first-order ETP model, a partial model of VES thermal parameters was established as shown in Figure 2.

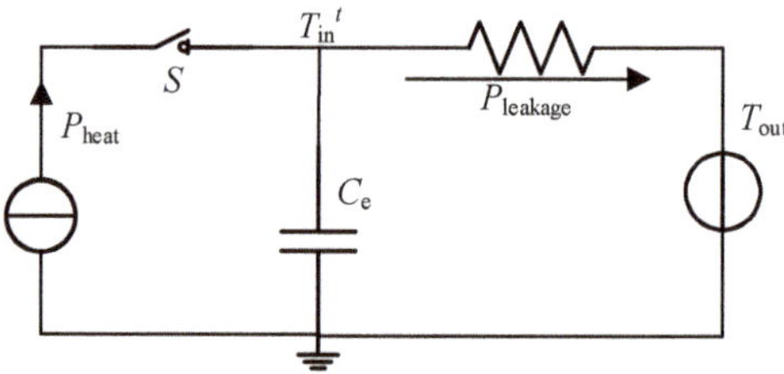

Figure 2. Thermal parameters part of the virtual energy storage (VES) model.

T_{in}^{t} represents the temperature (°C) of the room (chamber) at time t, S represents the switch state, with 0 representing disconnection (the device is closed) and 1 as closure (the device is open). $P_{heat} = \eta P_{ele}$ represents the VES power supply, whose specific form depends on the electric part. $P_{leakage} = (T_{in}^{t} - T_{out})/R_1$ represents the current leakage in the model, mainly represented by the energy loss caused by the temperature gap between the interior and exterior environment.

The electric power before control right transferring is set as the base power, P_{base}. Electric heating equipment that is not in use does not have any discharge capability, and P_{base} is 0. Charging/discharging power is closely related to the switching state, which directly reflects the real-time reserve energy resource requirements. If P_{ele} increases when the electric heating equipment participates in the demand response, and it can be considered that the VES is in charging state, and vice versa for the discharge state. Charging and discharging power can be expressed as:

$$P_{disc} = -S(t)P_{heat(t)} + P_{base}$$
$$P_{char} = S(t)P_{heat(t)} - P_{base}$$

(3)

where, the subscript *char* represents charging power and *disc* represents discharging power. $S(t)$ represents the switching state at time t. It can be seen that the charging/discharging capacity of VES comes from the change of state rather than the duration of the state of batteries. Charge/discharge time determines the sustainable response ability of demand resources. If $T_{in}(t_0) = C$, the solution of (2) is:

$$T_{in}(t) = T_{out}(t) + S\eta R_1 P_{ele}(t) - (T_{out}(t) + S\eta R_1 P_{ele}(t) - C)e^{-\frac{t}{R_1 C_e}} \tag{4}$$

and charge/discharge time is:

$$t_{on/off} = R_1 C_e \ln\left(\frac{C - T_{out}(t) - S\eta R_1 P_{ele}(t)}{T_{in}(t) - T_{out}(t) - S\eta R_1 P_{ele}(t)}\right) \tag{5}$$

Using the relationship between temperature and power in ETP model, the maximum capacitance of VES is:

$$Q_{capacity} = C_e(T_{max} - T_{min}) \tag{6}$$

where T_{max}, T_{min} are protocol maximum temperature and protocol minimum temperature after the control right is transferred, respectively. Based on the principle of energy conservation, the charge capacity at time t is

$$Q(t) = Q(t_0) + \int_{t_0}^{t} (\eta P_{ele}(\xi) - P_{leakage})d\xi \tag{7}$$

The ratio of residual energy to rated capacity under the same conditions can be expressed by:

$$VSOC = \frac{Q(t)}{Q_{capacity}} \tag{8}$$

Figure 3 shows the charge/discharge curves of two VES systems. Number 1 is denoted by solid lines and number 2 by dotted lines. They participate in the response at t_1, and their discharging powers are P_{char_1} and P_{char_2} respectively. Since it is convenient to control the equivalent VES of electric heating load, the power climbing state during the response process is neglected. t_{on_1} and t_{on_2} represent the discharge time, generally less than the maximum discharge time and limited by the VSOC state of virtual energy storage. For different VESs, their charging/discharging power and charge/discharge time are quite different, but their change modes are the same. The charging state is similar to the discharging state, so it is not necessary to elaborate.

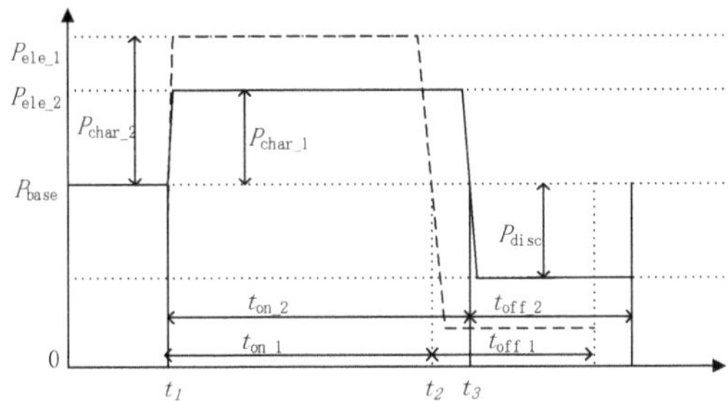

Figure 3. Charge-discharge curves of two VES systems.

2.4. Electrical Parameters Part of VES Model for An Electric Water Heater

Here we analyze the electrical parameters of a VES model of an electric heating load, and take an electric water heater as an example. An electric water heater keeps the chamber temperature within a specific temperature range by switching the on-off mode of resistance wire, and acts soon after receiving a control signal. As there is no delay when it starts up or turned off, its working state is single. There is no obvious power shock in the water heater. Under the working mode of rated voltage and rated current, temperature control is realized by switching on/off the devices. The relationship between heating power and electric power is as follows:

$$P_{\text{heatl}} = \eta P_{\text{rated}} \cdot S(t)$$

$$S(t) = \begin{cases} S(t-1) & T_{\min} < T_{\text{in}}(t) < T_{\max} \\ 0 & T_{\max} \le T_{\text{in}}(t) \\ 1 & T_{\min} \ge T_{\text{in}}(t) \end{cases} \tag{9}$$

where $S(t)$ represents the switching state of the electric water heater at time t, $S(t-1)$ represents the switching state at the last time step, 1 for running and 0 for stop; P_{heatl} represents heating power (kW); P_{rated} is the rated electric power of electric water heater (kW); $T_{\text{in}}(t)$ represents the chamber temperature at time t (°C).

As shown in Figure 4, the VES model mainly includes the electric power curve, the electrical part and the thermal part. The arrowed solid line indicates the energy flow, and the arrowed dotted line indicates the signal flow.

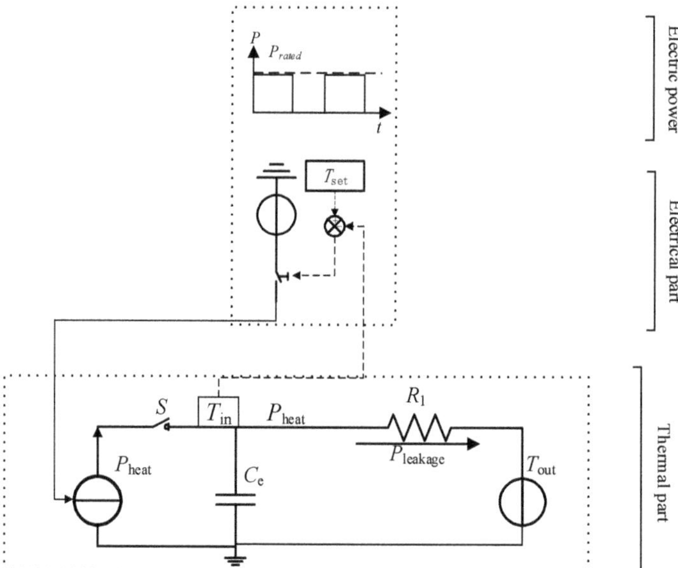

Figure 4. VES model of electric water heater.

3. Model Preprocessing

In advanced metering infrastructures (AMI), only discrete data are transmitted, and communication time intervals exist. It is necessary to discrete the original VES model. Furthermore, model linearization is also needed to simplify the calculation and reduce communications traffic.

3.1. Discretization

Assume that the time step is Δt, (3) can be expressed as:

$$T_{\text{in}}^{t+1} = T_{\text{out}}^{t+1} + S\eta R_1 P_{\text{ele}} - (T_{\text{out}}^{t} + S\eta R_1 P_{\text{ele}} - T_{\text{in}}^{t})e^{-\frac{\Delta t}{R_1 C_e}} \tag{10}$$

where T_{in}^{t} and T_{in}^{t+1} are the internal temperature of the room(chamber) at time t and $t+1$, respectively; T_{out}^{t} and T_{out}^{t+1} are the external ambient temperature at time t and $t+1$, respectively.

By connecting the discrete dots calculated by Equation (10) into a smooth curve, the relationship between time and temperature can be accurately described. Figure 5 shows the change of electric power and chamber temperature over time, where T_{set} is the setting temperature.

In $0\sim t_1$, the water heater is heated from the initial temperature to the protocol maximum temperature. In $t_1\sim t_2$, the water heater stops heating until the temperature drops to the protocol minimum temperature. The dot dash expresses the temperature drop curve after the unit is shut down.

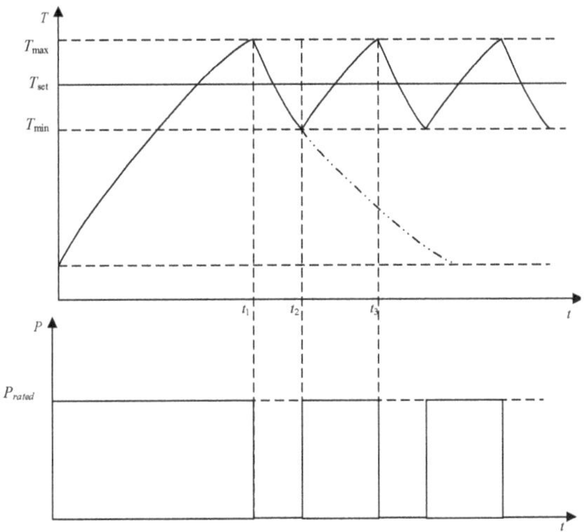

Figure 5. Variation curve of electric water heater temperature to power input.

3.2. Linearization

In order to simplify the calculation, we intend to linearize the temperature curve in Figure 5 and prove the rationality.

The process of temperature change is slow and the time for equipment to participate in demand response is relatively short, so it can be considered that the external temperature is constant. That is, $T_{out}{}^{t} = T_{out}{}^{t+1}$. The temperature range of the room(chamber) is set to $[T_{min}, T_{max}]$, the control cycle of P_{ele} is t_{cyc}, and the time interval(i.e., time step) is Δt. The length of time of electric power input is t_{on}, and the length of time without electric power input is t_{off}. Substitute T_{min}, T_{max} into Equation (10) for reception calculation, we get:

$$T_{max} = (T_{out} + \eta R_1 P_{ele})(1 - e^{-\frac{t_{on}\Delta t}{R_1 C_e}}) + T_{min}e^{-\frac{t_{on}\Delta t}{R_1 C_e}}$$

$$T_{min} = T_{out}(1 - e^{-\frac{t_{off}\Delta t}{R_1 C_e}}) + T_{max}e^{-\frac{t_{off}\Delta t}{R_1 C_e}} \tag{11}$$

$$t_{cyc} = t_{on} + t_{off}$$

After solution, t_{on} and t_{off} are described by:

$$t_{off} = \frac{R_1 C_e}{\Delta t}\ln\left(\frac{T_{max} - T_{out}}{T_{min} - T_{out}}\right)$$

$$t_{on} = \frac{R_1 C_e}{\Delta t}\ln\left(\frac{T_{min} - T_{out} - \eta R_1 P_{ele}}{T_{max} - T_{out} - \eta R_1 P_{ele}}\right) \tag{12}$$

To describe the temperature change of each iteration by the ratio of the time step Δt to t_{on} and t_{off}, we obtain:

$$\begin{cases} T_{in}{}^{t+1} = T_{in}{}^{t} + \frac{\Delta t}{t_{on}}(T_{max} - T_{min})s = 1 \\ T_{in}{}^{t+1} = T_{in}{}^{t} - \frac{\Delta t}{t_{off}}(T_{max} - T_{min})s = 0 \end{cases} \tag{13}$$

3.3. Rationality of Linearization

Take the actual running condition of an electric water heater as an example. Assume that an electric water heater is heated from 30 °C to 50 °C, and after that the temperature is controlled from $T_{\min}$ to $T_{\max}$; the operation parameters of the electric water heater are as below: t_{on} = 20 min, t_{off} = 20 min, P_{ele} = 2000 W, $T_{\min}$ = 50 °C, $T_{\max}$ = 60 °C, T_{out} ≡ 20 °C, T_{in}^{0} = 30 °C. It is available from (12) that: $R_1 = 3.5 \times 10^{-2}$ °C/W; $C_e = 1.192 \times 10^{5}$ J/°C. The shorter the communication time step, the more accurate the model and the more timely the control are. But at the same time, the communication pressure and the construction cost of AMI will increase. In this paper, the time step Δt is 1 min.

Simulation is conducted on MATLAB R2016a (MathWorks, Natick, MA, USA) and the variation of temperature along time is shown in Figure 6. The solid line represents the results of the first-order ETP model, and the dotted line represents the results of the linearized ETP model.

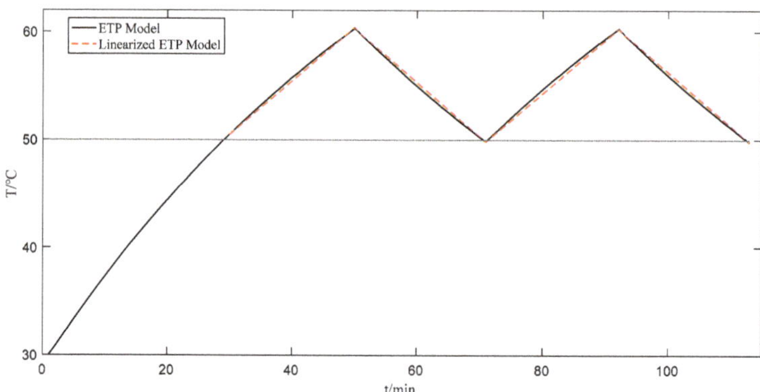

Figure 6. Comparison of temperature curves between linearized model and original ETP model.

Calculate the root mean square error (RMSE) of the two curves, and the smaller *RMSE is*, the less the influence of substitution will be.

$$RMSE(X, Y) = \sqrt{\frac{1}{N}\sum_{i=1}^{N}(X_i - Y_i)^2} \tag{14}$$

As the calculated RMSE between the linearized ETP model and the first-order ETP model within 113 min is only 0.2751 °C, it can be concluded that the linearized model fits well with the original model and will not bring significant change to the related results. Actually, the smaller the heating time t_{on} is or the greater the cooling time t_{off} is, the smaller RMSE is. Considering the actual situation of electric water heater, the duration of heating process of the equipment is generally much shorter than the duration of cooling process, that is, ton < toff.

4. Control Strategy of Electric Heating Loads Based on Virtual State of Charge (VSOC) Priority List

4.1. Proposal of the Control Strategy

Set the following assumptions:

The internal and external environment do not change when the control right is transferred; the energy conversion of electric heating equipment is 100%; the energy loss only comes from the difference between the chamber temperature and outside environment; the refresh time interval of communication data is Δt.

The following analysis still takes the electric water heater as an example. When a certain number of electric water heaters are controlled at the same point, the charge and discharge power of the *j*th one can be described by:

$$P^j_{disc}(t) = -S^j(t)P^j_{rated} + P_{base}{}^j$$
$$P^j_{char}(t) = S^j(t)P^j_{rated} - P_{base}{}^j \tag{15}$$

where the superscript *j* represents the *j*th VES.

At time *t*, the *j*th VES (VSOC^j) is like formula (8). After the model preprocessing of discretization and linearization, bring (13) into (7) to derive the formula (16):

$$VSOC^j(t) = \frac{Q(t)}{Q_{capacity}} = \frac{C^j(T^j(t) - T^j{}_{min})}{C^j(T^j{}_{max} - T^j{}_{min})} = \frac{T^j(t) - T^j{}_{min}}{T^j{}_{max} - T^j{}_{min}} \tag{16}$$

In (16), electrical parameters are described by thermal parameters, and interconversion from electrical parameters to thermal parameters is completed in time domain. The relationship between linearized temperature curve, VSOC curve and electric power is shown in Figure 7.

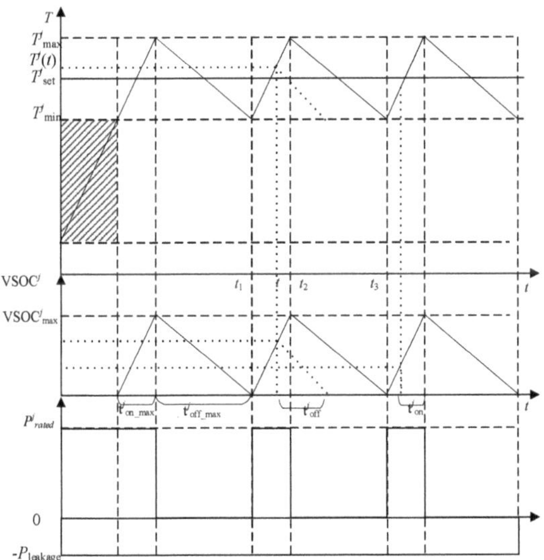

Figure 7. The relationship between linearized temperature curve, virtual state of charge (VSOC) curve and electric power.

The shaded portion indicates that the temperature of VES equipment has not reached the protocol value. $T^j(t)$ represents the temperature of *j*th VES at time *t*; t^j_{off} and t^j_{on} are remaining discharge and charge time of *j*th VES at time *t*; $t^j_{off\_max}$ and $t^j_{on\_max}$ are the maximum discharge and charge time of *j*th VES.

The recursion formula of VSOC^j is as follows:

$$VSOC^j(t+1) - VSOC^j(t) = \frac{T^j(t+1) - T^j{}_{min}}{T^j{}_{max} - T^j{}_{min}} - \frac{T^j(t) - T^j{}_{min}}{T^j{}_{max} - T^j{}_{min}} \tag{17}$$

By substituting (13) into (17), it can be concluded that VSOC^j is related to time step:

$$VSOC^j(t+1) = \frac{\Delta t}{S^j t^j_{on\_max} - (1 - S^j)t^j_{off\_max}} + VSOC^j(t) \tag{18}$$

Charge/discharge time are important constraint indexes. On the one hand, they can ensure that the VES will not charge or discharge excessively, which means that the temperature fluctuation of the electric water heater is within the set range. On the other hand, they can ensure that the state of the device will not change during the time step, which shuns affecting the control accuracy and bringing unnecessary grid side fluctuations. According to Figure 7, we can obtain:

$$\begin{aligned}
\frac{t^j_{\text{off}}}{t^j_{\text{off_max}}} &= \frac{\text{VSOC}^j(t) - \text{VSOC}^j_{\min}}{\text{VSOC}^j_{\max} - \text{VSOC}^j_{\min}} \\[2mm]
\frac{1 - t^j_{\text{on}}}{t^j_{\text{on_max}}} &= \frac{\text{VSOC}^j(t) - \text{VSOC}^j_{\min}}{\text{VSOC}^j_{\max} - \text{VSOC}^j_{\min}}
\end{aligned} \tag{19}$$

$$\begin{aligned}
t^j_{\text{off}} &= \frac{T^j(t) - T^j_{\min}}{T^j_{\max} - T^j_{\min}} t^j_{\text{off_max}} = \text{VSOC}^j(t) t^j_{\text{off_max}} \\[2mm]
t^j_{\text{on}} &= \frac{T^j_{\max} - T^j(t)}{T^j_{\max} - T^j_{\min}} t^j_{\text{on_max}} = (1 - \text{VSOC}^j(t)) t^j_{\text{on_max}}
\end{aligned} \tag{20}$$

To sum up, taking discharge as an example, the control strategy of VES are:

$$\begin{cases}
P^j_{\text{disc}}(t) = -S^j(t) P^j_{\text{rated}} + P_{\text{base}}{}^j \\[2mm]
\text{VSOC}^j(t+1) = \dfrac{\Delta t}{S^j t^j_{\text{on_max}} - (1 - S^j) t^j_{\text{off_max}}} + \text{VSOC}^j(t) \\[2mm]
t^j_{\text{off}} = \dfrac{T^j(t) - T^j_{\min}}{T^j_{\max} - T^j_{\min}} t^j_{\text{off_max}} = \text{VSOC}^j(t) t^j_{\text{off_max}}
\end{cases} \tag{21}$$

Taking the demand side response of discharge condition as an example, the principle of the control strategy for VESs is to control the switching state mainly based on the sequence of VSOC values, that is, the unit with higher VSOC is shut down preferentially. The main objective function is to meet the power shortage in each time step. The marginal limit conditions are that the discharge time is longer than the time step and VES do not overcharge or overdischarge.

The specific control function at time t are

$$\begin{cases}
P^t_s \leq \min_Q \sum P^t_{\text{disc}}(j_n) \\[2mm]
0 \leq \text{VSOC}^t(j_n) \leq 1 \\[2mm]
t^j_{\text{off}} \geq \Delta t
\end{cases} \tag{22}$$

where P^t_s represents the power shortage at time t; Q represents the set arranged from large to small according to VSOC; j_n is an element of set Q, and n represents the order of j in the new set.

4.2. Simulation of the Strategy

MATLAB is used as the simulation platform to verify the control effect of trunked dispatching of the electric heating loads with different parameters and working states. The program mainly includes the following steps: data refreshment, dealing with VSOC off-limit problem, generating control queue Q based on VSOC values, calculating whether the power shortage is satisfied, handling the switch state of controlled energy storage and updating the states of VES iteratively. The parameters of the example are: the amount of electric water heaters under control is 100; the initial value of VSOC is uniformly distributed from 0 to 1; the switching function is 0~1 integer distributed; the time step is 1min; the rated power of the equipment is uniformly distributed from 1.5 to 2.5 kW. For each water heater, the maximum charge and discharge time are 15~25 min and 30~50 min uniformly distributed, respectively; the protocol minimum and maximum temperature are 45~55 °C and 55~65 °C uniformly distributed, respectively.

Suppose the power shortage in the power system is 30 kW and the protocol control time is 30 min, and the control result is shown in Figure 8. After analysis, it can be found that the response power can satisfy the power shortage well in a short time, but there is an excessive response at 23 min, as shown in Figure 8a. At the same time, a large number of VSOC values reach the limit in Figure 8b, which shows that there is a certain relationship between the excessive response of power and a large number of VESs reaching the limits simultaneously.

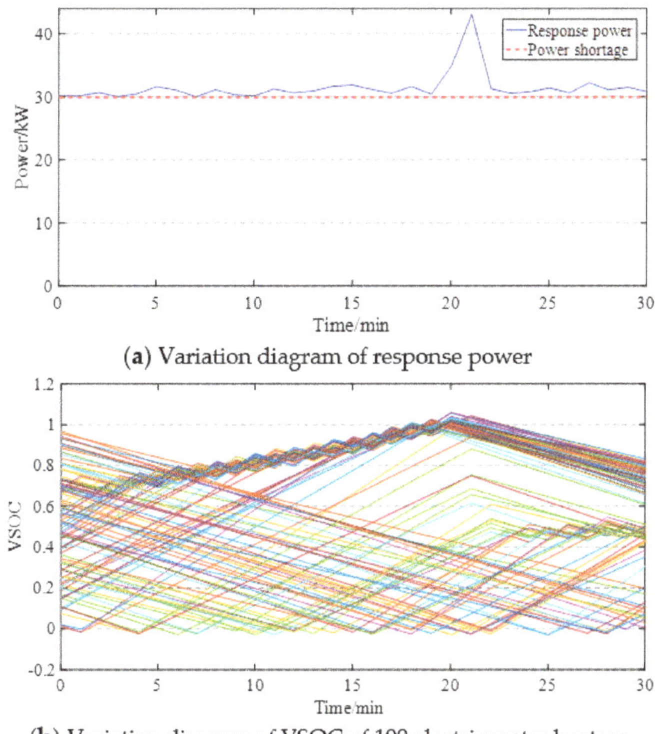

(a) Variation diagram of response power

(b) Variation diagram of VSOC of 100 electric water heaters

Figure 8. The response variation of power and VSOC in 30 min.

Suppose the power shortage is 30 kW and the protocol control time is 180 min. The control result is shown in Figure 9. It is found that the response power basically satisfies the power shortage in a long time, but at the same time there are more excessive responses and insufficient responses. By comparing Figure 9a,b, we come to the same conclusion as Figure 8: when a large number of VSOC values are concentrated and near the limit value, they will lead to excessive or insufficient response. Especially starting at 135 min, since most VESs are close to the limit of VSOC = 0, and in order to maintain the marginal condition VSOC > 0, a large number of VESs are forced to open and charge, resulting in insufficient response over a period of time.

Above all, the control strategy based on the VSOC priority list can achieve a relatively stable power response in a short time. However, limited by the state of VES, the control results over a long period of time are yet to be adjusted and optimized.

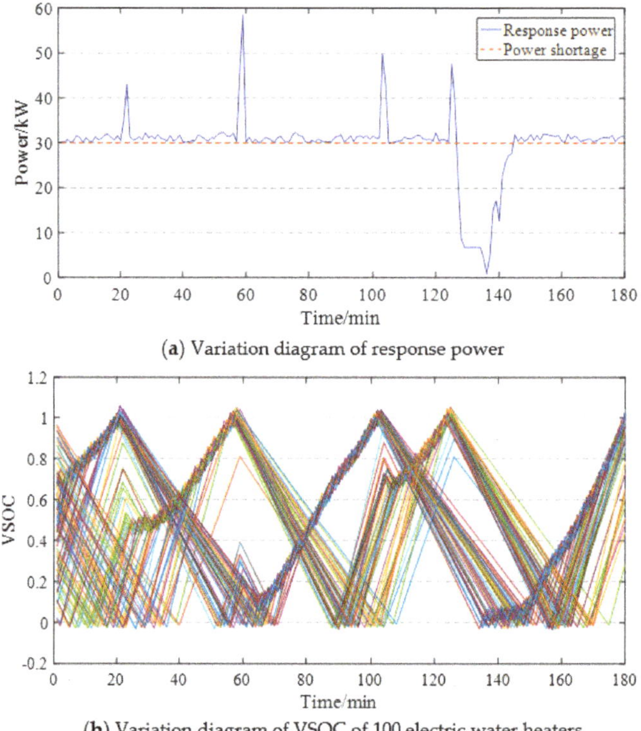

(**a**) Variation diagram of response power

(**b**) Variation diagram of VSOC of 100 electric water heaters

Figure 9. The response variation of power and VSOC in 180 min.

4.3. Discussion

After observing Figures 8 and 9, it would be found that when the power response is excessive, more electric water heaters are forced to close and their equivalent VESs discharge in advance because their VSOC values reach the upper limit, which leads to load peak of excessive response. When the response is inadequate, more electric water heaters are in uncontrolled state, namely, the relevant equipment is closed and there is inadequate load capacity for respond. At the same time, we found that before insufficient response of VES, there is always a large excessive response. The excessive response power is so large that the state of charge and discharge is changed in advance, causing insufficient discharge capacity and insufficient response.

By analyzing Figure 8 and the first 30 min of Figure 9, it is found that the control effect is better when the control right is transferred for a short time than that for a long period of time. By comparison, it is found that the states of the VESs are more dispersed in a short period of time, while more concentrated after long-term control.

As shown below, the control effect is affected by the diversity of VES.

The degree of distribution of virtual state of charge of virtual energy storage is defined. It is expressed by the standard deviation of VSOC. It reflects the diversity of virtual VES. The greater the standard deviation, the higher diversity of related VES.

By applying the relevant parameters in 4.2, the variation of VES diversity in 180 min can be obtained, as shown in Figure 10. It is obvious that with the increasing of transfer time of control right, the diversity of VES converges to a smaller value oscillatorily. The histograms of VES distribution at special time points 10 min, 55 min, 95 min and 115 min are shown in Figure 11. At 10 min, the distribution of VES is relatively uniform, but at 55 min, 95 min, and 115 min, it is relatively concentrated and the diversity is lower. Compared with Figure 9, 55 min, 95 min, and 115 min are the time points when excessive or insufficient response occurs.

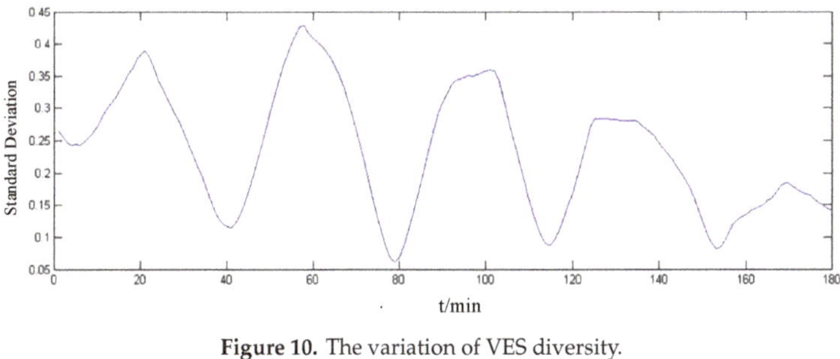

Figure 10. The variation of VES diversity.

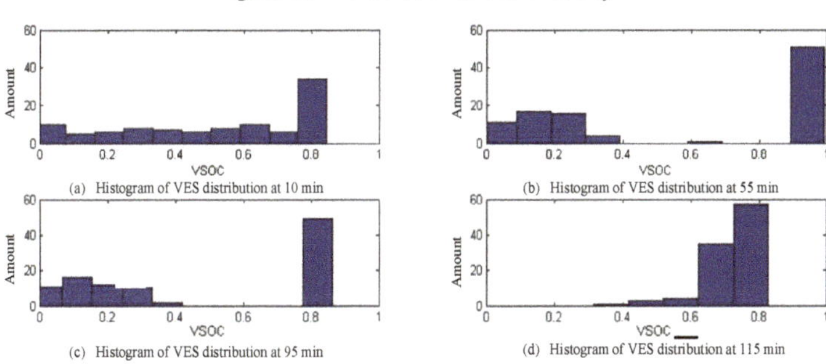

Figure 11. Histograms of VES distribution at different time points.

With increasing control time, VES diversity of electric water heaters is decreased, and that is before excessive response or insufficient response is low. Obviously, the diversity state of VES leads to excessive and insufficient response in the transfer process of control right.

In summary, in the response process, the above VSOC-priority VES control strategy makes the state of VESs convergent and thus reduces its diversity, which leads to worse results than expected, namely, excessive or insufficient response occurs frequently after long-term control right transfer.

5. Optimized Control Strategy of Electric Heating Loads Based on VSOC Priority List

5.1. Proposal of the Optimized Control Strategy

Based on the above analysis, the diversity of VES states directly affects the control results.

In (22), the switching state of set Q is refreshed within each communication time step in the original control strategy, which will inevitably lead to the discharge of VES with higher VSOC and charge of VES with lower VSOC, making the states of VESs synchronised, resulting in the reduction of the diversity of VESs and directly affecting the control effect.

In order to make the charge and discharge of each VES more complete, the improved control strategy is to reduce the switching times of VES, trying to change the active control to passive control according to limit value. To this end, the improved restrictive conditions are:

$$\begin{cases} P^t_s \leq \min\limits_{A} \sum P^t_{disc}(j_n) \\ P^t_s - \sum\limits_{A} P^t_{disc}(j_n) \leq \min\limits_{C_Q A} \sum P^t_{disc}(j_n) \\ 0 \leq VSOC^t(j_n) \leq 1 \\ t^j_{off} \geq \Delta t \end{cases} \tag{23}$$

where A represents the set of controlled VESs arranged from large to small according to VSOC; C_QA represents the complement set of A with Q as the complete set. The new restriction condition indicates that the VES discharging during the last time step is preferentially controlled to continue discharging, and then the power shortage is supplemented based on the order of VSOC values.

5.2. Simulations of the Optimized Strategy

The example parameters in Section 4 and the optimized control strategy based on VSOC priority list for electric water heaters are used to draw the VES diversity variation curve as shown in Figure 12.

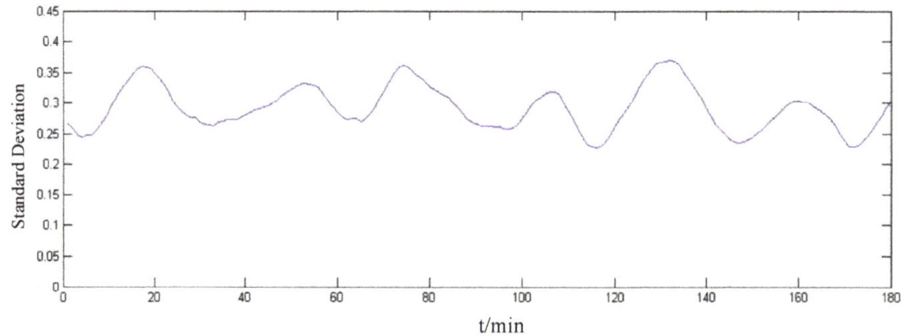

Figure 12. Variation of VES diversity of electric water heaters (after optimized).

Through quantitative calculation and comparison with Figure 10, it can be found that the diversity of VES obtained by the optimized control strategy is maintained well, the standard deviation oscillates within a stable range, and the amplitude is much smaller than that of the original control strategy. The histograms of VES distribution at 45 min, 90 min, 135 min and 180 min are shown in Figure 13. Obviously, the distribution of VES is more uniform and the diversity is better.

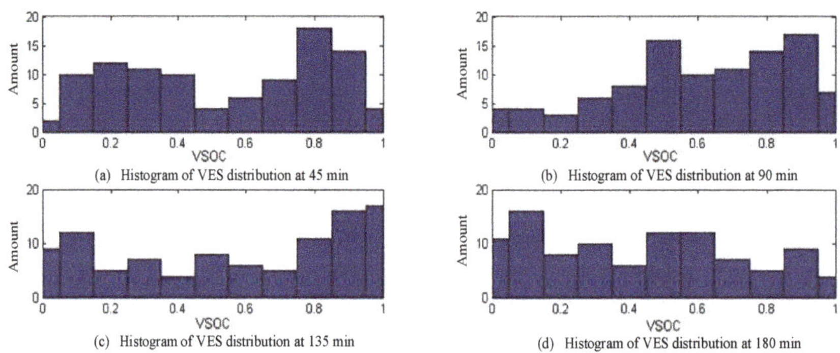

Figure 13. Histograms of VES distribution at different time points (after optimized).

The control effect is shown in Figure 14. The simulation result shows that the optimized control strategy has better control effect because long-term transfer of control rights does not result in insufficient or excessive response. The VSOC values of VESs are relatively uniform in the whole process, and there is no convergence of the states of VES. The power shortage can be reduced steadily by using the optimized control strategy to control electric heating loads.

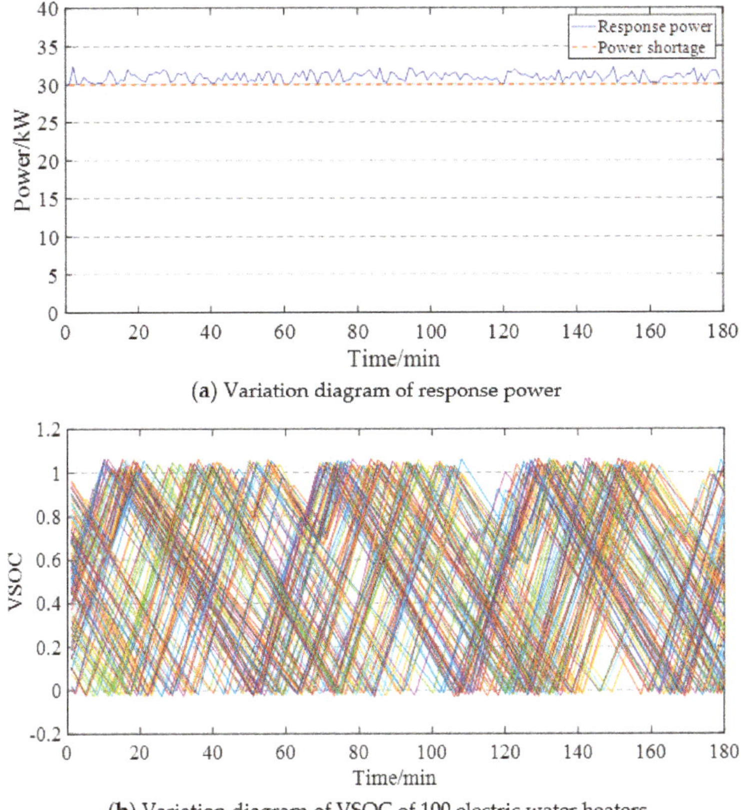

(**a**) Variation diagram of response power

(**b**) Variation diagram of VSOC of 100 electric water heaters

Figure 14. Control effect of optimized control strategy for electric water heaters.

The control effect of the optimized control strategy based on a VSOC priority list for electric water heaters under fluctuating power shortage is shown in Figure 15. As can be seen from the figure, the response power still tracks power shortage well, which shows that the control strategy proposed in this paper is also applicable to power grid with power shortage fluctuations.

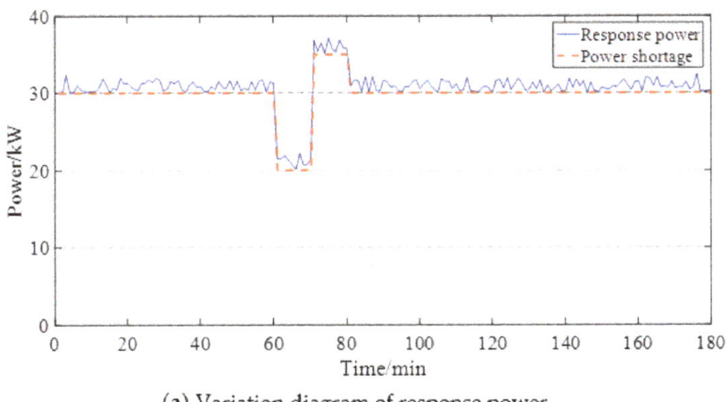

(**a**) Variation diagram of response power

Figure 15. *Cont.*

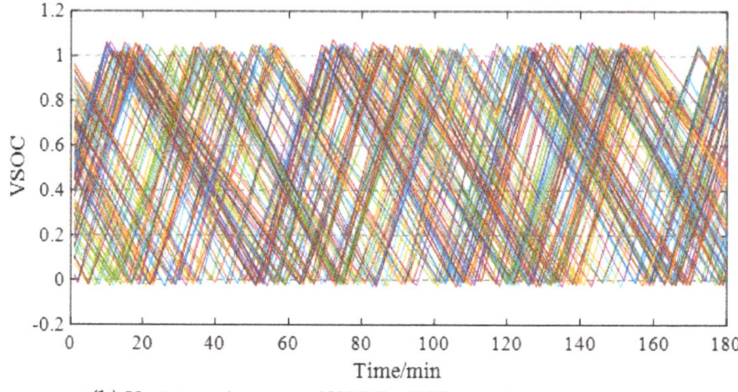

(**b**) Variation diagram of VSOC of 100 electric water heaters

Figure 15. Control effect of optimized control strategy under fluctuating power shortage.

6. Conclusions

In this paper, we establish a complete VES index system and propose a VES model which can reflect the practical electro-thermal exchange. The model is mainly divided into two parts: electrical parameters and thermal parameters, reflecting the impact of electric heating loads scheduling variation on the distribution network. The model is discretized to reduce communications traffic and linearized to simplify calculation. Taking the electric water heater as an example, a control strategy based on VSOC priority list is proposed, and simulation results show that this method can reduce the power shortage of the grid to a certain extent. By analyzing the insufficiency of the strategy, the optimized VSOC priority list control strategy, which optimizes the control effect of long-time scheduling, is put forward. The optimized control strategy can maintain the diversity of VES well and make it possible for the electric water heater to track the power shortage of the grid for a long time. Simulation examples are designed to verify the superiority and effectiveness of the proposed optimized control strategy.

The control strategy proposed in this paper is not only applicable to the electric water heater, but also to other electric heating loads that maintains the temperature of the room (chamber) in a specific temperature range by controlling the on-off mode of the resistance wire. For other types of electric heating loads, it is only necessary to modify the electrical parameters of the VES model and propose its control strategy in a targeted manner. The research in this paper is helpful to build a multi-VES system which can participate in the trunked dispatching of the power grid and promote the development of demand side response. The cost, benefit and pricing mechanism of demand response can be analyzed in the following studies.

Author Contributions: Conceptualization, Y.C.; Data curation, S.X.; Formal analysis, S.X.; Investigation, Y.Z.; Methodology, Y.C.; Project administration, Y.C.; Software, S.X.; Supervision, Y.Z.; Writing—original draft, Y.C., S.X.; Writing—review and editing, W.H. and R.Z.

Funding: This work is supported by SGCC program: Research on Extensive Application and Benefit Evaluation of Typical Power Substitution Technology Considering Power Quality Influence (52182018000H) and Lanzhou Jiaotong University-Tianjin University Innovation Fund Program: Research on Management Strategies of Power Quality Problems from Thermostatically Controlled Appliances (2018058).

Conflicts of Interest: The authors declare no conflicts of interest.

Abbreviation

P_{ele}	electric power
η	refrigeration or heating efficiency
T_{out}	ambient temperature
T_{in}	room (chamber) temperature
C_e	heat capacity of medium
R_1	heat transfer resistance of energy between the interior and the exterior environment of the room (chamber)
P_{heat}	VES power supply
$P_{leakage}$	leakage current
P_{base}	electric power before control right transfering
P_{disc}	discharging power
P_{char}	charging power
$S(t)$	switching state at time t
t_{on}	charge time
t_{off}	discharge time
$Q_{capacity}$	maximum capacitance of VES
T_{max}	protocol maximum temperature
T_{min}	protocol minimum temperature
$Q(t)$	charge capacity at time t
P_{heat1}	heating power
$S(t-1)$	switching state at the last time step
P_{rated}	rated electric power of electric water heater
Δt	time step
T_{in}^{t+1}	internal temperature of the room(chamber) at time $t+1$
$RMSE$	root mean square error of two curves,
$(Symbol)^j$	(Symbol) of jth VES
t^j_{off}	remaining discharge time of jth VES at time t
$t^j_{h\_on}$	remaining charge time of jth VES at time t
$t^j_{off\_max}$	maximum discharge time of jth VES
$t^j_{on\_max}$	maximum charge time of jth VES
P^t_s	power shortage at time t
Q	the set arranged from large to small according to VSOC
jn	the order of j in the new set
C_QA	the complement set of A with Q as the complete set

References

1. Zhang, Q.; Wang, X.F.; Wang, J.X.; Feng, C.; Liu, L. Survey of demand response research in deregulated electricity markets. *Autom. Electr. Power Syst.* **2008**, *32*, 97–106. [CrossRef]
2. Zhao, Y.C. Research on Cluster Scheduling for Thermostatically Controlled Load Based on Virtual Energy Storage Characteristic. Master's Thesis, Tianjin University, Tianjin, China, 2018.
3. Ruiz, N.; Cobelo, I.; Oyarzabal, J. A Direct Load Control Model for Virtual Power Plant Management. *IEEE Trans. Power Syst.* **2009**, *24*, 959–966. [CrossRef]
4. Gouveia, C.; Moreira, J.; Moreira, C.L.; Lopes, J.P. Coordinating Storage and Demand Response for Microgrid Emergency Operation. *IEEE Trans. Smart Grid* **2013**, *4*, 1898–1908. [CrossRef]
5. Li, P.; Han, Z.; Jia, X.; Mei, Z.; Han, X.; Wang, Z. Comparative analysis of an organic Rankine cycle with different turbine efficiency models based on multi-objective optimization. *Energy Convers. Manag.* **2019**, *185*, 130–142. [CrossRef]
6. Zhao, H.; Zhou, J.; Yu, E. Advanced metering infrastructure supporting effective demand response. *Power Syst. Technol.* **2010**, *34*, 13–20. [CrossRef]
7. Cao, B. Research on the Impacts of Climate and Built Environment on Human Thermal Adaptation. Ph.D. Thesis, Tsinghua University, Beijing, China, 2012.

8. Wang, D.; Fan, M.H.; Jia, H. User Comfort Constraint Demand Response for Residential Thermostatically-controlled Loads and Efficient Power Plant Modeling. *Proc. CSEE* **2014**, *34*, 2071–2077. [CrossRef]

9. Che, Y.; Xue, S.; He, W.; Li, Z.; Zhang, R. Frequency Regulation Strategy for Community Power Grid Based on improved Temperature State Priority List. *Ekoloji* **2019**, *107*, 2061–2068.

10. Zhou, L.; Li, Y.; Gao, C. Improvement of temperature adjusting method for aggregated air-conditioning loads and its control strategy. *Proc. CSEE* **2014**, *34*, 5579–5589. [CrossRef]

11. Yin, Z.J.; Che, Y.B.; Li, D.Z.; Liu, H.; Yu, D. Optimal Scheduling Strategy for Domestic Electric Water Heaters Based on the Temperature State Priority List. *Energies* **2017**, *10*, 1425. [CrossRef]

12. Nehrir, M.H.; Jia, R.; Pierre, D.A.; Hammerstrom, D.J. Power management of aggregate electric water heater loads by voltage control. In Proceedings of the 2007 IEEE Power Engineering Society General Meeting, Tampa, FL, USA, 24–28 June 2007.

13. Orphelin, M.; Adnot, J. Improvement of methods for reconstructing water heating aggregated load curves and evaluating demand-side control benefits. *IEEE Trans. Power Syst.* **1999**, *14*, 1549–1555. [CrossRef]

14. Molina, A.; Gabaldon, A.; Fuentes, J.A.; Alvarez, C. Implementation and assessment of physically based electrical load models: Application to direct load control residential programmes. *IEE Proc. Gener. Transm. Distrib.* **2003**, *150*, 61–66. [CrossRef]

15. Gao, C.; Li, Q.; Li, Y. Bi-level optimal dispatch and control strategy for air-conditioning load based on direct load control. *Proc. CSEE* **2014**, *34*, 1546–1555. [CrossRef]

16. Jia, H.; Qi, Y.; Mu, Y.F. Frequency response of autonomous microgrid based on family-friendly controllable loads. *Sci. China Technol. Sci.* **2013**, *56*, 693–702. [CrossRef]

17. Wang, D.; Zeng, R.; Mu, Y. An optimization method for new energy utilization using thermostatically controlled appliances. *Power Syst. Technol.* **2015**, *39*, 3457–3462. [CrossRef]

18. Lu, N. An Evaluation of the HVAC Load Potential for Providing Load Balancing Service. *IEEE Trans. Smart Grid* **2012**, *3*, 1263–1270. [CrossRef]

Article

Artificial Neural Networks Approach for a Multi-Objective Cavitation Optimization Design in a Double-Suction Centrifugal Pump

Wenjie Wang [1], Majeed Koranteng Osman [1,2], Ji Pei [1,*], Xingcheng Gan [1] and Tingyun Yin [1]

[1] National Research Center of Pumps, Jiangsu University, Zhenjiang 212013, China; wenjiewang@ujs.edu.cn (W.W.); mjk@ujs.edu.cn (M.K.O.); ganxingcheng@gmail.com (X.G.); tingyun_yin@ujs.edu.cn (T.Y.)

[2] Mechanical Engineering Department, Wa Polytechnic, Wa, Upper West, Ghana

* Correspondence: jpei@ujs.edu.cn; Tel.: +86-137-7647-4939

Received: 4 April 2019; Accepted: 24 April 2019; Published: 27 April 2019

Abstract: Double-suction centrifugal pumps are widely used in industrial and agricultural applications since their flow rate is twice that of single-suction pumps with the same impeller diameter. They usually run for longer, which makes them susceptible to cavitation, putting the downstream components at risk. A fast approach to predicting the Net Positive Suction Head required was applied to perform a multi-objective optimization on the double-suction centrifugal pump. An L_{32} (8^4) orthogonal array was designed to evaluate 8 geometrical parameters at 4 levels each. A two-layer feedforward neural network and genetic algorithm was applied to solve the multi-objective problem into pareto solutions. The results were validated by numerical simulation and compared to the original design. The suction performance was improved by 7.26%, 3.9%, 4.5% and 3.8% at flow conditions $0.6Q_d$, $0.8Q_d$, $1.0Q_d$ and $1.2Q_d$ respectively. The efficiency increased by 1.53% $1.0Q_d$ and 1.1% at $0.8Q_d$. The streamline on the blade surface was improved and the vapor volume fraction of the optimized impeller was much smaller than that of the original impeller. This study established a fast approach to cavitation optimization and a parametric database for both hub and shroud blade angles for double suction centrifugal pump optimization design.

Keywords: multi-objective optimization; artificial neural network; NPSHr prediction; cavitation optimization; CFD

---

1. Introduction

Over recent years, pump manufacturers have intensified their quest to rapidly develop cost-competitive and high-performance pumps with compact and robust structures to meet consumer's limitless demands for high end centrifugal pumps, since they offer wide-ranging stable operation both in industrial and agricultural applications. As such, double-suction centrifugal pumps are widely used in various fields, since their flow rate is twice that of single-suction pumps with the same impeller diameter [1]. For systems with large capacity demands, the pumps are usually run for longer periods, making them very susceptible to cavitation; this puts downstream components at risk of being damaged, since, in most cases, the flow passage gets blocked by cavitation bubbles [2].

Recent advances in the design of centrifugal pump systems does not only require energy efficient and quieter systems, but also reductions in design time and lower costs. Engineers have therefore devised optimization methods and algorithms with numerical simulations so that optimization does not rely solely on the designer's experience [3]. Aside from the optimization algorithms there are other approaches to optimization, such as the experimental design known as the design of experiment (DOE) method, and the application of meta-models as surrogates for optimization [4]. Pei et al. [5] applied the L_9 (3^3) orthogonal design of experiments and computational methods to optimize the cavitation performance in a centrifugal pump to reduce the required Net Positive Suction Head (NPSHr) by 0.63 m. A cavitation performance optimization was also carried out on a large-scale axial flow pump using orthogonal DOE to reduce the cavity length and vapor volume fraction [6]. Nataraj et al. [7] used the Taguchi test design (ODOE) and numerical simulations to optimize the design of the centrifugal pump impeller. Xu et al. [8] conducted a multi-objective and multi-parameter optimization study of a using Taguchi method. The optimized impeller increased in efficiency by 3.09% and reduced critical cavitation point by 1.45 m. Also, the Response Surface Method, Kriging Model and Artificial Neural Networks are popular surrogate models that have been applied in centrifugal pump optimization design [9–11]. Jin et al. [12] compared various surrogate models based on several defined criteria, including prediction accuracy, efficiency, and robustness, and concluded that a neural network should be used for higher-order nonlinear problems. Artificial Neural Networks (ANNs) abstract the human brain's vegetative cell network from the angle of data process, establish an easy model, and form completely different networks which are in step with different affiliation strategies. ANNs have been applied in turbomachinery by various researchers [13–16]. Pei et al. [17] carried out a multi-objective optimization on the inlet pipe shape of a vertical inline pump and using an ANN and Multi-Objective Generic Algorithm (MOGA) to increase the efficiency over a wide range.

During cavitation optimization designs, a numerical process of optimizing a collection of individual objective functions is simultaneously performed [5,18]. The process involves the calculation of several test cases to determine the NPSHr for each case, which consumes a lot of computational time. Traditionally, during the use of Computational Fluid Dynamics (CFD) for cavitation simulations, the common practice has been to set the inlet boundary to total pressure, while the boundary at the outlet is fixed to the volume flow rate. Although this boundary set has been fruitfully implemented in numerical simulations with and without cavitation models, the choosing of total inlet pressure relies on guesswork if there is no foreknowledge of the NPSHr range [19–22]. This approach usually requires about 10–15 independent simulations before good accuracy is reached [21,23]. Moreover, this pure guessing game is likely to draw the simulation into severe cavitation conditions, which takes much longer times to converge, especially in situations where transient simulations are required to acquire the necessary accuracy. [24,25]. A recent approach proposed by Ding et al. [26] introduced a controllable and more predictable simulation procedure where the traditional boundary set were substituted by introducing a new boundary pair and an algorithm developed to estimate a good value for a static pressure that correlated to a 3% drop in pump head. The procedure predicted NPSHr at a head drop of 3.3% just in three simulation steps in an industrial centrifugal pump. Despite the fact that this novel approach proved numerically stable for simulating cavitation flows, the application of this method in cavitation flow simulations has been very rare, and has not been applied to the double-suction pump.

In this study, a fast approach to NPSHr prediction was used to perform a multi-objective cavitation design optimization first to improve hydraulic efficiency and secondly to improve the suction performance of the double-suction centrifugal pump. The choice of decision variables was centered on the impeller hub and shroud angles, since there have been several cavitation performance enhancement studies on the blade inlet angle. An L_{32} (8^4) orthogonal array was designed to evaluate 8 geometrical parameters at 4 levels each. A two-layer feedforward neural network and genetic algorithm was applied to solve the multi-objective problem into 2D pareto-frontier solutions that meet the objective functions. The study established a fast approach to cavitation optimization design and also created a parametric database of the impeller hub and shroud blade angles for the double-suction centrifugal pump optimization design.

2. Description of Computational Domain

The 250GS40 double-suction centrifugal pump (Figure 1) was developed with PTC Creo parametric 4.0 (Boston, MA, United States). It was designed according to a specific speed of 126.57. It has a shrouded impeller with six twisted blades. The spiral suction domain has flow directing baffles to de-swirl the flow towards the impeller eye. The pump has a single volute; Table 1 presents the design specifics of the pump.

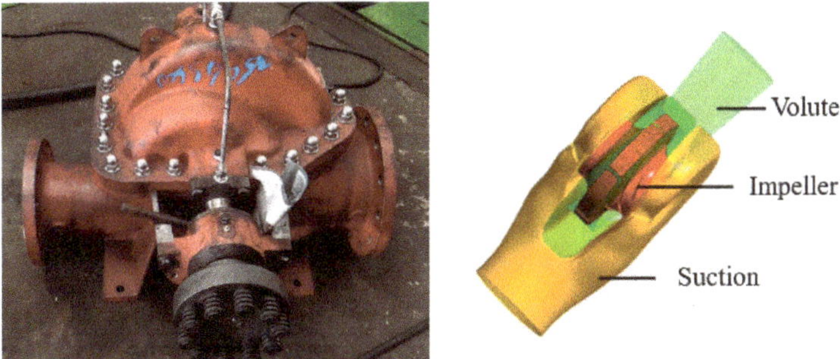

Figure 1. Tested pump and computational domain.

Table 1. Design specifications of model pump.

Design Parameters	Value
Flow rate, Q_d (m³/h)	500
Head, H (m)	40
Rotational speed, N (rpm)	1480
Number of blades, z	6
Suction diameter, D_s (mm)	250
Impeller inlet diameter, D_1 (mm)	192
Impeller outlet diameter, D_2 (mm)	365
Delivery diameter, D_d (mm)	200
Efficiency, η	84
NPSHr (m)	3.5

2.1. Computational Grid and Mesh Sensitivity

The flow passage for the tested pump was meshed with ANSYS ICEM 18.0 (ANSYS Inc., Canonsburg, PA, United States) mesh tool. High quality structural hexahedral mesh was built to achieve maximum calculation accuracy. The grid size and blocking method was used and the grids for the impeller and suction chamber were refined with large numbers for higher precision. The grids near the walls were refined further to attain boundary motion features. An overview of the computational mesh is shown in Figure 2.

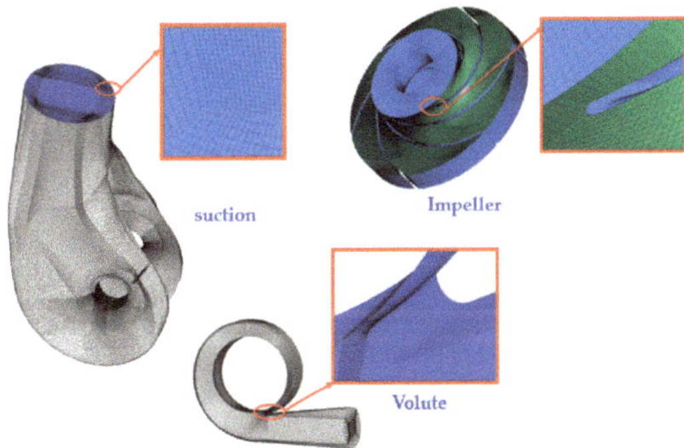

Figure 2. Mesh overview of the flow domains.

Based on similar works [17,27,28], 5 independent high quality structural hexahedral meshes were built and a grid sensitivity analysis executed. Simulations were performed at the design flow rate of to determine the mesh influence on pump head and efficiency. The effects of the mesh density on the hydraulic performance remained fairly constant when the number of grid elements increased past 4.26×10^6 elements, suggesting that numerical accuracy gradually stabilized as the grid number increased. The mesh density with elements 4,266,423 was therefore used for the computations to reduce calculation load and computation time. Statistics for the selected mesh density is presented in Table 2, and the grid sensitivity of the 5 independent mesh generated is shown in Figure 3.

Table 2. Grid cells of the selected mesh.

Domain	Impeller	Suction	Volute	All Domains
No of Elements	1,199,880	1,938,727	1,127,816	4,266,423
No of Nodes	1,285,686	2,019,162	1,159,980	4,464,828

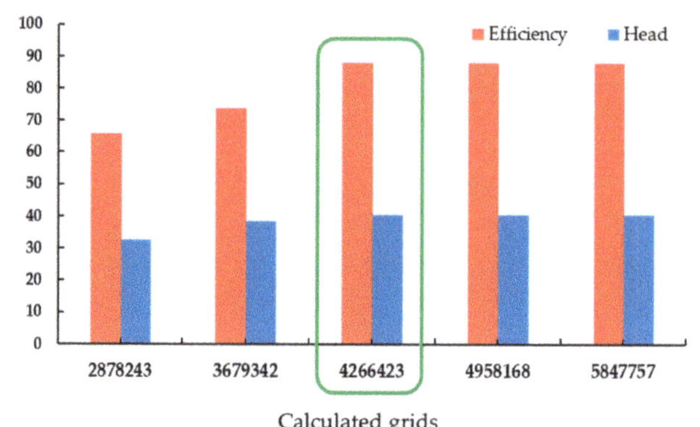

Figure 3. Performance comparison for 5 independent grids.

2.2. Governing Equations

The continuity equation, which is the basic equation governing two-phase flow is adopted from the Navier-Stokes equations [29] which is time dependent, and is given as:

$$\frac{\partial \rho}{\partial t} + \frac{\partial}{\partial x_i}\left(\rho u_j\right) = 0 \tag{1}$$

$$\frac{\partial\left(\rho u_i u_j\right)}{\partial x_j} = -\frac{\partial p}{\partial x_i} + \frac{\partial}{\partial x_j}\left[(\mu + \mu_t)\left(\frac{\partial u_i}{\partial x_j} + \frac{\partial u_j}{\partial x_i} - \frac{2}{3}\frac{\partial u_i}{\partial x_i}\delta_{ij}\right)\right] \tag{2}$$

Density and dynamic viscosity mixtures are represented by ρ and μ respectively. Velocity is denoted by u, p for pressure and turbulent viscosity is μ_t. Variables *i and j* are axis directions. The turbulence model chosen was SST k–ω, because it exhibits a combined advantage of both the k–ω and k–ε turbulence models [30,31]. The cavitation model used is the mass transport equation is deduced from Rayleigh-Plesset's equation. Transport equation for bubble formation and collapse is expressed as:

$$\frac{\partial(\rho_v \alpha_v)}{\partial_t} + \frac{\partial}{\partial x_j}\left(\rho_v \alpha_v u_j\right) = \dot{m} = \dot{m}^+ - \dot{m}^- \tag{3}$$

$$m^+ = C_{vap}\frac{3r_g(1-\alpha_v)\rho_v}{R_b}\sqrt{\frac{2}{3}\frac{max(p_v - p, 0)}{\rho_l}} \tag{4}$$

$$m^- = C_{cond} \frac{3\alpha_v \rho_v}{R_b} \sqrt{\frac{2}{3} \frac{max(p - p_v, 0)}{\rho_l}} \tag{5}$$

where α_v denotes the vapor fraction, m^+ and m^- stands for mass transfer terminology for evaporation and condensation respectively. The coefficients for condensation and evaporation phases are represented as C_{cond} and C_{vap}. r_g, is the nucleation site volume fraction, R_b, is the bubble radius, ρ_v, for vapor density, ρ_l, the liquid density and p_v, whch is also the saturation pressure. Standardized values according to literature are: $C_{vap} = 50$, $C_{cond} = 0.01$, $r_g = 5 \times 10^{-4}$, $R_b = 10^{-6}$ m, $\rho_v = 0.554$ kg/m^3, $\rho_l = 1000$ kg/m^3 and $p_v = 3169$ Pa [32,33].

2.3. NPSHr Prediction Procedure

For a given flow rate condition, three key procedures are required to calculate for the NPSHr. The approach used here is adapted from Ding et al. [26]. There are however two other optional procedural steps (pre and post procedural) that can be applied by discretionally where necessary. *Step 0 (optional):* The first step is an optional step of performing a quick simulation with the traditional boundary settings to give an idea of the pump head and can be skipped if the head is already known. *Step 1:* Recalculated pump head with new boundary pair to obtain clear-cut reference point in predicting the 3% head drop. The inlet is held fixed to the flow rate and static pressure at the outlet. The outlet static pressure, P_{Sout}, is estimated as follows.

$$P_{S_{OUT}}(1) = H(0) + P_{Tin}(0) - H_D(0) = P_{S_{OUT}}(0) \tag{6}$$

Step 2: Set outlet static pressure to 97% of head from step 1. This step prevents the simulation from running into severe cavitation since the estimated NPSHa is not far from the cavitation point.

$$P_{S_{OUT}}(2) = 0.97H_{100} - H_D(1) \tag{7}$$

Here, P_{Sout} (2), is used to expressed outlet static pressure for step 2, H_{100}, is the pump head at no head drop calculated from step 1, and H_D, is the dynamic head calculated from step 1. *Step 3:* This step is to correct the errors in the previous step by adjusting the outlet boundary condition. The predicted results are expected to be as near as possible to the NPSHr.

$$P_{S_{OUT}}(3) = 0.97H_{100} + P_{T_{IN}}(2) - H_D(2) \tag{8}$$

2.4. Numerical Simulation Setup

Three dimensional Reynolds-Averaged Navier Stokes (3-D RANS) equation for a fully developed flow was solved using the commercial CFD package ANSYS CFX (ANSYS Inc). Water at room temperature was selected as the working fluid for the homogeneous fluid model and the reference pressure was 0 atm. An isothermal heat transfer rate was selected to render the system in thermal equilibrium with its surroundings at 25 °C. In addition, all flow domains were considered to have a surface roughness of 25 μm with no-slip. Due to additional effects in the viscous sublayer, an automatic near-wall treatment was applied. An inlet viscosity ratio of 10 was chosen to correspond to a medium turbulence intensity level of 5% at the pump inlet. To guarantee consistency, convergence and accuracy during the simulations, a high-resolution upwind scheme was employed to solve both steady and unsteady equations. For no cavitation conditions, pressure opening and flowrate boundary conditions were specified at the inlet and outlet. Cavitation simulations were performed under steady state with boundary conditions specified as static pressure for the outlet and mass flowrate at the inlet. The volume fraction of water $(1 - \alpha)$ was set to 1, whereas the vapor volume fraction, α, set to 0 at the inlet of the pump. Steady state iterations were set to a maximum of 700 for no cavitation conditions and up to 3000 for cavitation conditions. However, iterations converged when maximum residual values were less than or equal to 10^{-5}, and this occured when the flow had reached its stable periodicity.

2.5. Test Setup to Validate Numerical Method

The pump performance tests were done at Shandong Shuanglun Company Ltd., China. The test rig schematics can be seen in Figure 4, and the test setup is shown in Figure 5. The flow rate was measured with LWGY-200A electromagnetic flowmeter from Shanghai Zijiu Instrument Co. Ltd., China, with an error margin of ±0.5%. The fluid pressure was measured with WT200 pressure transmitters from Shanghai Weiltai Instrument Co. Ltd., China, and the uncertainty is less than ±0.1%. Hydraulic performance and cavitation tests were performed and evaluation was done according basic pump theory [34].

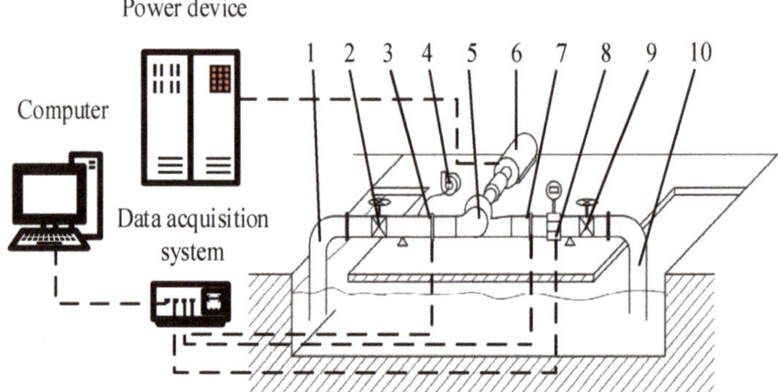

Figure 4. Scheme of the test rig. 1: Inlet pipe, 2(9): Valve, 3(7): Pressure transducer, 4: Vacuum pump, 5: Tested pump, 6: Driven motor, 8: Magnetic flow meter, 10: Outlet pipe.

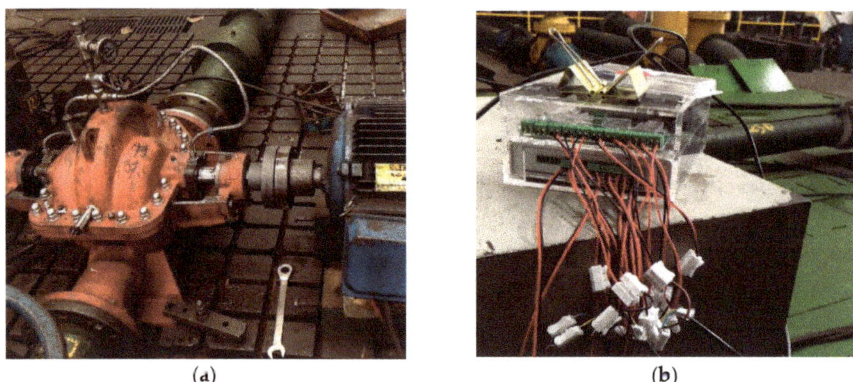

Figure 5. Test Pump (**a**) and Data Acquisition Device (**b**).

3. Optimization Procedure

The procedure for optimization is shown in Figure 6. The initial process was to sample the input bound variables based on an orthogonal design of experiment method. Secondly, series of single-suction impellers were designed by the hydraulic design software CFturbo (CFturbo® GmbH, Dresden, Germany). It was then mirrored with PTC Creo parametric 4.0 to obtain the 3D model of the double-suction impeller. ANSYS ICEM 18.0 and CFX 18.0 (ANSYS Inc.) were adopted for the numerical simulations to obtain the efficiencies and NPSHr, and the results were selected as objectives to train the surrogate models. The fourth part was to solve the models using multi-objective genetic algorithm to obtain pareto solutions. Finally, the optimized cases were verified by numerical simulation to improve the reliability of the results.

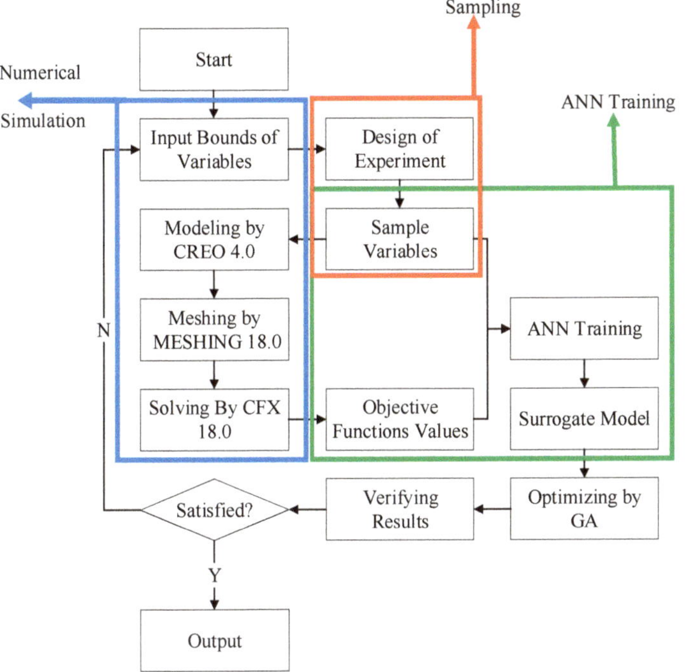

Figure 6. Flowchart of Optimization process.

3.1. Objective Functions

In the design process of centrifugal pump, the optimization goals and targets are considered as significant indexes for performance evaluation. For this study, the efficiency and NPSHr of the double-suction centrifugal pump at the design condition are selected as the optimization targets, and they are obtained by numerical model calculation. The expression for calculating efficiency is:

$$\eta = \frac{\rho g H Q}{P_s} \tag{9}$$

Here, ρ is density, H is the pump head, Q is the flow discharge, and P_S the shaft power. For pump cavitation, NPSHr, which is the minimum pressure required at the suction point to prevent cavitation is a significant performance index. The expression for NPSHr is given as:

$$NPSHr = \frac{P_s - P_v}{\rho g} + \frac{V_s^2}{2g} \tag{10}$$

Here, P_s and V_s are the reference pressure and velocity, and P_v is vapor pressure.

3.2. Decision variables and Array Orthogonal Design

There were space constraints due the structure of the suction and volute casing; therefore the shape of the impeller was maintained by holding constant the inlet diameter D_1, the impeller outlet diameter D_2, the hub diameter D_h, and the blade width at outlet b_2. Only the blade profile is optimized. Eight geometrical parameters namely hub inlet angle $\beta_{1\,hub}$, hub exit angle $\beta_{2\,hub}$, hub wrap angle $\varphi_{1\,hub}$, leading edge wrap angle at hub $\Delta\varphi_{0\,hub}$, shroud inlet angle $\beta_{3\,shroud}$, shroud exit angle $\beta_{4\,shroud}$, shroud wrap angle $\varphi_{2\,shroud}$ and the leading edge wrap angle at shroud $\Delta\varphi_{0\,shroud}$, were selected as optimization variables. Each parameter is given a set of four values. Table 3 shows the decision variables and their set of values used for the parameterization. Orthogonal Design of Experiment was applied here to design the experimental scheme. The main inputs to the design were the input variables (factors) and the number of values in each variable (levels). From Table 3, an orthogonal scheme of L_{32} (8^4) with n levels; p factors; and m schemes was designed in Table 4 according to the formula:

$$L_m\left(n^p\right) \tag{11}$$

Table 3. Range design of variables.

Trial No	A	B	C	D	E	F	G	H
	$\beta_{1hub}/^\circ$	$\beta_{2hub}/^\circ$	$\varphi_{1hub}/^\circ$	$\Delta\varphi_{0hub}/^\circ$	$\beta_{3shroud}/^\circ$	$\beta_{4shroud}/^\circ$	$\varphi_{2shroud}/^\circ$	$\Delta\varphi_{0shroud}/^\circ$
Original	17	29.43	143	0	15	29.43	143	0
1	15	26	139	−5	13	26	139	−5
2	17	28	143	−2.5	15	28	143	−2.5
3	19	30	145	2.5	17	30	145	2.55
4	21	32	148	5	19	32	148	5

Table 4. Orthogonal scheme.

Trial No	A	B	C	D	E	F	G	H
1	17	28	148	−2.5	13	26	148	−5
2	21	28	145	−2.5	19	28	139	−2.5
3	21	26	143	−2.5	17	26	145	5
4	21	28	139	5	19	32	145	−5
5	15	28	143	2.5	15	32	139	5
...	...	...	...	...	...	...	...	...
28	19	32	143	−2.5	15	30	139	−5
29	15	30	148	−2.5	19	30	143	5
30	15	32	145	5	17	28	143	−5
31	15	28	148	−5	15	28	145	2.5
32	21	32	143	−5	13	32	143	2.5

To eliminate the time of 2D hydraulic design and 3D modeling, the 3D hydraulic design software CFturbo 10.3 (CFturbo® GmbH) was used to carry out the 3D parametric design of the double suction impeller. In the parametric design, the focus is to control the blade profile. The meridional shape has 4 points for controlling the lines and the inclination of the angles of the hub and shroud (Figure 7). The position of the inlet edge of the blade is adjusted by a Bézier curve with five points. CFturbo is limited to single-suction impeller design only, and therefore a single-suction impeller was designed according to the hydraulic characteristics and mirrored into a double-suction impeller using PTC PTC Creo 4.0. The 32 design variables were written into the trail files using MATLAB R2017b (Mathworks Inc., Natick, MA, USA), then started PTC Creo with these trail files by BAT codes to complete the modeling process. ANSYS Workbench (ANSYS Inc.) was adopted to combine meshing with CFX solver since it can work in batch mode with a journal file, making it easier to automate meshing and solving using MATLAB and BAT codes.

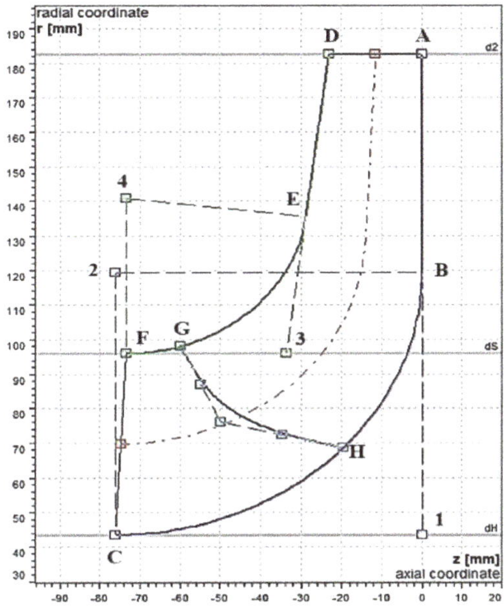

Figure 7. Meridional shape of impeller and leading edge of blade.

3.3. Artificial Neural Network Training

For this study, a dual-layer feed-forward artificial neural network with sigmoid hidden neurons and linear output neurons was adopted to fit the NPSHr and efficiency at the design flow condition and 8 design variables of the impeller. The Levenberg-Marquardt algorithm was adopted as the training algorithm to build the feedforward neural network. This is because compared with the disadvantages of traditional BPNNs, such as slow convergence speed and local minimum problems, the convergence rate of the Levenberg-Marquardt algorithm is the fastest of all traditional or improved networks, and it has been shown to achieve excellent evaluation and prediction results [35,36]. Also, to improve the prediction performance of ANN, it is important to use a much effective activation function in order to obtain a higher prediction accuracy. Therefore, the activation function used for the feedforward NN was the Tangent Hyperbolic Activation Function (*tanh*). This is because *tanh* has a much better recognition accuracy for multi-layer neural networks [37,38]. Its main advantage is the ability to produce zero centred output which aids the back-propagation process [39]. Figures 8 and 9 are the dual-layer feedforward neural network structures used for training the efficiency and NPSHr objectives. The mathematical relation for the ANN function is written as Equation (12), the activation function, *tanh* is written as Equation (13) and the linear function as Equation (14).

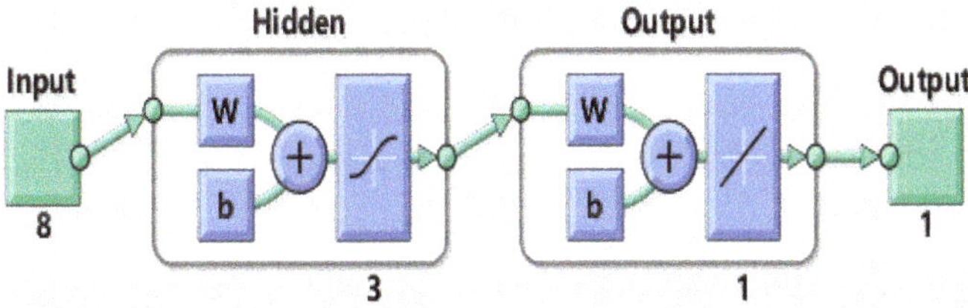

Figure 8. ANN structure for efficiency objective.

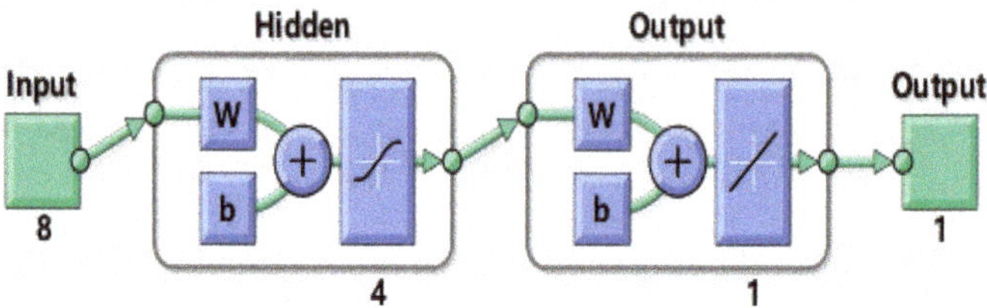

Figure 9. ANN structure for NPSHr objective.

$$y = g\left(\sum_{j=1}^{n} w_j^2 \times f\left(\sum_{k=1}^{m} w_{k,j}^1 x_k + b_n^1\right)\right) + b^2 \tag{12}$$

$$f(x) = \left[\frac{2}{(1+e^{-2x})}\right] - 1 \tag{13}$$

$$g(x) = ax + b \tag{14}$$

Here, weight coefficient is expressed as w, threshold is b. Superscripts 1 and 2 represent the coefficients from the first layer to the second layer, and from the hidden layer to the output layer respectively.

3.4. Multi-objective Optimization Design

The multi-objective problem, which is defined as an *N-dimensional* problem with M objective functions is mathematically expressed as:

$$\begin{cases} \frac{max}{min}\,[\sigma_1(X), \sigma_2(X), \ldots, \sigma_M(X)] \\ subject\ to \\ \\ \delta_i(X) \le 0, i = 1, 2, \ldots, M \end{cases} \tag{15}$$

where $X = (x_1, x_2, \ldots, x_N)$ is the N-dimensional vector;

$\sigma_j = (X)(j = 1, 2, \ldots, M)$ are the objective functions;
$\delta_i(X) \le 0$ is the variables limit.

Hence, the problem for the multi-objective optimization can be described as follows.

$$\begin{cases} Find \\ maximize \quad \eta = f_1(A, B, C, D, E, F, G, H) \\ miximize \quad NPSHR = f_2(A, B, C, D, E, F, G, H) \\ subject\ to \\ \qquad 15° \le A \ge 21° \\ \qquad 26° \le B \ge 32° \\ \qquad 139° \le C \ge 148° \\ \qquad -5° \le D \ge 5° \\ \qquad 13° \le E \ge 80° \\ \qquad 26° \le F \ge 80° \\ \qquad 139° \le G \ge 148° \\ \qquad -5° \le H \ge 5° \end{cases} \tag{16}$$

The value of each function could not be used to evaluate the individuals since the objective functions were more than one. The expression in Equation (17) was adopted [17] to evaluate the individuals and solve the problem.

$$F(x) = \left[\frac{1}{1+\|x-y\|_2} \right] \tag{17}$$

where x is any single population individual; y is the Pareto efficient individual closet to x; $\|x-y\|$ is the Euclidean distance between x and y.

To solve this problem, a Pareto-optimal solution was applied to determine the optimal parameter combinations that would best solve the problem. To obtain the global Pareto frontier of the two objective functions, the Multi-objective Generic Algorithm (MOGA) was applied [40]. To invoke the algorithm, *gamultiobj* a customized MATLAB function [41] was used. This function has the ability to use a controlled elitist individual, which gives it an advantage over the simple genetic algorithm, and has been successfully applied in optimization works [42]. In construction of the Pareto-optimal solutions, the following input parameters were used. Population size of 100, Pareto-front population of 0.8, crossover fraction of 0.85, 1000 generations, and function tolerance of 10.

4. Discussion of results

4.1. Numerical Model Validation

The efficiency calculated from the simulation results is the hydraulic efficiency only. To compare the simulation results with the experimental results, the overall efficiency of the pump, which comprises the hydraulic efficiency η_h, the mechanical efficiency η_m, and the volumetric efficiency η_v has to be considered. This is because the experimental efficiency includes the hydraulic efficiency, volumetric efficiency and mechanical efficiency. Thus, following the examples from [24,43], the efficiency obtained from the simulation is further processed, before comparing with the test results. The mechanical efficiency is expressed as:

$$\eta_m = 1 - 0.07 \frac{1}{\left(\frac{n_s}{100}\right)^{\frac{7}{6}}} - 0.02 \tag{18}$$

The Volumetric efficiency is expressed as:

$$\eta_v = \frac{1}{1 + (0.68 n_s)^{\frac{-2}{3}}} \tag{19}$$

The relationship between the efficiencies is expressed as:

$$\eta = \eta_h \eta_m \eta_v \tag{20}$$

Computational results from the numerical simulations were matched with the experiments in Figure 10. The pump head dropped gradually as the flow rate was increased, and the trend between the numerical and experimental heads were quite similar. The experimental head was, however, higher than the numerical head whereas the efficiency from the test results was lower than the calculate efficiency. The experimental head at the best efficiency point (BEP) was 41.49 m, and the corresponding efficiency was 86.63%, at a flow rate of 518.57 m³/h, whereas the BEP for the numerical predictions occurred at head 40.24 m and efficiency 88.39 %. The relative errors at the BEP were 3.01% and 2.03% respectively. The experimental head at the design point was 41.83 m and efficiency 85.20%. Overall, the agreement between the results from the experiments and the simulation was very good, and this affirmed the reliability of the numerical approach.

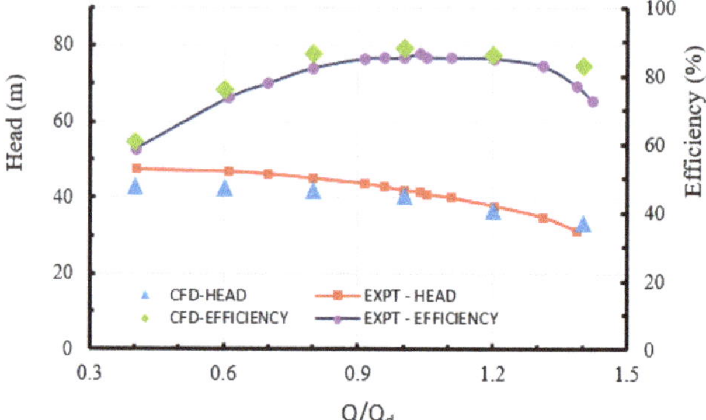

Figure 10. Hydraulic performance comparison of test and numerical results.

A comparison of the cavitation characteristics at the pump design point, Q_d, between the simulation and the experimental results is shown in Figure 11a. The calculated NPSHr was 2.532 m at a corresponding head drop of 3.12%. The deviation between the unsteady simulation and the experimental NPSHr for the design flow rate was calculated as 4.05%. It can be noted that both calculated and the test measurements for the NPSHr did not exceed the design NPSHr, and the characteristic "sudden" head-drop was very well predicted by the numerical simulation. The suction performance at different flow rates for the numerical and experimental results are compared in Figure 11b. The computational results were relatively higher than the experimental test values at part load conditions of $0.6Q_d$ and comparatively lower at design and overload conditions. That notwithstanding, the deviations were minimal and therefore the numerical results agreed well with the experimental results to established the suitability of the numerical approach for simulating and predicting cavitation flow in centrifugal pumps.

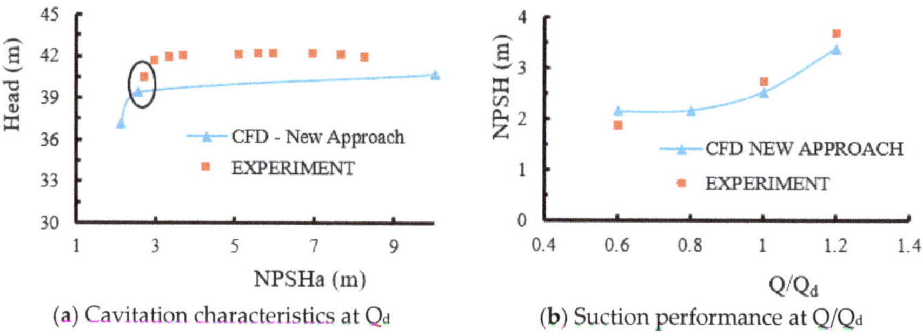

(**a**) Cavitation characteristics at Q_d (**b**) Suction performance at Q/Q_d

Figure 11. Suction performance prediction and comparison.

4.2. Optimization Results

4.2.1. Comparison of CFturbo Impeller to Original Impeller

The hydraulic performance characteristics of the design impeller built with CFturbo (CFturbo® GmbH) is compared with the original model. This is to validate the new design schemes to be produced by the CFturbo hydraulic software. For all three characteristics compared, the design condition recorded the lowest deviation of 0.4% for power and head, whereas the deviation in efficiency was 0.03%. This is shown in Figure 12. Looking at the margin of deviation, there was a perfect agreement between the two models, indicating that the model built by Cfturbo hydraulic design software could be used for the design of 3D models according to the design variables.

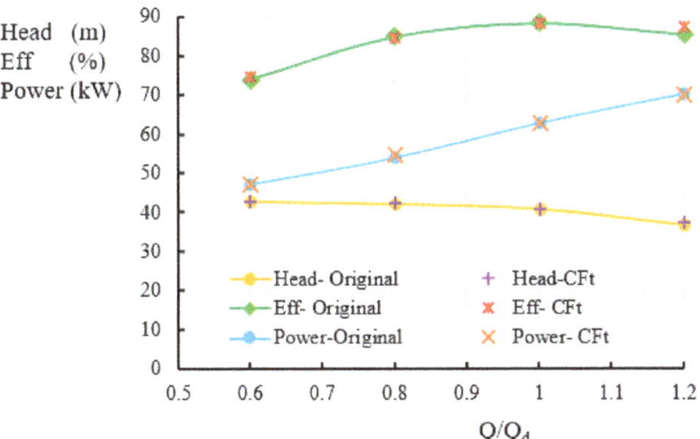

Figure 12. Comparison of Cfturbo built impeller to original model.

4.2.2. Orthogonal Test Results

The 32 impellers designed from Table 4 were simulated with ANSYS-CFX 19.2 (ANSYS Inc.). Investigations were done at the design flow rate as per the objectives of the optimization. The results from the orthogonal test are presented in Table 5.

Table 5. Orthogonal scheme results.

Trial No	A	B	C	D	E	F	G	H	η (%)	NPSHr (m)
1	17	28	148	−2.5	13	26	148	−5	88.03	2.420
2	21	28	145	−2.5	19	28	139	−2.5	88.69	2.338
3	21	26	143	−2.5	17	26	145	5	89.21	2.364
4	21	28	139	5	19	32	145	−5	88.16	2.374
5	15	28	143	2.5	15	32	139	5	87.41	2.647
6	15	32	139	−2.5	17	32	148	−2.5	87.26	2.464
...	...	...	...	...	...	...	...	...	...	...
27	19	32	148	5	15	26	145	−2.5	88.93	2.488
28	19	32	143	−2.5	15	30	139	−5	87.79	2.450
29	15	30	148	−2.5	19	30	143	5	87.87	2.442
30	15	32	145	5	17	28	143	−5	88.43	2.517
31	15	28	148	−5	15	28	145	2.5	88.18	2.513
32	21	32	143	−5	13	32	143	2.5	87.17	2.521

4.2.3. ANN Metamodeling and Validation

Artificial neural networks were adopted as the metamodel to build the relationship between the two objective functions and the design variables for optimization. To determine the suitability of the surrogates for further optimization, the R-square analysis was carried out to measure the strength of the relationship between the linear model and the dependent variable. Figure 13 shows the R-square of the ANN models of the objective functions. The R-square values for both efficiency and NPSHr were calculated as 0.96758 and 0.96083 respectively, indicating that the ANN models for can be applied to the multi-objective optimization since the prediction accuracy is high enough. As from previous works, validation of a metamodel is a requirement before it can be used as a surrogate [17,44]. A comparison of the ANN prediction and the CFD simulation results from is drawn in Figure 14. From the graph, the predicted ANN model values are in agreement with the CFD simulation values. Hence, both NPSHr and efficiency would be used as objective functions in the multi-objective optimization.

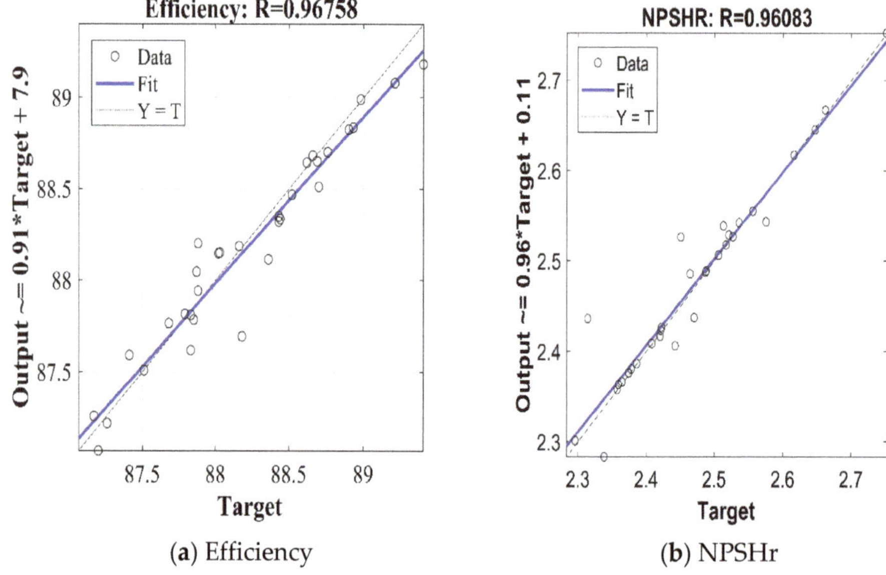

(a) Efficiency

(b) NPSHr

Figure 13. R-Square analysis of efficiency and NPSHr.

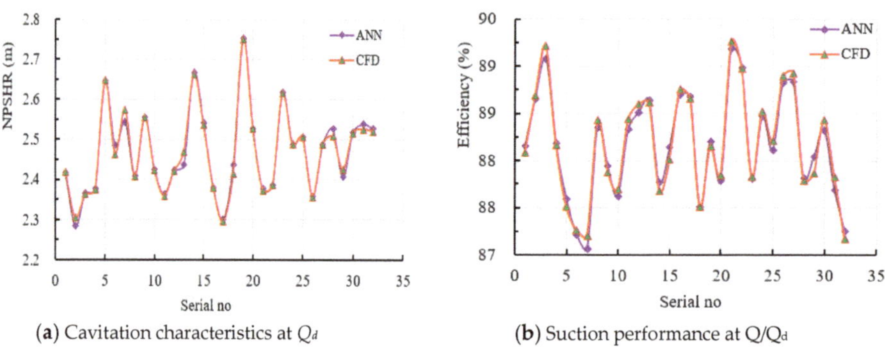

(a) Cavitation characteristics at Q_d

(b) Suction performance at Q/Q_d

Figure 14. Validation of ANN prediction with CFD.

4.2.4. Optimal Solution Solving-MOGA Solutions

The multi-objective generic algorithm was applied to establish an optimization model with MATLAB. The Pareto frontiers from ANN for the two objective functions have been presented in Figure 15. The pareto solutions presented a set of 100 optimal schemes that satisfied both objective functions. In real pump applications, it is necessary to maximize economic benefits; hence, a maximum efficiency point was considered. Consequently, the lower the suction performance the more resilient the pump is to cavitation, and so the minimum NPSHr point was also considered. A third point, the middle point is also considered. Three impeller schemes were then built according to the optimum decision variables in Table 6 and simulated with ANSYS 19.2 (ANSYS Inc.).

The comparison of the original case and the calculation results from the three optimal cases at the design condition is presented in Table 7. Although the head is not an optimization objective, it shouldn't be made worse. For all the 3 cases, the head of case 1 reduced by 5% even though it had the highest increase in efficiency of about 2.6%. There increase in efficiency of case 3 is 0.08%, which is almost negligible. Case 2 however increased its efficiency by 1.53%. For NPSHr, the cavitation performance for case 2 improved tremendously by 7.26%, whereas that of case 3 recorded the best cavitation performance by improving 8.21%. Despite case 3 performing slightly better than case 2 for cavitation performance, optimized case 2 would best suit the multi-objective optimization since its increase in efficiency is much significant. The selected best case was modelled with the original suction and volute units and numerically simulated for the optimization objectives.

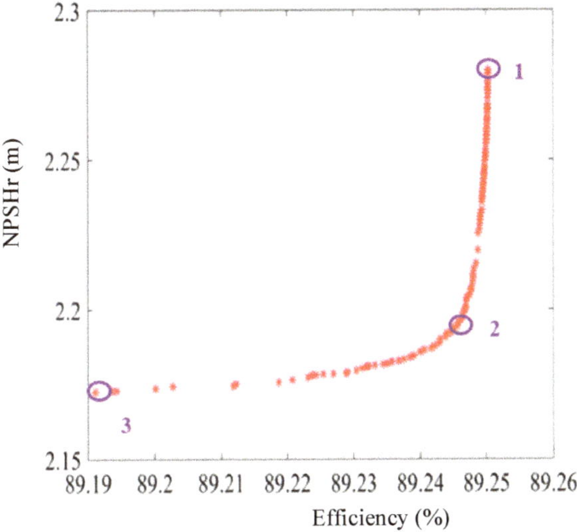

Figure 15. Pareto-frontiers from ANN.

Table 6. Variables for Optimum Cases.

Cases	β1	β2	φ1	Δφ1	β 3	β4	φ2	Δφ2	η (%)	NPSHr (m)
1	21.000	26.00	148.00	5.000	19.000	26.00	139.93	5.00	89.25	2.280
2	20.994	26.006	145.69	4.999	18.999	26.01	143.34	3.16	89.247	2.205
3	19.931	26.00	139.00	4.999	18.999	26.001	146.95	−5.00	89.191	2.173

Table 7. Comparison of original case and best optimal cases.

Name	η (%)	NPSHr (m)	Head (m)
Original case	88.28017	2.532	40.721
Case 1	90.64207	2.392	38.650
Case 2	89.65017	2.348	40.019
Case 3	88.35048	2.324	40.070

4.2.5. Comparison of External Characteristics—Optimized with Original Design

Figure 16 is a comparison of the predicted NPSHr for the optimized impeller and the original model at different flow conditions. It can be seen from the graph that the optimized impeller had improved cavitation performance as compared to the original impeller. There was a 7.26% improvement in suction performance at the design point. Comparatively, the improvement in suction performance decreased slightly as the flow moved from the design point. At 1.2 Q_d, the suction performance was improved by 3.9%. At part loads however, the improvement in suction performance was 3.8% and 4.5% for 0.6 Q_d and 0.8 Q_d flow conditions respectively. The predicted efficiencies for the optimized and original impellers were also compared in Figure 17 at off design conditions. The maximum increase in efficiency which is 1.53% occurred at the nominal flow rate. At 0.8 Q_d, there was in increase in efficiency of 1.1%. At overload and deep part loads, the effects of the optimization of hydraulic efficiency was very little, with 0.4 Q_d improving by only 0.2%.

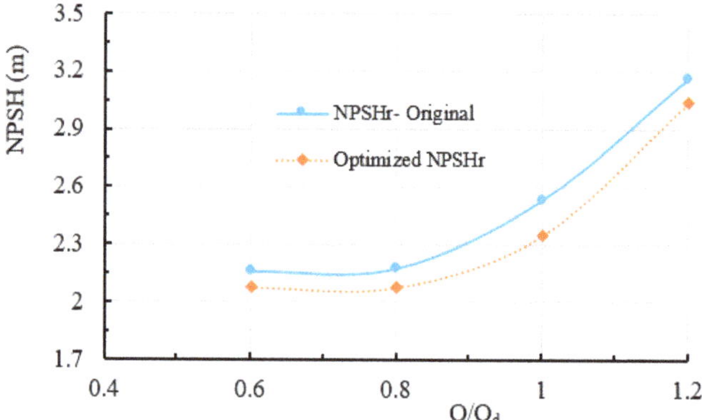

Figure 16. Predicted NPSHr comparison at different flow rates.

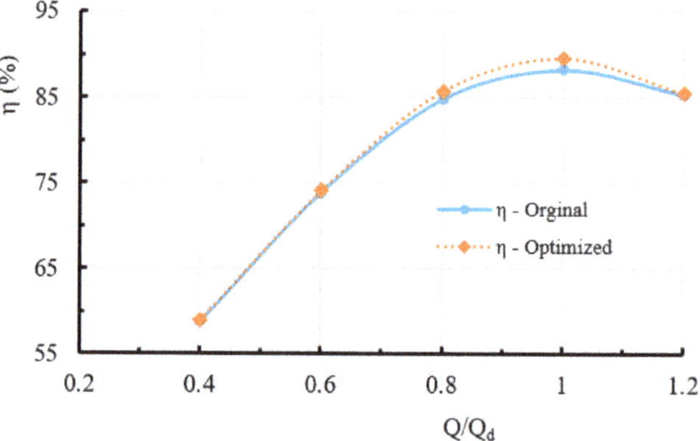

Figure 17. Predicted efficiency comparison at different flow rates.

4.2.6. Internal Flow Analysis at 1.0 Q_d

The static pressure distribution on the blade suction and pressure blade surfaces were compared at the design condition in Figure 18. For both the original impeller and the optimized impeller, the static pressure increased steadily with a smooth transition along the impeller's radial direction. Generally, low pressure gradients existed in the blade leading regions, which moved towards the blade suction surface (SS). The optimized impeller however had relatively higher pressure distributions along the blade surface with original impeller having lower distributions in the leading regions. For the original impeller, the span of the low pressure regions went as far as the middle of the blade for the suction side. At the pressure side (PS) of the blade, the trend was same. Static pressure was much higher than the suction side which is very usual of the blade. The optimized blade saw much improvements in the pressure distribution, with the pockets of low pressure regions disappearing.

Figure 19 compares the of blade streamlines of the optimum and original designs at no cavitation conditions to illustrate the improvement of pump performance. The streamline distribution along the blade surface is uniform and symmetric. There was some flow distortion observed very close to the leading edge of one of the blades. This however disappeared and the flow distribution became uniform when as the flow moved towards the middle of the blade. In the optimized impeller, no flow distortion was seen. The velocity direction was along with the blade, indicating the ability of the optimized blade to control the fluid much better than the original blade.

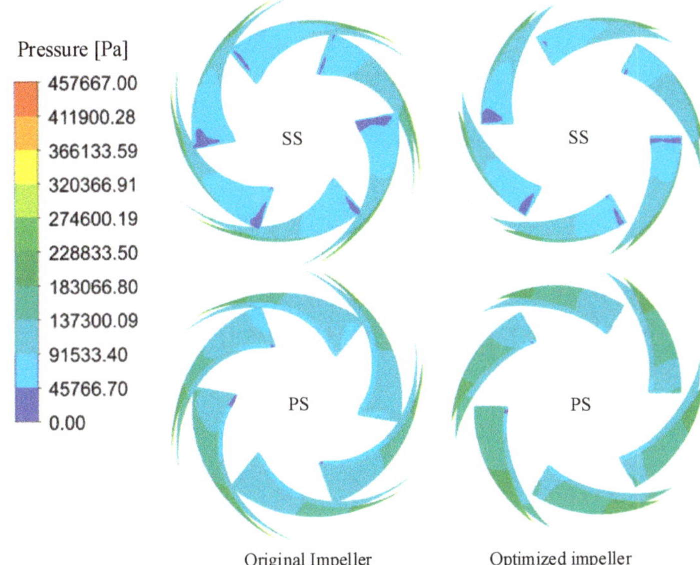

Figure 18. Comparison of static pressure distribution in impeller blade surface.

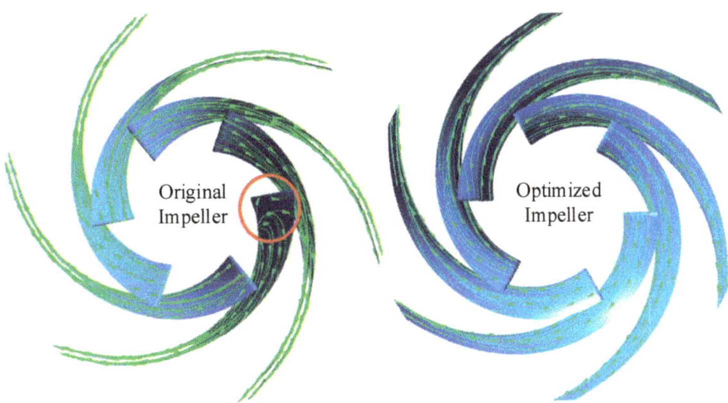

Figure 19. Comparison of streamline on blade surface at no cavitation condition.

At critical cavitation conditions, flow distortions and reverse flow were observed on the blade surface for both designs. As in Figure 20, the suction side of both blades had a smooth distribution at the leading edge. Distortions appeared along the blade surface on different blades up to the middle of the blade where the flow became uniform, and by this, the static pressure would have increased and bubble cavities disappearing. The distortions on the optimized impeller however covered a smaller region as compared to the original impeller; this is also an indication that flow distribution had improved in cavitation conditions. The pressure side of the blade had flow distortions that appeared similar for both the original and optimized schemes. However, a critical look revealed some vortex and reverse flow appearing closer to the leading edge of some blades of the original impeller.

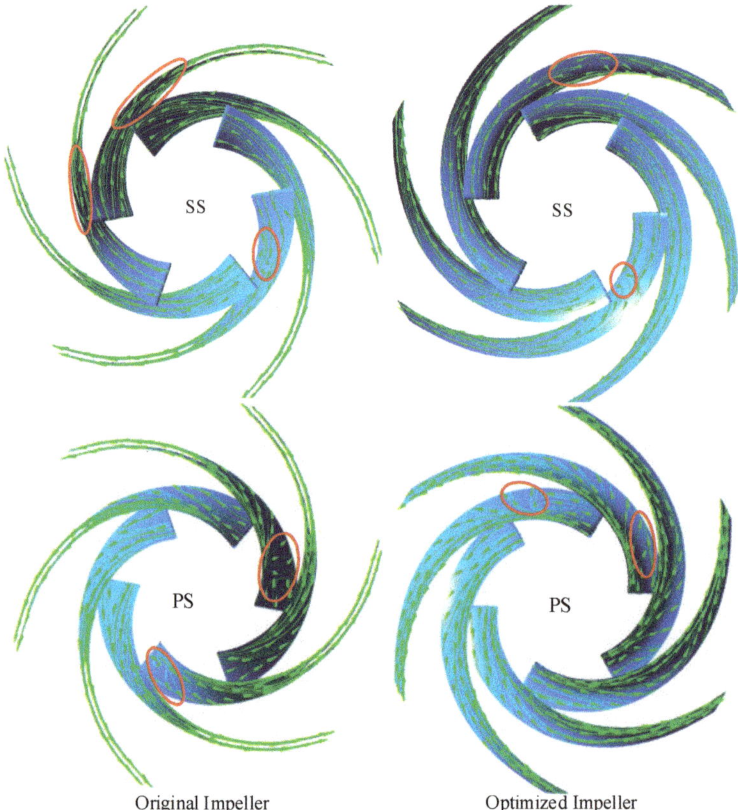

Figure 20. Compare streamline on blade surface at cavitation condition.

4.2.7. Vapor volume fraction analysis at 1.0 Q_d

To compare the vapor and bubble distribution within the two schemes, the optimized impeller was first compared with the original impeller at the nominal operating condition. From Figure 21, pockets of vapor above 5% volume fraction could be seen at the right corner of the leading edge for all the blades in the original impeller. The case was rather different for the optimized impeller with vapor cavities appearing on only two blades. This indicates the improvement in cavitation performance by the optimized impeller. For the case of critical cavitation, Figure 22 shows the volume fraction distribution in the impeller for water vapor above 10% volume fraction.

On comparison, the vapor volume fraction of the optimized impeller was shown to be much smaller than that of the original impeller, whereas for most of the blades, the bubble cavities spread up to the middle of the impeller before collapsing, and the bubble distribution was much lower in the optimized impeller, mostly covering just a quarter of the blade length. For the suction domain, vapor cavities formed around the wear rings which indicated the presence of suction ring cavitation for both design schemes. Figure 23 is the bubble distribution in suction domain at critical cavitation condition. The cavity formed around on the suction ring closer to the suction tongue region was not surprising, due to the geometry configuration. However, the bubble distribution in the suction unit was relatively smaller in the optimized impeller other than the original impeller which also proves that the optimized impeller has improved cavitation performance.

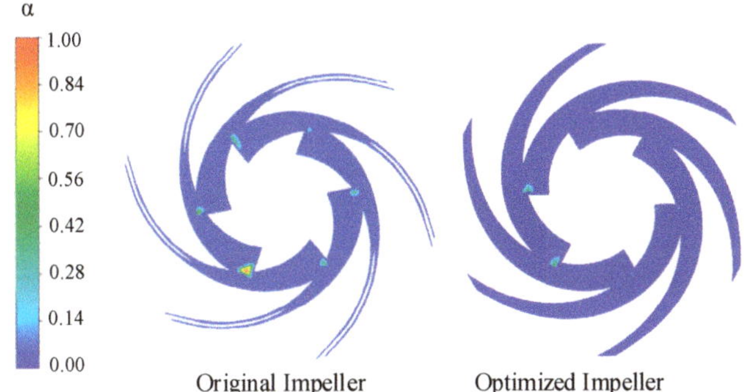

Figure 21. Volume fraction distribution in impeller at NPSH = 8.77 m.

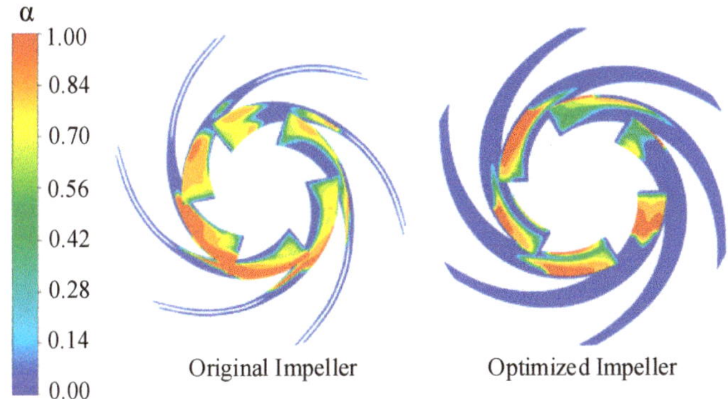

Figure 22. Volume fraction distribution in impeller at critical cavitation.

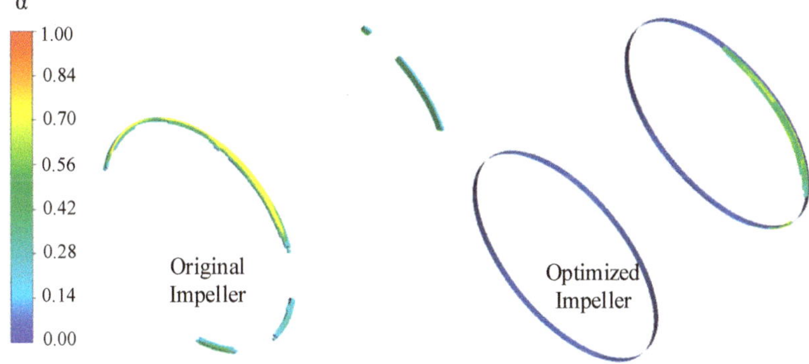

Figure 23. Bubble distribution in suction domain at critical cavitation.

5. Conclusion

A multi-parameter and multi-objective optimization was performed on an axially-split double-suction centrifugal pump first to increase efficiency and second to improve cavitation performance. An L_{32} (8^4) orthogonal array was designed to evaluate 8 geometrical parameters at 4 levels each. A two-layer feedforward neural network and genetic algorithm was applied to solve the multi-objective problem into pareto solutions that meets the objective functions. Three best cases from the pareto solutions were validated by numerical simulation and compared to the original design. The results from the study are as follows:

1. There was a 7.26% improvement in suction performance at the design point. At 1.2 Q_d, the suction performance was improved by 3.9%. At part loads, however, the suction performance improved by 3.8% and 4.5% for 0.6 Q_d and 0.8 Q_d flow conditions respectively.

2. The efficiency increased by 1.53% at the nominal flow rate and 1.1% at 0.8 Q_d. For overload and deep part loads, the effects of the optimization of hydraulic efficiency was very low.

3. For the optimized design, the pressure distribution at the leading regions were comparatively higher and the streamline on the blade surface was improved.

4. By comparison, the vapor volume fraction of the optimized impeller was much smaller than that of the original impeller. Also, the bubble distribution in the suction unit was relatively smaller in the optimized impeller.

5. This study provides a theoretical reference and a parametric database for both hub and shroud blade angles for double suction centrifugal pump optimization design.

Author Contributions: Conceptualization, M.K.O. and W.W.; Methodology, W.W. AND X.G; Software, X.G.; Validation, M.K.O., W.W. and J.P.; Formal analysis, M.K.O. and J.P.; Writing—original draft preparation, M.K.O.; Writing—Review and Editing, T.Y.; Supervision, J.P.

Funding: This work is supported by the National Key Research and Development Program (Grant No. 2018YFB0606103), National Natural Science Foundation of China (Grant No. 51879121, 51779107, 51809121), Six Talent Peaks Project (GDZB - 047) and Qing Lan Project of Jiangsu Province.

Acknowledgments: The authors are much grateful for the financial support provided by the National Natural Science Foundation of China and the kind help of 4CPump research group.

Nomenclatures

Latin symbols

b_2	blade width, mm
C_{cond}	Condensation coefficient
C_{vap}	Evaporation coefficient
D	Impeller diameter, mm
H	Head, m
H_D	Dynamic Head, m
k	Kinetic energy of turbulence, m^2/s^2
m^+	Evaporation rate
m^-	Condensation rate
N	Rotational speed, r/min
p	Pressure, Pa
P_S	Shaft power, kW
Q	Flow rate, m^3/h
R_b	Bubble radius, m
r_g	Nucleation site volume fraction
u	Velocity, m/s
z	Number of blades

Greek symbols

α_v	Vapor volume fraction
$\beta_{1\,hub}$	Inlet angle at hub, °
$\beta_{2\,hub}$	Exit angle at hub, °
$\beta_{3\,shroud}$	Inlet angle at shroud, °
$\beta_{4\,shroud}$	Exit angle at shroud, °

ϵ	Turbulence dissipation rate, m^2/s^3
η	Efficiency, %
ρ	Density, kg/m^3
μ	Dynamic viscosity, Pa.s
μ_t	Turbulent viscosity, m^2/s
ω	Specific dissipation of turbulent kinetic energy, s^{-1}
$\varphi_{1\,hub}$	wrap angle at hub, °
$\varphi_{2\,shroud}$	wrap angle at shroud, °
$\Delta\varphi_{0\,hub}$	leading edge wrap angle at hub, °
$\Delta\varphi_{0\,shroud}$	leading edge wrap angle at shroud, °

Abbreviations

2D	Two-Dimensional
3D	Three-Dimensional
ANN	Artificial Neural Network
BEP	Best Efficiency Point NPSHr
CFD	Computational Fluid Dynamics
DOE	Design of Experiment
MOGA	Multi-objective Generic Algorithm
NPSHr	Net positive suction head required
ODOE	Orthogonal Design of Experiment
SST	Shear Stress Transport
RANS	Reynolds-averaged Navier-Stokes

References

1. Zou, Z.; Wang, F.; Yao, Z.; Tao, R.; Xiao, R.F.; Li, H.C. Impeller Radial Force Evolution in a Large Double-Suction Centrifugal Pump During Startup at the Shut-Off Condition. *Nucl. Eng. Des.* **2016**, *310*, 410–417. [CrossRef]

2. Azizi, R.; Attaran, B.; Hajnayeb, A.; Ghanbarzadeh, A.; Changizian, M. Improving Accuracy of Cavitation Severity Detection in Centrifugal Pumps Using a Hybrid Feature Selection Technique. *Measurement* **2017**, *108*, 9–17. [CrossRef]

3. Li, Z.; Zheng, X. Review of Design Optimization Methods for Turbomachinery Aerodynamics. *Prog. Aerosp. Sci.* **2017**, *93*, 1–23. [CrossRef]

4. Zhou, L.; Shi, W.; Wu, S. Performance Optimization in a Centrifugal Pump Impeller by Orthogonal Experiment and Numerical Simulation. *Adv. Mech. Eng.* **2013**, *5*, 385809. [CrossRef]

5. Pei, J.; Yin, T.; Yuan, S.; Wang, W.J.; Wang, J.B. Cavitation Optimization for a Centrifugal Pump Impeller by Using Orthogonal Design of Experiment. *Chin. J. Mech. Eng.* **2017**, *30*, 103–109. [CrossRef]

6. Yan, H.; Liu, M.; Liang, X.; Shan, Z.B. Numerical Simulation on Cavitation Characteristics of Large-Scale Axial-Flow Pumps Based on Orthogonal Experiment. *J. Huazhong Univ. Sci. Technol. (Nat. Sci. Ed.)* **2014**, *29*. (In Chinese) [CrossRef]

7. Nataraj, M.; Arunachalam, V.P. Optimizing Impeller Geometry for Performance Enhancement of a Centrifugal Pump Using the Taguchi Quality Concept. *Proc. Inst. Mech. Eng. Part A J. Pow. Energy* **2006**, *220*, 765–782. [CrossRef]

8. Xu, Y.; Tan, L.; Cao, S.; Qu, S.H. Multiparameter and Multiobjective Optimization Design of Centrifugal Pump Based on Orthogonal Method. *Proc. Inst. Mech. Eng. Part C J. Mech. Eng. Sci.* **2017**, *231*, 2569–2579. [CrossRef]

9. Kaveh, A.; Laknejadi, K. A New Multi-Swarm Multi-Objective Optimization Method for Structural Design. *Adv. Eng. Softw.* **2013**, *58*, 54–69. [CrossRef]

10. Nourbakhsh, A.; Safikhani, H.; Derakhshan, S. The Comparison of Multi-Objective Particle Swarm Optimization and Nsga Ii Algorithm: Applications in Centrifugal Pumps. *Eng. Optim.* **2011**, *43*, 1095–1113. [CrossRef]

11. Safikhani, H.; Khalkhali, A.; Farajpoor, M. Pareto Based Multi-Objective Optimization of Centrifugal Pumps Using Cfd, Neural Networks and Genetic Algorithms. *Eng. Appl. Comput. Fluid Mech.* **2011**, *5*, 37–48. [CrossRef]

12. Jin, R.; Wei, C.; Simpson, T.W. Comparative Studies of Metamodelling Techniques under Multiple Modelling Criteria. *Struct. Multidiscip. Optim.* **2001**, *23*, 1–13. [CrossRef]

13. Tao, R.; Xiao, R.; Zhu, D.; Wang, F.J. Multi-Objective Optimization of Double Suction Centrifugal Pump. *Proc. Inst. Mech. Eng. Part C J. Mech. Eng. Sci.* **2018**, *232*, 1108–1117. [CrossRef]

14. Wang, G.G.; Shan, S. Review of Metamodeling Techniques in Support of Engineering Design Optimization. *J. Mech. Des.* **2006**, *129*, 370–380. [CrossRef]

15. Mengistu, T.; Ghaly, W. Aerodynamic Optimization of Turbomachinery Blades Using Evolutionary Methods and Ann-Based Surrogate Models. *Optim. Eng.* **2008**, *9*, 239–255. [CrossRef]

16. Derakhshan, S.; Mohammadi, B.; Nourbakhsh, A. Incomplete Sensitivities for 3d Radial Turbomachinery Blade Optimization. *Comput. Fluids* **2008**, *37*, 1354–1363. [CrossRef]

17. Pei, J.; Gan, X.; Wang, W.; Yuan, S.Q.; Tang, Y.J. Multi-Objective Shape Optimization on the Inlet Pipe of a Vertical Inline Pump. *J. Fluids Eng.* **2019**, *141*. [CrossRef]

18. Tao, R.; Xiao, R.; Yang, W.; Wang, F.J.; Liu, W.C. Optimization for Cavitation Inception Performance of Pump-Turbine in Pump Mode Based on Genetic Algorithm. *Math. Probl. Eng.* **2014**, *2014*, 234615. [CrossRef]

19. Goncalves, E.; Patella, R.F. Numerical Simulation of Cavitating Flows with Homogeneous Models. *Comput. Fluids* **2009**, *38*, 1682–1696. [CrossRef]

20. Bakir, F.; Rey, R.; Gerber, A.; Belamri, T.; Hutchinson, B. Numerical and Experimental Investigations of the Cavitating Behavior of an Inducer. *Int. J. Rotating Mach.* **2004**, *10*, 15–25. [CrossRef]

21. Li, X.; Yu, B.; Ji, Y.; Lu, J.X.; Yuan, S.Q. Statistical Characteristics of Suction Pressure Signals for a Centrifugal Pump under Cavitating Conditions. *J. Therm. Sci.* **2017**, *26*, 47–53. [CrossRef]

22. Li, X.; Yuan, S.; Pan, Z.; Yuan, J.P.; Fu, Y.X. Numerical Simulation of Leading Edge Cavitation within the Whole Flow Passage of a Centrifugal Pump. *Sci. China Technol. Sci.* **2013**, *56*, 2156–2162. [CrossRef]

23. Zhang, F.; Yuan, S.; Fu, Q.; Pei, J.; Böhle, M.; Jiang, X.S. Cavitation-Induced Unsteady Flow Characteristics in the First Stage of a Centrifugal Charging Pump. *J. Fluids Eng.* **2017**, *139*, 011303. [CrossRef]

24. Tang, X.L.; Zou, M.D.; Wang, F.J.; Li, X.Q.; Shi, X.Y. Comprehensive Numerical Investigations of Unsteady Internal Flows and Cavitation Characteristics in Double-Suction Centrifugal Pump. *Math. Probl. Eng.* **2017**, *2017*, 5013826. [CrossRef]

25. Lu, J.; Yuan, S.; Luo, Y.; Yuan, J.P.; Zhou, B.L.; Sun, H. Numerical and Experimental Investigation on the Development of Cavitation in a Centrifugal Pump. *Proc. Inst. Mech. Eng. Part E J. Proc. Mech. Eng.* **2016**, *230*, 171–182. [CrossRef]

26. Ding, H.; Visser, F.; Jiang, Y. A Practical Approach to Speed up Npshr Prediction of Centrifugal Pumps Using Cfd Cavitation Model. In Proceedings of the ASME 2012 Fluids Engineering Division Summer Meeting collocated with the ASME 2012 Heat Transfer Summer Conference and the ASME 2012 10th International Conference on Nanochannels, Microchannels, and Minichannels, Rio Grande, PR, USA, 8–12 July 2012; American Society of Mechanical Engineers: New York, NY, USA, 2012.

27. Gu, Y.; Pei, J.; Yuan, S.; Wang, W.J.; Zhang, F.; Wang, P.; Appiah, D.; Liu, Y. Clocking Effect of Vaned Diffuser on Hydraulic Performance of High-Power Pump by Using the Numerical Flow Loss Visualization Method. *Energy* **2019**, *170*, 986–997. [CrossRef]

28. Pei, J.; Zhang, F.; Appiah, D.; Hu, B.; Yuan, S.Q.; Asomani, S.N. Performance Prediction Based on Effects of Wrapping Angle of a Side Channel Pump. *Energies* **2019**, *12*, 139. [CrossRef]

29. Medvitz, R.B.; Kunz, R.F.; Boger, D.A.; Lindau, J.W.; Yocum, A.M.; Pauley, L.L. Performance Analysis of Cavitating Flow in Centrifugal Pumps Using Multiphase Cfd. *J. Fluids Eng.* **2002**, *124*, 377–383. [CrossRef]

30. Menter, F.R. Two-Equation Eddy-Viscosity Turbulence Models for Engineering Applications. *AIAA J.* **1994**, *32*, 1598–1605. [CrossRef]

31. Bardina, J.; Huang, P.; Coakley, T.; Bardina, J.; Huang, P.; Coakley, T. Turbulence Modeling Validation. In Proceedings of the 28th Fluid Dynamics Conference, Snowmass Village, CO, USA, 29 June–2 July 1997.

32. Zwart, P.J.; Gerber, A.G.; Belamri, T. A Two-Phase Flow Model for Predicting Cavitation Dynamics. In Proceedings of the Fifth International Conference on Multiphase Flow, Yokohama, Japan, 30 May–3 June 2004.

33. Mejri, I.; Bakir, F.; Rey, R.; Belamri, T. Comparison of Computational Results Obtained from a Homogeneous Cavitation Model with Experimental Investigations of Three Inducers. *J. Fluids Eng.* **2006**, *128*, 1308–1323. [CrossRef]

34. Gülich, J.F. *Centrifugal Pumps*; Springer Science & Business Media: Berlin, Germany, 2010.

35. Yu, H. Levenberg-Marquardt Training. In *Intelligent Systems*; Springer: Berlin, Germany, 2011.
36. Zhou, R.; Wu, D.; Fang, L.; Xu, A.J.; Lou, X.W. A Levenberg–Marquardt Backpropagation Neural Network for Predicting Forest Growing Stock Based on the Least-Squares Equation Fitting Parameters. *Forests* **2018**, *9*, 757. [CrossRef]
37. Karlik, B.; Olgac, A.V. Performance Analysis of Various Activation Functions in Generalized Mlp Architectures of Neural Networks. *Int. J. Artif. Intell. Expert Syst.* **2011**, *1*, 111–122.
38. Neal, R.M. Connectionist Learning of Belief Networks. *Artif. Intell.* **1992**, *56*, 71–113. [CrossRef]
39. Nwankpa, C.; Ijomah, W.; Gachagan, A.; Marshall, S. Activation Functions: Comparison of Trends in Practice and Research for Deep Learning. *arXiv* **2018**, arXiv:1811.03378.
40. Gen, M.; Cheng, R. *Genetic Algorithms and Engineering Optimization*; John Wiley & Sons: Hoboken, NJ, USA, 2000; Volume 7.
41. MATLAB. *The Language of Technical Computing-Release 14*; The MathWorks Inc.: Natick, MA, USA, 2004.
42. Shim, H.-S.; Kim, K.-Y.; Choi, Y.-S. Three-Objective Optimization of a Centrifugal Pump to Reduce Flow Recirculation and Cavitation. *J. Fluids Eng.* **2018**, *140*, 091202. [CrossRef]
43. Pei, J.; Wang, W.-J.; Yuan, S.-Q. Statistical Analysis of Pressure Fluctuations During Unsteady Flow for Low-Specific-Speed Centrifugal Pumps. *J. Cent. South Univ.* **2014**, *21*, 1017–1024. [CrossRef]
44. Zhang, Y.; Hu, S.; Wu, J.; Zhanga, Y.Q.; Chen, L.P. Multi-Objective Optimization of Double Suction Centrifugal Pump Using Kriging Metamodels. *Adv. Eng. Softw.* **2014**, *74*, 16–26. [CrossRef]

Article

Power Transmission Congestion Management Based on Quasi-Dynamic Thermal Rating

Yanling Wang [1], Zidan Sun [1], Zhijie Yan [2], Likai Liang [1,*], Fan Song [1] and Zhiqiang Niu [3]

[1] School of Mechanical, Electrical and Information Engineering, Shandong University (Weihai), Weihai 264209, China; wangyanling@sdu.edu.cn (Y.W.); szd940921@163.com (Z.S.); songfan@mail.sdu.edu.cn (F.S.)

[2] China Telecom, Nanjing 210000, China; yzj0308605@163.com

[3] State Grid Weihai Power Supply Company, Weihai 264200, China; nujusinu@126.com

* Correspondence: lianglikai@sdu.edu.cn; Tel.: +86-182-6319-6315

Received: 20 March 2019; Accepted: 24 April 2019; Published: 26 April 2019

Abstract: Transmission congestion not only increases the operation risk, but also reduces the operation efficiency of power systems. Applying a quasi-dynamic thermal rating (QDR) to the transmission congestion alarm system can effectively alleviate transmission congestion. In this paper, according to the heat balance equation under the IEEE standard, a calculation method of QDR is proposed based on the threshold of meteorological parameters under 95% confidence level, which is determined by statistical analysis of seven-year meteorological data in Weihai, China. The QDR of transmission lines is calculated at different time scales. A transmission congestion management model based on QDR is established, and the transmission congestion alarm system including conductor temperature judgment is proposed. The case shows that transmission congestion management based on QDR is feasible, which improves the service life and operation flexibility of the power grid in emergencies and avoids power supply shortages caused by unnecessary trip protection.

Keywords: transmission line; meteorological parameter; quasi-dynamic thermal rating (QDR); transmission congestion

---

1. Introduction

In view of the challenges of renewable energy, load growth and obsolete distribution facilities, it is imperative to improve transmission capacity [1]. At the same time, the reliability and safety of power supplies are always primary problems. Transmission congestion aggravates the power supply crisis. In the event of transmission congestion, the use of electricity by enterprises and residents has to be limited, or electricity supplies have to be cut off altogether [2]. Therefore, it is of great significance to alleviate transmission congestion and improve the service life and operation flexibility of power grids in emergency situations.

There are several generator units, transmission lines and loads in power systems. The active power flow on each branch is determined by the system structure and the output of generator unit. The absolute value of the active power flow on each branch is set to a safety limit in order to leave sufficient safety margin for the system to be adjusted in emergency. Once the absolute value of the active power flow exceeds the safety limit, the system will overload or violate voltage safety constraint, resulting in the electricity demand cannot be satisfied, which is called transmission congestion [3].

Managers adopt some protection schemes, such as limiting the use of electricity or planned outage, to regulate power flow and node voltage. It is an effective method to alleviate transmission congestion to maintain the stability and connectivity of the system, which is called transmission congestion management. Considering the system stability and management cost, generator regulation and load shedding are usually adopted [4]. When the transmission congestion is serious, it is necessary to regulate the generation and load side simultaneously. In most cases, load shedding is a remedial measure, and certain compensation should be paid according to the cost of generator rescheduling [5]. Therefore, preventive measures can reduce the management costs which arise due to transmission congestion.

Static thermal rating (STR) is used as the maximum ampacity of transmission lines in traditional preventive measures. If a branch fails, the current on other branches will exceed STR. At this point, some measures, such as trip protection, output reduction and load shedding, will be adopted to regulate the generation or load side, thereby greatly reducing the current of fault-free branch and avoiding thermal overload [6]. STR is a conservative method based on severe weather conditions. The thermal load capacity of transmission lines is often underestimated using STR as a reference for power dispatching, resulting in unreasonable utilization of transmission capacity [7,8]. Compared with STR, dynamic thermal rating (DTR) determines the ampacity by real-time meteorological data. In the favorable conditions of wind speed and ambient temperature, the maximum ampacity of transmission lines is significantly improved [9,10].

At present, the application of DTR technology has been studied in transmission congestion management. In [11], the classical method considering DTR and voltage stability limit was used to solve the optimal power flow in congestion management. A congestion management model based on DTR for distributed robust optimization was proposed in [12]. The thermal overload risk in short-term load forecasting was studied and the possibility of multi-line overload was evaluated to control the overload risk within the system security. In [2], an alternative solution based on DTR to alleviate transmission congestion was presented. Smart adaptations based upon varying weather conditions provided a feasible scenario for DTR of transmission lines. In order to ensure the safety of the system operation, the most perfect scheme is to install thermal sensors on each span of the transmission lines, which aims to obtain the running state and the surrounding meteorological data accurately [13]. However, such a scheme leads to the installation of redundant sensors, and the high cost makes it difficult to implement [14]. At the same time, large-scale deployment of thermal sensors may further lead to calculation complexity and dimensionality reduction in DTR evaluation [15]. Finally, the time variant of DTR increases the complexity of system operation and control.

In addition to above problems, there are some errors between meteorological forecast data and real-time data due to the large fluctuation of meteorological parameters and the short variation interval, which leads to the results of DTR often deviate from the actual value [16]. The concept of quasi-dynamic thermal rating (QDR) was proposed in [17]. QDR is an ideal solution to solve above problems effectively. QDR uses statistical method to determine the thermal rating with a certain confidence level based on the meteorological data defined in time scale. Compared with DTR, as slightly conservative as QDR is, it is more reliable and lower cost. In [18], a market-based real time transmission congestion management algorithm taking QDR into account was proposed, which fully exploit the capability of conductors to withstand different current flows when the system is faced with an emergency situation. The results show that the congestion mitigation, reduction in the congestion costs and load shedding is possible. On the basis of [18], the ampacity calculated by QDR is used as an important reference for overheating alarm in power grid, and the conductor temperature is the core basis for judging thermal overload of lines in this paper. The trip protection, power reduction or load shedding can be adopted to improve the accuracy of system congestion management.

The rest of this paper is organized as follows. In Section 2, the thermal rating model of overhead transmission lines based on IEEE standard is introduced. In Section 3, the key meteorological parameters based on historical meteorological data in Weihai are analyzed, the threshold of meteorological parameters under 95% confidence level is determined, and the QDR at different time scales is obtained by steady-state heat balance equation. A transmission congestion management method based on QDR is proposed in Section 4, which integrates conductor temperature judgment module into transmission congestion alarm system. In Section 5, an improved 14-bus case is given to analyze the variation of currents in fault according to the characteristics of conductor electrothermal under dynamic thermal balance. The trip protection scheme and congestion management decision based on QDR are also given. Conclusions are given in Section 6.

2. Thermal Rating Calculation Model of Transmission Line

The thermal load capacity is determined by the physical properties of conductor. The solution of thermal load capacity is an important issue after determining conductor material, geometric section and maximum allowable operating temperature. The calculation principle is derived from the heat balance equation of conductor. The conductor temperature is affected by its current carrying value and ambient conditions. The main factors are the joule heat caused by the current passing through the line and the heat absorbed from solar radiation. The cooling effects of the transmission line are mainly the convection heat generated by the wind and the radiation heat due to the temperature difference between the conductor temperature and ambient temperature. Therefore, the conductor temperature is a function of current, wind speed, illumination and ambient temperature. According to the IEEE standard, the heat balance equation representing the dynamic change of conductor temperature is shown in Equation (1):

$$q_c + q_r + mC_p\frac{dT_c}{dt} = q_s + I^2R(T_c) \tag{1}$$

where q_c is the convection heat caused by the wind speed, W/m; q_r is the radiation heat caused by temperature differences, W/m; q_s is the absorption heat from solar radiation, W/m; I is the current carrying, A; T_c is the conductor temperature, °C; $R(T_c)$ is the conductor resistance per unit length of the conductor at the temperature of T_c, Ω/m; m is the mass per unit length of the conductor, kg/m; C_p is the specific heat capacity of the conductor, J/(kg·°C); t is the time, s.

When the current and the weather conditions are constant, the absorption and loss of heat will be in equilibrium. The heat balance equation in the steady state is shown in Equation (2):

$$q_c + q_r = q_s + I^2R(T_c) \tag{2}$$

It is assumed that the conductor operates at the maximum allowable temperature T_{max} and the meteorological parameters are known, the ampacity of the transmission lines can be deduced from Equation (2), as shown in Equation (3):

$$I = \sqrt{\frac{q_c + q_r - q_s}{R(T_{max})}} \tag{3}$$

The type of the line in this paper is LGJ-400/50. STR is a conservative method based on severe weather conditions. The wind speed is 0.5 m/s. The ambient temperature is 40 °C, and the sunshine intensity is 1000 W/m². The STR calculated by Equation (3) is 592 A.

3. Quasi-dynamic Thermal Rating of Transmission Line

DTR determines the ampacity of overhead transmission lines according to the real-time meteorological parameters such as wind speed, ambient temperature and sunshine, which can effectively improve the utilization ratio of transmission lines. However, the time-varying increases the complexity of system. QDR is used to analyze the historical data of key meteorological parameters to determine the threshold values under different confidence levels and time scales (monthly, seasonally and yearly). The thermal ratings in different time scales are calculated by the threshold of each parameter. Theoretically, the selection of time scale for thermal rating can be arbitrary. However, for regions with obvious seasonal variations in meteorological conditions, time scales based on seasonal variations can greatly improve the ampacity of transmission lines. Monthly, seasonally and yearly rating of QDR are mainly studied in this paper. Firstly, the meteorological parameter thresholds under 95% confidence level are calculated based on wind speed and ambient temperature with the interval of

10 min. The meteorological data are from November 24, 2008 to November 23, 2015 in Weihai, China. Next, the QDRs of year, season and month are solved.

3.1. Statistical Analysis of Key Meteorological Parameters

The climate in Weihai belongs to marine monsoon climate, with four distinct seasons, clear monsoon and large wind-force. The seven-year meteorological data of Weihai are analyzed. The statistics of the key meteorological parameters are as follows.

3.1.1. Statistical Analysis of Wind Speed

From November 24, 2008 to November 23, 2015, the maximum wind speed in Weihai is 22.7 m/s. The maximum differences of the maximum frequency wind speed in different years is 0.4 m/s. Similarly, the wind speed difference of the same season in different years is 1.9 m/s, and the difference of the same month in different years is 1.8 m/s. The above differences illustrate that it is necessary to use statistical analysis of meteorological data for many years to drive the thermal rating of the line. In the same years, the maximum differences of the maximum frequency wind speed in different seasons and months of the same year are 3.4 m/s and 3.3 m/s respectively. The differences indicate that different time scales for thermal rating will directly affect the QDR.

The frequency distribution histogram of the seven-year wind speed is shown in Figure 1. It can be seen that wind speed are mainly concentrated in the interval of 0-5 m/s. If the confidence level is set to 95%, there are 368,116 × 95%wind speed data is higher than the threshold and the annual wind speed threshold is 2.2 m/s.

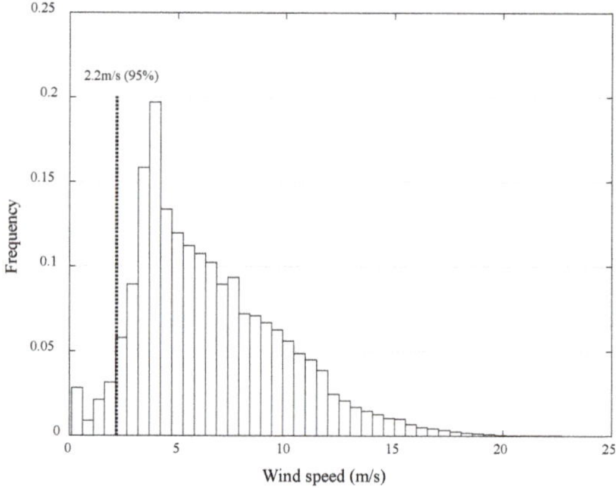

Figure 1. The frequency distribution histogram of wind speed.

3.1.2. Statistical Analysis of Ambient Temperature

According to statistics, the maximum and minimum temperatures are 42.0 °C and −13.5 °C, respectively. The maximum annual temperature difference is 48.9 °C. The maximum difference of the average temperatures in different years is 1.9 °C. Similarly, the temperature differences of the same seasons in different years, and the same months in different years are 3.5 °C and 4.3 °C respectively. In the same year, the maximum differences of the average temperatures of different seasons and months are 27.8 °C and 30.4 °C, respectively. The above temperature differences show the necessity of thermal rating in different time scales based on meteorological data. The frequency distribution histogram of the seven-year ambient temperature is shown in Figure 2. The conductor ampacity decreases with the increase of ambient temperature, so a high temperature threshold is set. The annual temperature threshold under the confidence level of 95% is 27.2 °C.

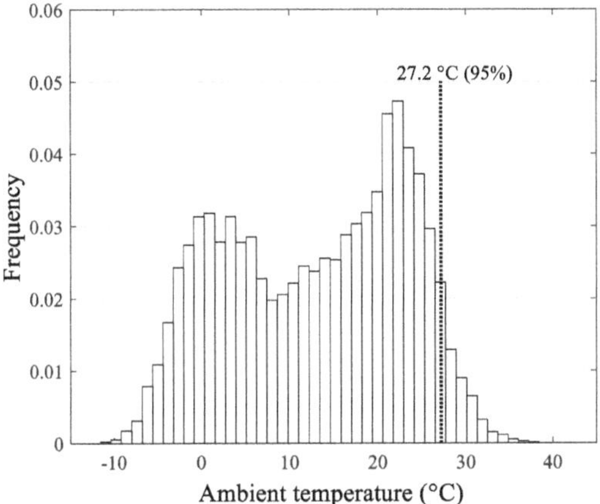

Figure 2. The frequency distribution histogram of ambient temperatures.

In addition, according to the statistical analysis of the wind direction, the wind direction shows a high variability from 0° to 180°, and especially the low wind speed is non-directional. Considering the strong randomness of wind direction and the change of line direction, the distribution of wind incidence angle along the line changes greatly. Therefore, the average value of long-term incidence angle of 45° is used to calculate the thermal rating of the line.

3.2. Quasi-dynamic Thermal Rating

The type of the line in this paper is LGJ-400/50, the typical overhead lines in Weihai of China, whose diameter is 27.63 mm and the maximum allowable operating temperature is 70 °C. With the condition of overhead lines is unknown, the absorptivity and emissivity of sunshine are usually 0.5. The seven-year meteorological data are divided into several subsets according to above time scales, and the frequency distributions of wind speed and ambient temperature in each subset are obtained. Next, the threshold of each parameter is calculated based on confidence level. Using the steady state heat balance equation and the threshold values of meteorological parameters, the thermal ratings of the line in different time scales are calculated. The flow chart of the method is shown in Figure 3.

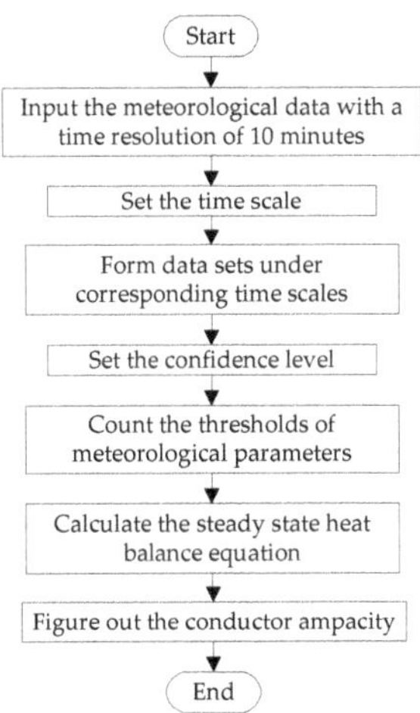

Figure 3. The flow chart of quasi-dynamic thermal rating driven by meteorological data.

According to Equation (3) and the meteorological parameter threshold, the wind speed and ambient temperature thresholds at different time scales under the confidence level of 95% are shown in Table 1. The yearly, seasonally and monthly ratings are shown in Figure 4.

Table 1. Meteorological parameter thresholds and quasi-dynamic thermal ratings under confidence level of 95%.

Time Scale	Wind Speed (m/s)	Ambient Temperature (°C)	QDR (A)	Time Scale	Wind Speed (m/s)	Ambient Temperature (°C)	QDR (A)
year	2.2	27.2	1109.9				
				March	2.4	14.4	1341.3
spring	2.5	23.2	1236.2	April	2.4	20.5	1258.7
				May	2.7	26.5	1216.4
				June	2.2	28.3	1113.1
summer	1.8	30.2	1016.5	July	1.8	30.4	1013.6
				August	1.6	30.9	979.3
				September	2.0	26.8	1101.6
autumn	2.1	24.7	1149.0	October	2.4	21.8	1240.1
				November	2.2	15.4	1292.6
				December	2.3	7.6	1409.3
winter	2.2	6.8	1399.7	January	2.2	5.4	1416.3
				February	2.3	6.9	1417.8

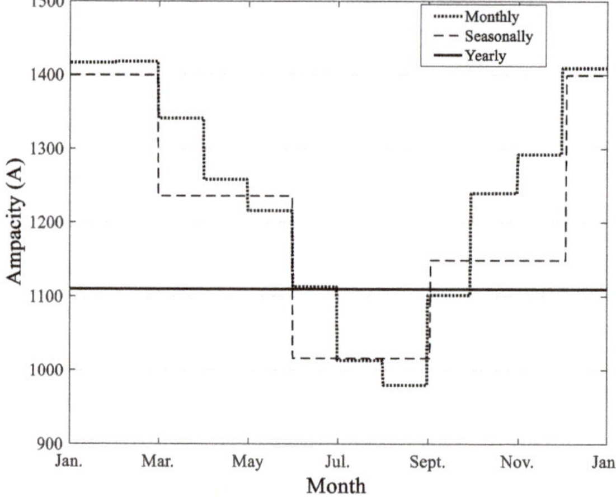

Figure 4. The results of quasi-dynamic thermal ratings under confidence level of 95%.

As shown in Table 1 and Figure 4, there are obvious differences between meteorological parameters in different months, which makes large difference of carrying capacity. There is much room for improvement of the ampacity with favorable meteorological parameters. The thermal carrying capacity reaches the maximum in the winter of February and the minimum in the summer of August, respectively. The difference between them is 438.5 A. The thermal load capacity of overhead lines is seasonally dependent, which can greatly improve the ampacity in winter. In addition, the average of rating monthly is higher than that of rating seasonally, indicating that shortening the time scale can significantly improve the ampacity. The confidence level and time scale are important parameters affecting the ampacity. With the increase of confidence level, the ampacity is increased. Even if the confidence level is 99%, the rating yearly is 745.5 A, which is much higher than the traditional STR of 592 A, as well as the seasonally and monthly thermal rating under different confidence levels. It is further illustrated that QDR can effectively improve the utilization ratio of lines.

4. Design of Transmission Congestion Management Based on Quasi-dynamic Thermal Rating

STR is replaced by QDR as the maximum ampacity to relax the thermal restriction of transmission lines. When the line fails, real-time meteorological data are used to determine the occurrence and time of thermal overload, so as to regulate the generator.

The method is used to deal with the transmission congestion caused by line fault. The conductor temperature is an important signal to adjust the generation side. The logical design diagram is shown in Figure 5. The current value, the real-time meteorological data, and the circuit breaker state are the input signals of the system. If a fault occurs, the circuit breaker on the fault line will cut off and a breaking signal will feed back to the managers. The regulatory signal will be sent out only the system satisfies the following conditions simultaneously: the open circuit signal is detected; a warning signal is sent out when the current value of the fault-free line exceeds the safety limit of QDR; the real-time conductor temperature exceeds the maximum allowable operating temperature.

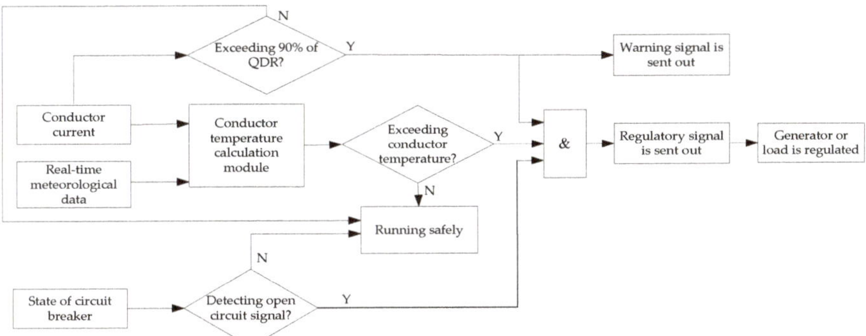

Figure 5. Design of transmission congestion management based on quasi-dynamic thermal rating.

The tripping signal is sent out when the conductor overheats. Because the thermal time constant is larger than the electrical time constant, the system security will not be affected as long as the trip protection or load reduction is completed within a short time after receiving the regulatory signal. The delay time in this system is 0.1 s.

5. Case

A 14-bus system is given to analyze the transmission congestion. The base voltage and capacity are 220 kV and 100 MVA, respectively. There are 17 high-voltage transmission lines with the type of LGJ-400/50 and the voltage of 220 kV whose transmission distance is between 100 km and 300 km. Conductor resistance, reactance and susceptance are 0.07875 Ω/km, 0.405 Ω/km and 2.815 $\times$ 10^{-6} S/km, respectively. The structure of the system is shown in Figure 6.

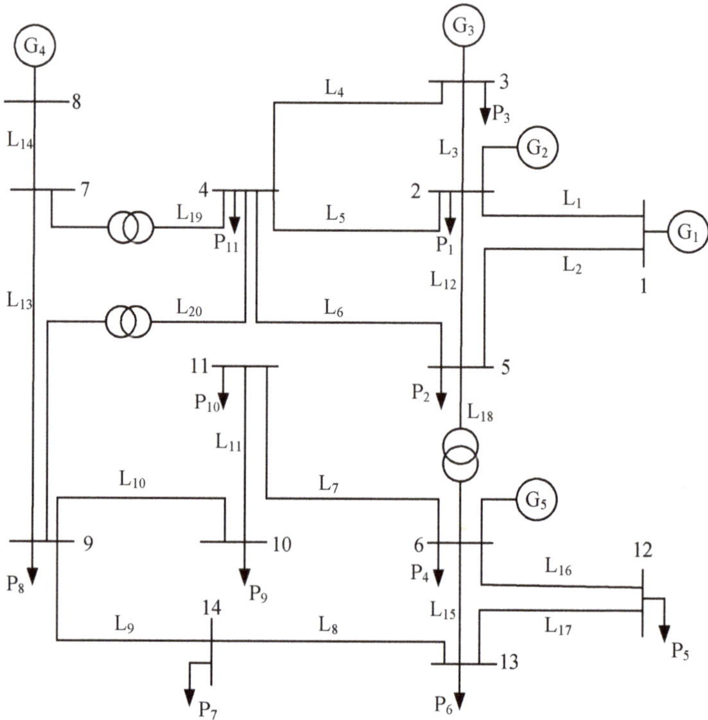

Figure 6. The structure of 14-bus system.

The parameters of generator, transmission line and load are shown in Tables 2–4 respectively, where the voltage, resistance, power and other parameters are expressed as per-unit value.

Table 2. Generator parameters.

Number	Type	Node Voltage (p. u.)	Active Power (p. u.)
1	slack	1.06	3.41
2	PV	1.045	0.86
3	PV	1.01	1.8
4	PV	1.025	1.25
5	PV	1.07	0.95

Under normal operation, the current values of all branches in the system are lower than STR of 592 A. Compared with other branches, the currents of L_1, L_2, L_{13}, L_{15} and L_{18} are larger, but they don't exceed 592 A. The current values of the five branches are shown in Table 5.

Table 3. Transmission line and transformer parameters.

Branch	Type	Resistance (p. u.)	Reactance (p. u.)	Susceptance (p. u.)
L_1		0.0163	0.0837	0.0681
L_2		0.0244	0.1255	0.1022
L_3		0.0408	0.2092	0.1703
L_4		0.0326	0.1674	0.1362
L_5		0.0489	0.2511	0.2043
L_6		0.0163	0.0837	0.0681
L_7		0.0978	0.5022	0.4086
L_8	Transmission line	0.1565	0.8035	0.6538
L_9		0.1252	0.6428	0.5230
L_{10}		0.0326	0.1674	0.1362
L_{11}		0.0782	0.4018	0.3269
L_{12}		0.0587	0.3013	0.2452
L_{13}		0.0326	0.1674	0.1362
L_{14}		0.0326	0.1674	0.1362
L_{15}		0.0626	0.3214	0.2615
L_{16}		0.1252	0.6428	0.5230
L_{17}		0.1565	0.8035	0.6538
L_{18}		0	0.2520	0
L_{19}	Transformer	0	0.2091	0
L_{20}		0	0.5562	0

Table 4. Load parameters.

Number	Node	Active Power (p. u.)	Reactive Power (p. u.)
1	2	1.36	0.13
2	3	0.94	0.15
3	4	0.48	−0.3
4	5	0.9	0.2
5	6	0.9	0.35
6	9	1.26	0.35
7	10	0.09	0.04
8	11	0.04	0.01
9	12	0.15	0.03
10	13	1.13	0.25
11	14	0.14	0.05

Table 5. Branch current.

Branch	Current (A)
L_1	402.8
L_2	528.6
L_{13}	390.9
L_{15}	548.6
L_{18}	434.3

Analysis of Transmission Congestion

Any actions violating the power grid restrictions may cause transmission congestion. The power flow is analyzed in 14-bus system with transmission congestion caused by line fault. With the occurrence of overload, short circuit or undervoltage, the circuit breaker will automatically cut off, resulting in a sharp increase in active power flow on the fault-free lines. Once the current exceeds the rating, the line may be thermal overload, resulting in transmission congestion to endanger the operation of power system.

It is assumed that three classical line faults occur when the system runs to 300 s, at the same time the circuit breaker on the line starts and cuts off the branch.

Fault 1: only L_4 fails;

Fault 2: L_5 and L_{12} fail (five seconds apart);

Fault 3: L_2, L_7 and L_{17} fail (five seconds apart).

Generally, there are few simultaneous faults of three branches in the system. In order to ensure the stability of the system, it is assumed that the time interval of the line fault is 5s. The power flow of each branch under the three kinds of faults is monitored. The conductor temperature after occurrence of faults is analyzed.

Only the branch with the greatest current change is analyzed after the circuit breaker is cut off. The currents on other branches change insignificantly. The circuit breaker starts up with a break signal at the same time. The current variation of the three faults are shown in Figures 7–9, respectively.

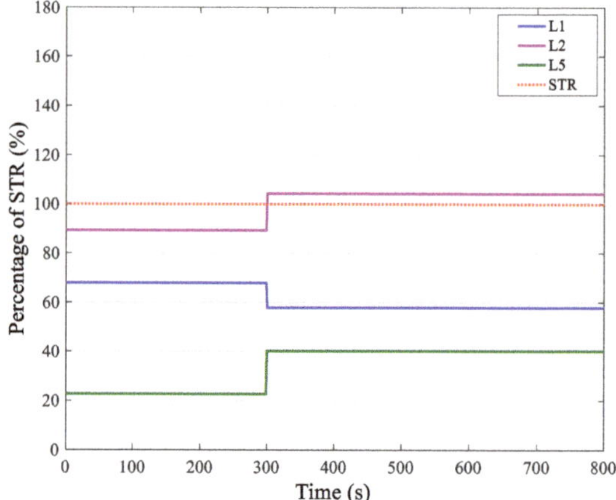

Figure 7. Current variation of fault 1.

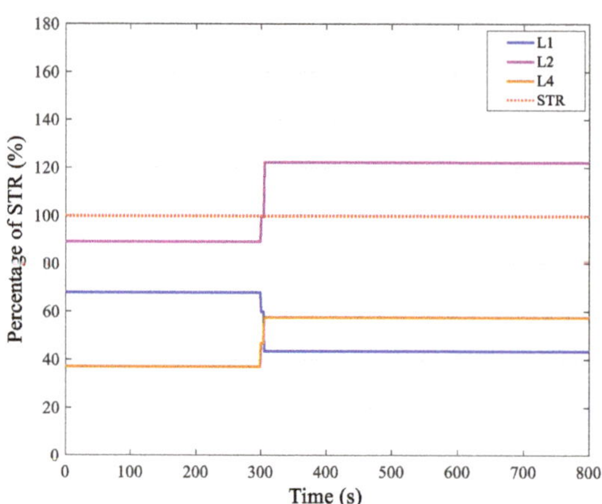

Figure 8. Current variation of fault 2.

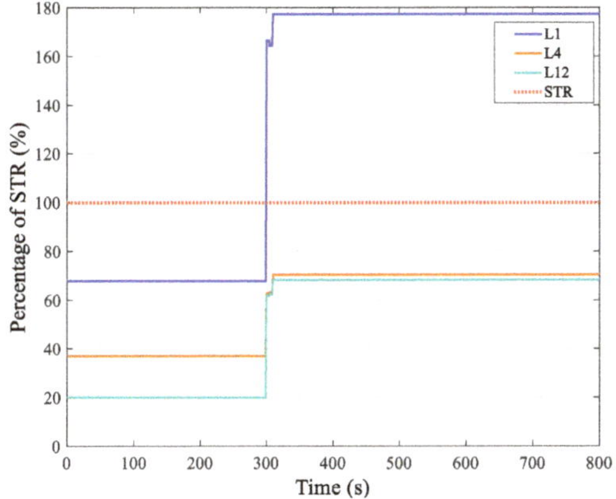

Figure 9. Current variation of fault 3.

For fault 1, L_4 fails when the system runs to 300 s, at the same time the circuit breaker on L_4 starts. Consequently, as shown in Figure 7, the currents on L_1, L_2 and L_5 change obviously. The current on L_1 decreases to lower than 60% of STR, however, the currents on L_2 and L_5 increase. The current on L_5 rises to about 40% of STR still lower than STR. The current on L_2 rises to 617.2 A, which is 4.3% higher than STR.

For fault 2, L_5 and L_{12} fail when the system run to 300 s and 305 s respectively. The circuit breakers on L_5 and L_{12} start at 300 s and 305 s, respectively. As shown in Figure 8, at 300 s, the current on L_1 decreases and the currents on L_2 and L_4 rise, and they are all lower than STR. From 300 s to 305 s, only the current on L_2 is close to STR. After 305 s, the current on L_2 is up to 723.4 A which is 22.2% higher than STR.

For fault 3, it is assumed that the circuit breakers on L_2, L_7 and L_{17} start at 300 s, 305 s and 310 s, respectively. As shown in Figure 9, from 300 s to 305 s, the current on L_1 rises to 986.6 A. Although the currents on L_4 and L_{12} rise sharply, it is much smaller than STR. From 305 s to 310 s, L_2 and L_7 are cut off. At the same time, the current on L_1 reduces to 974.1 A. After 310 s, all fault branches are cut off. The current on L_1 rises to 1065.6 A, reaching 180% of STR. The current on L_4 and L_{12} is still lower than STR. The circuit breaker signals in fault 3 are shown in Figure 10.

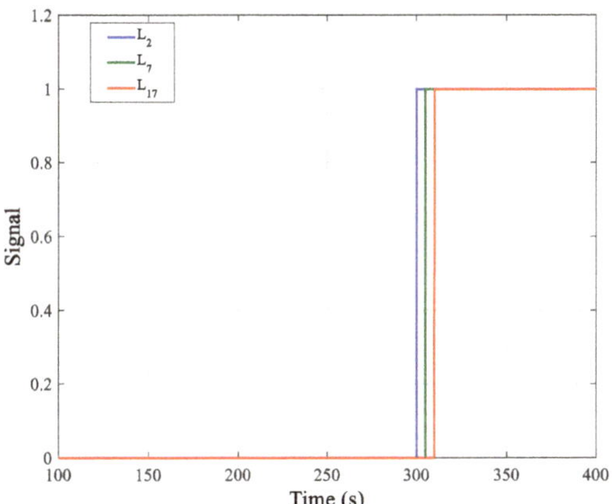

Figure 10. Circuit breaker signals of fault 3.

The three faults lead to the currents on critical fault-free branches exceed STR. In the case of transmission congestion management based on STR, it is necessary to adopt corrective measures and cut power or load in serious case. A transmission congestion management scheme based on QDR is presented in this paper. In order to ensure the system security, the safety limit of the active power and the confidence level of QDR are set to 90% and 99%, respectively. It is necessary to analyze the conductor temperature and decide to take preventive measures when the current exceeds the safety limit of QDR.

The confidence level of QDR is set at 99% and the safety threshold of judgment module is 90% to keep enough safety margin to prevent conductor temperature from exceeding the thermal limit. The yearly, seasonally, monthly rating under the confidence level of 99% and safety limit of 90% can be obtained using the method proposed in Section 3, as shown in Table 6.

Table 6. Safety limit of quasi-dynamic thermal rating.

Time Scale	Rating (A)	Safety Limit (A)	Time Scale	Rating (A)	Safety Limit (A)
year	745.5	671.0	STR	592	
			March	866.1	779.5
spring	833.9	750.5	April	872.0	784.8
			May	867.6	780.8
			June	744.8	670.3
summer	685.2	616.7	July	681.5	613.4
			August	672.9	605.6
			September	762.4	686.2
autumn	780.8	702.7	October	809.6	728.6
			November	884.9	796.4
			December	977.0	879.3
winter	1017.5	915.8	January	1136.9	1023.2
			February	1041.1	937.0

It can be seen that the QDR significantly improves the ampacity of lines. Compared with STR, the yearly safety limits, the average seasonally and monthly safety limits increased by 13.3%, 26.1%, and 30.7%, respectively.

The 14-bus power system with three kinds of transmission congestions is studied. Meteorological data around overhead transmission lines are from the observatory of Shandong University (Weihai). The interval of meteorological data is 5 min. It is assumed that there is no significant change in meteorological data in 5 min. The yearly rating and the winter rating are used as QDR to verify the effectiveness of the method based on QDR.

In order to verify that yearly rating can avoid unnecessary regulation after fault occurs, the ambient parameters are set to the severe value in 2016 (the minimum average wind speed and the maximum average ambient temperature in one hour). According to the statistics of meteorological data, the meteorological data at 11 a.m. on July 24, 2016 are chosen as the most conservative in the whole year. The sampling of meteorological data around the line at 5 min is shown in Table 7.

Table 7. Meteorological data around the line in one hour.

Time (s)	Wind Speed (m/s)	Ambient Temperature (°C)
0–300	1.8	31.3
301–600	2.2	31.1
601–900	1.8	30.9
901–1200	1.8	30.6
1201–1500	1.8	30.7
1501–1800	0.4	31.1
1801–2100	0.9	31.6
2101–2400	3.1	31.5
2401–2700	4.5	30.6
2701–3000	3.1	30.1
3001–3300	4.0	29.3
3301–3600	3.6	28.5

According to the time resolution of meteorological data, the conductor temperatures are divided. The conductor temperatures are analyzed with a line fault. Figure 11 shows the changes of conductor temperature after the occurrence of the fault.

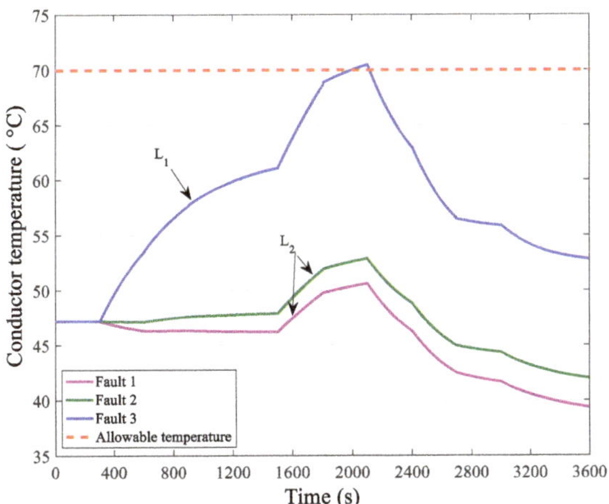

Figure 11. Change of conductor temperature after fault.

In fault 1, although the current value of L_2 is 617.2 A which exceeds STR, it never exceeds the safety limit of yearly rating of 671.0 A. Therefore, the warning signal will not be sent out and the conductor temperature is much smaller than the thermal limit. In fault 2, the current value of L_2 is 723.4 A, which exceeds the safety limit of yearly rating. A warning signal will be sent out after 300s. At the same time, the conductor temperature module shows that the conductor temperature is always in allowable range. Therefore, the regulatory signal will not be sent out and it is non-essential to adjust the generator or load. In fault 3, the conductor temperature of L_1 is 1065.6 A after 310s, which is much greater than the safety limit of yearly rating. In addition, it can be seen form Figure 11 that the conductor temperature of L_1 exceeds the maximum limit at 1995s to cause overload. Meanwhile, the warning and regulatory signals will be sent out at 300s and 1995s, respectively, as shown in Figure 12.

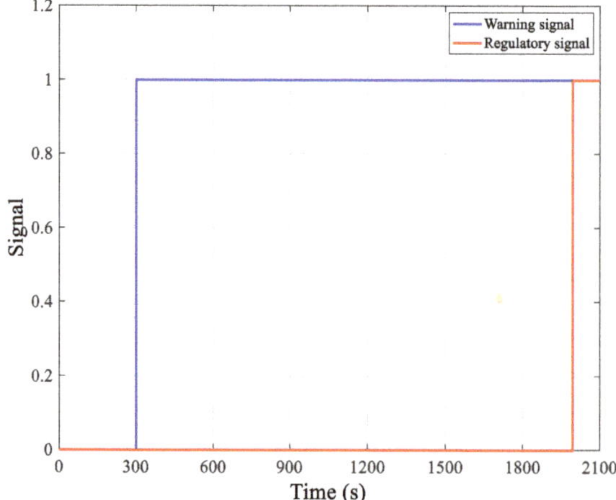

Figure 12. Warning and regulatory signals.

Due to the hysteresis of conductor temperature, the warning and regulatory signals are 28 min apart, which means the managers have enough time to make decision after receiving the warning signal. If the system returns to normal operation within 28 min, there is no need to issue a regulatory signal. If the fault cannot be effectively removed within 28 min, the following methods are adopted according to the power flow after receiving the regulatory signal at 1995s. First, regulate the generator. Open L_{16} in 0.1s to complete trip protection of generator G_5. The current on L_1 reduces to 617.5 A, however, it is still the largest. Second, cut the load. After receiving the regulatory signal, the load P_1 is cut off and the current on L_1 reduces to 663.1 A. The two methods make the current value of L_1 fall within the safety limit of the yearly rating to ensure the system security. The current value in the second method is slightly higher than that in the first one. The current and conductor temperature change of L_1 are shown in Figure 13.

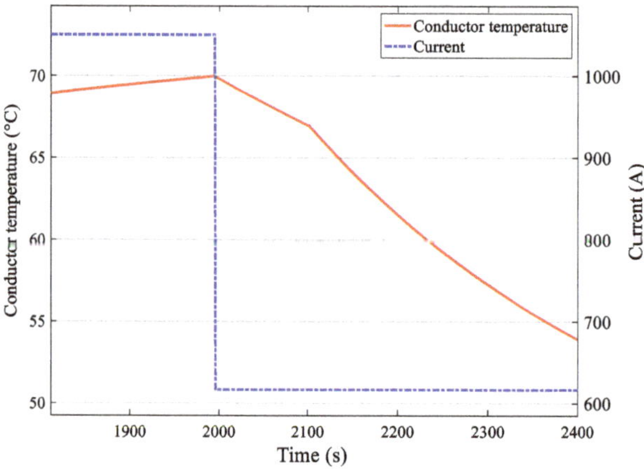

Figure 13. Change of conductor temperature after load shedding.

As shown in Figure 13, the conductor temperature is always lower than 70 °C and runs in safety after regulating the generation side or the load side after fault 3, indicating the congestion management based on yearly rating is effective. The analysis is based on the severe meteorological conditions in

2016. It is feasible to use the yearly rating under the severe meteorological conditions, indicating that the yearly rating can be applied to other meteorological conditions.

Because the difference between the winter rating and the yearly rating is the largest, the winter rating is used to verify the effective of QDR. The meteorological data at 8 a.m. on February 26, 2016 are chosen as the most conservative in winter. The change of meteorological parameters is shown in Table 8.

Table 8. Meteorological data of rating in winter.

Time (s)	Wind Speed (m/s)	Ambient Temperature (°C)
0–300	1.3	11.1
301–600	1.3	10.7
601–900	1.8	10.2
901–1200	0.9	9.9
1201–1500	0.4	10.7
1501–1800	0.4	11.6
1801–2100	0.9	12.1
2101–2400	0.9	12.0
2401–2700	0.9	11.8
2701–3000	0.9	11.6
3001–3300	1.3	11.2
3301–3600	0.9	10.9

The ampacity of transmission lines is higher in winter. As shown in Figure 14, the conductor temperature in three faults does not exceed the safety limit in this case. The current value of L_1 is 1065.6 A in fault 3, which exceeds the safety limit of the winter rating of 915.8 A. The conductor temperature in three faults does not exceed the safety limit in this case. As shown in Figure 14, the current value of L_1 in fault 3 exceeds the safety limit of the winter rating. The conductor temperature peaks at 1806s and later, there is no significant change in conductor temperature. The conductor temperature does not reach the thermal limit after the fault has occurred, so a warning signal is sent out without other actions. Therefore, it is completely feasible to take QDR as the reference for overheating alarm, which not only overcomes the conservation of STR, but also improves the accuracy of the trip protection. Taking the conductor temperature as the main basis for judging the thermal overload of transmission lines can provide safety operation time for fault repairing and improve the accuracy of congestion management.

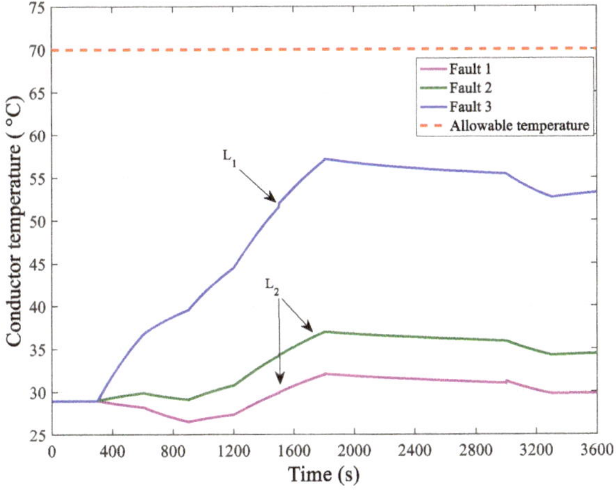

Figure 14. Change of conductor temperature after fault (winter).

6. Conclusions

In this paper, a method of transmission congestion management based on QDR is proposed, which integrates QDR technology into the trip protection scheme. The trip protection, based on the characteristic that the thermal time constant is greater than the electrical time constant, can provide sufficient dispatching time for managers without affecting the safety of system. This method predictably regulates the load pressure of line faults by trip protection to alleviate transmission congestion. A warning signal will be sent out to inform managers the overload. The method can improve the accuracy of transmission congestion alarm judgment and can be widely used in power system detection, protection and control. Increasing transmission capacity under real-time weather conditions, providing sufficient time for decision-making, and reducing unnecessary trip protection and load shedding are conducive to improve the utilization of existing transmission lines and economic benefits.

Author Contributions: Conceptualization, Y.W.; Methodology, Z.S.; Writing-original draft, Z.Y.; Writing-review and editing, L.L., F.S. and Z.N.

Funding: This research was funded by the National Natural Science Foundation of China, grant number 51641702, the Science and Technology Development Project of Shandong Province, China, grant number ZR2015ZX045, and the Science and Technology Development Project of Weihai City, China, grant number 2018DXGJ05.

Acknowledgments: We gratefully acknowledge the technical assistance of DL850E ScopeCorder.

Conflicts of Interest: The authors declare no conflict of interest.

References

1. Teh, J.; Ooi, C.A.; Cheng, Y.H. Composite reliability evaluation of load demand side management and dynamic thermal rating systems. *Energies* **2018**, *11*, 466. [CrossRef]
2. Zafran, M.; Arbab, M.N.; Ahmad, I. A Case Study on Alleviating Electric Transmission Congestion Using Dynamic Thermal Rating Methodology. In Proceedings of the IEEE International Conference on Energy Systems and Policies, Islamabad, Pakistan, 24–26 November 2014; pp. 1–6.
3. Anusha, P.; Prabhakar, K.S.; Kothari, D.P. Congestion management in power systems—A review. *Int. J. Elec. Power* **2015**, *70*, 83–90. [CrossRef]
4. Sun, Q.; Peng, J.C.; Pan, J.T.; Li, P. Congestion management considering multi-time interval demand response. *Power Syst. Technol.* **2010**, *34*, 139–143. [CrossRef]
5. Madani, V.; Adamiak, M.; Imai, S. IEEE PSRC report on global industry experiences with system integrity protection schemes. *IEEE Trans. Power Deliv.* **2010**, *25*, 2143–2155. [CrossRef]
6. Cong, Y.; Regulski, P.; Wall, P. On the use of dynamic thermal-line ratings for improving operational tripping schemes. *IEEE Trans. Power Deliv.* **2016**, *31*, 1891–1900. [CrossRef]
7. Heckenbergerová, J.; Musilek, P.; Filimonenkov, K. Quantification of gains and risks of static thermal rating based on typical meteorological year. *Int. J. Elec. Power* **2013**, *44*, 227–235. [CrossRef]
8. Jiashen, T.; Ching-Ming, L.; Nor, A.M.; Chia, A.O.; Yu-Huei, C. Prospects of using the dynamic thermal rating system for reliable electrical networks: A review. *IEEE Access* **2018**, *6*, 26765–26778. [CrossRef]
9. Zhang, H.; Du, M.; Zhao, Q.; Wei, Z.; Zhang, Q. Security Constrained Economic Dispatch with Dynamic Thermal Rating Technology Integration. In Proceedings of the IEEE International Conference on Power and Renewable Energy, Shanghai, China, 21–23 October 2016; pp. 709–713.
10. Ying, Z.F.; Chen, Y.S.; Feng, K. New DTR estimation method without measured solar and wind data. *J. Electr. Eng. Technol.* **2017**, *12*, 576–585. [CrossRef]
11. Capitanescu, F.; Cutsem, T.V. A unified management of congestions due to voltage instability and thermal overload. *Electr. Power Syst. Res.* **2007**, *77*, 1274–1283. [CrossRef]
12. Qiu, F.; Wang, J. Distributionally robust congestion management with dynamic line ratings. *IEEE Trans. Power Syst.* **2015**, *30*, 2198–2199. [CrossRef]
13. Jiang, J.A.; Wan, J.J.; Zheng, X.Y. A novel weather information-based optimization algorithm for thermal sensor placement in smart grid. *IEEE Trans. Smart Grid* **2018**, *9*, 911–922. [CrossRef]
14. Teh, J.; Cotton, I. Critical span identification model for dynamic thermal rating system placement. *IET Gener. Transm. Distrib.* **2015**, *9*, 2644–2652. [CrossRef]

15. Matus, M.; Saez, D.; Favley, M. Identification of critical spans for monitoring systems in dynamic thermal rating. *IEEE Trans. Power Deliv.* **2012**, *27*, 1002–1009. [CrossRef]

16. Kosec, G.; Maksić, M.; Djurica, V. Dynamic thermal rating of power lines—model and measurements in rainy conditions. *Int. J. Elec. Power* **2017**, *91*, 222–229. [CrossRef]

17. EPRI. *Increased Power Flow through Transmission Circuits: Overhead Line Case Studies and Quasi-Dynamic Rating*; EPRI Technical Report; EPRI: Palo Alto, CA, USA, 2006.

18. Esfahani, M.M.; Yousefi, G.R. Real time congestion management in power systems considering quasi-dynamic thermal rating and congestion clearing time. *IEEE Trans. Ind. Inf.* **2016**, *12*, 745–754. [CrossRef]

Article

Modeling of Future Electricity Generation and Emissions Assessment for Pakistan

Abdullah Mengal [1,*], Nayyar Hussain Mirjat [2], Gordhan Das Walasai [3], Shoaib Ahmed Khatri [2], Khanji Harijan [4] and Mohammad Aslam Uqaili [2]

1   Department of Mechanical Engineering, Balochistan University of Engineering & Technology, Khuzdar 89100, Pakistan
2   Department of Electrical Engineering, Mehran University of Engineering & Technology, Jamshoro 76062, Pakistan; nayyar.hussain@faculty.muet.edu.pk (N.H.M.); shoaib.ahmed@faculty.muet.edu.pk (S.A.K.); aslam.uqaili@faculty.muet.edu.pk (M.A.U.)
3   Department of Mechanical Engineering, Quaid-e-Awam University of Engineering, Sciences & Technology, Nawabshah 67480, Pakistan; valasai@quest.edu.pk
4   Department of Mechanical Engineering, Mehran University of Engineering & Technology, Jamshoro 76062, Pakistan; khanji.harijan@faculty.muet.edu.pk
*   Correspondence: abdullah.mengal@buetk.edu.pk; Tel.: +92-300-890-2886

Received: 31 December 2018; Accepted: 8 April 2019; Published: 12 April 2019

Abstract: Electricity demand in Pakistan has consistently increased in the past two decades. However, this demand is so far partially met due to insufficient supply, inefficient power plants, high transmission and distribution system losses, lack of effective planning efforts and due coordination. The existing electricity generation also largely depends on the imported fossil fuels, which is a huge burden on the national economy alongside causing colossal loss to the environment. It is also evident from existing government plans that electricity generation from low-cost coal fuels in the near future will further increase the emissions. As such, in this study, following the government's electricity demand forecast, four supply side scenarios for the study period (2013–2035) have been developed using Long-range Energy Alternatives Planning System (LEAP) software tool. These scenarios are Reference scenario (REF) based on the government's power expansion plans, and three alternative scenarios, which include, More Renewable (MRR), More Hydro (MRH), and More Hydro Nuclear (MRHN). Furthermore, the associated gaseous emissions (CO_2, SO_2, NO_X, CH_4, N_2O) are projected under each of these scenarios. The results of this study reveal that the alternative scenarios are more environmentally friendly than the REF scenario where penetration of planned coal-based power generation plants would be the major sources of emissions. It is, therefore, recommended that the government, apart from implementing the existing plans, should consider harnessing the renewable energy sources as indispensable energy sources in the future energy mix for electricity generation to reduce the fossil-fuel import bill and to contain the emissions.

Keywords: electricity demand; emissions; LEAP model; fossil fuels; renewable energy

---

1. Introduction

Electricity is considered one of the most important vectors for economic growth and development of any country. Industrial processes, transportation, education systems, construction activities, household appliances, as well as large businesses and small commercial services heavily rely on the electricity supply. The global electricity consumption is growing rapidly with population growth, and with the change of lifestyle across the world. The generation of electricity is a serious challenge, especially for developing countries. Globally, fossil fuels such as coal, oil, and natural gas are the major primary energy sources [1]. About 68% of electricity in the world is generated from these fossil fuels.

Electricity generation from fossil fuels produces the Greenhouse Gases (GHGs) along with air pollutant emissions such as oxides of nitrogen (NO_x) and sulfur dioxide (SO_2). These air pollutants and GHG emissions have negative impacts on the environment such as causing global warming, climate change, and health problems of all living organism [2–4].

Pakistan is a developing country having a low Gross Domestic Product (GDP) of 232 billion USD, the growth rate of GDP remained 4.4% during 2013. In terms of population, Pakistan ranks as the sixth largest nation in the world, and second populous country in South Asia with 182 million people recorded in 2013 [3,5]. The per capita, electricity consumption in Pakistan is ~449 kWh, which is also well below the world average energy consumption. Electricity shortage, which was estimated to be 1000–2000 MW in 2007, reached ~7000 MW in 2015 [5]. Many industries have been forced to shut down or slow down their production, and residential consumers in urban as well as rural areas are facing power cuts for about 12 h on a daily basis during the summer [6,7]. The electricity crises, without major steps taken, may worsen in the coming years due to the increase in demand and inconsistency in supply. In order to cope-up this challenge, the Government of Pakistan (GOP) announced a new power policy in 2013 [8]. The goal of this power policy was to build a power generation capacity that can meet electricity demand in the country. To achieve this goal on a long-term basis, the government made a plan to ensure the generation of electricity by focusing on shifting the country's electricity generation mix towards low-cost indigenous sources such as coal, hydro, gas, nuclear, and biomass, with a major share from coal [1]. To complement the government's efforts of overcoming electricity shortage, a careful planning exercise needs to be undertaken to devise future strategies.

Fossil fuel consumption, luxurious lifestyles, population and industrial growth are key drivers of climate change. GHG emissions from various sectors of the economy are also to blame. Pakistan's GHG emissions include 158.10 Mt of CO_2 (54%), 111.60 Mt of CH_4 (36%), 27.90 Mt of N_2O (9%), 2.17 Mt of CO (0.75), and 0.93 Mt of volatile organic carbon (VOC) (0.3%) [9]. Pakistan's accumulative CO_2 emissions are likely to reach 250 Mt by 2020, which may grow to 650 Mt if subsidies continue on fossil fuels. Energy and transport sectors contribute the largest share of approximately half the national GHG emissions of Pakistan, while the agricultural sector contributes 39%, according to a 2008 national greenhouse gas inventory [8].

Pakistan mainly uses natural gas and furnace oil for power generation. The country has also some abundant reserves of coal at Tharparkar but has not used these towards power generation in 2013 although some projects on coal based power generation are at advanced stages of realization. At present, Pakistan is on the path of an environmentally damaging energy mix with various government-planned projects in hand or to be realized in the near future. Pakistan's overall GHG emissions are projected to increase from 347 million tons of CO_2 equivalents (Mt CO_2-eq) in 2011 to 4621 Mt CO_2-eq in 2050 [1].

Energy modeling using various computer-based tools is attaining greater importance, and is now essentially used for energy planning [2]. Globally various energy planning models are available, which have been developed with different modeling approaches to address the energy-planning requirement on a case-to-case bases. Some of the well-known modeling tools are MARket ALlocation/The Integrated MARKAL-EFOM System (MARKAL/TIMES), EnergyPlan, Model for Energy Supply Strategy Alternatives and their General Environmental Impact (MESSEGE), and LEAP. However, amongst these and various other tools, the LEAP energy model is a freely available tool for academia with sufficient capabilities and easy to use features for energy scenario planning and emission analysis. LEAP is a scenario-based energy and environmental modeling tool used in several countries for energy and environmental planning [10–12]. It is a user-friendly energy-modeling tool, which facilitates the tracking of energy resource extraction, production, and consumption in all sectors of the economy. Lower data requirement of the LEAP model, and the built-in technology and environmental data base suit this study's requirements [3].

There are only a few studies, in Pakistan's case, in the contemporary literature which have taken into account the government's existing plans and policies, while further considering the energy resource potential to develop scenarios for electricity generation with emission projections. For example, GHG emissions of the electric power sector are estimated by Usama Perwaiz over the period of (2012–2030) with different scenarios [13], and a review study of GHG emissions for Pakistan from various sectors [1].

In this study, following electricity demand projection, environmental emissions have been estimated for the period 2013–2035, in accordance with the government's plans and policies to meet the project demand using the LEAP modeling tool. The study also considers the other environmentally friendly electricity generation options. To evaluate the diversification of the future electricity generation system, a Reference scenario (REF) and three alternative scenarios, More Renewable (MRR), More Hydro (MRH), and More Hydro Nuclear (MRHN) were developed. As such, this study not only provides insight into existing government plans for power generation alongside estimating the associated emission, but at the same time also propose three alternative scenarios based on the country's energy resources. These alternative scenarios for electricity generation propose utilizing indigenous resources, and as such, would lower the emissions compared to the REF scenario.

The next section of the paper provides an overview of the electricity supply and demand situation in Pakistan. A detailed analysis of future electricity generation plans is discussed in Section 3. Section 4 describes the development of Pakistan's LEAP energy model development for the expansion of the electricity generation sector, and emission assessment based on scenario analysis. In Section 5, the simulation results of the developed energy model are analyzed and discussed. The final section of the paper provides conclusions and recommendations for energy policy makers based on this study.

2. Electricity Supply and Demand Situation in Pakistan

Over the years, the growth in electricity demand in Pakistan has been witnessed owing to increase in the population, rapid urbanization, improved living standards, and some level of industrial growth. The major sectors consuming electricity are domestic, commercial, agricultural, and industrial. The gap between generation and demand is widening, and reached about 7 GW in 2013 as shown in Figure 1. Because of this demand–supply gap, economic development in the country has been widely compromised. The electricity crises of Pakistan are thus reflective of failed energy planning and policy regime of the country [2].

Pakistan heavily relies on fossil fuels like imported oil, natural gas, and a small share from coal, for electricity generation. The huge potential of hydro, other renewables, and indigenous coal is available in the country, but these resources have not been significantly exploited due to several constraints including technical, economic, and political. The installed electricity generation capacity in the country has increased only (14.86%) from 19,420 MW in 2008 to 22,812 MW in 2013 as shown in Figure 2 [5].

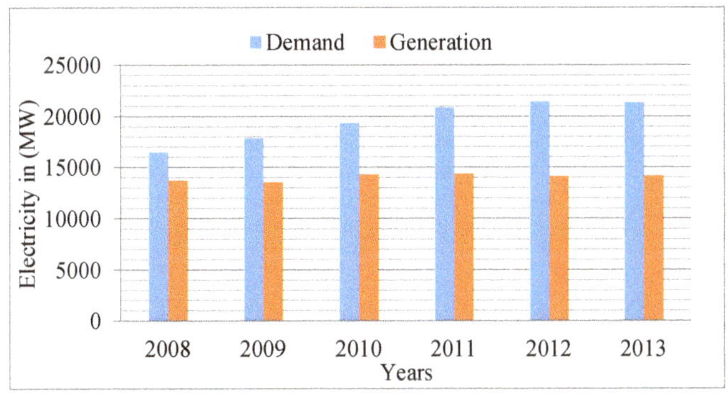

Figure 1. Electricity demand and generation in Pakistan [3,5].

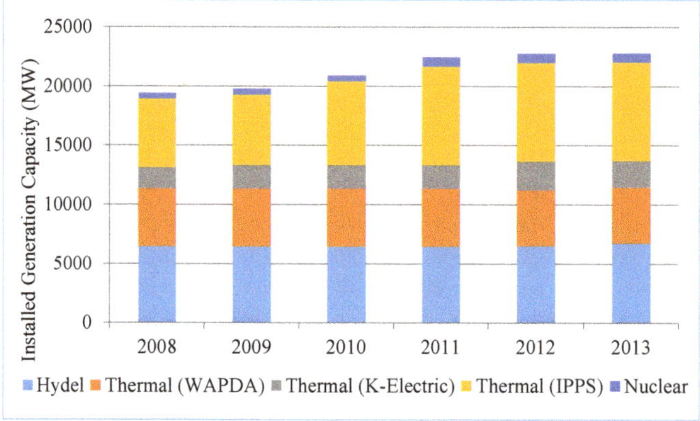

Figure 2. Electricity installed capacity in Pakistan from 2008 to 2013 [5].

Electricity generation in the country is mainly dominated by thermal power, and during 2012–2013, ~36% of electricity was produced using expensive imported oil. The costly electricity generation using imported oil is the major cause of the present electricity crisis, as the dwindling economy of the country cannot provide adequate fuel to these power plants [14]. The total electricity generation in the country during 2012–2013 was 96,122 GWh, the shares of thermal (oil, gas, coal), hydro, and nuclear generation were 61,711 GWh (64.2%), 29,857 GWh (31%) and 4553 GWh (5%), respectively [5]. Shares of different fuels in the total electricity generation during 2012–2013 are shown in Figure 3 [5]. Due to this over dependence of electricity generation on fossil fuels, CO_2 emissions have also increased to ~792.32 Mt with a growth rate of 12.62% [6].

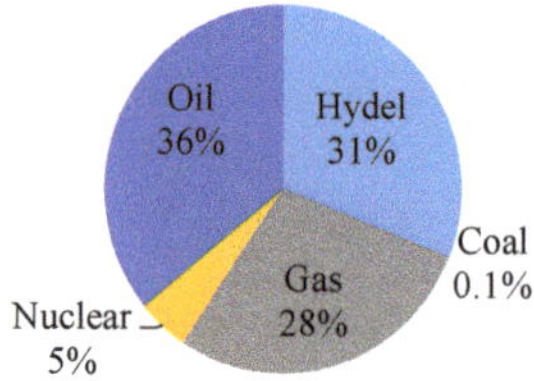

Figure 3. Shares of electricity generation from different sources during 2012–2013 [5].

The major concern pertaining electricity generation is that around half of the total installed generation capacity is underutilized due to the inappropriate fuel mix, inefficient power plants, and lack of proper maintenance of power plants [7]. In order to meet the augmenting demand of electricity, besides planning and developing the conventional sources of energy, Pakistan should also focus on exploiting the potential of hydro and other renewable energy resources, mainly biomass, biogas, wind, and solar, to develop a substantial share of sustainable energy sources in the electricity generation fuel mix [8].

Environment-related emissions from conventional fossil fuel based power plants have severally adverse effects on human societies, animals habitats, and woodlands by hot and cold waves, flash floods, glaciar melts, acidic rains, droughts, and other unwanted effects. Pakistan is among the top ten countries worst hit by climate change. Hundreds of people die every year because of heat waves. Natural habitat is at the destruction level, on the other hand, forests have dried due to dieback during 1998–2005 [1]. A proper strategic energy mix is required to fulfill present and future energy demand keeping in view of climate change.

Pakistan has abundant sources of indigenous energy, it is estimated that 185 Gt of coal reserves are available, of which around 7 Gt are technically feasible and could be utilized for power generation using clean technologies [10]. The estimated hydro power capacity of Pakistan is 100 GW, of which ~55 GW are technically feasible for power generation. The technical potential of Solar, Wind, and Biomass is 240, 62 and 8.6 MTOE/Year, respectively, which could be effectively used for power generation [11]. The energy potential of renewable only is 4.4 times the current energy demand of Pakistan, however, because of poor governance and unavailability of effective policies, only 33% of renewable energy resources have been utilized so far.

3. Future Electricity Generation Plans

In order to overcome the electricity crises in the country, the GOP has announced the development of various power projects based on coal, hydro, and other renewable energy sources under the Power Policy 2013. As a result of these efforts, it is expected that ~16–20 GW of the power generation capacity will be added to the national grid during the next 5–10 years, which will reduce the load shading duration [15]. In this context, a brief assessment of the future power projects under referred policy is undertaken as under:

3.1. Hydroelectric Power Plants

Hydroelectric-based power generation has a huge potential in the country and such plants can produce low-cost electricity. The government under Power Policy 2013 has proposed and initiated medium and long-term projects for hydroelectric capacity expansion. In this context, a total capacity of 384 MW power projects has been completed and connected to the grid in 2016. The Gulpur and Patrind hydel projects of smaller capacity, are also expected to be completed in the near future, which will add 247 MW to the national grid. Furthermore, around 969 MWs of electricity is expected from the Neelum–Jhelum project in the near future too. A number of hydroelectric projects are anticipated to come online during this plan period, which include the fourth and fifth Tarbela expansions. These expansions add the generating capacity to the tune of 1910 MW. The detailed engineering design for projects at Patan 2800 MW, Dasu of 2160 MW, and Thakot 2800 MW are also anticipated to be undertaken under the current Power Policy. Some other long-term projects are as Bunji 7100 MW, Kohala 1100 MW, and Diamer–Bahasha 4500 MW. Completion of these projects may save the country from the vulnerable conditions of electricity crises [15,16]. Brief detail of these projects is given in Table 1 below.

Table 1. Proposed hydroelectric power plants [17].

Name of Power Plant	Capacity (MW)
Hydro Neelum–Jhelum Hydroelectric	969
Tarbela 4 th,5th extension	1910
Patrind Hydroelectric	147
Akhori dam project	600
Sehra Hydroelectric Project	130
Dasu Hydroelectric Project	4320
Diamer–Basha Dam	4500
Suki Kinari Hydroelectric	870
Karot Hydroelectric Project	720
Bunji Hydroelectric	7100
Azad Pattan Hydroelectric	640
Lower Palas Hydroelectric	665
Lower Spat Gah Hydroelectric	496
Kohala Hydroelectric Project	1100
Mahl Hydroelectric Project	590
Thakot Hydroelectric	2800
Patan Hydroelectric	2800
Munda dam project	740
Mohmand dam Hydroelectric	800
Shyok dam project	690
Chakoti–Hattan Hydroelectric	500

3.2. Coal Power Plants

Realizing the low cost-based power generation from electricity, the government decided to install coal-based power plants at different locations of the country. In this context, various power plants based on imported coal having a total capacity of 5,580 MW are under construction with the financial support of China–Pakistan Economic Corridor (CPEC). One of these project of 1320 MW at Sahiwal, Punjab province was completed recently, and also commenced the commercial operation in July 2017 [18]. Furthermore, additional coal based power plants are also proposed under the Power Policy for installation at various locations inPunjab based on imported coal. Sindh Engro Coal Mining Company (SECMC) in a joint venture with Sindh Government and Engro power is also developing coal mines in Block-I and II in district Tharparkar. SECMC is expected to complete construction of a 660 MW power plant in the first phase, while in the second phase another 660 MW power plant would be commissioned by the end of 2019. In Block III of Thar coal mines, 5000 MW power plants are expected to be installed by various companies, whereas 7500 MW of power plants by Sino Sindh Resources (Pvt.) Limited (SSRL) China will be established in Block-I of Thar Coal Mines in different phases. Three power plants, based on coal, each of 1320 MW are proposed for installation at Jamshoro, Lakhra, and Port Qasim by PEPCO and K-Electric [19–22]. Table 2 summarizes the proposed coal-based power plants under current power policy of the government.

Table 2. Proposed coal power plants [17].

Name of Power Plant	Capacity (MW)
Imported Coal	
Coal power plants at Punjab	2×660
Coal power plants at Punjab	5280
Coal power plants at Jamshoro	2×660
Coal power plants at Hub	2×660
Coal power plants at Gawadar	300
Coal power at Port Qasim	2×660
Conversion of Jamshoro Power Plant from Oil to Coal	850
Conversion of Muzaffargarh Power Plant from Oil to Coal	1350
Conversion of Guddu Power Plant from Oil to Coal	640
Conversion of K-Electric Power Plant from Oil to Coal	1260
Conversion of HUBCO Power Plant from Oil to Coal	1292
Local coal	
Sino Sindh Resources (Pvt.) Limited (SSRL) (China	7500
Thar Power Company Ltd. (THARCO) SECM	5000
Oracle Coalfields UK	1400
GENCOS	1320
Sindh/ETON Japan Power	3960

3.3. Oil and Natural Gas Power Plants

The steep increase in furnace oil prices in the international market has rendered the electricity generation mix highly unsustainable and costly in Pakistan. Therefore, the government has decided in principle not to install oil fuel-based power plants in future, as such, in this study, the existing oil fuel power plants are gradually decreased whereas no new power plant is considered for during the study period.

Natural Gas contributes ~45% of the total primary energy supply mix in the country. Pakistan has a widespread gas network of pipelines to cater the requirement of more than 8.4 million consumers across the country by providing about 4 Billion Cubic Feet of natural gas per day [23]. The GOP is also implementing a multi-pronged approach which include import of piped natural gas from neighboring countries like Iran and Turkmenistan, or LNG from Qatar towards meeting its energy needs, especially for power generation [24]. In this context, two power plants based on natural gas Uch-II and Guddu with an installed capacity of 404 MW and 747 MW, respectively, were completed and added to the national grid in 2014. Four power plants having a total capacity of 4,883 MW on Re-gasified Liquefied Natural Gas (RLNG) are under development in different locations of the Punjab province. Table 3 below summarizes the natural gas-based power plants which have started commercial operation or expected to start commercial operation in the future [25].

Table 3. Proposed natural gas power plants [17].

Name of Power Plant	Capacity (MW)
Uch-II power plant	404
Guddu power plant	747
RLNG based power plant Bhikki Punjab	1180
RLNG based power plant Balloki Punjab	1223
RLNG based power plant Haveli Punjab	1230
RLNG based power plant Jhang Punjab	1250

3.4. Nuclear Power Plants

Pakistan Atomic Energy Commission (PAEC) is currently operating three nuclear power plants: Karachi Nuclear Power Plant (KANUPP), Chashma Nuclear Power Plant Unit-1 (C-1), and Unit-2 (C-2). The construction of two more power plants at Chashma, C-3 and C-4, of 340 MW each are in progress and expected to be commissioned in different phases in the near future. The ground-breaking ceremony of two Karachi-based Coastal Nuclear Power Plants (K-2) and (K-3) of 1100 MW each was also held in November 2013, and they are expected to be completed in 2020. Another 2200 MW nuclear power plant is also proposed at the coastal belt of Balochistan near Hub [26]. A summary of the proposed nuclear power plants under the Power Policy 2013 is given in Table 4.

Table 4. Proposed nuclear power plants [17].

Name of Power Plant	Capacity (MW)
Chashma Nuclear Power Plant Units-3 (C-3) and Unit-4 (C-4)	680
Karachi Nuclear Power Plants (K-2) and (K-3)	2200
Chashma Nuclear Power Plant Unit-5	1000
Coastal Nuclear Power Hub Balochistan	2200

3.5. Renewable Energy Power Plants

Pakistan has immense potential for renewable energy (RE) resources. These resources, if harnessed, can play a significant role towards the nation's energy security. In this context, GOP has tasked the Alternative Energy Development Board (AEDB) to ensure that 15% of total power generation should be from renewable energy (other than hydro) by 2030 [27–29]. Solar, wind, and biomass are the leading renewable energy sources of the country. The potential of each of these renewable energy resources is discussed as under:

3.5.1. Solar PV Power Plants

The solar radiation map of Pakistan is given in Figure 4. Realizing this potential, the government has taken steps to harness power from solar energy. A 1000 MW of solar PV-based power plant development, as such, has been undertaken at "Quaid-e-Azam Solar Power Park" in district Bahawalpur, Punjab province. Another 500 MW solar based power plant is also on the cards to be set up by a Chinese Company at "Quaid-e-Azam Solar Power Park" which is expected to be completed in near future [29,30].

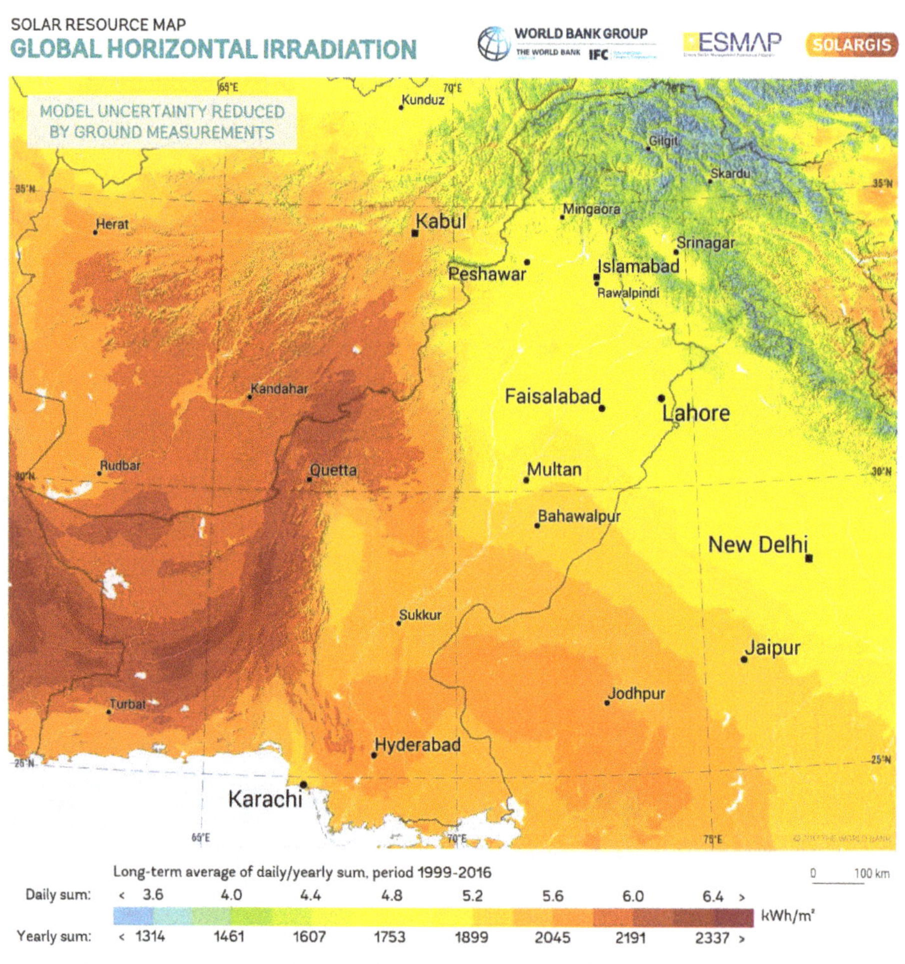

Figure 4. Solar radiation map of Pakistan [31].

3.5.2. Wind Power Plants

The coastal belt of Pakistan is blessed with a 60 km wide (Gharo-Keti Bandar) and 180 km long (up to Hyderabad) wind corridor. In addition to this wind corridor, there are other wind sites available in the coastal area of Balochistan and some northern areas of the country. The wind map of Pakistan is shown in Figure 5, which illustrates enormous wind energy potential in the country [32,33].

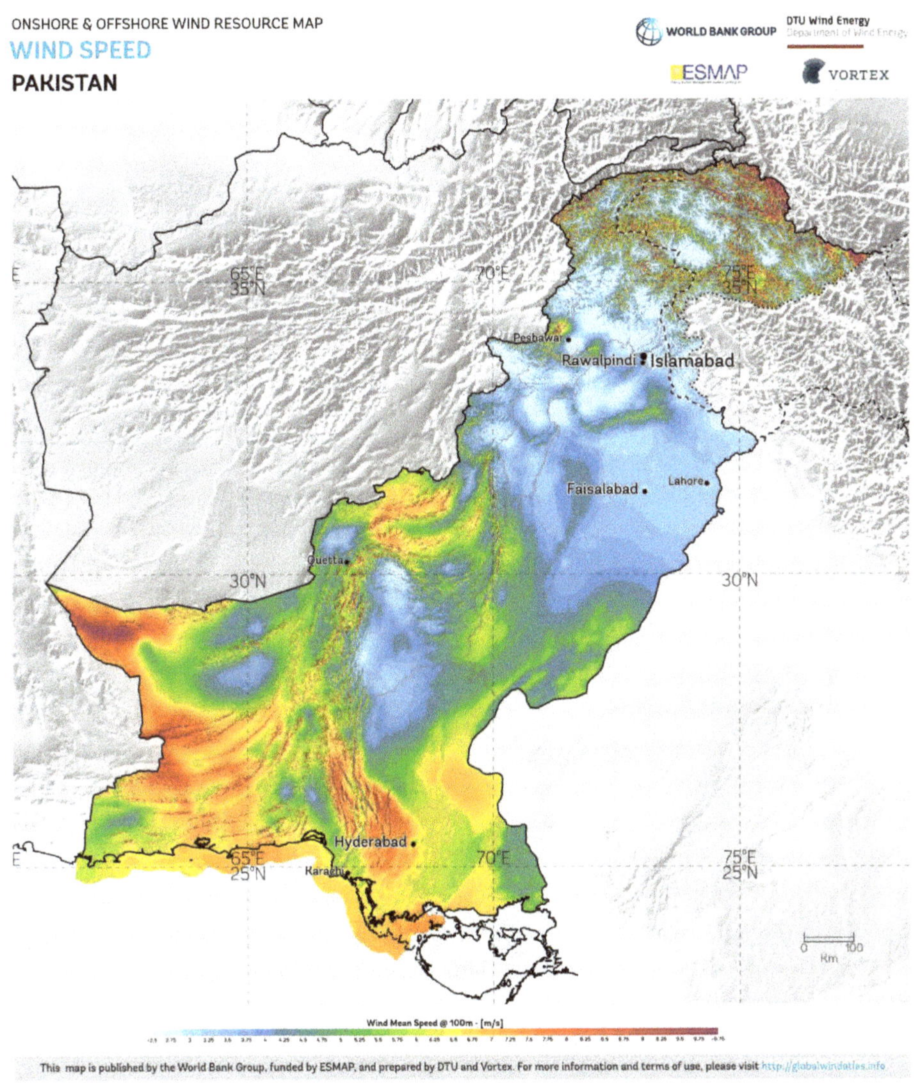

Figure 5. Wind map of Pakistan [34].

However, despite this huge potential, wind power is not utilized optimally in Pakistan, only two power plants, "Fauji Fertilizer Energy Company Limited (FFCEL)" and "Zorlu Wind Energy" with the cumulative installed capacity of 106 MW could only be connected to the national grid in the early phase of implementing the Power Policy 2013. Other wind power plants (50 MW Foundation Wind energy I, 50 MW Foundation Wind energy II, 50 MW Sapphire Power, 50 MW Metro and 50 MW China Three Gorges) are at the final stages. In addition, Letters of Support (LOS) have been issued for projects up to 450 MW, and an additional 2276 MW of wind power projects are currently in the feasibility evaluation process. Thus a cumulative 2726 MW of wind electricity could come online in the near future in different phases [9,15]. The brief of renewable energy projects already undertaken or to be undertaken as per government's plans are given in Table 5.

Table 5. Proposed renewable (other than hydro) power plants [17].

Name of Power Plant	Capacity (MW)
Solar PV power park Punjab	1000
Chinese solar Company	500
Zorlu Wind Energy Sindh	56
Fuji Fertilizer Energy Sindh	50
Capacity of wind power to be Commissioned	2726
Bagasse based power plant	83

3.5.3. Biomass Power Plants

Pakistan is the fifth largest producer of sugarcane in the world, with an average production of about 50 million tons annually. This huge amount of sugarcane is crushed in 80 sugar mills across the country produce, which in turn produce 10 million tons of bagasse. This huge resource of bagasse can be turned into an immense source of energy by producing 3000 MW of electricity. Another 5000 MW of electricity can also be produced from the livestock. The National Power Policy 2013 stipulated to produce at a minimum 83 MW of electricity from bagasse. In this context, Letters of Intent (LOIs) have also been issued to different companies by the Alternative Energy Development Board (AEDB) to develop a bagasse-based power plant. As such, biomass to energy power plants at Jhang and Faisalabad, Punjab, at Mirpurkhas, Sindh and Mardan, Khyber Pakhtunkhwa provinces are at different stages of the development [11,35,36].

4. Methodology

Power generation expansion planning is very crucial in the overall planning for a country or region. It provides insight into long-term alternative strategies to meet the future power requirements. It can also be adapted to search for minimum cost solutions which meet the present and future power demand [37,38]. However, sustainability-related concerns now emphasize optimal solutions instead of least-cost solutions. Various computer-based energy modeling tools are available for power generation planning, including MARKAL/TIMES, EnergyPlan, and LEAP. Each of these tools has its advantages and limitations. In this study, an energy scenario modeling exercise was undertaken using the computer-based modeling tool LEAP. The accounting platform in LEAP matches the energy demand through supply-side energy generation technologies and updates the system impacts consisting of electricity generation by type, system electricity generation cost, depletion of resources, and emission estimates. Subsequent sections elaborate the modeling exercise undertaken in this study.

4.1. LEAP Energy Model Framework Development

The LEAP modeling efforts adopt the bottom-up approach to meet the electricity demand by considering capacity factors of different fuel-based power plants, energy intensity of the power plants, and emissions intensity of the fuels. The LEAP model was developed by the Stockholm Environment Institute, and is widely used around the globe. The essential concept of LEAP is user-driven scenario analysis [39]. The scenario manager in the model sets the Base/Reference/Business As Usual (BAU) scenarios. The input data set for LEAP consists of various modules such as key assumptions, demand, transformation, and resources. The key assumption module includes Gross Domestic Product (GDP), GDP growth, total population, population growth, number of consumers and their growth, alongside other relevant parameters. In the key transformation module of LEAP, energy transformed from energy sources as input to energy product is modeled using a range of electricity technologies, including those in operation during the base year, and others anticipated in the government's plans, as well as those considered in accordance with the emerging trends of the generation technologies in the future. Figure 6 shows the LEAP model framework developed under this study focusing on electricity generation for the modeling period 2013–2035. The reference scenario was initially developed based on the governments' plan and policy followed by three alternative policy scenarios, which were then evaluated by comparing their obtained values with those of the reference scenario.

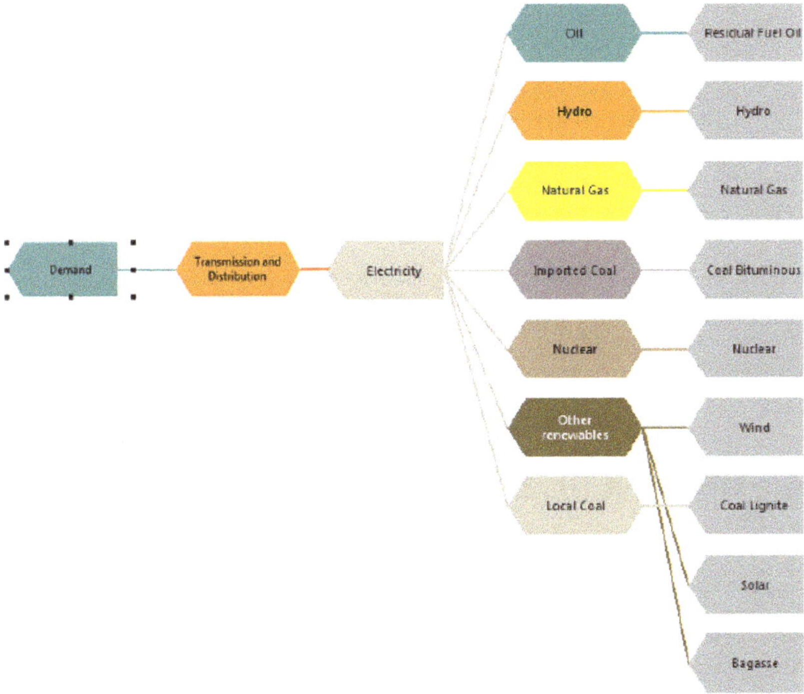

Figure 6. LEAP modeling framework for Electricity Generation 2013–2035.

4.2. Electricity Demand Forecast

Electricity demand is an important component, which must be estimated for the power generation expansion planning. The National Transmission and Dispatch Company (NTDC), a subsidiary of Water and Development Authority (WAPDA) of the GOP, has forecasted the electricity demand in the country using a multiple regression analysis. These estimates suggest an increase at a growth rate of 5.4% from 2013 to 2035 [40]. The main variables considered during the electricity demand forecast by the NTDC include the contribution of various sectors in the overall GDP, the population of the country, electricity consumption by various sectors and a number of consumers. Utilizing NTDC determined electricity demand growth rate of 5.4%, LEAP model estimated that that electricity demand is likely to increase from 139 TWh in 2013 (base year) to 442 TWh in 2035 (end year). Electricity demand forecast by NTDC and that of LEAP model of this study are illustrated and compared in Figure 7. It shows that demand forecast estimated by the LEAP model is generally in parallel to that of the estimated by NTDC.

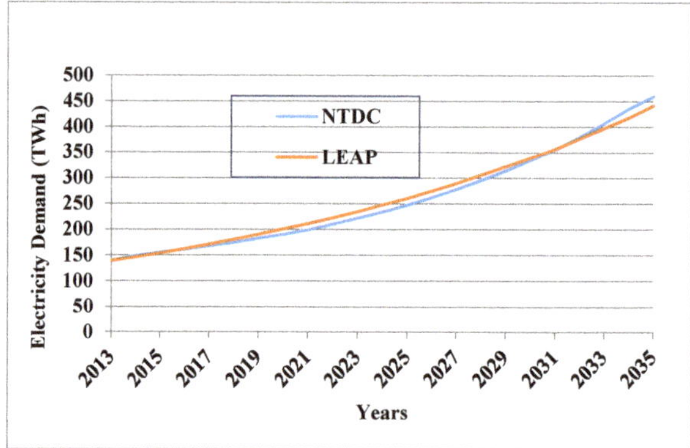

Figure 7. Comparison of electricity demand forecast from the LEAP Model and NTDC.

4.3. Basic Assumptions for LEAP Modeling

The basic assumptions considered while developing the LEAP modeling framework of this study are given in Table 6. In addition, the overall Transmission and Distribution (T&D) losses are assumed to decrease from 17.37% in 2013 to 12.5% in 2020, and about 11.7% by 2035 as projected by the NTDC [40]. Taking into account the GOP's current plant, the currently working oil and natural gas-based power plants are assumed to continue with existing efficiency and availability while the power plants being converted from oil- to coal-based power generation are assumed to operate on new efficiency and availability.

The reserve margin is described as the fraction of additional power generation capacity, which is available to meet the peak load in case the sudden increase in demand can happen. Keeping in view the situation of the power system in Pakistan, the reserve margin in this study is assumed as 12% for the modeling period. The Load Duration Curve (LDC) provides the variation in power demand during a specified period and under this study, the LDC is considered on the hourly basis throughout the year. The GHGs emissions from all the fuels used in the power generation system are assessed by their associated emission factors available in the Technology and Environment Database (TED) of the LEAP energy model. For all the fuels, the Tier 1 default emission factors published by Intergovernmental Panel on Climate Change (IPCC) are selected. The dispatch rule for the power generation plants is followed as per merit order according to the GOP plan and policy. The capacity credit, is defined as the fraction of availability of the power plant to the standard availability of thermal power plant, and for this study is assumed by following the precedent literature [13,41].

Table 6. Summary of model development assumptions.

Parameters	Assumptions
Transmission and Distribution (T&D) losses	Assumed to reduce from 17.37% in 2013 to 12.5% in 2020 and about 11.7% by 2035 as published by NTDC.
Emission factors of fuels	Default emission factors published by IPCC are available in LEAP TED database.
Reserve Margin	Assumed 12% according to the previous studies
Dispatch rule	Followed as merit order according to the government of Pakistan's regulating body.
Load Duration Curve (LDC)	Calculated according to the published data in the "Pakistan energy yearbook 2013".
Capacity credit	Capacity credit is assumed by previous studies.
Plant development	For presently operational oil and gas steam turbines existing technologies whereas for coal, combined cycle gas turbines, solar, wind and biomass new mature technologies are assumed.

Furthermore, in this study, four supply side scenarios have been developed to meet the demand forecast for the study period (2013–2035). A summary description of the four scenarios of the study is given in Table 7.

Table 7. Summary of scenario alternative of this study.

Scenario	Objective	Main Resources
Reference (REF)	In this scenario, the government's current plan and policy is followed.	As per the government plan and policy.
More Renewable Energy (MRR)	Under this scenario, the share of renewable energy resources (other than hydro) is increased while share of coal is decreased.	Renewable energy resources, solar, wind, and biomass.
More Hydro Energy (MRH)	Under this scenario, share of hydroelectric is increased while the share of coal is decreased.	Renewable energy resources, hydro, solar, wind, and biomass.
More hydro Nuclear Energy (MRHN)	Under this scenario, the share of hydro and nuclear is increased while reducing the share of coal.	Energy resources, hydro, nuclear, solar, wind, and biomass.

The detailed description of each of these scenarios is discussed in the following sections of this paper.

4.4. Reference or Base Scenario

Reference or Base scenario of this study was developed according to the future electricity generation plans of the GOP as discussed in Section 3, and is based on the existing National Power Policy. The new installed capacity in the reference scenario is projected to be based on different fuel sources like imported coal (20 GW), local coal (20 GW), hydro (36 GW), natural gas (8 GW), nuclear (8 GW), and renewables (wind, solar, biomass) other than hydro (15 GW). The base year electricity generation data are given in Table 8, which were also used as input to the LEAP scenarios manager.

Table 8. Base year electricity generation data [14,41,42].

Fuel	Efficiency (%)	Output (GWh)	Capacity (GW)	Maximum Availability (%)
Hydro	80	29,857	68	60
Coal	45	61	0.01223	75
Natural Gas	50	27,116	6.59	70
Nuclear	34	4553	0.75	85
Oil	40	34,534	8.2	50
Other Renewable	34	0	0	34

4.5. Alternative Scenarios

Three alternative scenarios were further developed using the scenario manager to compare the results with the reference scenario over the modeling period 2013–2035. The total installed capacity under these scenarios increased in such a way that electricity generation remained the same in the alternative scenarios as in the reference scenario, but only the shares of fuels in each alternative scenario alongside installed capacity varied accordingly. The brief description of the alternative scenarios developed in this study is provided in the following sections.

4.5.1. More Renewable Energy (MRR) Scenario

According to the government's existing power policy, a major share of electricity, which would be generated from the imported and indigenous coal, this results in more GHG and other air pollutant emissions. Following this path poses serious threats to regional and global environments. In order to counter this situation, the MMR scenario in this study takes into account Pakistan's enormous renewable energy potential, which is so far untapped. It is estimated that the exploitable potential of solar, wind, and biomass is 169 GW, 65 GW, and 15 GW, respectively in the country [13]. In this scenario, the share of renewables (other than hydro)increased while the share of coal for power generation decreased. The increment in renewable energy, in this scenario, is mainly from the wind, solar, and biomass resources.

4.5.2. More Hydro Energy (MRH) Scenario

Hydroelectric is a major source of electricity generation for the base-load requirements, and can meet the peak and unexpected electricity demands in the power system. The electricity generation cost of hydroelectric also cheap since no any fuel is consumed during the generation of electricity which makes it a competitive source of renewable energy [43]. The hydroelectric plants, during operation, also do not produce any emission and industrial waste like fossil fuel-based power plants. Pakistan is blessed with ~100 GW of hydroelectric potential out of which so far 59 GW has been identified. Talking into account the enormous potential of hydroelectric, in this scenario, the share increased, reducing the share of coal thus altering the reference scenario.

4.5.3. More Hydro Nuclear Energy (MRHN) Scenario

Nuclear is also among the group of energy resources and technologies that are available, and capable of dealing the environmental challenge faced due to the GHG and other air pollutant emissions due to combustion of other fossil fuels [44]. Nuclear and hydroelectric energies emit negligible amounts of CO_2 and other GHG, with the emissions considered for the total life cycle fewer than 15 CO_2-eq/kWh [45]. Nuclear-based power plants are base-load electricity generation options, which are particularly suitable for large-scale, uninterrupted electricity supply to meet the demand. Extended penetration of nuclear power in the energy supply mix can also help address the higher electricity generation costs. Keeping in view both the environment and cost related benefits of nuclear and hydroelectric energies in this scenario the shares of nuclear and hydroelectric are increased by cutting the share of coal.

The above-discussed scenarios were simulated in the LEAP energy model. The results of these simulations are discussed in the following section.

5. Results and Discussions

5.1. Reference or Base Scenario

Simulation results of the developed model for the Reference scenario of existing and planned electricity generations are shown in Figure 8a. According to these results, about 107 GW of new installed capacity is required from various available sources up to 2035 in order to meet the growing electricity demand. This newly installed capacity raises the overall capacity of the reference scenario by 124 GW in 2035. The maximum shares are from the low-cost electricity generation sources of coal and hydro with the capacities of 40 GW and 36 GW, respectively. In the new power policy, it has been decided by the government that more than 60% of the existing low-efficiency high fuel cost oil-fired power plants will be converted to low-cost coal based power plants. Therefore, instead of adding the oil-based power plants, the existing share of oil will be reduced from 8.6 GW to 3 GW. For the diversification of electricity generation fuel mix, government also plans to install new power plants based on natural gas. As such, another 8 GW of newly installed capacity from natural gas has been added to the total installed capacity under this scenario. The supply of natural gas for new power plants is planned for importation through LNG and gas pipeline. This scenario also takes into account the renewable energy resources and ~15 GW of renewable energy sources other than hydro have been included in the installed capacity by the year 2035. These developments are in line with existing GOP plans, which should be realized by the year 2035.

Electricity generation in the reference scenario, with additional installed capacity, increases to meet the growing electricity demand. The total electricity generation, therefore, increased from 96 TWh in 2013 to 442 TWh in 2035 as shown in Figure 8b. The major shares in the electricity generation were from coal and hydroelectric as per government plans. The share of oil decreased from 36% to only 2% in the year 2035, as there would be no new installations from oil-based sources during the study period. Reductions in oil shares were replaced by hydro- and coal-based power generation. The overall share in electricity generation from hydroelectric plants are anticipated to increases from 31%

in the base year to 37% in the end year, whereas the share of natural gas for electricity generation is expected to significantly decreases from 28.2% to 13% during the same period.

The share of nuclear-based electricity generation is anticipated to increases from 5% in the base year to 9% in the end year. The share of other renewables in 2013 was 0%, increased in 2015 by 1.2%, and is expected to reach 6% in 2035. Electricity generation from coal-fired power plants is more cost-competitive than oil-fired power plants thus the share of coal for electricity generation, under this scenario, increased from 0.1% to 34% during the modeling time period.

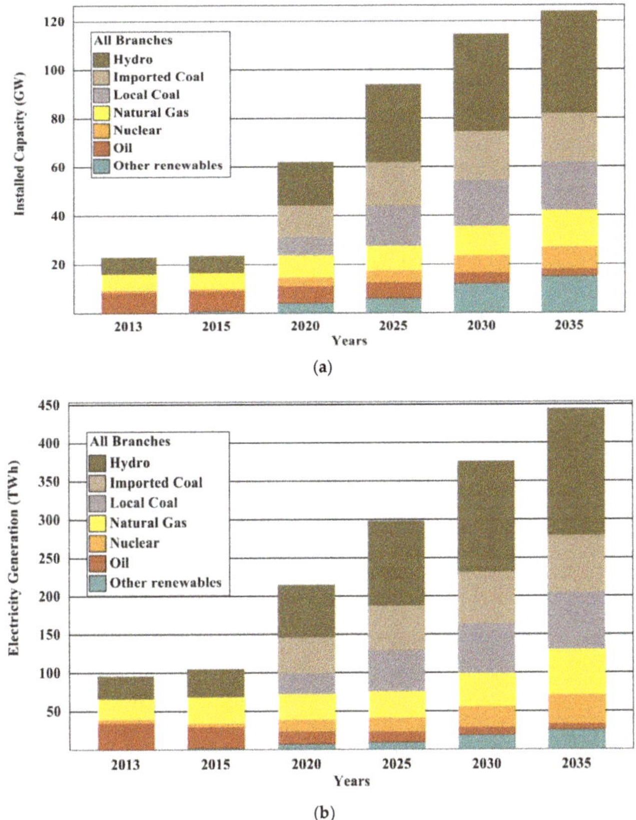

Figure 8. (**a**) Electricity installed capacity by source under reference/base scenario; and (**b**) total electricity generation by a source under reference/base Scenario.

The electricity output shares from various fuel sources for the modeling period under the reference scenario are shown in Table 9.

Table 9. Electricity output share from various fuel sources for the modeling period.

Fuel Source	Percentage Share 2013	Percentage Share 2035
Oil	36	2
Natural Gas	28.2	13
Nuclear	5	9
Coal	0.1	34
Hydro	31	37
Renewable	0.0	6

It is apparent that significant variation in fuel share for electricity generation under reference scenario may follow if GOP remain wedged on its plans.

5.1.1. MRR Scenario

Renewable energy sources such as solar, wind, and biomass are expected to play a significant role in the mitigation of emissions to counter the climate change and address the sustainability cause. Therefore, under this scenario, the share of installed capacity of renewables (other than hydroelectric) increased from 15 GW to 30 GW, while the installed capacity of coal decreased from 40 GW to 20 GW compared to the REF scenario as shown in Figure 9a. The installed capacities of other sources like hydro, nuclear, and oil remained the same as in the reference or base scenarios. The increased share of renewables shall reduce the reliance on coal, but it needs flexible power sources such as the natural gas power to compensate electricity demand due to the lower capacity factors of renewable energy based plants. The newly installed capacity from natural gas was 20 GW in the end year under this scenario, which enhanced the overall installed capacity from all sources to 132 GW to meet the electricity demand.

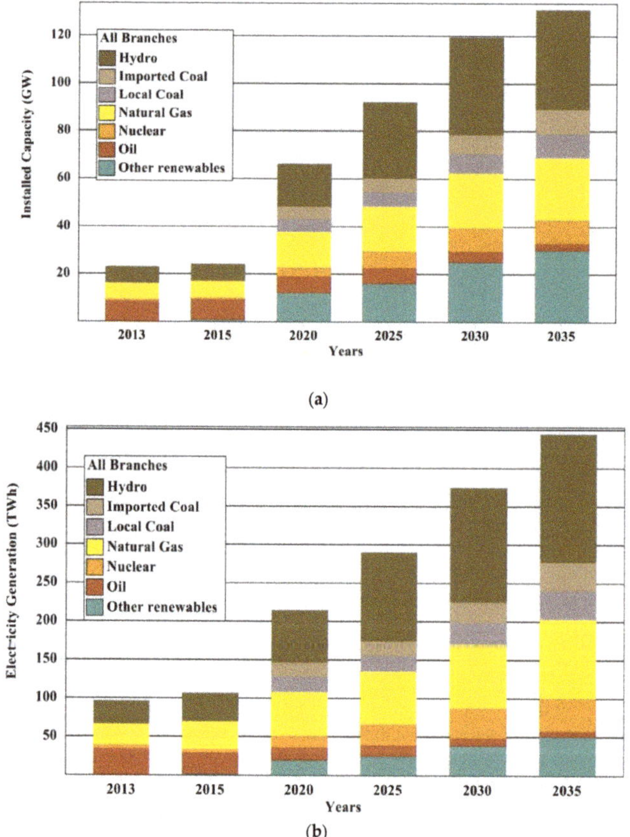

Figure 9. (**a**) Electricity installed capacity by source under MRR scenario; and (**b**) total electricity generation by source under MRR scenario.

The overall electricity generation output in this scenario was same as in the reference or base scenario, but only the share of renewables (other than hydroelectric) significantly increased due to the increment in their capacities. The share of other renewable energy sources, which was 0% in 2013, is anticipated to increase 11% in 2035. The share of coal is expected to decrease from 34% as in the reference scenario to 16% in the MRR scenario in 2035 as shown in Figure 9b. The penetration of renewable energy sources in the total electricity generation shall results in better environmental and economic impacts since these are indigenously available and emissions-free resources.

MRR scenario considerations and results are very important, and following this pattern would greatly help country to reduce oil import bill as well as help in containing the climate change.

5.1.2. MRH Scenario

In this scenario, which utilizes the country's hydro potential, the installed capacity of hydroelectric should increase from 36 GW to 40 GW by the end year 2035. The installed capacity shares of all other sources for power generation are assumed to remain the same as in the reference or base scenario except for the installed capacity of coal which should be reduced from 40 GW as in reference or base scenario to 34 GW in this scenario by the end year as shown in Figure 10a. Increasing the installed capacity of hydroelectric presents a viable option for electricity generation from cost-effective and green power generation sources. The total installed capacity under this scenario, should increase by ~123 GW in the year 2035, with substantial share of the hydroelectric plant in this scenario.

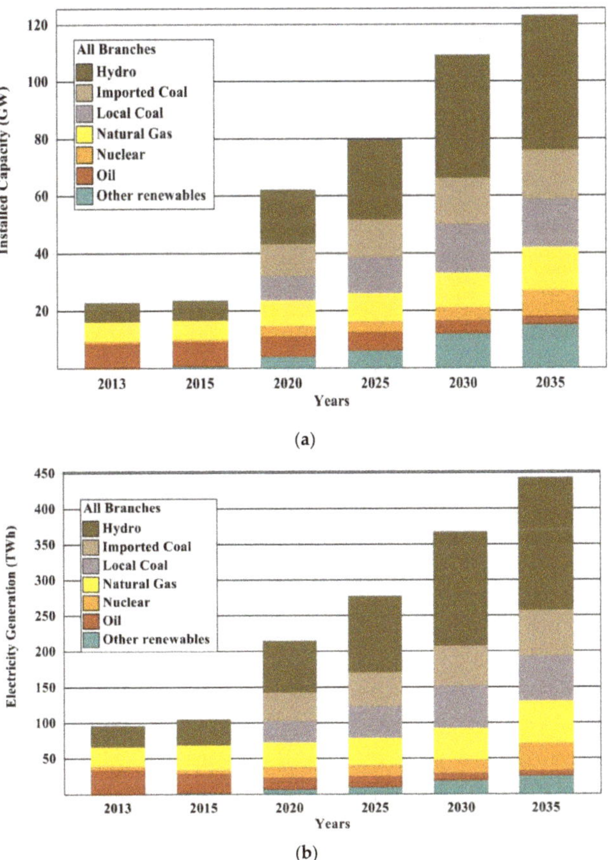

Figure 10. (**a**) Electricity installed capacity by source under MRH scenario; and (**b**) total electricity generation by a source under MRH scenario.

The electricity generation share under this scenario for hydroelectric is expected to increase from 37% in the reference scenario to 42% by the year 2035 as shown in Figure 10b. Increase in the share of hydroelectric depicts the low-cost, secure and sustainable electricity generation option. The share of coal should also decrease from 34% in the reference scenario to 28% in this scenario by 2035 to balance the increase of hydroelectric share.

The MRH scenario is focused on exploiting the indigenous hydroelectric resources and thus could significantly contribute towards reducing oil-based power generation as well containing emissions.

5.1.3. MRHN Scenario

In this scenario, alongside hydroelectric, nuclear power increment is considered to offer cost-effective, sustainable, technologically superior, and long-term energy supply option for electricity generation. The installed capacity of nuclear and hydroelectric, as such, is expected to increase from 8 GW and 36 GW as of in reference scenario to 10 GW and 40 GW under MRHN scenario, respectively by 2035. In response to an increase in installed capacity of these plants, the installed capacity of coal-based plant reduces from 40 GW in the reference scenario to 31 GW under MRHN scenario as shown in Figure 11a. The overall installed capacity in this scenario is expected to reach 123 GW in 2035 to meet the increased demand.

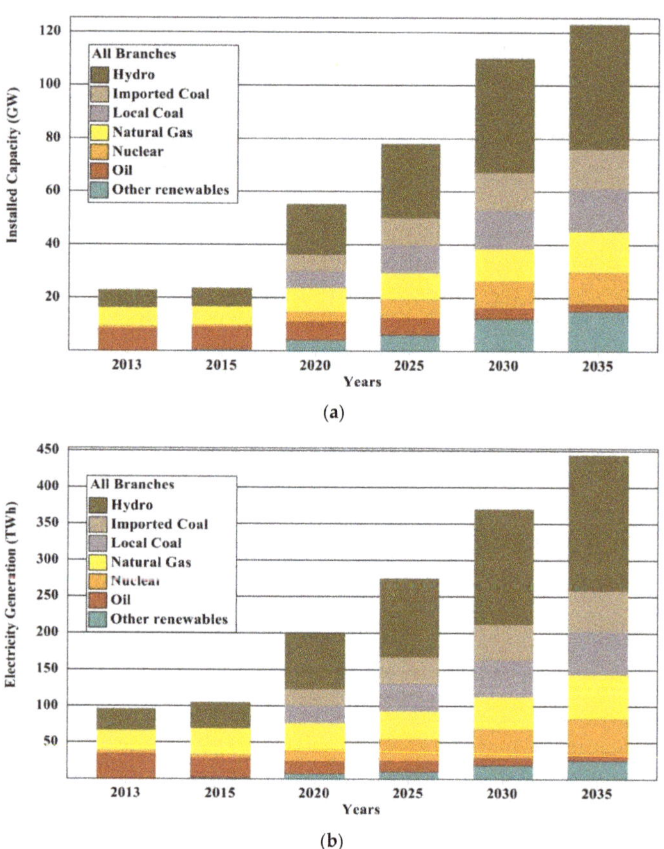

Figure 11. (**a**) Electricity installed capacity by source under MRHN scenario; and (**b**) total electricity generation by a source under MRHN scenario.

The electricity generation shares from nuclear and hydroelectric are subsequently higher in this scenario as shown in Figure 11b. These shares of electricity from nuclear and hydroelectric are expected to increase by 2% and 5%, respectively, under MRHN compared to the reference scenario in 2035. The share of coal would also decrease by 8% compared to the reference scenario. The reduction in electricity generation shares from coal is an optimistic situation to mitigate the effects of emissions, which are harmful for the environment.

MRHN scenario in addition to exploiting the hydroelectric potential also taken into account the nuclear-based power generation, which would require care full planning, and execution of such projects.

5.2. Environmental Emissions in Reference and All Alternative Scenarios

The LEAP energy model estimates the emissions from power plants using the emission factors and other technical characteristics integrated within TED, which are based on IPCC Tier 1 database. The simulated results of GHG emissions for the reference and alternative scenarios of this study are shown in Table 10.

Table 10. Air pollution and GHG emissions in different scenarios from 2013 to 2035.

Emissions	Units	2013	2035			
			REF	MRR	MRH	MRHN
CO_2	million tons	34	143	103	126	117
SO_2	kilo tons	253	553	372	487	454
NOx	kilo tons	92	157	167	146	140
CH_4	kilo tons	1	1.8	2.3	1.7	1.6
N_2O	kilo tons	0.2	82	41	70	63

Each of these tabulated emission components are further discussed with respect to the reference and other three alternative scenarios as under.

5.2.1. CO_2 Emission

CO_2 emissions are estimated to increase from 34 million tons in the base year 2013 to 143 million tons in 2035 under the reference scenario, and from 34 million tons to 103 million tons in the MRR scenario. In case of the MRH scenario, these emissions increase from 34 million tons in the base year to 126 million tons by the year 2035, while in MRHN scenario the CO_2 emissions are estimated to increase from 34 million tons to 117 million tons from 2013 to 2035, respectively. The CO_2 emissions results pertaining all four scenarios of the study are illustrated in Figure 12.

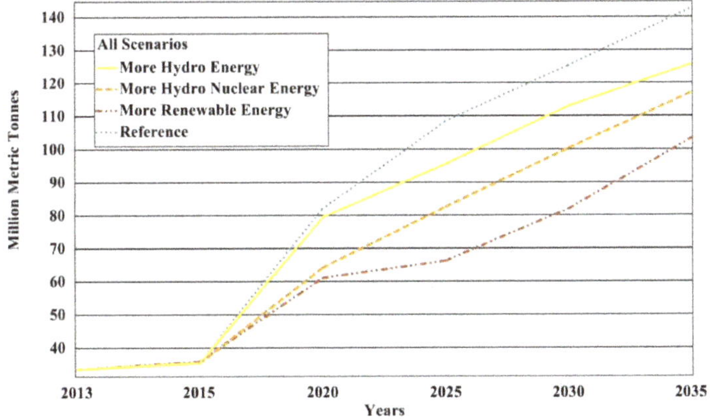

Figure 12. Annual CO_2 emissions in all scenarios.

CO_2 emissions are estimated to be 118% less in the MRR scenario compared to the reference scenario. This would be a significant reduction in emissions due to the addition of renewable energy sources. The higher CO_2 emissions in the reference scenario are evidently due to increased coal-based generation, which emits a huge amount of emissions. Under the MRHN and MRH scenarios, CO_2 emissions were 76% and 51%, respectively, less than the reference scenario due to hydroelectric and nuclear power-based capacity additions.

5.2.2. SO_2 Emissions

The utilization of both imported and indigenous coal as a fuel for electricity generation in all scenarios would produce a great amount of emissions. However, with highly efficient critical pressure boilers and advanced Fluidized Bed Combustion (FBC) combustion technologies considered in this study, hazardous emissions such as SO_2 could be reduced to a minimum level. Nevertheless, a substantial amount of emissions would be from the combustion of coal. In addition, SO_2 emissions from the existing oil-based power plants would also be present. In these, existing plants' no emission control systems are in place. Therefore, SO_2 emissions are projected to increase 300 thousand tons during the whole study period under the REF scenario, which is the highest increase as compared to the other alternative scenarios. The amount of SO_2 emissions in other alternative scenarios is projected to be 234 thousand Mt in MRH scenario, 201 thousand Mt in MRHN, and only 120 thousand Mt in MRR scenario by the end of modeling period. The projected SO_2 emissions under REF, and the three alternative scenarios, are shown in Figure 13.

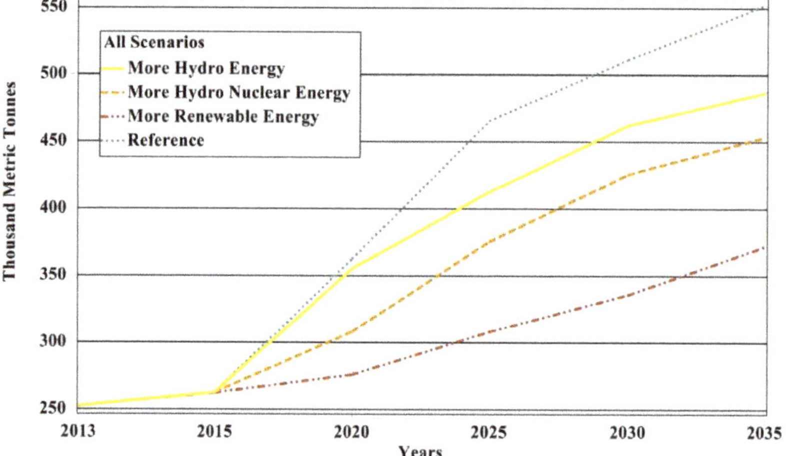

Figure 13. Annual SO_2 emissions in all scenarios.

5.2.3. NO_X Emissions

NO_X emissions from existing and planned thermal power plants in the reference and alternative scenarios for the modeling period are shown in Figure 14. These estimated projections illustrate that NO_X emissions are estimated to increase 72% in the REF scenario from 2013 till 2035, while an increase of 59%, 53%, and 83% is projected under MRH, MRHN, and MRR scenarios, respectively. The thermal power plants—particularly coal-based—proposed by the GOP during the study period, would be a major source of increased NO_X emissions in the reference scenario. NO_X emissions were also higher in the MRR scenario compared to the other alternative scenarios, due to the large installed capacity of natural gas to compensate the low capacity factor renewable power plants. The power plants based on natural gas produce more NO_X than coal-based power plants.

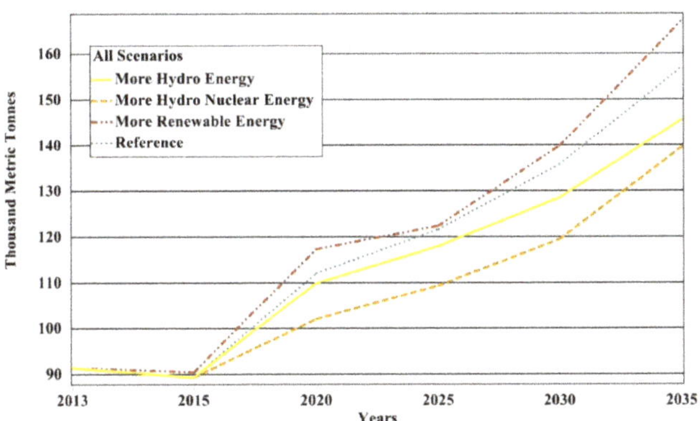

Figure 14. Annual NO$_x$ emissions in all scenarios.

5.2.4. CH$_4$ Emissions

Methane-like CO$_2$ is also another major component of GHGs. It is projected in this study that methane (CH$_4$) emissions from the existing and planned biomass, natural gas, and oil-based power generations plants would increase considerably. CH$_4$ emissions are projected to increase from 1 thousand Mt to 2.3 Mt from the base year (2013) to the end year (2035) in MRR scenario, which is a 137% increase compared to the base year emission of CH$_4$. These emissions are also projected to increase 80% in the REF scenario, 71% in the MRH scenario, and 65% in the MRHN scenario from the base year level in 2013 to the end year of study period as illustrated in Figure 15. The increasing share of CH$_4$ emissions are noticeably high in the MRR scenario compared to the other alternative scenarios due to high shares of biomass and natural gas fuels in the MRR scenario. Biomass fuels like crop residues—bagasse, rice husk, and corn straw—discharge large amounts of CH$_4$ emissions during combustion. CH$_4$ is also the main component of natural gas fuels, and as such, power plants based on natural gas would also emit unburnt CH$_4$.

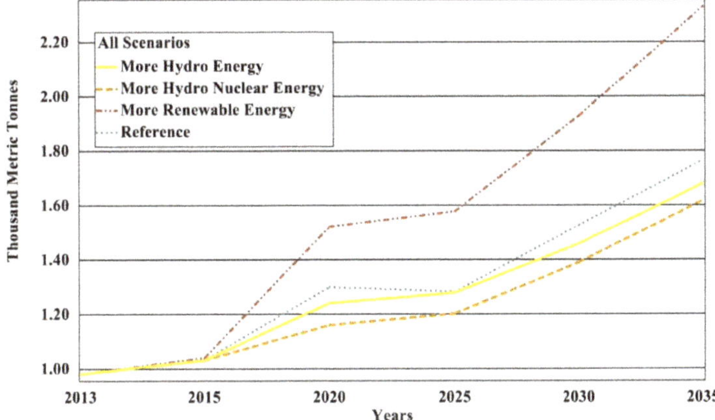

Figure 15. Annual CH$_4$ emissions in all scenarios.

5.2.5. N$_2$O Emissions

N$_2$O is also a GHG and mainly emitted due to the combustion of coal with minimum amount of such emissions from oil and natural gas fuel-based power plants. N$_2$O emissions under reference as well as all alternative scenarios as estimated in this study are shown in Figure 16. N$_2$O emissions are estimated to increase from 0.2 thousand Mt in 2013 to 82 thousand Mt in 2035 under the REF scenario, 0.2 thousand Mt to 70 thousand Mt under the MRH, 0.2 thousand Mt to 63 thousand Mt under the MRHN, and 0.2 thousand Mt to 41 thousand Mt under the MRR scenario for the same period. These emissions are highest in the REF scenario owing to the highest share of coal. In all alternative scenarios, N_2O emissions were comparatively less than the REF scenario. These emissions were lowest in the MRR scenario since a large amount of electricity is generated from renewable energy sources, which do not produce any N_2O emissions.

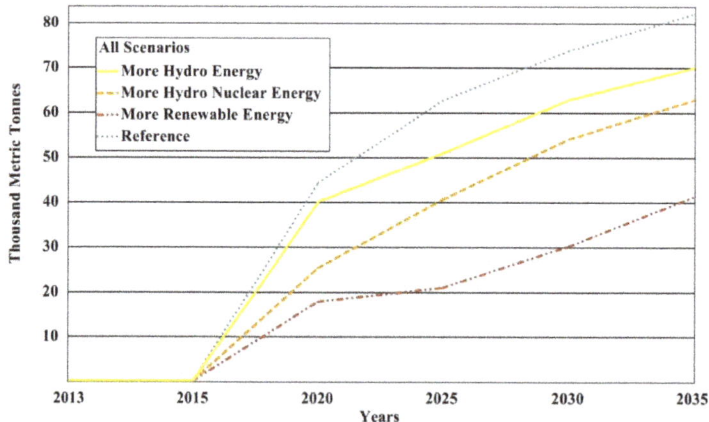

Figure 16. N_2O emissions in all scenarios.

It can be concluded from these results pertaining emission—under different scenarios of this study—that all the alternative scenarios have low overall emissions compared to the reference scenario. However, the MRR scenario has the lowest CO_2 (28% less then reference scenario), SO_2 (33% less than reference scenario), and N_2O (50% less than reference scenario) emissions compared to the reference scenario, whereas the MRHN scenario has the lowest NO_x (11% less than the reference scenario) and CH_4 (11% less than the reference scenario) emissions.

6. Conclusions and Recommendations

The electricity demand in Pakistan is increasing with the increase in population, load growth, economic development, and improved lifestyle. This growing demand cannot be met from existing electricity generating plants. Furthermore, generating electricity from costly imported fossil fuels have plunged the country into existing power crises owing to huge oil import bills and issues of circular debt. In order to improve the electricity demand–supply balance, the government has initiated some positive steps to resolve the electricity crises on priority basis as discussed in Section 3. Modeling the planned power generation capacity of the country was undertaken via the REF scenario and three alternative scenarios: MRR, MRH, and MRHN, which developed using LEAP model. The emphasis of this energy modeling exercise has been to determine and suggest the environment friendly sources of electricity generation in Pakistan for the study period 2013–2035.

The electricity generation output from existing power plants in 2013 remained 96 TWh, which according to this study is projected to be 442 TWh in 2035, based on demand growth rate determined by the government. As such, the addition of new plants in order to meet the increased demand by the year 2035 is inevitable. In summary, the results of the model simulations of this study reveal that:

- The REF scenario based on government plans suggests a reduction of oil-based power generation, however, conversion of certain power plants to coal and addition of new coal-based power plants in this scenario lead to more emission which are highest compared to other alternative scenarios.

- The MRR scenario mainly emphasized the renewable sources for electricity generation wherein the share of renewables increases to be 11% of total electricity output in 2035. This scenario has minimum emissions resulting from power generation during the modeling period.
- The MRH scenario prefers hydroelectric-based power generations to the tune of 42% in the total energy mix by the year 2035. The coal-based generation forms about 28% in this scenario, as such, environmental emissions in the scenario are found to next highest to the REF scenario.
- In the MRHN scenario, the share of electricity output of hydroelectric and nuclear power is projected to increase from 31% to 42% and 5% to 11%, respectively, by the end year 2035, as such, environmental emissions, in this case, are minimum but second to those of the MRR scenario.

It is evident from the above-summarized conclusions of this study that:

- All alternative scenarios (MRR, MRH, and MRHN) are more environmentally friendly and acceptable compared to the base or REF scenario, and
- Coal-based power generation is the major sources of emission with local coal (indigenous lignite) emitting more emissions than imported (bituminous) coal.

Pakistan has a huge potential of renewable energy resources such as wind, solar, and biomass energy alongside hydroelectric potential, which, if effectively utilized can play a major role in meeting the electricity demand and would reduce the GHGs and other air pollutant emissions. It is, therefore, recommended that in order to address the current gap in supply and demand of electricity, decision makers at the government level apart from implementing the existing plans should also consider harnessing indigenous renewable energy resources for electricity generation. Renewable energy resources of country such as wind, solar, and biomass together with hydro and nuclear energy can form a substantial part of the future energy mix of the country to minimize oil import bills, reduce GHG emissions and air pollution appropriately.

Author Contributions: All the authors contributed to this work. A.M., K.H., and M.A.U. conceived and structured the study. A.M. developed the model. N.H.M., G.D.W. and S.A.K. analyzed the results and prepared the preliminary manuscript. A.M., K.H., and M.A.U. reviewed and finalized the manuscript.

Funding: This research received no external funding.

Acknowledgments: The authors highly acknowledge PPSP-USAID, Higher Education Commission Pakistan, Mehran University of Engineering & Technology Jamshoro, Quaid-e-Awam University of Engineering, Sciences & Technology, Nawabshah and Balochistan University of Engineering & Technology, Khuzdar for their financial and technical support for completing this research work.

Conflicts of Interest: The authors declare no conflict of interest.

References

1. Abas, N.; Kalair, A.; Khan, N.; Kalair, A.R. Review of GHG emissions in Pakistan compared to SAARC countries. *Renew. Sustain. Energy Rev.* **2017**, *80*, 990–1016. [CrossRef]
2. Mirjat, N.H.; Uqaili, M.A.; Harijan, K.; Valasai, G.D.; Shaikh, F.; Waris, M. A review of energy and power planning and policies of Pakistan. *Renew. Sustain. Energy Rev.* **2017**, *79*, 110–127. [CrossRef]
3. Mirjat, N.H.; Uqaili, M.A.; Harijan, K.; Walasai, G.D.; Mondal, M.A.H.; Sahin, H. Long-term electricity demand forecast and supply side scenarios for Pakistan (2015–2050): A LEAP model application for policy analysis. *Energy* **2018**, *165*, 512–526. [CrossRef]
4. Perwez, U.; Sohail, A. GHG Emissions and Monetary Analysis of Electric Power Sector of Pakistan: Alternative Scenarios and it's Implications. *Energy Procedia* **2014**, *61*, 2443–2449. [CrossRef]
5. HDIP. *Pakistan Energy Year Book, 2016*; Government of Pakistan: Islamabad, Pakistan, 2016.
6. Lin, B.; Raza, M.Y. Analysis of energy related CO_2 emissions in Pakistan. *J. Clean. Prod.* **2019**, *219*, 981–993. [CrossRef]
7. Kessides, I.N. Chaos in power: Pakistan's electricity crisis. *Energy Policy* **2013**, *55*, 271–285. [CrossRef]
8. Rauf, O.; Wang, S.; Yuan, P.; Tan, J. An overview of energy status and development in Pakistan. *Renew. Sustain. Energy Rev.* **2015**, *48*, 892–931. [CrossRef]

9. Bhutto, A.W.; Bazmi, A.A.; Zahedi, G. Greener energy: Issues and challenges for Pakistan—Wind power prospective. *Renew. Sustain. Energy Rev.* **2013**, *20*, 519–538. [CrossRef]

10. National Electric Power Regulatory Authority (NEPRA). *Pakistan Coal Power Generation Potential, National Electric Power and Regulatory Authority*; NEPRA: Islamabad, Pakistan, 2004.

11. Farooqui, S.Z. Prospects of renewables penetration in the energy mix of Pakistan. *Renew. Sustain. Energy Rev.* **2014**, *29*, 693–700. [CrossRef]

12. Yousuf, I.; Ghumman, A.; Hashmi, H.; Kamal, M. Carbon emissions from power sector in Pakistan and opportunities to mitigate those. *Renew. Sustain. Energy Rev.* **2014**, *34*, 71–77. [CrossRef]

13. Perwez, U.; Sohail, A.; Hassan, S.F.; Zia, U. The long-term forecast of Pakistan's electricity supply and demand: An application of long range energy alternatives planning. *Energy* **2015**, *93*, 2423–2435. [CrossRef]

14. HDIP. *Pakistan Energy Year Book, 2013*; Government of Pakistan: Islamabad, Pakistan, 2013.

15. MOW&P. *National Power Policy 2013*; Ministry of Water and Power, Government of Pakistan: Islamabad, Pakistan, 2013.

16. Water and Power Development Authority (WAPDA). *Hydro Potential Pakistan*; Water and Power Development Authority: Lahore, Pakistan, 2011.

17. Ministry of Finance, Revenue and Economic Affairs. *Economic Survey of Pakistan 2016–2017*; Ministry of Finance, Revenue and Economic Affairs: Islamabad, Pakistan, 2017.

18. China-Pakistan Economic Corridor (CPEC). *China-Pakistan Economic Corridor, Updated Energy Projects*; CPEC, 2016. Available online: http://cpec.gov.pk/energy (accessed on 6 June 2016).

19. BOI. *Coal Based Projects*; Board of Investment Prime Minister's Office, 2014. Available online: http://boi.gov.pk/ (accessed on 14 September 2014).

20. Coutinho, M.; Butt, H.K.; Rauf, S. *Environmental Impact Assessment Guidance for Coal Fired Power Plants in Pakistan, IUCN Pakistan (National Impact Assessment Programme)*; International Union for Conservation of Nature and Natural Resources (IUCN): Gland, Switzerland, 2014.

21. Thar Coal & Energy Board (TCEB). *Projects, Thar Coal Energy Board, Sindh Governement*; TCEB, 2015. Available online: http://sindhcoal.gos.pk/ (accessed on 15 April 2016).

22. Uqaili, M.A.; Hussain, N.; Mengal, A.; Harijan, K. Supply side initiatives for overcoming electricity crisis in Pakistan. In Proceedings of the 14th International Conference on Sustainable Energy Technologies, SET 2015, Nottingham, UK, 25–27 August 2015; pp. 572–580.

23. GOP. *Pakistan Economic Survey 2013–2014*; Ministry of Finance, Finance Division, Government of Pakistan, 2015. Available online: http://www.finance.gov.pk/survey_1314.html (accessed on 15 June 2017).

24. Farooq, M.K.; Kumar, S. An assessment of renewable energy potential for electricity generation in Pakistan. *Renew. Sustain. Energy Rev.* **2013**, *20*, 240–254. [CrossRef]

25. Private Power and Infrastructure Board (PPIB). 2016. Available online: www.ppib.gov.pk (accessed on 16 August 2017).

26. PAEC. Nuclear Power: A Viable Option for Electricity Generation. Available online: http://www.paec.gov.pk/NuclearPower/ (accessed on 26 March 2014).

27. Javaid, M.A.; Hussain, S.; Maqsood, A.; Arshad, Z.; Arshad, A.; Idrees, M. Electrical energy crisis in Pakistan and their possible solutions. *Int. J. Basic Appl. Sci.* **2011**, *11*, 38–52.

28. Alternative Energy Development Board (AEDB), Government of Pakistan. 2014. Available online: http://www.aedb.org/ (accessed on 13 June 2016).

29. PM launches solar power plant in Bahawalpur. *Dawn*, 9 May 2014.

30. Harijan, K.; Uqaili, M.A.; Mirza, U.K. Assessment of solar PV power generation potential in Pakistan. *J. Clean Energy Technol.* **2015**, *3*, 54–56. [CrossRef]

31. Energy Sector Management Assistance Program (ESMAP). *Global Solar Atlas*; ESMAP: Washington, DC, USA, 2018.

32. Harijan, K.; Uqaili, M.A.; Memon, M.; Mirza, U.K. Assessment of centralized grid connected wind power cost in coastal area of Pakistan. *Renew. Energy* **2009**, *34*, 369–373. [CrossRef]

33. Mengal, A.; Uqaili, M.A.; Harijan, K.; Memon, A.G. Competitiveness of wind power with the conventional thermal power plants using oil and natural gas as fuel in Pakistan. *Energy Procedia* **2014**, *52*, 59–67. [CrossRef]

34. Energy Sector Management Assistance Program (ESMAP). *Renewable Energy Resource Mapping in Pakistan*; ESMAP: Washington, DC, USA, 2017; Available online: http://pubdocs.worldbank.org/en/344131489521020746/Solargis-Presentation-Pakistan-Solar-Mapping-Final-Results-07Mar2017.pdf (accessed on 4 April 2017).
35. Zuberi, M.J.S.; Hasany, S.Z.; Tariq, M.A.; Fahrioglu, M. Assessment of biomass energy resources potential in Pakistan for power generation. In Proceedings of the 4th International Conference on Power Engineering, Energy and Electrical Drives, Istanbul, Turkey, 13–17 May 2013; pp. 1301–1306.
36. Perwez, U.; Sohail, A. Forecasting of Pakistan's net electricity energy consumption on the basis of energy pathway scenarios. *Energy Procedia* **2014**, *61*, 2403–2411. [CrossRef]
37. SEI. *Long-range Energy Alternatives Planning System—LEAP*; Stockholm Environment Institute: Stockholm, Sweden, 2014.
38. Meryem, S.S.; Ahmad, N.A.; Aziz, N. Evaluation of biomass potential for renewable energy in Pakistan using LEAP model. *Int. J. Emerg. Trends Eng. Dev.* **2013**, *1*, 243–249.
39. COMMEND. An Introduction to LEAP. 2015. Available online: http://www.energycommunity.org/default.asp?action=47 (accessed on 23 March 2016).
40. NTDC. *Electricity Demand Forecast Based on Multiple Regression Analysis*; National Transmission and Distribution Company, Government of Pakistan: Lahore, Pakistan, 2013.
41. Mariam Gul, W.A.Q. Modeling Diversified Electricity Generation Scenarios for Pakistan. In Proceedings of the IEEE Power and Energy Society General Meeting, San Diego, CA, USA, 22–26 July 2012.
42. IRG. *Pakistan Integrated Energy Model (Pak-IEM) Policy Analysis Report*; International Resources Group: Washington, DC, USA, 2011.
43. Taliotis, C.; Bazilian, M.; Welsch, M.; Gielen, D.; Howells, M. Grand Inga to power Africa: Hydropower development scenarios to 2035. *Energy Strategy Rev.* **2014**, *4*, 1–10. [CrossRef]
44. Van der Zwaan, B. The role of nuclear power in mitigating emissions from electricity generation. *Energy Strategy Rev.* **2013**, *1*, 296–301. [CrossRef]
45. IAEA. *Climate Change and Nuclear Power 2014*; International Atomic Energy Agency: Vienna, Austria, 2014.

Article

Numerical Investigation of Influence of Reservoir Heterogeneity on Electricity Generation Performance of Enhanced Geothermal System

Yuchao Zeng [1,2], Liansheng Tang [1,2,*], Nengyou Wu [3,4], Jing Song [1,2] and Zhanlun Zhao [1,2]

[1] School of Earth Science and Engineering, Sun Yat-sen University, Guangzhou 510275, China; zengyuc@126.com (Y.Z.); songj5@mail.sysu.edu.cn (J.S.); zhaozhlun@mail2.sysu.edu.cn (Z.Z.)
[2] Guangdong Provincial Key Laboratory of Mineral Resources & Geological Processes, Guangzhou 510275, China
[3] Laboratory for Marine Mineral Resources, Qingdao National Laboratory for Marine Science and Technology, Qingdao 266071, China; wuny@ms.giec.ac.cn
[4] Qingdao Institute of Marine Geology, China Geological Survey, Qingdao 266071, China
*   Correspondence: eestls@mail.sysu.edu.cn; Tel.: +86-20-8411-2113

Received: 22 February 2019; Accepted: 5 April 2019; Published: 9 April 2019

Abstract: The enhanced geothermal system (EGS) reservoir consists of a heterogeneous fracture network and rock matrix, and the heterogeneity of the reservoir has a significant influence on the system's electricity generation performance. In this study, we numerically investigated the influence of reservoir heterogeneity on system production performance based on geological data from the Gonghe Basin geothermal field, and analyzed the main factors affecting production performance. The results show that with the increase of reservoir heterogeneity, the water conduction ability of the reservoir gradually reduces, the water production rate slowly decreases, and this causes the electric power to gradually reduce, the reservoir impedance to gradually increase, the pump power to gradually decrease and the energy efficiency to gradually increase. The fracture spacing, well spacing and injection temperature all have a significant influence on electricity generation performance. Increasing the fracture spacing will significantly reduce electric power, while having only a very slight effect on reservoir impedance and pump power, thus significantly decreasing energy efficiency. Increasing the well spacing will significantly increase the electric power, while having only a very slight effect on the reservoir impedance and pump power, thus significantly increasing energy efficiency. Increasing the injection temperature will obviously reduce the electric power, decrease the reservoir impedance and pump power, and thus reduce energy efficiency.

Keywords: reservoir heterogeneity; enhanced geothermal system; electricity generation; performance; influence

---

1. Introduction

1.1. Background

The enhanced geothermal system (EGS) adopts artificial circulating water to extract heat from the fractured hot dry rock (HDR) at a depth of 3–10 km, and is an effective approach to exploiting the high-temperature geothermal energy stored deep in the earth [1]. All over the world, the total EGS resource reserves within a 10km depth amounts to about 40–400 MEJ (1 EJ = 10^{18} J), approximately 100–1000 times the quantity of fossil energy [2]. In China, total EGS resource reserves within a 3–10 km depth amounts to 20.90 MEJ; if the recoverable fraction is taken as 2%, the recoverable EGS resource amounts to 4400 times the total annual energy consumption of China in 2010 [3]. Compared with other renewable energy sources, the EGS resource is very suitable for generating base-load electric power, with nearly no pollution emissions and with a high utilization efficiency [1]. It is predicted that there will be commercial exploitation of EGS in the next 15 years, with large scale utilization of EGS to generate electricity by 2030 [1]. It is suggested that EGS will provide about 100,000 MW electric power by 2050 in the USA and this will occupy about 10% of total electricity generating capacity [1].

The field tests and experimental studies of EGSs are time-consuming, expensive and very difficult, while numerical studies are very fast, lower cost and easy, thus numerical simulation studies of EGS have received more and more attention all over the world and have made important progress in recent years. Two issues need to be considered in the simulation of EGS reservoirs: fracture representation and simplification of the coupled hydraulic-thermal-mechanical-chemical interaction between the fractured rock and circulating water [4,5]. For fracture representation, there are two main types of method: the equivalent continuum method (ECM) and the discrete fracture network (DFN) method [4,5]. The ECM will regard the actually discrete and interconnected fracture network as continuous porous media and use the mature theories of fluid flow in porous media to describe the water seepage and heat transfer process in the fractured rocks, and this method is mainly used for densely fractured reservoirs [6–10]. The commonly used ECM includes the equivalent porous media (EPM) method, the double-porosity method (DPM) and the multiple interacting continua (MINC) method. The DFN method considers the fracture orientation, spacing and other mechanical properties to establish a fracture network model [4–10]. For simplification of the multiphysics field, recently, the coupling between water flow and heat transfer has been most important and most considered, while the models considering the coupling among the hydraulic, thermal and mechanic effects are increasing [4–10].

The EPM method is mainly used to model densely fractured reservoirs where fracture spacing is small and fracture density is high, especially for average fracture spacing less than 2–3 m [4–10]. Birdsell et al. used the EPM method to develop a three-dimensional model of fluid, heat and tracer transport in the Fenton Hill HDR reservoir [11]. McDermott et al. used the EPM method to analyze the influence of coupled processes on differential reservoir cooling in heat extraction from crystalline rocks [12]. Watanabe et al. used the EPM method to study the uncertainty of thermo-hydro-mechanical coupled processes in heterogeneous porous media [13]. Zeng et al. adopted the EPM method to analyze the electricity generation potential from the EGS reservoirs at Desert Peak geothermal and Yangbajing geothermal field [5,14–19]. Cheng et al. employed the EPM method to analyze the influencing factors of heat extraction from EGSs considering water losses [20]. Based on the EPM method, Hu et al. established a novel fully-coupled flow and geomechanics model in EGS reservoirs [21].

When the average fracture spacing is higher than 10 m, we must consider the temperature difference between rock and water, and the DPM or MINC method is more reasonable for this [4,5]. Sanyal et al. employed the DPM method to analyze the power generation prospects of EGS at the Desert Peak geothermal field [22]. Taron et al. used the DPM method to study the hydrologic-thermal-mechanical-chemical processes in the EGS reservoir [23,24]. Gelet et al. adopted the DPM method to establish a hydro-thermo-mechanical coupled model in local thermal non-equilibrium for fractured HDR reservoirs and found that fluid loss is high initially and decreases over time [25,26]. Benato et al. used the DPM simulator TFReact to analyze the mechanisms influencing permeability evolution during the reservoir stimulation and circulation at Desert Peak geothermal field [27]. Pruess et al. used the MINC method to evaluate the heat extraction rate from EGS reservoirs where the heat transmission fluid is either CO_2 or water [28,29]. Spycher et al. used the MINC method to establish a phase-partitioning model for CO_2-Brine mixtures at elevated temperatures and pressures and applied it to CO_2-EGSs [30]. Borgia et al. used the MINC method to analyze salt precipitation in the fractures of a CO_2-EGS [31]. Xu et al. used the MINC method to calculate the power generation potential of an EGS by water circulating through two horizontal wells in the Gonghe Basin geothermal field [32]. Zeng et al. employed the MINC method to compute the electricity generation potential at the Yangbajing geothermal field [33].

If the data from reservoir fracture distribution are adequate, the DFN model can be adopted, and the use of the DFN method has been increasing recently [4–10]. Baujard et al. used the DFN model to study the impact of fluid density on the pressure distribution and stimulated volume in the Soultz HDR reservoir and found that the density difference between the in situ reservoir fluid and the injected fluid might play a significant role in the hydraulic stimulation of the reservoir [34]. Kolditz et al. used the DFN model to study fluid flow and heat transfer in fractured crystalline rocks in Rosemanowes HDR reservoir and they also made long-term predictions of the thermal performance of HDR systems [35,36]. Jing et al. adopted the stochastic DFN model to study the heat extraction performance of EGS and found that rock thermoelasticity has an obvious effect on the production temperature [37–39]. Based on the DFN model, Sun et al. studied heat extraction in EGS with the hydraulic-thermal-mechanical coupling method and the results show the significance of taking into account the hydraulic-thermal-mechanical (HTM) coupling effect when investigating the performance and efficiency of EGS [40,41].

Though much important progress has been made in recent years, most fractured reservoirs represented by the ECM are homogeneous and the reservoir heterogeneity is not taken into account [4–10]. In fact, because the formation is usually layered and the hydrofracture effect is commonly heterogeneous, the EGS reservoirs are generally heterogeneous [1]. A report from Huang et al. [42] has indicated that reservoir heterogeneity has a significant influence on the heat production performance of EGS, but they only discussed the quantitative relation between the heat extraction ratio and the reservoir heterogeneity. There are quantitative relationships between the electric power, flow impedance, energy efficiency and the reservoir heterogeneity, however recently there is a lack of deep and systematic studies on these quantitative relations [4–10]. In order to analyze the influence of reservoir heterogeneity on the electricity generation performance of the EGSs, in this work we established the numerical model of the EGSs and discussed the influence of reservoir heterogeneity on electricity generation performance in detail based on the geological data of the Gonghe Basin geothermal field [32]. These will lay a good foundation for future research and development of EGSs at the Gonghe Basin geothermal field.

1.2. Research Objectives

The research objectives of this work are to establish a numerical model of EGS with heterogeneous reservoir and to reveal the influence of reservoir heterogeneity on system electricity generation performance.

The novelty of this work is in the following three features. First, we used the MINC method to represent the fractured reservoir and the temperature difference between circulating water and rock matrix was taken into consideration. Second, the layered EGS reservoir with heterogeneous permeability was considered and the corresponding numerical model was established. Third, through a comparison with a homogenous reservoir we examined the impact mechanism of the reservoir heterogeneity on system production performance.

2. Electricity Generation Method and Well Design

2.1. Heat Production Method

In order to deeply analyze the influence of reservoir heterogeneity on system production performance, in this work we considered a five-spot well configuration at the Gonghe Basin geothermal field to mine the heat—namely four production wells at corners and one injection well in the center, as shown in Figure 1. The distance is 1000 m between two adjacent production wells. Only one quarter of the domain needs to be simulated due to symmetry. This kind of well configuration is usually used to analyze heat mining performance [28–31]. Based on the geological data at Gongbe Basin geothermal field, in this work we aimed to mine the heat at a depth of 2700–3200 m [32]. The vertical wells are perforated over the whole reservoir height of 500 m to obtain maximum water production rate and thermal power. As shown in Figure 2, along the circumference of the vertical wells there are 8 grooves evenly distributed and previous studies have shown that this kind of well design can obtain a much higher mining efficiency [14–19]. The whole injection rate or production rate is distributed across the four gridblocks where the well is located, thus one gridblock will only represent the production rate through two grooves, namely one quarter of the whole production rate [14]. We used the constant water production rate method to mine the heat in the fractured rocks. For the production well, we installed a downhole pump to maintain the bottomhole production pressure P_{pro} at a constant; for the injection well, we installed an injection pump to maintain the injection rate q at a constant. Field tests and numerical simulations all show that this kind of injection and production method can greatly reduce reservoir impedance and water losses [4,5]. When injecting cold water into the fractured reservoir, the formation pressure will rise. To avoid second reservoir growth and water losses, P_{inj} must be lower than the minimum reservoir principal stress [5]. Based on the experience from the oil and gas industry, in order to avoid the second fracture growth P_{inj} must be kept below an upper limit P_{max} [5]:

$$P_{inj} < P_{max} \tag{1}$$

where $P_{max} = fP_{W0}$; P_{W0} is the initial pressure of the wellbore; $f = 1.2$ is the safety factor and is determined by the actual geologic conditions [5]. At the Gonghe Basin geothermal field, the initial wellbore pressure of the injection well at the intermediate depth of 2950 m is $P_{W0} = 29.20$ MPa, so the $P_{max} = 35.04$ MPa in this work. The initial wellbore pressure of the production well at the intermediate depth of 2950 m is 29.20 MPa. Based on the engineering data at Desert Peak geothermal field, under current pump technology the maximum pressure drawdown in the bottomhole production well is 3.40 MPa [5]. Therefore, in this work the minimum bottomhole production pressure P_0 at the production well is (29.20 − 3.40) = 25.80 MPa, and the production pressure P_{pro} is decreased to 25.80 MPa to maintain continuous production.

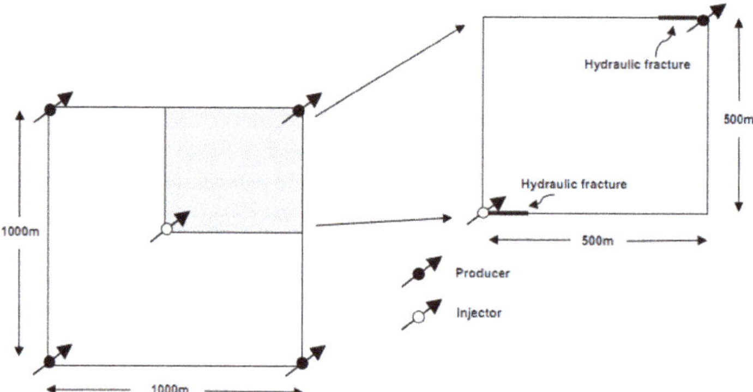

Figure 1. Five-spot well pattern for the 2700–3200 m fractured reservoir.

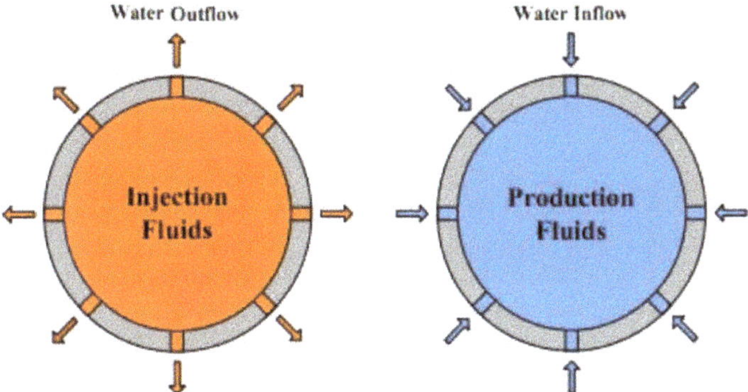

Figure 2. Well design used in the five-spot enhanced geothermal system (EGS) at the Gonghe Basin geothermal field.

2.2. Electricity Generation Method

So far, the commonly used methods for electricity generation in the geothermal industry include the dry steam system, the flash system and the binary system, and the applicable conditions and efficiency of each system are all different [1]. The dry steam system is mainly used for high-temperature geothermal resources where the geothermal fluid is mainly in the form of steam. The flash system and binary system are mainly used for medium and low temperature geothermal resources where the geothermal fluid is mainly in the form of hot liquid. The factors affecting the conversion efficiency of the geothermal power plant includes production temperature, system design, heat loss from equipment, non condensable gases (NCG) content, turbines and generator efficiency and other factors [43]. At the Gonghe Basin geothermal field, the average temperature of the fractured formation at a depth of 2700–3200 m is about 180 °C, the most suitable method is the binary system according to reports from Xu et al., so in this numerical study we used the binary system to calculate the production performance [32].

The scheme of the basic binary geothermal power plant at Gonghe Basin geothermal field was reported by Xu et al. in Reference [32]. Based on the studies from Zeng et al., the optimized injection temperature for the circulation system is 60 °C [5]. Neglecting the temperature drop when the water flows from the production well to the power plant, the production temperature T_{pro} is regarded as the inlet temperature for the power plant. The mean annual temperature in the Gonghe Basin is 4.1 °C [32], thus the heat rejection temperature of $T_0 = 277.25$ K is used for electric power calculation. Based on the second law of thermodynamics, the fraction of the total produced heat that can be converted to the maximum mechanical work f_R can be calculated as Equation (2), in which the T_0 and T_{pro} are all absolute temperature.

$$f_R = 1 - \frac{T_0}{T_{pro}} \tag{2}$$

3. Numerical Method

3.1. Mathematical Model

Because the pressure in the fractured reservoir is great enough, the water remains in the liquid state when temperature is at 180 °C, thus it is water saturated single liquid flow in the fractured formation. The mass conversation equation is (3), where ρ is water density, ϕ reservoir porosity, V the velocity vector.

$$\frac{\partial(\rho\phi)}{\partial t} + \nabla \cdot (\rho V) = 0 \tag{3}$$

The momentum conservation equation for single liquid flow is the classical Darcy's law, as Equations (4)–(6), where V_x, V_y and V_z is velocity component along the x, y and z direction, K_x, K_y and K_z is reservoir permeability along the x, y and z direction, μ the dynamic viscosity, P the pressure, $g = 9.80$ m/s^2 the acceleration of gravity.

$$V_x = -\frac{K_x}{\mu}\frac{\partial p}{\partial x} \tag{4}$$

$$V_y = -\frac{K_y}{\mu}\frac{\partial p}{\partial y} \tag{5}$$

$$V_z = -\frac{K_z}{\mu}\left(\frac{\partial p}{\partial y} + \rho g\right) \tag{6}$$

Assuming that the rock matrix and circulating water is in local thermodynamic equilibrium, namely the rock temperature is equal to the water temperature, in the rock the heat transfer is conduction, and in the water the heat transfer is convection and conduction, thus the energy conservation equation for single liquid flow is Equation (7):

$$[\phi(\rho c_p)_f + (1-\phi)(\rho c_p)_s]\frac{\partial T}{\partial t} + (\rho c_p)_f(V\cdot\nabla)T = [\phi k_f + (1-\phi)k_s]\nabla^2 T \tag{7}$$

where T is the temperature, $(\rho c_p)_f$ is product of water density and water specific heat capacity, $(\rho c_p)_s$ is product of solid density and solid specific heat capacity. c_p is specific heat capacity, k_f is water heat conductivity, and the k_s is solid heat conductivity. Because in this work the variation of reservoir pressure and temperature is very great and its influence on water density and viscosity is significant, we considered that the water density and viscosity are functions of pressure and temperature, as Equations (8) and (9). In this paper, the ρ and the μ are calculated from steam table equations as given by the International Formulation Committee [44]. For more information about the state equations, the reader can refer to Reference [44].

$$\rho = \rho(p, T) \tag{8}$$

$$\mu = \mu(p, T) \tag{9}$$

3.2. The MINC Method

For EGS reservoirs with large fracture spacing, the MINC method is an effective approach for modeling fluid flow and heat transfer. In the fractured reservoirs, matrix blocks with low permeability are embedded in the fracture network, and fluid flow mainly occurs through the fractures [44]. For describing the fluid and heat transport process in the fractured media, it is necessary to resolve the driving temperature, pressure and mass fraction gradients at the matrix/fracture interface. In the MINC approach, the pressure and temperature changes in the matrix are controlled locally by the distance from the fractures. Based on Xu et al. [32], in this work the matrix blocks are divided into four subgrids with volume fractions of 0.08, 0.2, 0.35 and 0.35. The fracture domain occupies a volume fraction of 0.02. In this work, the TOUGH2-EOS1 codes are employed to carry out the simulation. For more information about the MINC method and the TOUGH2 codes, the reader can refer to Reference [44].

3.3. Domain, Grid and Parameters

As stated above, in this work we mainly aim to exploit the heat stored in the fractured reservoir at a depth of 2700–3200 m and the distance between two adjacent production wells is 1000 m, as shown in Figure 1. Because of symmetry, only one quarter of the whole domain needs to be simulated, thus the actual calculated domain is 500 m $\times$ 500 m $\times$ 500 m. As shown in Figure 3, in the horizontal direction the grid that is within 50 m of the wells is refined, the width of every gridblock is 5 m and there are

10 gridblocks near the wells. For the other subdomain that is not near the wells, the 400 m length is evenly divided into 8 gridblocks, and the width of every gridblock is 50 m. Therefore, there are a total of 28 gridblocks in the x direction and the y direction. In the vertical direction, the 500 m height is evenly divided into 10 gridblocks, and the height of every gridblock is 50 m. With this grid arrangement, in the simulated domain there are total $28 \times 28 \times 10 = 7840$ gridblocks. Based on Section 3.2, the reservoir domain is further divided into 5 continua for the MINC method, thus there are total $7840 \times 5 = 39{,}200$ gridblocks for the whole calculated domain. Under reference condition, we assumed that after hydrofracture, the fracture spacing is 50 m [32]. According to Pruess et al. [29], the conductive heat transfer between the impermeable cap rock or base rock and the permeable reservoir can be neglected for the fracture spacing of 50 m, thus in this work the conductive heat transfer between the confined rocks and the reservoir is neglected and only the heat transfer process within the reservoir is considered in this simulation. Assuming the surrounding rocks are impermeable, the water loss can be neglected [5], and the water injection rate q_{inj} is equal to the water production rate q_{pro}:$q_{inj} = q_{pro}$. Neglecting the water loss can greatly simplify the calculation of reservoir performance, and this has been adopted from previous studies by Zeng et al. and Xu et al. [4,5,32].

In order to investigate the influence of reservoir heterogeneity on system electricity generation performance, under reference conditions we considered three permeability distribution patterns, as shown in Figure 4. In this work, we assumed that the permeability is independent of the porosity, namely the relationship between permeability and porosity is not taken into account. There are 10 layers in the vertical direction, and the average permeability of the 10 layers is maintained at 50 mD ($1 \text{ mD} = 1.0 \times 10^{-15} \text{ m}^2$). For the first case R1, we considered a homogenous reservoir of uniform permeability, and the permeability of every layer is 50 mD. For the second case R2, we considered a heterogeneous reservoir, in which the permeability of the 5th and 6th layer is 200 mD, and the permeability of the remaining 8 layers is 12.50 mD. For the third case R3, we considered a heterogeneous reservoir, in which the permeability of 3rd layer, 4th layer, 7th layer and 8th layer is 100 mD, and the permeability of the remaining 6 layers is 16.67 mD. We can easily find that the R1 reservoir is homogenous, the R2 reservoir is most heterogeneous, and the heterogeneity of the R3 reservoir is in between. Namely, the ranking for the heterogeneity of the three cases is: R2 > R3 > R1. Based on the studies from Xu et al. at the Gonghe Basin geothermal field, the fracture porosity is taken as 0.5 [32]. The other model parameters are listed in Table 1, in which most are referred to work of Xu et al. [32].

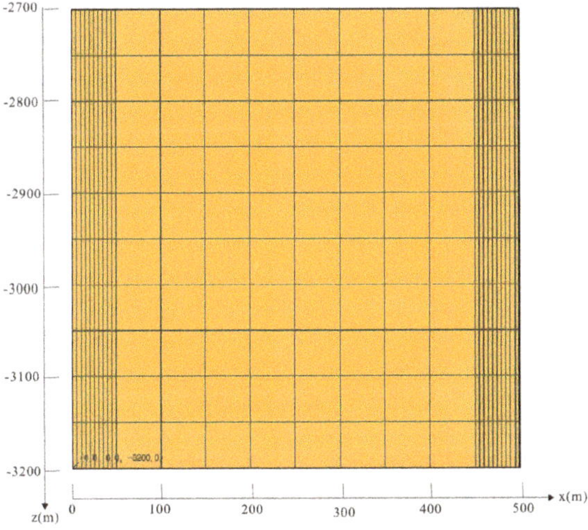

Figure 3. Simulation domain and grid used in this work.

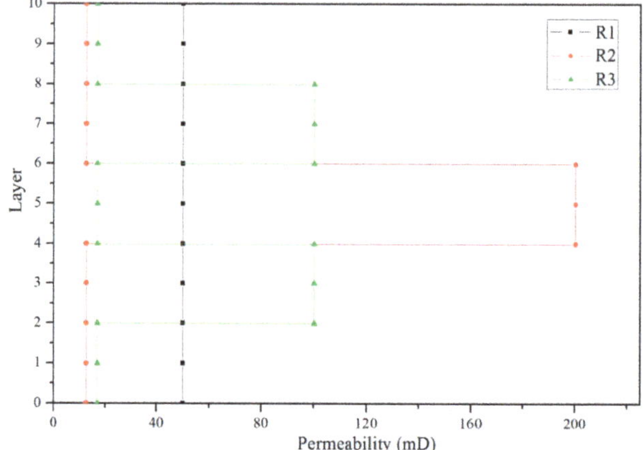

Figure 4. The permeability distribution pattern for the three cases: R1, R2 and R3.

Table 1. EGS reservoir properties and conditions at the Gonghe Basin geothermal field [32].

Parameter	Value
Rock grain density	2650 kg/m^3
Rock specific heat	1000 J/(kg·K)
Rock heat conductivity	2.50 W/(m·K)
Fracture system volume fraction	2%
Fracture spacing	50 m
Porosity in fracture system	0.5
Porosity in matrix	1.0×10^{-5}
Permeability in matrix	1.0×10^{-18} m^2
Injection temperature	60 °C (275.571 kJ/kg)
Bottomhole production pressure	25.80 MPa
Productivity index	5.0×10^{-12} m^3

3.4. Boundary and Initial Conditions

Neglecting the conductive heat transfer between the cap rock or base rock and the reservoir and assuming that the surrounding rocks are impermeable, the topmost and bottommost boundaries in Figure 3 are all no-flow for mass and heat. Because of symmetry, the lateral boundaries in Figure 3 are also all no-flow for mass and heat. The initial pressure is $P = -0.0088z + 3.24$(MPa), and the initial temperature is 180 °C [32].

4. Influence of Reservoir Heterogeneity on the Electricity Generation Performance

4.1. The Determination of Water Production Rate

In previous studies, we have clearly stated the determination method for the water production rate [4]. A lower water production rate will decrease electric power and reduce the practical application value of the system. With an increase of the water production rate, the injection pressure gradually increases; however, for given reservoir conditions, the tolerable maximum injection pressure is finite. Based on (1), we gradually increased the water production rate and calculated the corresponding injection pressure under various conditions. With the increase of water production rate, the production temperature T_{pro} rapidly declines [4,5]. For engineering applications, during the exploiting period the drop of the production temperature T_{pro} should be less than 10%, or the decline of reservoir temperature will be too great, and this will affect the regenerability of the geothermal resource [4,5]. As stated above, based on the two principles that the maximum injection pressure must be lower than

P_{max} = 35.04 MPa and the production temperature drop must be less than 10%, we can determine the available maximum water production rate, and under this water production rate the system can obtain maximum electric power. Because of symmetry the total water production rate and thermal power of the five well system is 4 times that of the simulated domain. Figure 5 shows the change of the injection pressure P_{inj} corresponding to the three reference cases and Figure 6 shows the change of the production temperature T_{pro} corresponding to the three reference cases. For the R1 reservoir, the water production rate for the simulated domain is 40 kg/s, thus the total water production rate of the five well system is 160 kg/s. For the R2 reservoir, the water production rate for the simulated domain is 17.50 kg/s, thus total water production rate of the five well system is 70 kg/s. For the R3 reservoir, the water production rate for the simulated domain is 30 kg/s, thus total water production rate of the five well system is 120 kg/s. It can be easily found that with the increase of the reservoir heterogeneity, namely R1 < R3 < R2, the corresponding water production rate gradually decreases. Though the average permeabilities of the three reference cases are the same, for the heterogeneous reservoir, the permeabilities of the various layers are different and with the increase of the reservoir heterogeneity, the fluid conduction ability of the reservoir is decreasing, thus the available water production rate gradually declines. It can be seen in Figure 5 that with the increase of the reservoir heterogeneity—namely R1 < R3 < R2—the corresponding injection pressure gradually decreases. As stated above, the injection pressure is mainly determined by the water production rate and a higher water production rate means a higher injection pressure. Because with the increase of reservoir heterogeneity—namely R1 < R3 < R2—the water production rate gradually decreases, thus the corresponding injection pressure gradually declines. It can also be seen in Figure 5 that during the period that the injection pressure is gradually increasing, the maximum injection pressure under the three reference cases are all lower than P_{max} = 35.04 MPa. This is mainly due to the decline of the reservoir temperature, causing an increase of the water viscosity and the rise of reservoir impedance [4,5]. This is in agreement with previous studies by Zeng et al. [4,5]. Figure 6 shows that changes in the production temperature under three cases are consistent for the determined water production rate, and the production temperature decreases to about 162 °C, a reduction of about 10%. This indicates that the reservoir heterogeneity has only a very slight influence on the production temperature for the determined water production rate. Though reservoir heterogeneity increases local ability to conduct water, it decreases the global ability to conduct water and heat, thus it has only a very slight influence on the production temperature when reducing the water production rate.

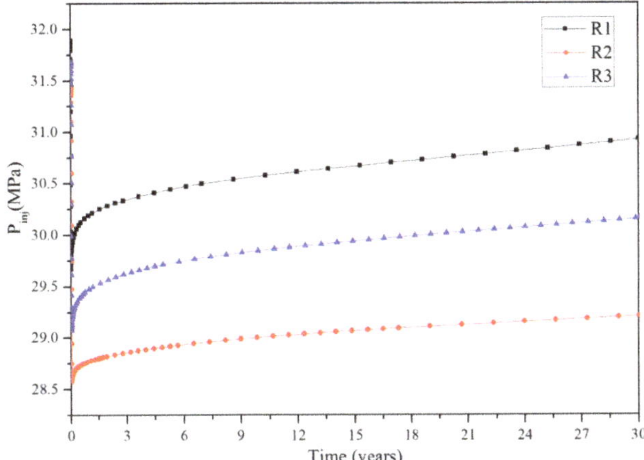

Figure 5. Change of the injection pressure P_{inj} corresponding to the three reference cases.

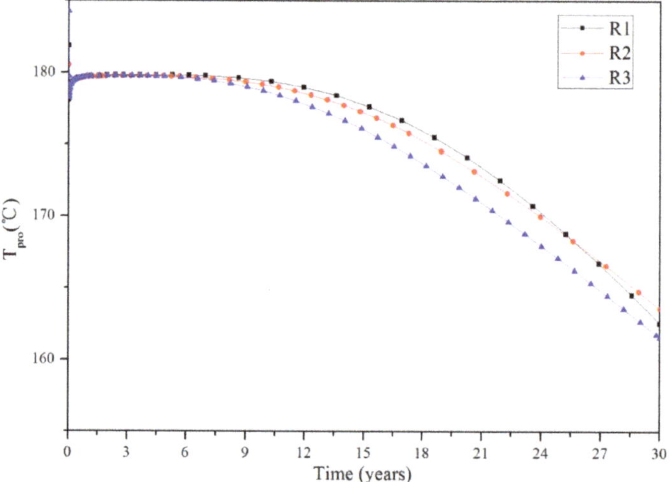

Figure 6. Change of the production temperature T_{pro} corresponding to the three reference cases.

4.2. Influence of Reservoir Heterogeneity on the Electric Power

For the binary system, if the water production rate of the simulated domain is q, the total water production rate for the five well system is Q = 4q. In a more realistic environment, it is very likely that there will be a short circuit between the injection well and one of the production wells; therefore, that production well will produce more than the rest. For simplification, in this work we have assumed there is no short circuit between the injection well and production wells, and that the water production rate of each production well is equal. If the injection specific enthalpy of the injection water is h_{inj}, the production specific enthalpy is h_{pro}, then the thermal power of the system W_{h} can be calculated by Equation (10), where the temperature drop, when flowing from production well to power plant, is neglected [5]. The h_{pro} is calculated according to the bottomhole production temperature and pressure: $h_{\text{pro}} = h_{\text{pro}}(P_{\text{pro}}, T_{\text{pro}})$. As stated above, h_{inj} = 275.571 kJ/kg and is corresponding to T_{inj} = 60 °C.

$$W_{\text{h}} = Q(h_{\text{pro}} - h_{\text{inj}}) = 4q(h_{\text{pro}} - h_{\text{inj}}) \tag{10}$$

Based on Equation (2), the fraction of the total produced heat that can be converted to the maximum mechanical work is f_R. If the utilization efficiency of the maximum mechanical work transferred to electric power is 0.45 [4,5], the electric power W_{e} of the EGS power plant can be calculated as Equation (11). As stated above, at the Gonghe Basin geothermal field, T_0 = 277.25 K.

$$W_{\text{e}} = 0.45 f_R W_{\text{h}} = 0.45 Q(h_{\text{pro}} - h_{\text{inj}})\left(1 - \frac{T_o}{T_{\text{pro}}}\right) = 1.8q(h_{\text{pro}} - h_{\text{inj}})\left(1 - \frac{T_o}{T_{\text{pro}}}\right) \tag{11}$$

Figure 7 shows the change in electric power over 30 years under three reference cases. Based on previous studies, the change of the electric power of the EGS power plant can be divided into two stages: a stable stage and a declining stage [4,5]. During the stable stage, T_{pro} maintains an initial reservoir temperature and the corresponding electric power is also maintained unchanged; during the declining stage, the T_{pro} gradually declines and the corresponding electric power also gradually reduces based on Equation (11). It can be easily seen in Figure 7 that for the R1 reservoir, the electric power W_{e} is highest, and over the 30 years, the W_{e} gradually decreases from 13.97 MW to 11.20 MW. For the R3 reservoir, the W_{e} is in between, and it gradually decreases from 10.45 MW to 8.29 MW. For the R2 reservoir, the W_{e} is lowest, and it slowly reduces from 6.10 MW to 4.97 MW during the 30 years. According to Figure 6, the T_{pro} are very close in the three reference cases, thus the main factor affecting the W_e is the water production rate q based on Equation (11). For the R1, R3 and R2 reservoir, the q is 40 kg/s, 30 kg/s and 17.50 kg/s, respectively. The q gradually decreases, thus the corresponding W_e also reduces. This indicates that with the increase of the reservoir heterogeneity, namely R1 < R3 < R2, the water conduction ability of the reservoirs decreases, the corresponding water production rate reduces and the electric power declines. It is clear that during the reservoir stimulation stage, we should control the reservoir to be uniformly stimulated and make the permeability of every layer the same and this will greatly improve the water conduction ability of the reservoir and increase the electric power.

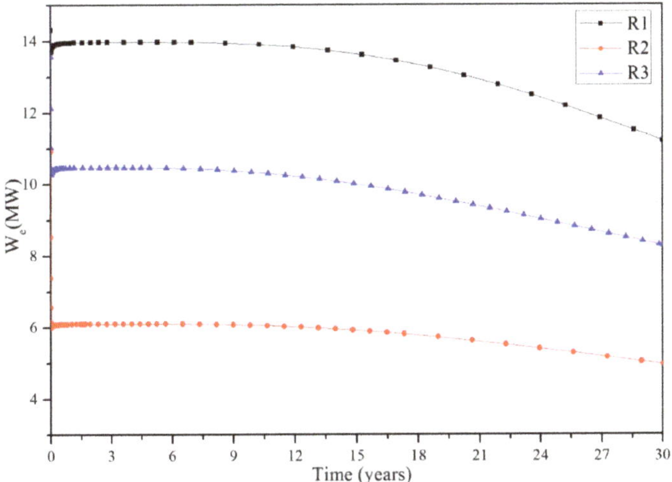

Figure 7. Change of the electric power during the 30 years under the three reference cases.

4.3. Influence of Reservoir Heterogeneity on the Reservoir Impedance

The water flow impedance of the system can be calculated as Equation (12), where P_{pro} = 25.80 MPa. In previous studies, Zeng et al. has analyzed the main factors influencing the flow impedance [4,5], and found that with heat mining the reservoir temperature gradually declines, the water viscosity slowly increases, and this causes the gradual increase of the reservoir impedance.

$$I_R = \frac{P_{inj} - P_{pro}}{q} \tag{12}$$

Figure 8 shows the change of the reservoir impedance over the 30 years under the three reference cases. For the R1 reservoir, over the 30 years the reservoir impedance is lowest, and it gradually increases from 0.097 MPa/(kg/s) to 0.128 MPa/(kg/s). For the R3 reservoir, the reservoir is in between, and it slowly increases from 0.110 MPa/(kg/s) to 0.145 MPa/(kg/s) over the 30 years. For the R2 reservoir, the reservoir impedance is highest, and it increases from 0.160 MPa/(kg/s) to 0.195 MPa/(kg/s) over the 30 years. These indicate that with the increase of the reservoir heterogeneity, namely R1 < R3 < R2, the reservoir impedance increases. As stated above, this is mainly because the reservoir heterogeneity significantly reduces the global water conduction ability of the reservoir; when the average permeability of the various layers is constant, the water conduction ability of the reservoir decreases with the increase of the reservoir heterogeneity, thus the reservoir impedance increases with an increase in the reservoir heterogeneity. In this work, the rock deformation due to mechanics and thermoelasticity is not taken into consideration in the simulation. However, in factual EGS reservoir, as the reservoir cools, the fractures may dilate, and this increases permeability and reduces inter-well impedance.

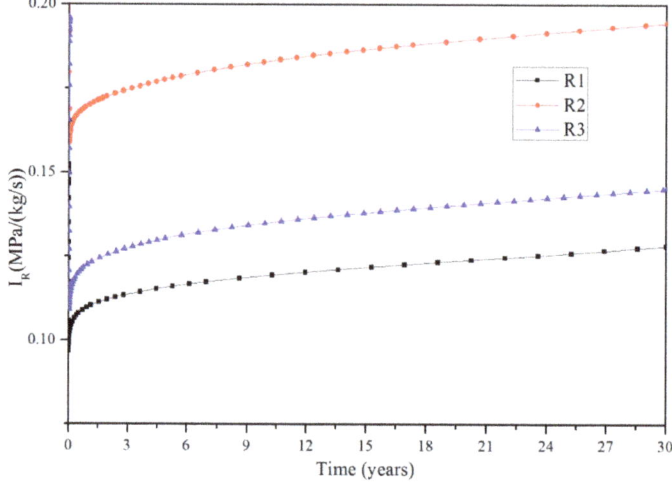

Figure 8. Change of the reservoir impedance during the 30 years under the three reference cases.

4.4. Influence of Reservoir Heterogeneity on the Pump Power

The internal energy consumption $W_p = W_{p1} + W_{p2}$, includes mainly the energy consumed by the injection pump W_{p1} and the production pumps W_{p2} [4,5]:

$$W_{p1} = \frac{4q(P_{inj} - \rho g h_1)}{\rho \eta_p} \tag{13}$$

$$W_{p2} = \frac{4q(\rho g h_2 - P_{pro})}{\rho \eta_p} \tag{14}$$

where h_1 is the depth of the injection well, h_2 is the depth of production well, and $\eta_p = 80\%$ is the pump efficiency [4,5]. Based on these the internal energy consumption W_p is Equation (15):

$$W_p = W_{p1} + W_{p2} = \frac{4q(P_{inj} - P_{pro}) - 4\rho q g(h_1 - h_2)}{\rho \eta_p} \tag{15}$$

In this work, $h_1 = h_2 = 3200$ m, thus the Equation (15) of W_p can be reduced into Equation (16):

$$W_p = W_{p1} + W_{p2} = \frac{4q(P_{inj} - P_{pro})}{\rho \eta_p} \tag{16}$$

In Equation (16), the water density ρ is determined by the reservoir pressure and temperature. When the pressure is within 25.80–31.00 MPa and the temperature is within 60–180 °C, the maximum value of the water density is 996.25 kg/m^3, the minimum value of the water density is 902.61 kg/m^3, thus the average value of the water density is $\rho = 949.43$ kg/m^3. Based on previous studies from Zeng et al. [4,5], adopting the average value of the water density in Equation (16) is accurate and reliable, thus in this work we used the average density of $\rho = 949.43$ kg/m^3 for calculation and analysis. Figure 9 shows the change of the pump power W_p during the 30 years under the three reference cases. Based on Equation (16), during the mining period the P_{inj} is increasing, thus the W_p is also gradually rising, and this is in accordance with Figure 9. For the R1 reservoir, the W_p is highest and it gradually increases from 0.82 MW to 1.08 MW. For the R3 reservoir, the W_p is in between and it gradually increases from 0.52 MW to 0.69 MW. For the R2 reservoir, the W_p is lowest and it gradually rises from 0.26 MW to 0.31 MW. So, we can see that with an increase of reservoir heterogeneity—namely R1 < R3

< R2—the W_p gradually decreases. Based on Equation (16), this is mainly because, with the increase of reservoir heterogeneity, the available water production rate declines. Therefore, the water conduction ability reduction caused by the increase of the reservoir heterogeneity can significantly influence the pump power.

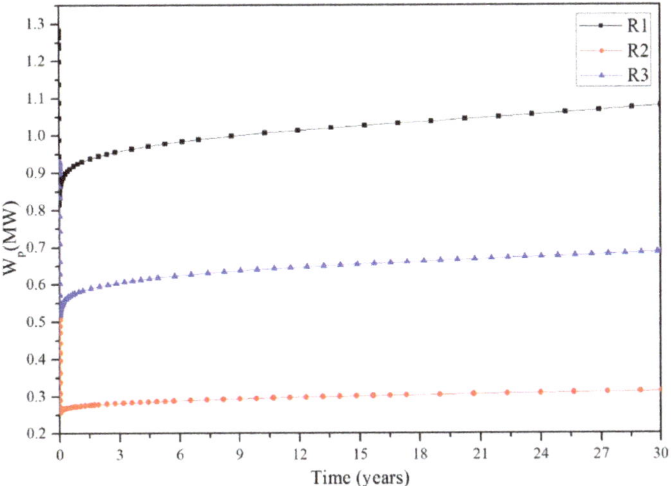

Figure 9. Change of the pump power during the 30 years under the three reference cases.

4.5. Influence of Reservoir Heterogeneity on the Energy Efficiency

The energy efficiency η of the system is defined as the ratio of the total produced electric energy to the internal energy consumption, and can be calculated as Equation (17):

$$\eta = \frac{W_e}{W_p} = \frac{0.45\rho\eta_p(h_{pro} - h_{inj})(1 - T_o/T_{pro})}{(P_{inj} - P_{pro}) - \rho g(h_1 - h_2)} \tag{17}$$

In the calculation of Equation (17), the water density is still taken as the average value of ρ = 949.43 kg/m^3. Figure 10 shows the change of the energy efficiency during the 30 years under the three reference cases. Based on Equation (17), during the heat mining because the T_{pro} and h_{pro} gradually decreases while the P_{inj} gradually increases, the energy efficiency η gradually reduces, and this is in agreement with Figure 10. For the R1 reservoir, the η is lowest and it decreases from 20.62 to 10.36. For the R3 reservoir, the η is in between and it reduces from 23.13 to 12.04. For the R2 reservoir, the η is highest and it decreases from 24.05 to 15.83. So, it can be found that with the increase of the reservoir heterogeneity—namely R1 < R3 < R2—the η rises. Though the reservoir heterogeneity significantly reduces the water conduction ability of the reservoir, under a lower water production rate the system obtains a higher energy efficiency. These are in accordance with previous studies from Zeng et al. [4,5]. Though the heterogeneous reservoir can obtain higher energy efficiency, it decreases the water production rate and also electric power is reduced, thus the economic benefit of the heterogeneous reservoir is still lower than that of the homogenous reservoir.

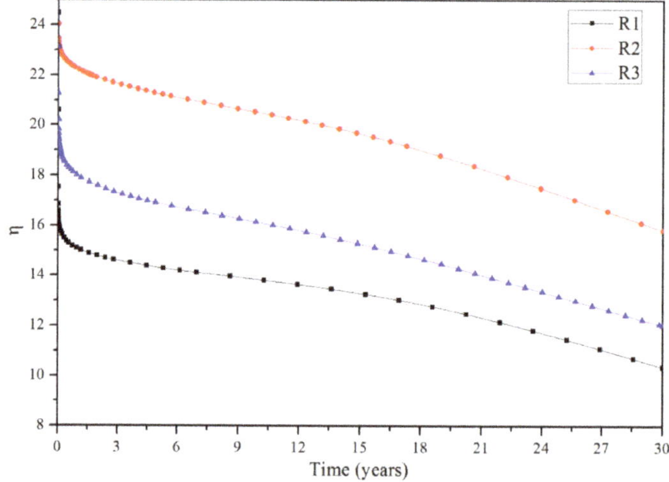

Figure 10. Change of the energy efficiency during the 30 years under the three reference cases.

4.6. Influence of Reservoir Heterogeneity on the Pressure Field

Figure 11 shows the evolution of the spatial distribution of the fracture pressure over the 30 years under the three reference cases. High pressure annular regions gradually form near the injection well, and the fracture pressure declines from the injection well to the production well. With heat mining, the high pressure regions gradually expand, the reservoir pressure gradually increases, and this represents the thermal energy in the reservoir is being gradually extracted out. These are in accordance with previous studies from Zeng et al. [16]. From the comparison among the three reference cases, the distribution and evolution of the fracture pressure are basically identical and this shows that the reservoir heterogeneity has only a very slight influence on the pressure field.

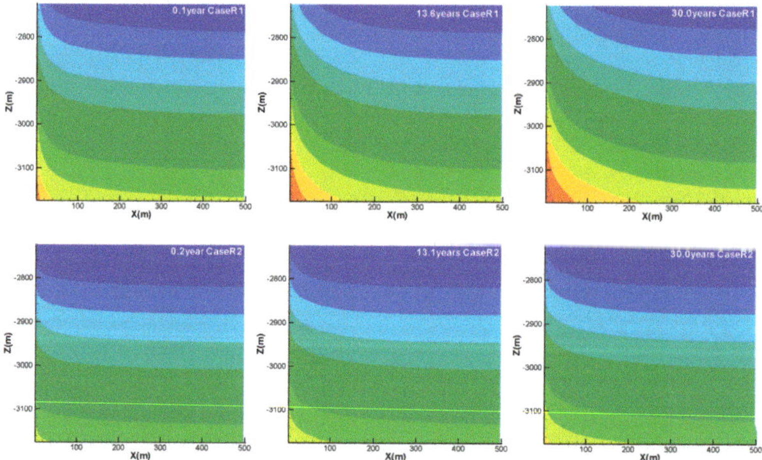

Figure 11. *Cont.*

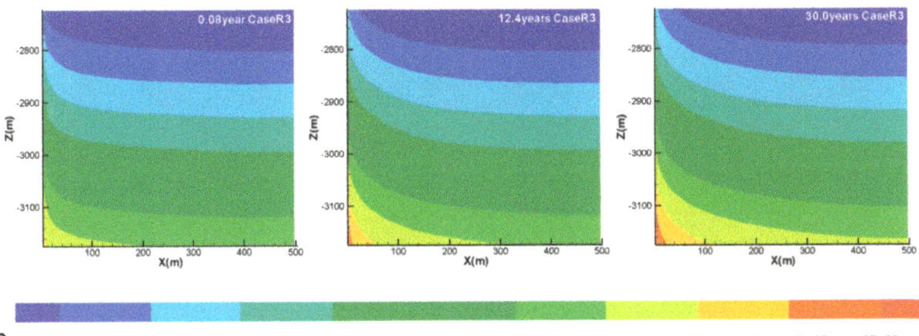

Figure 11. Evolution of spatial distribution of the fracture pressure (Pa) during the 30 years under the three reference cases.

4.7. Influence of Reservoir Heterogeneity on the Temperature Field

Figure 12 shows the evolution of the spatial distribution of the fracture temperature during the 30 years under the three reference cases. Annular low temperature regions gradually form near the injection well and the fracture temperature gradually increases from the injection well to the production well. With heat mining, the low temperature regions gradually expand, the reservoir temperature gradually declines, and this represents the result of the thermal energy in the reservoir being gradually extracted out. These are in accordance with previous studies from Zeng et al. [16]. Comparing the three reference cases, we find that a cold front forms in the high permeability layer of the heterogeneous reservoir. For the R1 reservoir, the temperature distribution of all the layers are basically identical and a cold front does not develop in which the temperature field is uneven. For the R2 reservoir, there forms one cold front in the two high permeability layers, because in these layers the water seepage velocity is much higher than that in the lower permeability layers. For the R3 reservoir, there forms two cold fronts in the four high permeability layers of the reservoir, also because the water velocity in these layers is much higher than that in the rest layers. These indicate that the reservoir heterogeneity has a significant influence on the fracture temperature field. In higher permeability layers, the water seepage velocity is increased, and there will form cold front, which means the temperature distribution along depth is uneven. The figure also shows that the system can benefit from buoyancy drive due to the temperature difference in the production and injection wells, and these are in agreement with studies from Huang et al. [42].

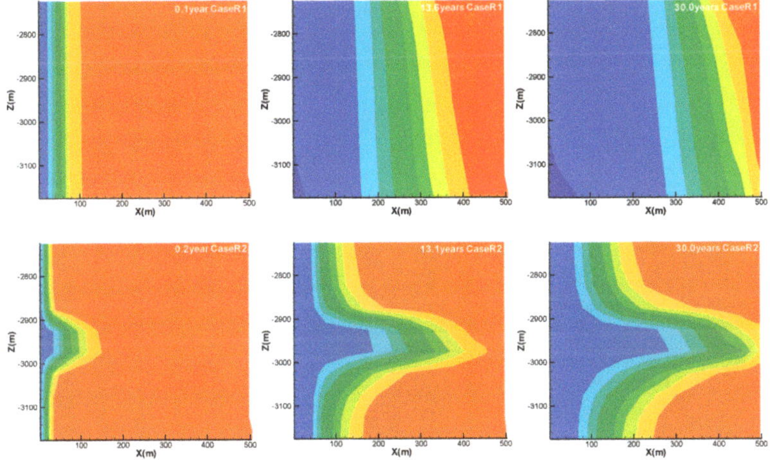

Figure 12. *Cont.*

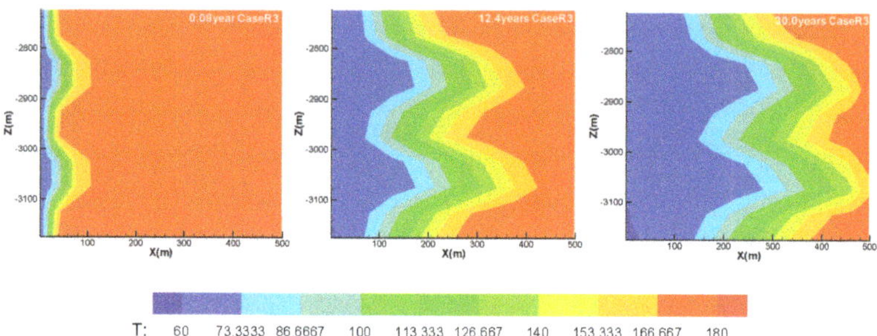

Figure 12. Evolution of the spatial distribution of the fracture temperature (°C) during the 30 years under the three reference cases.

4.8. Influence of Reservoir Heterogeneity on the Water Density Field

Figure 13 shows the evolution of the spatial distribution of the fracture water density over the 30 years under the three reference cases. Because near the injection well, it is high pressure and low temperature, annular high water density regions form. The water density reduces from the injection well to the production well. With heat mining, the high density regions expand toward the production well, the water density in the reservoir gradually increases, and this represents the result of the thermal energy being gradually extracted from the reservoir. Comparing the three reference cases we can find that in the high permeability layers there forms low density front, while in the low permeability layers the density distribution are basically even. For the R1 reservoir, the water density in each layer is basically identical, and a low density front does not form. For the R2 reservoir, there forms one low density front in the two high permeability layers. For the R3 reservoir, there forms two low density fronts in the four high permeability layers. As stated above, this is mainly because in the high permeability layers the horizontal velocity is much greater than that in the lower permeability layers, thus the low density contours are fronted in the high permeable layers. These indicate that the reservoir heterogeneity has a significant influence on the water density field. Higher permeability will increase the water conduction ability in the layers and there will form low density front, making the density distribution uneven along depth. This is in agreement with the studies by Huang et al. [42].

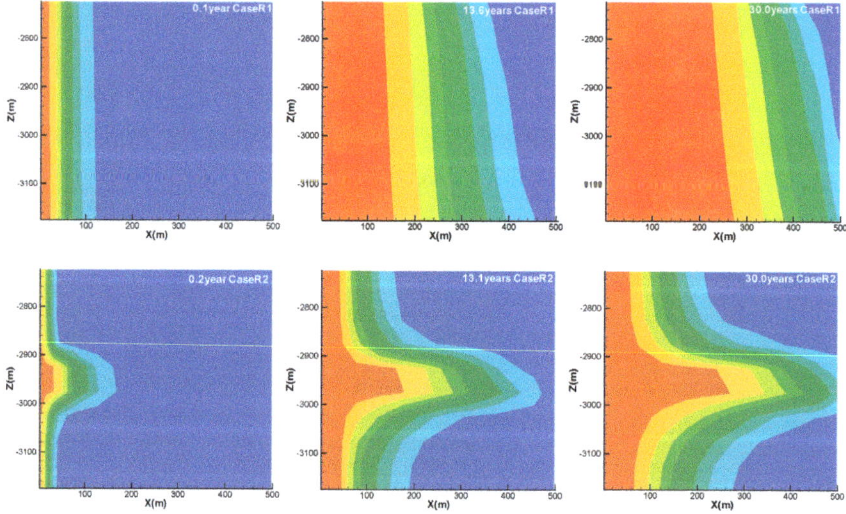

Figure 13. *Cont.*

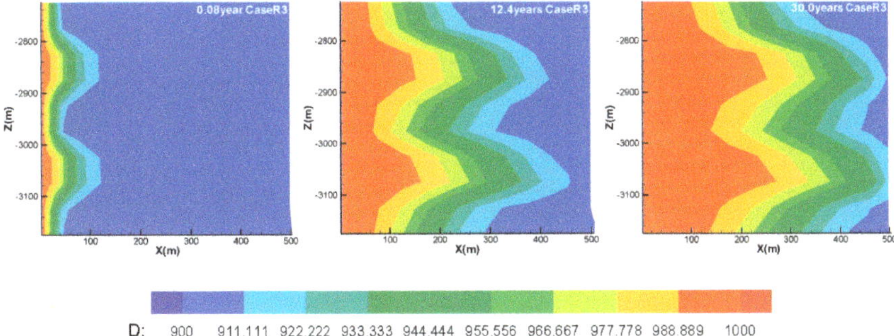

Figure 13. Evolution of the spatial distribution of the facture water density (kg/m^3) during the 30 years under the three reference cases.

5. Sensitivity Analysis

Many factors have significant influence on the production performance of the EGSs and previous studies can be found in References [4,19,20,45]. Most important parameters that can be controlled and adjusted are fracture spacing, well spacing and injection temperature [20,45], thus in this study we mainly investigated the influence of the above three parameters. Based on the above three reference cases, we further investigated the sensitivity of electricity generation to the three key parameters: fracture spacing D, well spacing WS and injection temperature T_{inj}. In detail we researched the performance and efficiency characteristics of the following 3 scenarios: (1) increasing D to D = 75 m; (2) increasing WS to WS = 600 m; (3) increasing T_{inj} to T_{inj} = 80 °C. Figures 14–17 show the sensitivity of electric power, reservoir impedance, pump power and energy efficiency to the above three parameters, respectively.

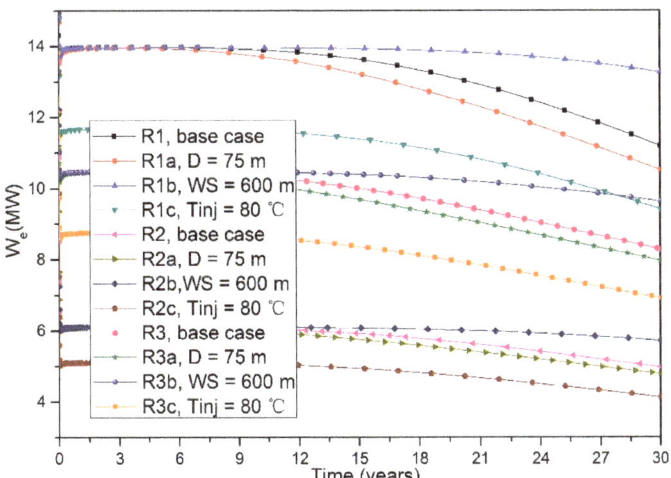

Figure 14. Sensitivity of electric power to main parameters.

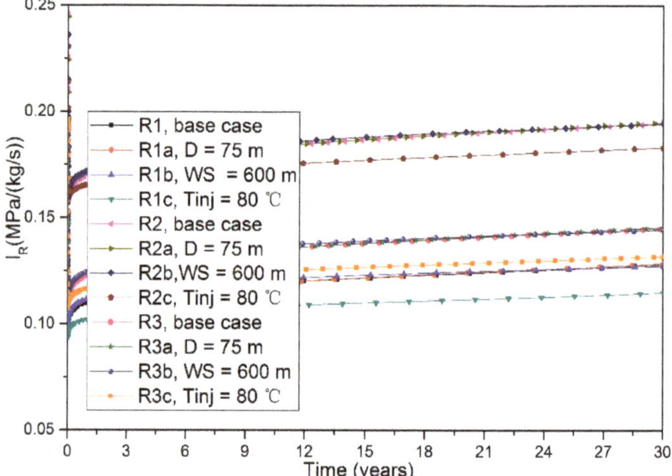

Figure 15. Sensitivity of reservoir impedance to main parameters.

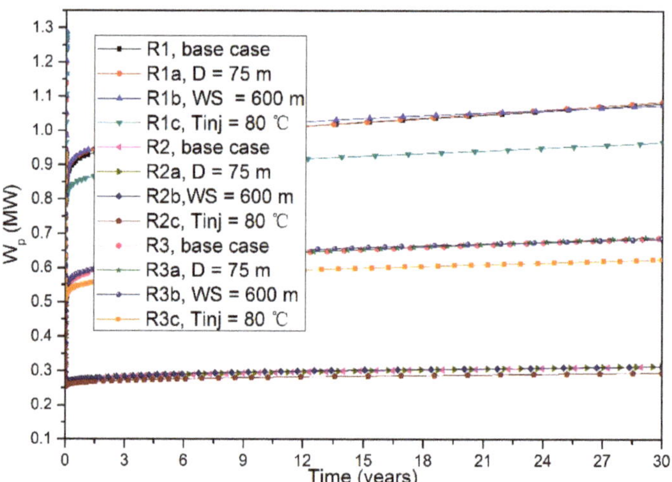

Figure 16. Sensitivity of pump power to main parameters.

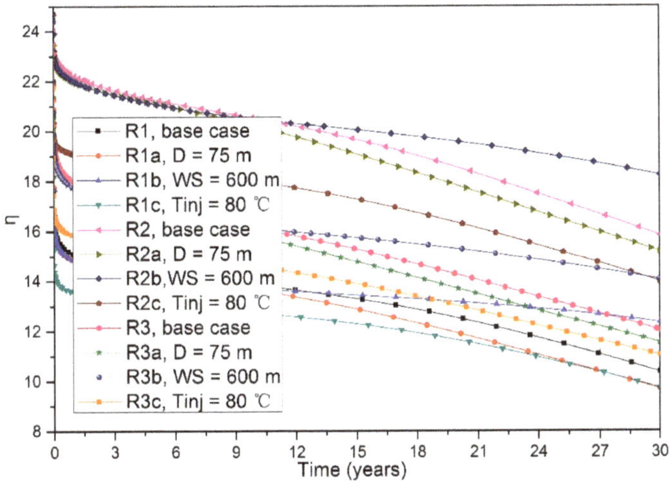

Figure 17. Sensitivity of energy efficiency to main parameters.

5.1. Sensitivity to Fracture Spacing

Figure 14 R1a shows that increasing D from 50 m to 75 m results in a decrease of W_e from 13.97–11.20 MW to 13.83–10.52 MW. For Figure 14 R2a and R3a, we can also find that an increase of D will cause the reduction of W_e. This is mainly because the fracture spacing determines the heat transfer area between the fractured rock and circulating water, higher fracture spacing will decrease the ratio between surface area and reservoir volume, finally decrease the heat transfer area, and this will significantly reduce the W_e according to the heat transfer formula. These are in agreement with studies from Sanyal et al. [22]. Figure 15 R1a shows that increasing D from 50 m to 75 m results in only very slight influence on the I_R. For Figure 15 R2a and Figure 15 R3a, similarly we can find that the increase of D has only very slight influence on the I_R. This is because the I_R is mainly determined by water viscosity and reservoir permeability [4,5]. In this study, the viscosity and reservoir permeability are all independent of the D, thus the change of the D has only a very slight effect on the I_R. This is in accordance with previous studies from Zeng et al. [33]. Figure 16 R1a shows that increasing D from 50 m to 75 m results in only a very slight influence on the W_p. For Figure 16 R2a and Figure 16 R3a, similarly we can find that the increase of D has only a very slight influence on the W_p. Because the increase of D has only a very slight influence on the I_R, based on Equations (12) and (16), the increase of D also has only a very slight effect on the P_{inj} and thus has only a slight effect on the W_p according to Equation (16). This is in accordance with previous studies by Zeng et al. [33]. Figure 17 R1a shows that increasing D from 50 m to 75 m results in a decrease of η from 20.62–10.36 to 19.31–9.69. For Figure 17 R2a and R3a, similarly we can find that the increase of D will cause the reduction of the η. This is because the increase of D will reduce the W_e, while only has a very slight influence on the W_p, based on Equation (17) this will obviously decrease the η. This is in accordance with previous studies from Zeng et al. [33]. Overall, the D has a significant influence on the electricity generation performance, within a certain range, increasing D will reduce the W_e, while have only a very slight effect on the I_R and W_p, thus significantly decrease the η.

5.2. Sensitivity to Well Spacing

Figure 14 R1b shows that increasing WS from 500 m to 600 m results in an increase of W_e from 13.97–11.20 MW to 19.72–13.27 MW. For Figure 14 R2b and R3b, similarly we can find that the increase of WS will cause the rise of W_e. This is because higher WS increases the reservoir volume between the injection well and production well, thus increases the T_{pro} and h_{pro}, according to Equation (11) this obviously increases the W_e. This proves that the WS is an important design parameter for EGS construction, it is directly related to the W_e, thus the determination of WS should be based on an accurate analysis of the geological data of the geothermal field. Figure 15 R1b shows that increasing WS from 500 m to 600 m results in only a very slight influence on the I_R. For Figure 15 R2b and Figure 15 R3b, similarly we can find that the increase of WS has only a very slight effect on the I_R. As stated above, the I_R is mainly determined by water viscosity and reservoir permeability—the viscosity and permeability are independent of WS, thus it has only a very slight influence on the I_R. Figure 16 R1b shows that increasing WS from 500 m to 600 m results in only a very slight influence on the W_p. For Figure 16 R2b and R3b, similarly we can find that the increase of WS has only a very slight effect on the W_p. This is mainly because the increase of WS has only a very slight on the I_R, based on Equations (12) and (16), the increase of WS has only a very slight influence on the P_{inj} and W_p. Figure 17 R1b shows that increasing WS from 500 m to 600 m results in an increase of η from 20.62–10.36 to 22.56–12.35. For Figure 17 R2b and R3b, similarly we can find that an increase of WS significantly increases the η. As stated above, this is because the increase of WS obviously increases the W_e, while it has only a very slight effect on the W_p, based on Equation (17) this will increase the η. Overall, the WS has a significant influence on the electricity generation performance, within a certain range, increasing the WS will significantly increase the W_e, while have only a very slight effect on the I_R and W_p, thus significantly increase the η.

5.3. Sensitivity to Injection Temperature

Figure 14 R1c shows that increasing the T_{inj} from 60 °C to 80 °C results in a decrease of W_e from 13.97–11.20 M to 12.06–9.41 MW. For Figure 14 R2c and R3c, similarly we can find that the increase of the T_{inj} reduces the W_e. This is in accordance with previous studies from Zeng et al. [4,5]. This is because the increase of T_{inj} increases the h_{inj}, based on Equation (11) when the other conditions are unchanged, this will significantly decrease the W_e. Figure 15 R1c shows that increasing the T_{inj} from 60 °C to 80 °C results in a decrease of I_R from 0.097–0.128 MPa/(kg/s) to 0.095–0.115 MPa/(kg/s). For Figure 15 R2c and R3c, similarly we can find that the increase of the T_{inj} will significantly reduce the I_R. As mentioned above, the I_R is mainly determined by the water viscosity and reservoir permeability [4,5]. The increase of T_{inj} will increase the reservoir temperature, decrease the water viscosity, thus reducing the I_R. This is in accordance with previous studies by Zeng et al. [33]. Figure 16 R1c shows that increasing the T_{inj} from 60 °C to 80 °C results in a decrease of W_p from 0.82–1.08 MW to 0.80–0.97 MW. For Figure 16 R2c and R3c, similarly we can see that the increase of the T_{inj} will significantly decrease the W_p. This is because the increase of the T_{inj} reduces the I_R, according to Equations (12) and (16), this will significantly reduce the P_{inj} and thus further decrease the W_p. Figure 17 R1c shows that increasing the T_{inj} from 60 °C to 80 °C results in a decrease of the η from 20.62–10.36 to 18.20–9.73. For Figure 17 R2c and R3c, similarly we can find that the increase of the T_{inj} will significantly reduce the η. This is a comprehensive result of both reduction of W_e and W_p. Overall, the T_{inj} has a significant influence on the electricity generation performance, within a certain range, increasing the T_{inj} will reduce the W_e, decrease the I_R and W_p, and thus reduce the η.

6. Conclusions

In this study, we numerically investigated the influence of reservoir heterogeneity on the electricity generation performance of an EGS reservoir and analyzed the main factors affecting the production performance. The conclusions are as follows:

(1) With increasing of the reservoir heterogeneity, the water conduction ability of the reservoir gradually decreases, the available water production rate gradually reduces, thus the electric power gradually decreases.

(2) With increasing of the reservoir heterogeneity, the reservoir impedance gradually increases.

(3) With increasing of the reservoir heterogeneity, the pump power gradually reduces.

(4) With increasing of the reservoir heterogeneity, the energy efficiency gradually increases.

(5) The reservoir heterogeneity has a significant influence on the fracture temperature field. In higher permeability layers, a cold front will form and make the temperature distribution along depth uneven.

(6) The fracture spacing has a significant influence on the electricity generation performance, within a certain range, increasing the fracture spacing will obviously reduce the electric power, while having only very slight effect on the reservoir impedance and pump power, thus significantly decreasing the energy efficiency.

(7) The well spacing has a significant influence on the electricity generation performance, within a certain range, increasing the well spacing will obviously increase the electric power, while having only very slight effect on the reservoir impedance and pump power, thus significantly increasing the energy efficiency.

(8) The injection temperature has a significant influence on the electricity generation performance, within a certain range, increasing the injection temperature will obviously reduce the electric power, decrease the reservoir impedance and pump power, thus reducing the energy efficiency.

Author Contributions: The author contributed equally to the research and writing of this manuscript.

Acknowledgments: The authors gratefully appreciate the financial support from the Young Teacher Training Program of Sun Yat-sen University (Grant 17lgpy44); the Natural Science Foundation of China (Grant 41572277, 41877229); the Joint Foundation of NFSC and Guangdong Province (Grant U1401232); the Natural Science Foundation of Guangdong Province (Grant 2014A030308001).

Conflicts of Interest: The authors declare no conflict of interest.

Nomenclature

D	fracture spacing, m
g	gravity, 9.80 m/s^2
h	well depth, m
h_1	depth of injection well, m
h_2	depth of production well, m
h_{inj}	injection specific enthalpy, kJ/kg
h_{pro}	production specific enthalpy, kJ/kg
I_R	reservoir impedance, MPa/(kg/s)
k	reservoir permeability, m^2
k_f	fracture permeability, m^2
k_m	matrix permeability, m^2
k_x	intrinsic permeability along x, m^2
k_y	intrinsic permeability along y, m^2
k_z	intrinsic permeability along z, m^2
P	pressure, MPa
P_{max}	critical pressure, MPa
P_{inj}	injection pressure, MPa
P_{pro}	production pressure, MPa
P_0	bottomhole production pressure, MPa
q	water production rate, kg/s
Q	total water production rate, kg/s
T	temperature, °C
T_0	mean heat rejection temperature, 282.15 K
T_{pro}	production temperature, °C
T_{inj}	injection temperature, °C
W_p	electric power of pump, MW
WS	well spacing, m
W_e	electric power, MW

x, y, z	cartesian coordinates, m
ϕ	reservoir porosity
η	energy efficiency
η_p	pump efficiency, 80%
ρ	water density, kg/m^3

References

1. Tester, J.W.; Livesay, B.; Anderson, B.J.; Moore, M.C.; Bathchelor, A.S.; Nichols, K.; Blackwell, D.D.; Petry, S.; Dipoppo, R.; Toksoz, M.N.; et al. *The Future of Geothermal Energy: Impact of Enhanced Geothermal Systems (EGS) on the United States in the 21st Century. An assessment by an MIT-Led Interdisciplinary Panel*; MIT: Cambridge, MA, USA, 2006.

2. Zhao, Y.S.; Wan, Z.J.; Kang, J.R. *Introduction to HDR Geothermal Development*; Science Press: Beijing, China, 2004. (In Chinese)

3. Wang, J.; Hu, S.; Pang, Z.; He, L.; Zhao, P.; Zhu, C.; Rao, S.; Tang, X.; Kong, Y.; Luo, L.; et al. Estimate of geothermal resources potential for hot dry rock in the continental area of China. *Sci. Technol. Rev.* **2012**, *30*, 25–31. (In Chinese)

4. Zeng, Y.C.; Tang, L.S.; Wu, N.Y.; Cao, Y.F. Analysis of influencing factors of production performance of enhanced geothermal system: A case study at Yangbajing geothermal field. *Energy* **2017**, *127*, 218–235. [CrossRef]

5. Zeng, Y.; Su, Z.; Wu, N. Numerical simulation of heat production potential from hot dry rock by water circulating through two horizontal wells at Desert Peak geothermal field. *Energy* **2013**, *56*, 92–107. [CrossRef]

6. Pruess, K. Modelling of geothermal reservoirs: Fundamental processes, computer simulation, and field applications. In Proceedings of the 10th New Zealand Geothermal Workshop, Auckland, New Zealand, 2–4 November 1988.

7. Willis-richards, J.; Wallroth, T. Approaches to the modeling of HDR reservoirs: A review. *Geothermics* **1995**, *24*, 307–332.

8. Hayashi, K.; Willis-Richards, J.; Hopkirk, R.J.; Niibori, Y. Numerical models of HDR geothermal reservoirs—A review of current thinking and progress. *Geothermics* **1999**, *28*, 507–518. [CrossRef]

9. Sanyal, S.K.; Butler, S.J.; Swenson, D.; Hardeman, B. Review of the state-of-the-art of numerical simulation of enhanced geothermal system. In Proceedings of the World Geothermal Congress, Kyushu-Tohoku, Japan, 28 May–10 June 2000.

10. O'Sullivan, M.J.; Pruess, K.; Lippmann, M.J. State of the art of geothermal reservoir simulation. *Geothermics* **2001**, *30*, 395–429. [CrossRef]

11. Birdsell, S.; Robinson, B. A three-dimensional model of fluid, heat, and tracer transport in the Fenton Hill hot dry rock reservoir. In Proceedings of the Thirteenth Workshop on Geothermal Reservoir Engineering, Stanford, CA, USA, 19–21 January 1988.

12. Mcdermott, C.I.; Randriamanjatosoa, A.R.; Tenzer, H.; Kolditz, O. Simulation of heat extraction from crystalline rocks: The influence of coupled processes on differential reservoir cooling. *Geothermics* **2006**, *35*, 321–344. [CrossRef]

13. Watanabe, N.; Wang, W.Q.; McDermott, C.I.; Taniguchi, T.; Kolditz, O. Uncertainly analysis of thermo-hydro-mechanical coupled processes in heterogeneous porous media. *Comput. Mech.* **2010**, *45*, 263–280.

14. Yuchao, Z.; Nengyou, W.; Zheng, S. Numerical simulation of heat production potential from hot dry rock by water circulating through a novel single vertical fracture at Desert Peak geothermal field. *Energy* **2013**, *63*, 268–282.

15. Yuchao, Z.; Nengyou, W.; Zheng, S. Numerical simulation of electricity generation potential from fractured granite reservoir through a single horizontal well at Yangbajing geothermal field. *Energy* **2014**, *65*, 472–487.

16. Yuchao, Z.; Jiemin, Z.; Nengyou, W. Numerical simulation of electricity generation potential from fractured granite reservoir through vertical wells at Yangbajing geothermal field. *Energy* **2016**, *103*, 290–304.

17. Yuchao, Z.; Jiemin, Z.; Nengyou, W. Numerical investigation of electricity generation potential from fractured granite reservoir by water circulating through three horizontal wells at Yangbajing geothermal field. *Appl. Therm. Eng.* **2016**, *104*, 1–15.

18. Yuchao, Z.; Jiemin, Z.; Nengyou, W. Numerical investigation of electricity generation potential from fractured granite reservoir through a single vertical well at Yangbajing geothermal field. *Energy* **2016**, *114*, 24–39.

19. Yuchao, Z.; Liansheng, T.; Nengyou, W.; Jing, S.; Yifei, C. Orthogonal test analysis on conditions affecting electricity generation performance of an enhanced geothermal system at Yangbajing geothermal field. *Energies* **2017**, *10*, 2015.

20. Cheng, W.L.; Wang, C.L.; Nian, Y.L.; Han, B.B.; Liu, J. Analysis of influencing factors of heat extraction from enhanced geothermal systems considering water losses. *Energy* **2016**, *115*, 274–288. [CrossRef]

21. Hu, L.T.; Winterfeld, P.H.; Fakcharoenphol, P.; Wu, Y.S. A novel fully-coupled flow and geomechanics model in enhanced geothermal reservoirs. *J. Petrol. Sci. Eng.* **2013**, *107*, 1–11. [CrossRef]

22. Sanyal, S.K.; Butler, S.J. An analysis of power generation prospects from enhanced geothermal systems. In Proceedings of the World Geothermal Congress 2005, Antalya, Turkey, 24–29 April 2005; pp. 1–6.

23. Taron, J.; Elsworth, D.; Min, K.B. Numerical simulation of thermal-hydrologic-mechanical-chemical processes in deformable, fractured porous media. *Int. J. Rock Mech. Min. Sci.* **2009**, *46*, 842–854. [CrossRef]

24. Taron, J.; Elsworth, D. Thermal-hydrologic-mechanical-chemical processes in the evolution of engineered geothermal reservoirs. *Int. J. Rock Mech. Min. Sci.* **2009**, *46*, 855–864. [CrossRef]

25. Gelet, R.; Loret, B.; Khalili, N. A thermal-hydro-mechanical coupled model in local thermal non-equilibrium for fractured HDR reservoir with double porosity. *J. Geophys. Res.* **2012**, *117*, 1–23. [CrossRef]

26. Gelet, R.; Loret, B.; Khalili, N. Thermal recovery from a fractured medium in local thermal non-equilibrium. *Int. J. Numer. Anal. Method Geomech.* **2013**, *37*, 2471–2501. [CrossRef]

27. Benato, S.; Taron, J. Desert Peak EGS: Mechanisms influencing permeability evolution investigated using dual-porosity simulator TFReact. *Geothermics* **2016**, *63*, 157–181. [CrossRef]

28. Pruess, K. Enhanced geothermal system (EGS) using CO_2 as working fluid-A novel approach for generating renewable energy with simultaneous sequestration of carbon. *Geothermics* **2006**, *35*, 351–367. [CrossRef]

29. Pruess, K. On production behavior of enhanced geothermal systems with CO_2 as working fluid. *Energy Convers. Manag.* **2008**, *49*, 1446–1454. [CrossRef]

30. Spycher, N.; Pruess, K. A phase-partitioning model for CO_2-brine mixtures at elevated temperatures and pressures: Application to CO_2-enhanced geothermal systems. *Transp. Porous Media* **2010**, *82*, 173–196. [CrossRef]

31. Borgia, A.; Pruess, K.; Kneafsey, T.J.; Oldenburg, C.M.; Pan, L. Numerical simulation of salt precipitation in the fractures of a CO_2-enhanced geothermal system. *Geothermics* **2012**, *44*, 13–22. [CrossRef]

32. Xu, T.F.; Yuan, Y.L.; Jia, X.F.; Lei, Y.D.; Li, S.T.; Feng, B.; Hou, Z.Y.; Jiang, Z.J. Prospects of power generation from an enhanced geothermal system by water circulation through two horizontal wells: A case study in the Gonghe Basin, Qinghai Province, China. *Energy* **2018**, *148*, 196–207. [CrossRef]

33. Zeng, Y.; Tang, L.; Wu, N.; Cao, Y. Numerical simulation of electricity generation potential from fractured granite reservoir using the MINC method at the Yangbajing geothermal field. *Geothermics* **2018**, *75*, 122–136. [CrossRef]

34. Baujard, C.; Bruel, D. Numerical study of the impact of fluid density on the pressure distribution and stimulated volume in the Soultz HDR reservoir. *Geothermics* **2006**, *35*, 607–621. [CrossRef]

35. Kolditz, O.; Clauser, C. Numerical simulation of flow and heat transfer in fractured crystalline rocks: Application to the hot dry rock site in Rosemanowes. *Geothermics* **1998**, *27*, 1–23. [CrossRef]

36. Kolditz, O. Modelling flow and heat transfer in fractured rocks: Conceptual model of a 3-D deterministic fracture network. *Geothermics* **1995**, *24*, 451–470. [CrossRef]

37. Jing, Z.; Willis-Richards, J.; Hashida K, W. A three-dimensional stochastic rock mechanics model of engineered geothermal systems in fractured crystalline rock. *J. Geophys. Res.* **2000**, *105*, 23663–23679. [CrossRef]

38. Jing, Z.; Watanabe, K.; Willis-Richards, J.; Hashida, T. A 3-D water/rock chemical interaction model for prediction of HDR/HWR geothermal reservoir performance. *Geothermics* **2002**, *31*, 1–28. [CrossRef]

39. Jing, Y.N.; Jing, Z.Z.; Willis-Richards, J.; Hashida, T. A simple 3-D thermoelastic model for assessment of the long-term performance of the Hijiori deep geothermal reservoir. *J. Volcanol. Geotherm. Res.* **2014**, *269*, 14–22. [CrossRef]

40. Sun, Z.X.; Zhang, X.; Xu, Y.; Yao, J.; Wang, H.X.; Lv, S.H.; Sun, Z.L.; Huang, Y.; Cai, M.Y.; Huang, X.X. Numerical simulation of the heat extraction in EGS with thermal-hydraulic-mechanical coupling method based on discrete fractures model. *Energy* **2017**, *120*, 20–33. [CrossRef]

41. Yao, J.; Zhang, X.; Sun, Z.X.; Huang, Z.Q.; Liu, J.R.; Li, Y.; Xin, Y.; Yan, X.; Liu, W.Z. Numerical simulation of the heat extraction in 3D-EGS with thermal-hydraulic-mechanical coupling method based on discrete fractures model. *Geothermics* **2018**, *74*, 19–34. [CrossRef]

42. Huang, W.; Cao, W.; Jiang, F. Heat extraction performance of EGS with heterogeneous reservoir: A numerical evaluation. *Int. J. Heat Mass Transfer* **2017**, *108*, 645–657. [CrossRef]

43. Zarrouk, S.J.; Moon, H. Efficiency of geothermal power plants: A worldwide review. *Geothermics* **2014**, *51*, 142–153. [CrossRef]

44. Pruess, K.; Oldenburg, C.; Moridis, G. *TOUGH2 User's Guide, Version 2.0*; Lawrence Berkeley National Laboratory: Berkeley, CA, USA, 1999.

45. Asai, P.; Panja, P.; Mclennan, J.; Moore, J. Performance evaluation of enhanced geothermal system (EGS): Surrogate models, sensitivity study and ranking key parameters. *Renew. Energy* **2018**, *122*, 184–195. [CrossRef]

Article

A Rotor-Sync Signal-Based Control System of a Doubly-Fed Induction Generator in the Shaft Generation of a Ship

Trong-Thang Nguyen

Faculty of Energy Engineering, Thuyloi University, 175 Tay Son, Dong Da, Hanoi 11398, Vietnam; nguyentrongthang@tlu.edu.vn; Tel.: +84-038-846-8555

Received: 20 February 2019; Accepted: 29 March 2019; Published: 1 April 2019

Abstract: A doubly-fed induction machine in generator-mode is popularly used for energy generation, particularly in the case of a variable speed, such as in the wind generator, the shaft generator of a ship, because the doubly-fed induction generator is able to maintain a stable frequency when changing the rotor speed. This paper aims to propose a novel method for controlling the shaft generation system of a ship using a doubly-fed induction generator. This method uses the rotor signals of a small doubly-fed induction machine as base components to create the control signal for the doubly-fed induction generators. The proposed method will be proven by both theory and a simulation model. The advantage of the proposed method is that the control system of the generator can be simply built, but it functions effectively. The generator voltage always coincides with the grid voltage, even when the grid voltage and the rotor speed are changed, and the reactive and active power of the generator fed into the grid can be separately controlled.

Keywords: shaft generator; DFIG; shipboard; power; control

---

1. Introduction

Today, fuel resources are increasingly becoming exhausted, so identifying and using renewable sources is very urgent and necessary. In order for these renewable energy sources to operate reliably, they must be able to operate together or connected to the grid. Some researches [1,2] have succeeded in solving these difficulties. However, on a ship, it is more difficult to connect the generator voltage to the grid, because the grid of the ship is a soft-grid, and so the voltage is frequently changed. Therefore, solving this problem is a challenge and an opportunity for scientists.

On a ship, the power station must be optimally exploited when cruising to reduce energy consumption, noise, environmental pollution and to avoid the negative impact on people and nature. When navigating on the sea, in steady climatic and weather conditions, the main propulsion engines of the ship's propeller often do not reach full capacity. To take advantage of this surplus capacity, the ship is designed with a shaft generator, which works together with the diesel generator.

The required power for a ship in cruise mode only accounts for 5–10% of the main engine power. Therefore, ships with a shaft generator use the main engine's surplus power to save the operation time of the diesel generators, reduce the consumption of materials, and improve the lifetime of the diesel generator. In particular, the production cost of a power unit, created by a shaft generator, is only equal to 50% of the cost, when using the diesel generator [3].

However, as the power station system adds the shaft generator, the ship's power system becomes more complicated, posing technical problems that need to be addressed. The most complicated problem is that the output voltage of the shaft generator must coincide with the unstable grid voltage of the ship, when changing the rotor speed of the main engine [4,5]. Further, the engine power is much larger than the electric power, so the rotor speed of the generator is independent of the electric power.

The rotor speed depends only on the operation of the ship, so the speed range of the rotor is very wide. Therefore, the most effective solution is to use the doubly-fed induction machine in generator mode because of its ability to maintain a stable frequency of this system.

The doubly-fed induction machine (DFIM) is an induction machine with both stator and rotor windings [6]. The doubly-fed induction machine plays the role of generation (DFIG), with inherent advantages, such as a small control circuit, suitable for the variable speed system, so DFIG has been applied in many energy generation systems, such as the shaft generator in a ship, the wind generator system. In these generator systems, the control circuit is located in the rotor, and the energy emitted in the stator is transmitted to the grid directly. Thus, the control circuit power is much lower than the grid-transmitted power.

The control circuit has two main parts: The first part is connected to the grid [7,8], used to regulate the DC voltage [9,10]. The second part is connected to the rotor of DFIG [11,12], used to control the reactive and active power of DFIG fed into the grid [13,14]. The rotor side control is more complicated than the grid side control, thus attracting many scientists around the globe. In this study, we are only interested in controlling the rotor side control, so we assume that the DC source already exists and is stable.

There is a lot of research on controlling the control system for the rotor side [15,16], most of which use the space vector modulation technique. Relying on that technique, in order for the controller to perform the mission, all parameters of the stator and rotor must be transformed in a synchronous reference frame [17,18], which coincides with the space vector of the grid voltage (grid voltage-orientated coordinates) or with the space vector of the stator flux (stator-flux-orientated coordinates). However, the final parameters for controlling the system are not within these coordinates. Thus, the control structure must consist of two coordinate conversion stages: First, all input and feedback parameters of the controller are transformed in a synchronous reference frame; and lastly, all output parameters of the controller are transformed in the rotor or stator frame. For these reasons, the control system calculations must be more complex.

In this research, the author proposes a simple and effective technique, which does not need the coordinate-conversion stages. Relying on this proposed technique, all signals of the control system are continuous. This technique uses the rotor signals of a small DFIM as a base to create the control signals of the DFIG rotor current.

2. The Proposed Model

The purpose of the control system is to make the DFIG stator voltage coincide with the grid voltage in the case of a changing rotor speed and grid voltage. Thus, the author proposes a method of using a small DFIM with the stator connected to the grid, so the change of the DFIM rotor voltage depends on two factors: the rotor speed and the grid voltage. The natural change in the DFIM rotor voltage is therefore suitable for controlling the DFIG rotor current. DFIM only acts as a sync signal generator, so it has a small capacity. In order for the sync signal to not be distorted, the output of this signal is connected to high resistance. The initially proposed model is shown in Figure 1.

The generation system includes:

- The main machine, which is used for pulling the propeller. In addition, it is also used for pulling the shaft generator through the gearbox.
- DFIM with the stator connected to the grid and the rotor connected to the high-resistance mode. The DFIM rotor signal is a base component for creating the control signals for DFIG.
- An isolation stage, which is a circuit that has a high-resistance input, so the DFIM rotor is connected to the high resistance.
- The current control circuit, which creates the output current value that is equal to the input voltage value.
- DFIG, which creates the power that is fed into the grid.

Rotors of DFIG and DFIM are connected to each other tightly so that the angles between the stator and rotor are equal.

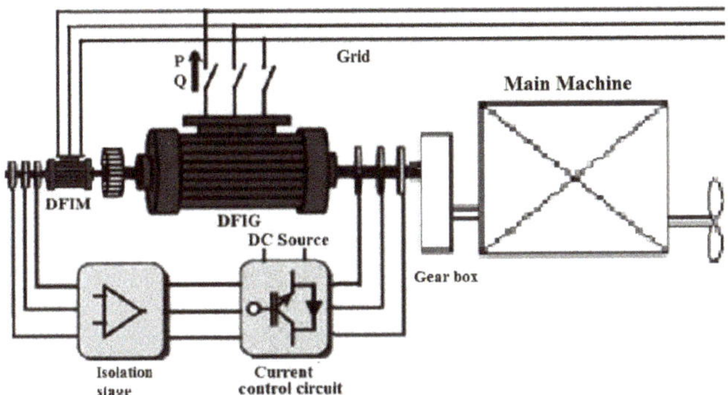

Figure 1. The initially proposed model.

3. Building the Equations of the Proposed Model

3.1. The Initial Equations

The system includes two machines, so the parameter symbols are as follows: ^{G}X for DFIG, and ^{M}X for DFIM. For example, $^{G}R_s$ is the stator resistance of DFIG, and $^{M}R_s$ is the stator resistance of DFIM. The DFIM equations in the grid voltage-orientated coordinates are as follows [19–21]:

$$\begin{cases} ^{M}\underline{u}_s^f = {}^{M}R_s \cdot {}^{M}\underline{i}_s^f + \dfrac{d(^{M}\underline{\psi}_s^f)}{dt} + j\cdot\omega_s\cdot {}^{M}\underline{\psi}_s^f & \text{(1a)} \\[2mm] ^{M}\underline{u}_r^f = {}^{M}R_r\cdot {}^{M}\underline{i}_r^f + \dfrac{d(^{M}\underline{\psi}_r^f)}{dt} + j\cdot\omega_r\cdot {}^{M}\underline{\psi}_r^f & \text{(1b)} \\[2mm] ^{M}\underline{\psi}_s^f = {}^{M}\underline{i}_s^f\cdot {}^{M}L_s + {}^{M}\underline{i}_r^f\cdot {}^{M}L_m & \text{(1c)} \\[2mm] ^{M}\underline{\psi}_r^f = {}^{M}\underline{i}_s^f\cdot {}^{M}L_m + {}^{M}\underline{i}_r^f\cdot {}^{M}L_r & \text{(1d)} \end{cases} ,$$

The DFIM rotor is connected to the high resistance, so $^{M}\underline{i}_r^f = 0$. Substituting $^{M}\underline{i}_r^f = 0$ for (1c), the DFIM stator flux $^{M}\underline{\psi}_s^f = {}^{M}\underline{i}_s^f\cdot {}^{M}L_s$. Substituting $^{M}\underline{\psi}_s^f$ for (1a), the DFIM stator voltage:

$$^{M}\underline{u}_s^f = {}^{M}R_s\cdot {}^{M}\underline{i}_s^f + {}^{M}L_s\cdot \frac{d(^{M}\underline{i}_s^f)}{dt} + j\cdot\omega_s\cdot {}^{M}L_s\cdot {}^{M}\underline{i}_s^f. \tag{2}$$

On the rotor shaft-orientated coordinates, the DFIM rotor voltage and DFIM rotor flux equations are as follows [20,21]:

$$\begin{cases} ^{M}\underline{u}_r^r = {}^{M}R_r\cdot {}^{M}\underline{i}_r^r + \dfrac{d(^{M}\underline{\psi}_r^r)}{dt} & \text{(3a)} \\[2mm] ^{M}\underline{\psi}_r^r = {}^{M}\underline{i}_s^r\cdot {}^{M}L_m + {}^{M}\underline{i}_r^r\cdot {}^{M}L_r & \text{(3b)} \end{cases} .$$

Substituting $^{M}\underline{i}_r^r = 0$ for (3b), $^{M}\underline{\psi}_r^r = {}^{M}\underline{i}_s^r\cdot {}^{M}L_m$, and substituting for (3a):

$$^{M}\underline{u}_r^r = {}^{M}L_m\cdot \frac{d(^{M}\underline{i}_s^r)}{dt}, \tag{4}$$

$^{M}\underline{u}_r^r$ is fed into the isolation stage. The first section of the isolation stage is the amplifier with the gain G_{ss}, so the output voltage vector of this section in the rotor shaft-orientated coordinate is:

$$\underline{u}_{ss}^r = G_{ss}\cdot {}^{M}L_m\cdot \frac{d(^{M}\underline{i}_s^r)}{dt}. \tag{5}$$

Next, $\underline{u}_{ss}^{r}$ is fed into the integral section, so the signal output vector of the integral section in the rotor shaft-orientated coordinate is:

$$\underline{u}_{is}^{r} = \int \underline{u}_{ss}^{r} = \int G_{ss}{}^{M}L_{m}\frac{d({}^{M}\underline{i}_{s}^{r})}{dt} = G_{ss} \cdot {}^{M}L_{m} \cdot {}^{M}\underline{i}_{s}^{r}. \tag{6}$$

Converting (6) into an equation in the grid voltage-orientated coordinates:

$$\underline{u}_{is}^{f} = G_{ss} \cdot {}^{M}L_{m} \cdot {}^{M}\underline{i}_{s}^{f}, \tag{7}$$

$\underline{u}_{is}^{f}$ is fed into the current control circuit. This circuit makes the output current value equal the input voltage value:

$$\underline{i}_{c}^{f} = G_{ss} \cdot {}^{M}L_{m} \cdot {}^{M}\underline{i}_{s}^{f}. \tag{8}$$

The DFIG equations in the grid voltage-orientated coordinates are as follows:

$$\left\{ \begin{array}{ll} {}^{G}\underline{u}_{s}^{f} = {}^{G}R_{s} \cdot {}^{G}\underline{i}_{s}^{f} + \frac{d({}^{G}\underline{\psi}_{s}^{f})}{dt} + j \cdot \omega_{s} \cdot {}^{G}\underline{\psi}_{s}^{f} & (9a) \\ {}^{G}\underline{u}_{r}^{f} = {}^{G}R_{r} \cdot {}^{G}\underline{i}_{r}^{f} + \frac{d({}^{G}\underline{\psi}_{r}^{f})}{dt} + j \cdot \omega_{r} \cdot {}^{G}\underline{\psi}_{r}^{f} & (9b) \\ {}^{G}\underline{\psi}_{s}^{f} = {}^{G}\underline{i}_{s}^{f} \cdot {}^{G}L_{s} + {}^{G}\underline{i}_{r}^{f} \cdot {}^{G}L_{m} & (9c) \\ {}^{G}\underline{\psi}_{r}^{f} = {}^{G}\underline{i}_{s}^{f} \cdot {}^{G}L_{m} + {}^{G}\underline{i}_{r}^{f} \cdot {}^{G}L_{r} & (9d) \end{array} \right. .$$

The DFIG rotor receives the current $({}^{G}\underline{i}_{r}^{f})$ from the current control circuit. We control ${}^{G}\underline{i}_{r}^{f}$ to satisfy Equation (10):

$$ {}^{G}\underline{i}_{r}^{f} = \underline{i}_{c}^{f} = G_{ss} \cdot {}^{G}L_{m} \cdot {}^{G}\underline{i}_{s}^{f} \tag{10}$$

3.2. Adjusting the System before the DFIG Stator is Connected to the Grid

The DFIG stator is disconnected from the grid, so the DFIG stator current is ${}^{G}\underline{i}_{s}^{f} = 0$. The DFIG rotor current is ${}^{G}\underline{i}_{r}^{f} = {}^{G}\underline{i}_{r0}^{f} = G_{ss} \cdot {}^{M}L_{m} \cdot {}^{M}\underline{i}_{s}^{f}$. Substituting ${}^{G}\underline{i}_{s}^{f} = 0$ for ${}^{G}\underline{i}_{r}^{f}$. with (9c), the DFIG stator flux is: ${}^{G}\underline{\psi}_{s}^{f} = G_{ss} \cdot {}^{M}L_{m} \cdot {}^{G}L_{m} \cdot {}^{M}\underline{i}_{s}^{f}$. Substituting ${}^{G}\underline{\psi}_{s}^{f}$ and ${}^{G}\underline{i}_{s}^{f} = 0$ for (9a), the DFIG stator voltage is:

$${}^{G}\underline{u}_{s}^{f} = G_{ss} \cdot {}^{M}L_{m} \cdot {}^{G}L_{m} \cdot \frac{d({}^{M}\underline{i}_{s}^{f})}{dt} + j \cdot \omega_{s} \cdot G_{ss} \cdot {}^{M}L_{m} \cdot {}^{G}L_{m} \cdot {}^{M}\underline{i}_{s}^{f} = G_{ss} \cdot {}^{M}L_{m} \cdot {}^{G}L_{m} \cdot \left(\frac{d({}^{M}\underline{i}_{s}^{f})}{dt} + j \cdot \omega_{s} \cdot {}^{M}\underline{i}_{s}^{f} \right). \tag{11}$$

Considering Equation (2), which is the DFIM stator voltage equation, the grid voltage equation also includes two parts:

- The first part creates the heat of the resistor: ${}^{G}\underline{u}_{sr}^{f} = {}^{G}R_{s} \cdot {}^{G}\underline{i}_{s}^{f}$, but this part is very small in ${}^{G}\underline{u}_{s}^{f}$, so it can be neglected.
- The second part creates the flux:

$${}^{M}\underline{u}_{s\psi}^{f} = {}^{M}L_{s} \cdot \left(\frac{d({}^{M}\underline{i}_{s}^{f})}{dt} + j \cdot \omega_{s} \cdot {}^{M}\underline{i}_{s}^{f} \right). \tag{12}$$

- The frequency of ${}^{G}\underline{u}_{s\psi}^{f}$ is equal to the frequency of the grid voltage.

Combining (12) with (11), it can be seen that the frequency of ${}^{G}\underline{u}_{s\psi}^{f}$ is equal to the frequency of the DFIG stator voltage.

Finally, the frequency of the DFIG stator voltage is equal to the frequency of the grid voltage, and the phase difference between the two voltages is very small and constant, so it can be ignored or can be compensated for in an easy way, that is, by rotating the shaft between DFIG and DFIM.

The amplitude of the DFIG stator voltage can be adjusted by G_{ss} in the isolation stage. In order for $^G\underline{u}_s^f = {}^M\underline{u}_{s\psi}^f$, G_{ss} is evaluated by balancing the equation:

$$^G\underline{u}_s^f = {}^M\underline{u}_{s\psi}^f \text{ so } G_{ss} = {}^ML_s/({}^ML_m \cdot {}^GL_m). \tag{13}$$

After adjusting, all of the above stages will be kept stable.

In the stator-fix-orientated coordinates, the DFIG stator voltage (ignoring the small voltage in the resistor) is [20,21]:

$$^G\underline{u}_s^s = \frac{d({}^G\underline{\psi}_s^s)}{dt} = j \cdot \omega_s \cdot {}^G\underline{\psi}_s^s. \tag{14}$$

The above equation shows that the phase of the stator voltage is equal to the phase of the stator flux plus $\pi/2$.

$^G\underline{i}_{r0}^f$ creates the flux, so the phase of $^G\underline{i}_{r0}^f$ is equal to the phase of the flux, so the phase of $^G\underline{i}_{r0}^f$ is equal to the phase of the stator voltage minus $\pi/2$. Thus, in the grid voltage-orientated coordinates, $(-{}^G\underline{i}_{r0}^f)$ is a base component to create the q axis component of the rotor current $(^Gi_{rq})$. The phase of the d axis component of the rotor current $(^Gi_{rd})$ is equal to the phase of $(-{}^G\underline{i}_{r0}^f)$ minus $\pi/2$, or is equal to the phase of $^G\underline{i}_{r0}^f$ plus $\pi/2$. The vector graph in the grid voltage-orientated coordinates is shown in Figure 2.

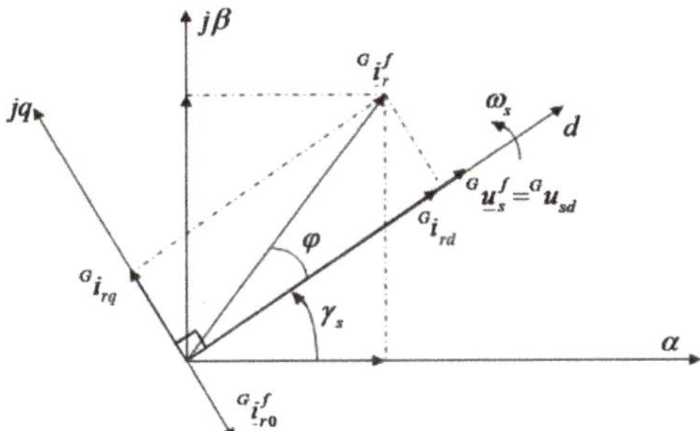

Figure 2. The doubly-fed induction machine plays the role of generation (DFIG) rotor current vector in the grid voltage-orientated coordinates.

3.3. Controlling the System in the Grid-Connected Mode

In the grid-connected mode, the DFIG stator feeds the current $^G\underline{i}_s^f$ to the grid. The DFIG rotor current is $^G\underline{i}_r^f = {}^G\underline{i}_{r0}^f + {}^G\underline{i}_{rt}^f$.

Where: $^G\underline{i}_{r0}^f$ is the DFIG rotor current in the grid-disconnected mode. $(^G\underline{i}_{rt}^f)$ is created in order that the DFIG stator feeds the current $^G\underline{i}_s^f$ to the grid. Substituting $^G\underline{i}_r^f = {}^G\underline{i}_{r0}^f + {}^G\underline{i}_{rt}^f$ and $^G\underline{i}_s^f$ for (9c), the DFIG stator flux is as follows:

$$^G\underline{\psi}_s^f = {}^G\underline{i}_s^f \cdot {}^GL_s + ({}^G\underline{i}_{r0}^f + {}^G\underline{i}_{rt}^f) \cdot {}^GL_m \tag{15}$$

Substituting $^G\underline{\psi}_s^f$ for (9a), the DFIG stator voltage is:

$$^G\underline{u}_s^f = {}^GR_s \cdot {}^G\underline{i}_s^f + {}^GL_s \cdot \frac{d({}^G\underline{i}_s^f)}{dt} + {}^GL_m \cdot \frac{d({}^G\underline{i}_{r0}^f)}{dt} + {}^GL_m \cdot \frac{d({}^G\underline{i}_{rt}^f)}{dt} + j \cdot \omega_s \cdot {}^G\underline{i}_s^f \cdot {}^GL_s + j \cdot \omega_s \cdot {}^G\underline{i}_{r0}^f \cdot {}^GL_m + j \cdot \omega_s \cdot {}^G\underline{i}_{rt}^f \cdot {}^GL_m \tag{16}$$

The voltage in ^GR_s is very small, so this voltage can be ignored. Thus, $^G\underline{u}_s^f$ is written as:

$$^G\underline{u}_s^f = {}^GL_s \cdot \frac{d({}^G\underline{i}_s^f)}{dt} + {}^GL_m \cdot \frac{d({}^G\underline{i}_{r0}^f)}{dt} + {}^GL_m \cdot \frac{d({}^G\underline{i}_{rt}^f)}{dt} + j \cdot \omega_s \cdot {}^G\underline{i}_s^f \cdot {}^GL_s + j \cdot \omega_s \cdot {}^G\underline{i}_{r0}^f \cdot {}^GL_m + j \cdot \omega_s \cdot {}^G\underline{i}_{rt}^f \cdot {}^GL_m \tag{17}$$

The DFIG stator voltage must coincide with the grid voltage and be steady. Combining ${}^{G}\underline{u}_{s}^{f}$ in Equation (17) with ${}^{G}\underline{u}_{s}^{f}$ in Equation (11):

$$ {}^{G}\underline{i}_{rt}^{f} = -({}^{G}L_{s}/{}^{G}L_{m})\cdot{}^{G}\underline{i}_{s}^{f}. \tag{18} $$

Analyzing the components of ${}^{G}\underline{i}_{rt}^{f}$:

$$ \begin{cases} {}^{G}i_{sd} = -({}^{G}L_{m}/{}^{G}L_{s})\cdot{}^{G}i_{rtd} & (19a) \\ {}^{G}i_{sq} = -({}^{G}L_{m}/{}^{G}L_{s})\cdot{}^{G}i_{rtq} & (19b) \end{cases} $$

3.4. Controlling the DFIG Stator Power

In the grid voltage-orientated coordinates, ${}^{G}u_{sq} = 0$, so the DFIG stator power is [14,15]:

$$ \begin{cases} P = (3/2)\cdot{}^{G}u_{sd}\cdot{}^{G}i_{sd} & (20a) \\ Q = (3/2)\cdot{}^{G}u_{sd}\cdot{}^{G}i_{sq} & (20b) \end{cases} . $$

Substituting ${}^{G}i_{sd}$ (19a) for (20a) and ${}^{G}i_{sq}$ (19b) for (20b):

$$ \begin{cases} P = (-3/2)\cdot{}^{G}u_{sd}\cdot{}^{G}i_{rtd}\cdot({}^{G}L_{m}/{}^{G}L_{s}) & (21a) \\ Q = (-3/2)\cdot{}^{G}u_{sd}\cdot{}^{G}i_{rtq}\cdot({}^{G}L_{m}/{}^{G}L_{s}) & (21b) \end{cases} , $$

where:

$$ \begin{cases} {}^{G}i_{rtd} = G_{p}\cdot{}^{G}i_{rd0} \\ {}^{G}i_{rtq} = G_{q}\cdot{}^{G}i_{rq0} \end{cases} , \tag{22} $$

where: ${}^{G}i_{rq0}$ is created by ${}^{G}i_{rq0} = -{}^{G}i_{r0}^{f}$; and ${}^{G}i_{rd0}$ is created by rotating the vector ${}^{G}\underline{i}_{r0}^{f}$ by angles $\pi/2$. Substituting ${}^{G}i_{rtd}$ and ${}^{G}i_{rtq}$ for (21a,b):

$$ \begin{cases} P = -(3/2)\cdot G_{p}\cdot{}^{G}u_{sd}\cdot{}^{G}i_{rd0}\cdot({}^{G}L_{m}/{}^{G}L_{s}) = G_{p}\cdot X & (23a) \\ Q = -(3/2)\cdot G_{q}\cdot{}^{G}u_{sd}\cdot{}^{G}i_{rq0}\cdot({}^{G}L_{m}/{}^{G}L_{s}) = G_{q}\cdot Y & (23b) \end{cases} . $$

In the grid voltage-orientated coordinates, ${}^{G}u_{sd}$, ${}^{G}i_{rd0}$, ${}^{G}i_{rq0}$ are steady, so X and Y are steady. Thus, the reactive and active power can be controlled by adjusting G_{q} and G_{p} separately. The system block diagram is shown in Figure 3.

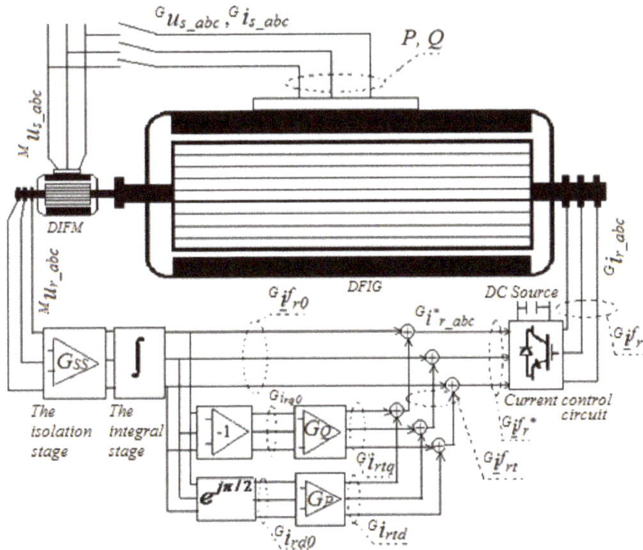

Figure 3. The system block diagram.

4. Building the Model to Demonstrate the Correctness of the Proposed Structure

In the above section, based on the DFIG equations, the author proposed a control structure and demonstrated its ability to control the system. For clarity, in this section, the author will build circuits and emulate systems using the Matlab-Simulink software (R2014b, MathWorks Inc., Natick, MA, USA).

Based on the system structure in Figure 3, building the system includes:

- DFIM with the stator connected to the grid and the rotor functioning in high-resistance mode.
- The isolation stage, which is a circuit with a high-resistance input, so the DFIM rotor is connected to the high resistance.
- The integral stage, which is a circuit in which output signals are created by integrating the input signal.
- The current control circuit, which creates the currents fed into the DFIG rotor.
- The $e^{j\pi/2}$ stage, which creates the output signals in order that the phases of the output signals are equal to the phases of the input signals plus $\pi/2$.
- The amplifying stage G_q: Based on (23b), the reactive power of the DFIG stator can be controlled by adjusting G_q.
- The amplifying stage G_p: Based on (23a), the active power of the DFIG stator can be controlled by adjusting G_p.
- DFIG, which is the doubly-fed induction generator that creates the currents fed into the grid.

All of the above stages are available in Matlab-Simulink. These stages' fabrication techniques in practice are very easy. The details are as follows:

- Four amplifier stages, such as G_{ss}, G_p, G_q, (-1), and one integral stage are available in the library of Matlab-Simulink.
- The $e^{j\pi/2}$ stage is the three-phase circuit, with a phase difference between each two input phases that is a $2\pi/3$. For example, working on the A-phase, the phase of the output signal (S_a') is equal to the phase of the input signal (S_a) plus $\pi/2$. The signal (S_a') is created by the formula (Figure 4a) [22]:

$$S_a' = (2/\sqrt{3}) \cdot (0.5 S_a + S_c). \tag{24}$$

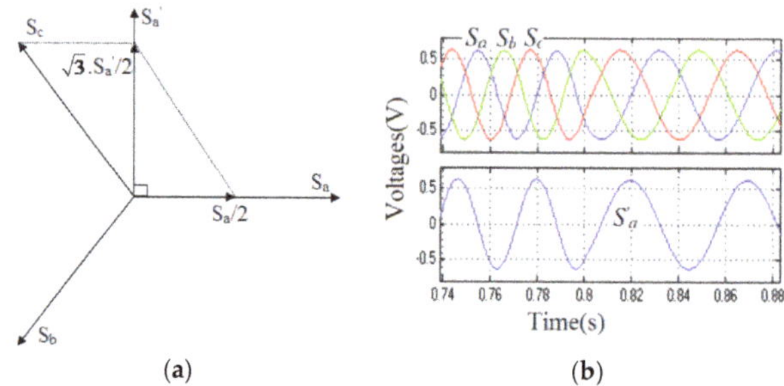

Figure 4. The $e^{j\pi/2}$ stage: (**a**) the vector graph; (**b**) the signals.

Running the $e^{j\Pi/2}$ stage with input signals (S_a, S_b, S_c), the A-phase output signal (S_a') is presented in Figure 4b.

The current control circuit has three phases (Figure 5a), and each phase is controlled by two IGBTs independently. For example, the A-phase with the desired current value (i^*_{ra}) and the actual current value (i_{ra}). If i^*_{ra} is less than i_{ra}, G1_1 is turned off, and G2_1 is turned on to decrease i_{ra}; otherwise, G1_1 is turned on, and G2_1 is turned off to increase i_{ra}. Running this circuit, the results are shown in Figure 5b.

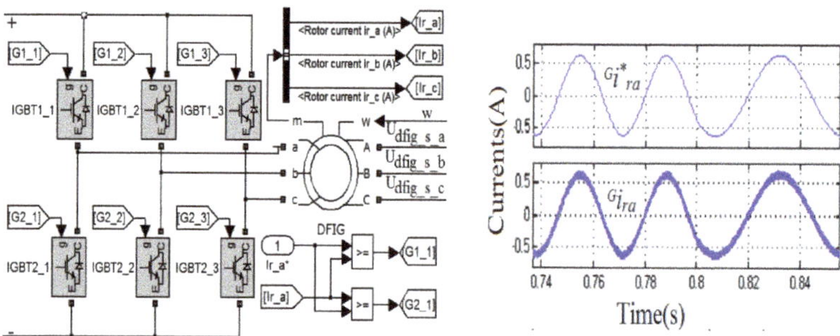

Figure 5. The current control stage: (**a**) the circuit; (**b**) the currents.

- The other stages: The other generator, 600V-1MVA, the transformer, TR1, the wire, L, the rectifier circuit with three level bridges, the voltage measurement device, the current measurement device, the breaker, etc.

Finally, the simulation model is shown in Figure 6, and the parameters of DFIG and DFIM are shown in Table 1.

Table 1. The parameters of DFIG and DFIM.

	U (V)	f (HZ)	S (VA)	L_s (H)	R_s (Ω)	L_m (H)	L_r (H)	R_r (Ω)	p
DFIG	400	60	1,000,000	3.9×10^{-4}	1.56×10^{-3}	0.0121	3.95×10^{-4}	1.62×10^{-3}	2
DFIM	400	60	1500	6.93×10^{-3}	0.512	0.0511	3.92×10^{-3}	0.690	2

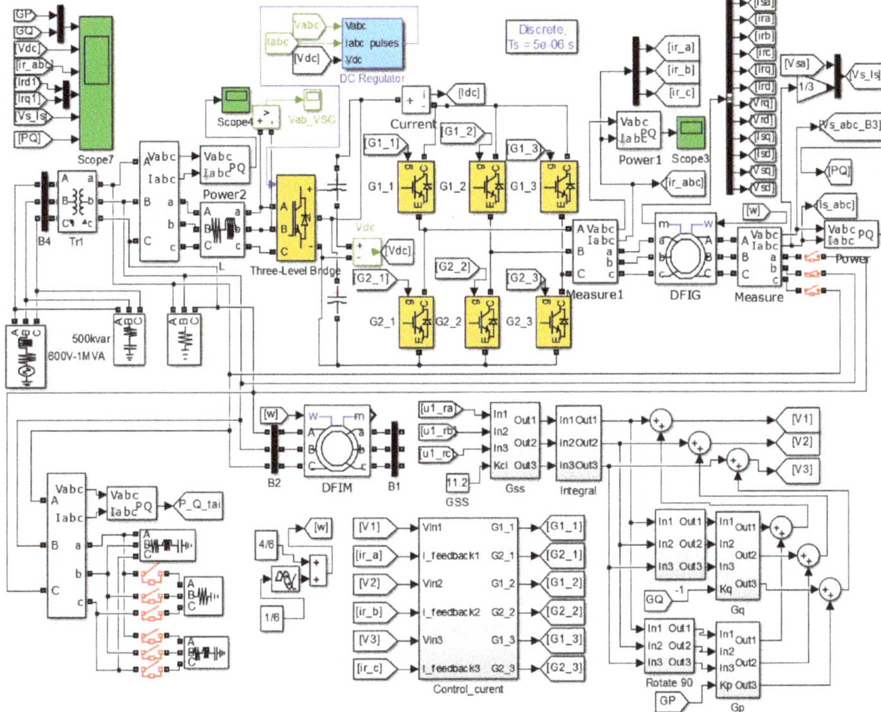

Figure 6. The simulation system model.

5. The Results and Discussion

5.1. Before the DFIG Stator Is Connected to the Grid (the Grid-Disconnected Mode)

G_p and G_q are set to zero. The multiplier G_{ss} is adjusted in order that the DFIG stator voltage coincides with the grid voltage. The signals during the adjustment of G_{ss} are shown in Figure 7.

The DFIG stator generates the voltages (example in the A-Phase: $^G u_{sa}$), with the phase and frequency being always equal to the phase and frequency of the grid voltage. Thus, in order for the DFIG stator to be connected to the grid, we only need to adjust the amplitude of $^G u_{sa}$. The simulation results show that if the value of G_{ss} increases, the amplitude of $^G u_{sa}$ increases. If the value of G_{ss} decreases, the amplitude of $^G u_{sa}$ decreases. At time=1.6s and setting $G_{ss} = 11.2$, the amplitude of $^G u_{sa}$ is equal to the amplitude of the grid voltage ($^M u_{sa}$), so the DFIG stator can be connected to the grid.

Next, the model was run to test the ability of the DFIG stator voltage ($^G u_{sa}$) to coincide with the grid voltage ($^M u_{sa}$) in the case of the changing of the rotor speed (ω) and the grid voltage.

The ability to coincide with the grid voltage of the generator in the case of the changing of ω is shown in Figure 8. When ω is closer to 1 pu (synchronous speed), the frequency and amplitude of DFIM rotor voltages ($^M u_{ra}, ^M u_{rb}, ^M u_{rc}$) are reduced, the DFIG rotor current ($^G i_{ra}, ^G i_{rb}, ^G i_{rc}$) and amplitudes are steady, and the frequencies are reduced. When ω is equal to 1 pu,$^M u_{ra}, ^M u_{rb}, ^M u_{rc}$ are zero, and $^G i_{ra}, ^G i_{rb}, ^G i_{rc}$ become the steady ones. Finally, $^G u_{sa}$ and $^M u_{sa}$ always have an equal frequency, equal amplitude, and equal phase. Thus, in the case of the changing of the rotor speed, the DFIG stator voltage always coincides with the grid voltage.

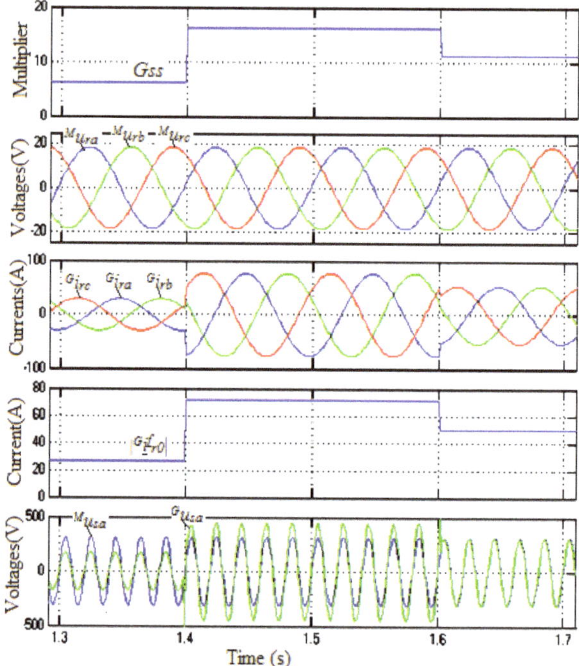

Figure 7. The signals during an adjustment of G_{ss}.

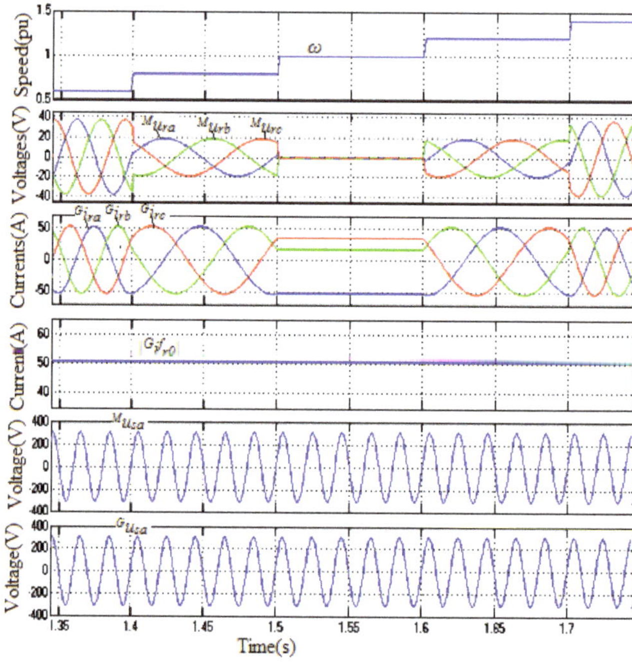

Figure 8. The ability to coincide with the grid-voltage of the generator when changing the rotor speed.

In addition, the simulation results in Figure 8 show that when changing the rotor speed (ω), the DFIG rotor current ($^{G}i_{r0}^{f}$) in the grid voltage-orientated coordinates is constant, and this result coincides with the above conclusions. $^{G}i_{r0}^{f}$ is constant, so $^{G}i_{r0}^{f}$ is the base component for modulating the d, q axis component of the DFIG rotor current ($^{G}i_{rd}$, $^{G}i_{rq}$) in the grid voltage-orientated coordinates.

When the grid voltage ($^{M}u_{sa}$) is reduced, the ability to coincide with the grid-voltage of the generator is shown in Figure 9. The DFIM rotor voltages ($^{M}u_{ra}$, $^{M}u_{rb}$, $^{M}u_{rc}$) and the DFIM rotor currents ($^{G}i_{ra}$, $^{G}i_{rb}$, $^{G}i_{rc}$) are changed, but the DFIG stator voltage ($^{G}u_{sa}$) always coincides with the grid voltage ($^{M}u_{sa}$).

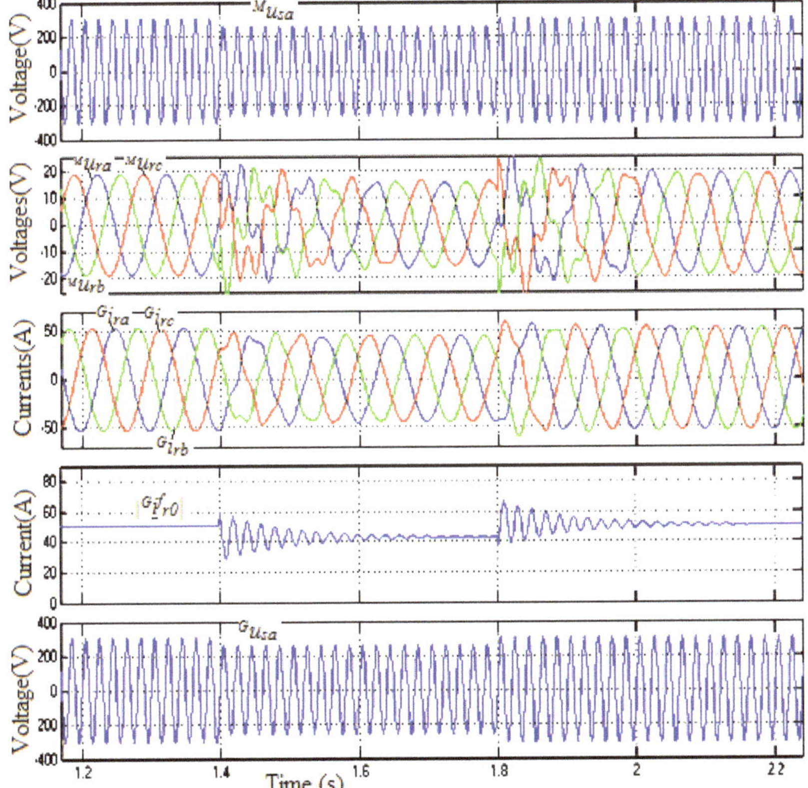

Figure 9. The ability to coincide with the grid-voltage of the generator when changing the grid-voltage.

In grid-disconnected mode, the conclusion is that, after adjusting G_{ss}, the DFIG stator voltage always coincides with the grid voltage, even when the grid voltage and the rotor speed are changing. This is a very good condition for connecting the DFIG stator to the grid.

5.2. Controlling the System in the Grid-Connected Mode

When the DFIG stator is connected to the grid, it is very easy to control Q and P separately by adjusting G_q and G_p. The simulation results are shown in Figure 10.

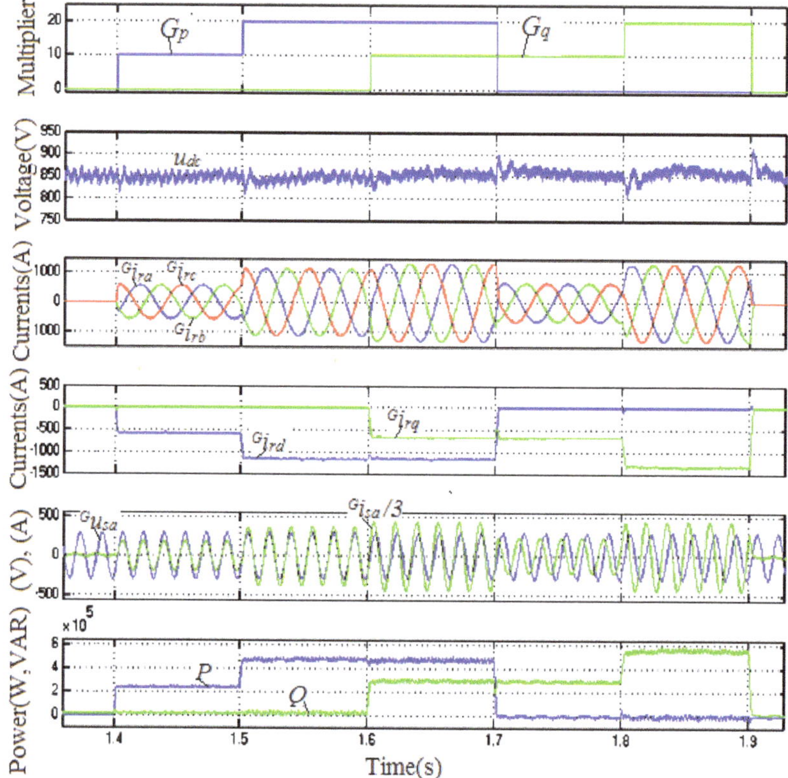

Figure 10. The process of adjusting P and Q through G_p and G_q.

In the period before 1.4 s and after 1.9 s, setting G_p and G_q to zero, the result is $^{G}i_{sa} = 0$ (A), so $P = 0$ and $Q = 0$.

In the period from 1.4 s to 1.6 s, $G_p \neq 0$ and $G_q = 0$. The simulations show that the phase of the DFIG stator current is equal to the phase of the grid voltage (example in the A Phase: $^{G}i_{sa}$, $^{G}u_{sa}$), so the DFIG stator feeds the active power to the grid. In addition, if G_p increases by double, the DFIG stator current amplitude increases by double, and the active power increases by double.

From 1.7 s to 1.9 s, $G_q \neq 0$ and $G_p = 0$. The simulations show that the phase of the DFIG stator current is faster than the phase of the grid voltage at an angle of $\pi/2$, so the DFIG stator feeds the reactive power to the grid. In addition, if G_q increases by double, the stator current amplitude increases by double, and the reactive power increases by double.

Thus, these simulation results show that it is very easy to control Q and P by adjusting G_q and G_p separately.

Next, when changing the rotor speed (ω), the stability of the system is shown in Figure 11. Setting $G_p = 10$ and $G_q = 0$, the simulation results show that the DFIG rotor currents ($^{G}i_{ra}$, $^{G}i_{rb}$, $^{G}i_{rc}$) are changed, but the DFIG stator current (example in the A Phase: $^{G}i_{sa}$) is not changed, so P and Q are not changed. Thus, it is confirmed that the system is stable when the rotor speed is changed.

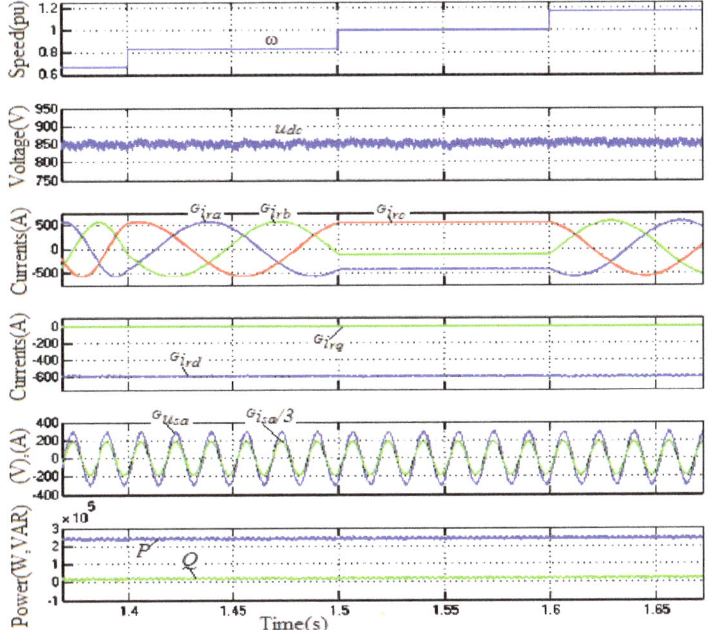

Figure 11. The stability of the system when the rotor speed is changed.

The reaction of the system in the case of the changing of the grid voltage is shown in Figure 12.

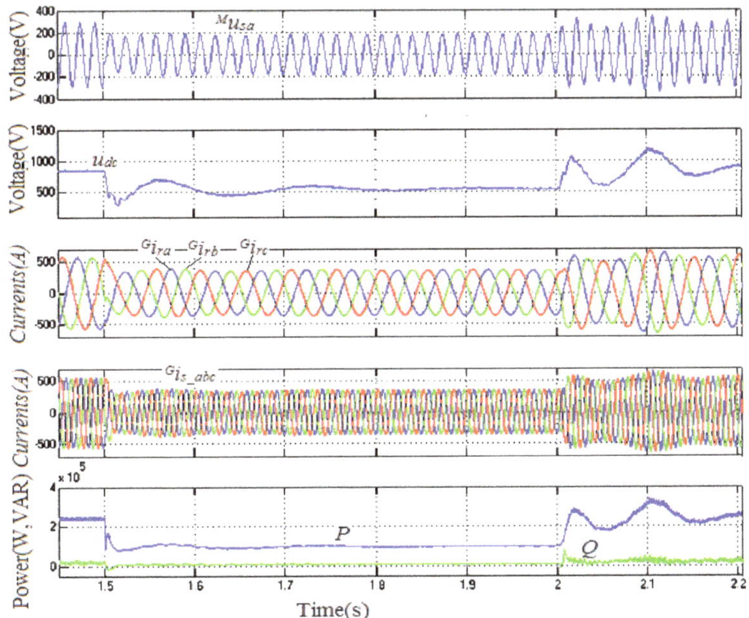

Figure 12. The reaction of the system when the grid voltage is changed.

With simple conventional power generation systems, when the DFIG stator is connected to the grid, if the grid voltage is reduced, the voltage difference between the generator terminal and the grid is increased rapidly, so the generator is over-current. However, in this proposed system, the simulation results show that, if the grid voltage is reduced, the DFIG stator current fed into the grid is decreased, so the DFIG stator power fed into the grid is decreased. Thus, the natural reaction of the system is suitable for the case of dropping the grid voltage, because the generator is not over-current.

In summary, the simulation results show that, when the DFIG is not connected to the grid, the natural characteristics of the system are such that the DFIG stator voltages and the grid voltages have an equal phase and equal frequency. We only need to adjust the amplitude of the DFIG stator voltage to make it equal to that of the grid-voltage by adjusting G_{ss}. After this, the DFIG stator voltages always coincide with the grid voltages, even when the grid voltage and the rotor speed are changed. This is a very good condition for connecting the stator DFIG to the grid. In the grid-connected mode, it is very easy to control P and Q by adjusting G_p and G_q separately. Therefore, controlling the power of the generator fed into the grid will be convenient and effective.

6. Conclusions

This paper presents a novel method for controlling DFIG connected to the grid. This new method has been fully demonstrated by the author in both theory and simulation. Compared with the previous system, the proposed system needs small DFIM more to generate the rotor signal, but it ignores the encoder. The natural feature of the proposed system is that the output voltage always coincides with the grid voltage. All stages are very simple, and the control system does not need the coordinate conversion stages, so it is easy and cost-effective to fabricate the generator system.

When the DFIG stator is disconnected from the grid, the DFIG stator voltages and the grid voltages have an equal phase and equal frequency. Thus, in order for the DFIG stator to be connected to the grid, we only need to adjust the amplitude of the DFIG stator voltage by adjusting G_{ss}. After this, the DFIG stator voltages always coincide with the grid voltages, even when the grid voltage and the rotor speed are changed. This is a very good condition for connecting the stator DFIG to the grid.

When the DFIG stator is connected to the grid, it is very easy to control P and Q by adjusting G_p and G_q separately.

Based on this novel method, the structure of the control system is very simple, and the generation system operates easily and effectively. This system is suitable for energy generation system applications with a variable speed, particularly on ships with an unstable grid voltage.

However, the limitation of this study is that it does not offer a solution to the problem of the generator supplying loads independently when the generator is not connected to the grid. Further studies will overcome the above limitation and present practical application results for ships. Therefore, it will be possible to show more clearly the effectiveness of the proposed method.

Funding: This research received no external funding.

Conflicts of Interest: The author declares no conflict of interest.

Nomenclature

Symbol	Unit	Definition
$\underline{u}_s^f, \underline{u}_r^f$	V	Stator, rotor voltage vector in the grid voltage-orientated coordinates
$\underline{i}_s^f, \underline{i}_r^f$	A	Stator, rotor current vector in the grid voltage-orientated coordinates
$\underline{\psi}_s^f, \underline{\psi}_r^f$	Wb	Stator, rotor flux vector in the grid voltage-orientated coordinates
$\underline{i}_s^r, \underline{i}_r^r$	A	Stator, rotor current vector in the rotor shaft-orientated coordinates
$\underline{u}_r^r$	V	Rotor voltage vector in the rotor shaft-orientated coordinates
$\underline{u}_s^s$	V	Stator voltage vector in the stator-fix-orientated coordinates
$\underline{\psi}_s^s$	Wb	Stator flux vector in the stator-fix-orientated coordinates
R_s, R_r	Ω	Stator, rotor resistance

$L_s, L_r L_m$	H	Stator, rotor, mutual inductance
ω_s, ω_r	pu	Stator, rotor electrical angular velocity
$\omega\omega$	pu	Rotor speed
P	W	Active power
Q	VA	Reactive power
i_{sd}, i_{sq}	A	d, q components of the stator current
i_{rd}, i_{rq}	A	d, q components of the rotor current
u_{rd}, u_{rq}	V	d, q components of the rotor voltage
u_{sd}, u_{sq}	V	d, q components of the stator voltage

References

1. Rana, M.M.; Li, L.; Su, S.W. Controlling the renewable microgrid using semidefinite programming technique. *Int. J. Electr. Power Energy Syst.* **2017**, *84*, 225–231. [CrossRef]
2. Rana, M.M.; Li, L.; Su, S.W. Distributed dynamic state estimation over a lossy communication network with an application to smart grids. In Proceedings of the 2016 IEEE 55th Conference on Decision and Control (CDC), Las Vegas, NV, USA, 12–14 December 2016; pp. 6657–6662.
3. MAN B&W Diezel A/S, Shaft Generators for the MC and ME Engines, Denmark. 2004. Available online: http://marineengineering.co.za/education/information/electrical (accessed on 30 March 2018).
4. Xia, K.; Zhang, Z.; Wang, N.; Zhang, P. Operation control and simulation research of the variable-speed constant-frequency system of the ship shaft generator. In Proceedings of the 2016 IEEE Region 10 Conference (TENCON), Singapore, 22–25 November 2016; pp. 301–304.
5. Gao, J.; Wan, S.; Liu, L. The research of marine shaft generator system based on brushless doubly-fed machine. In Proceedings of the 17th International Conference on Electrical Machines and Systems, Hangzhou, China, 22–25 October 2014; pp. 151–155.
6. Bodson, M. Speed Control for Doubly Fed Induction Motors with and Without Current Feedback. *IEEE Trans. Control Syst. Technol.* **2019**. [CrossRef]
7. Boutoubat, M.; Mokrani, L.; Zegaoui, A. Power quality improvement by controlling the Grid Side Converter of a wind system based on a DFIG. In Proceedings of the 2017 6th International Conference on Systems and Control (ICSC), Batna, Algeria, 7–9 May 2017; pp. 360–365.
8. Tang, H.; He, W.; Chi, Y.; Tian, X.; Li, Y.; Wang, Y. Impact of grid side converter of DFIG on sub-synchronous oscillation and its damping control. In Proceedings of the 2016 IEEE PES Asia-Pacific Power and Energy Engineering Conference (APPEEC), Xi'an, China, 25–28 October 2016; pp. 2127–2130.
9. Yao, J.; Li, H.; Chen, Z.; Xia, X.; Chen, X.; Li, Q.; Liao, Y. Enhanced control of a DFIG-based wind-power generation system with series grid-side converter under unbalanced grid voltage conditions. *IEEE Trans. Power Electron.* **2013**, *28*, 3167–3181. [CrossRef]
10. Martinez, M.I.; Tapia, G.; Susperregui, A.; Camblong, H. Sliding-mode control for DFIG rotor-and grid-side converters under unbalanced and harmonically distorted grid voltage. *IEEE Trans. Energy Convers.* **2012**, *27*, 328–339. [CrossRef]
11. Mahalakshmi, R.; Viknesh, J.; Ramesh, M.G.; Vignesh, M.R.; Sindhu Thampatty, K.C. Fuzzy Logic based Rotor Side Converter for constant power control of grid connected DFIG. In Proceedings of the 2016 IEEE International Conference on Power Electronics, Drives and Energy Systems (PEDES), Trivandrum, India, 14–17 December 2016; pp. 1–6.
12. Srirattanawichaikul, W.; Premrudeepreechacharn, S.; Kumsuwan, Y. A comparative study of vector control strategies for rotor-side converter of DFIG wind energy systems. In Proceedings of the 2016 13th International Conference on Electrical Engineering/Electronics, Computer, Telecommunications and Information Technology (ECTI-CON), Chiang Mai, Thailand, 28 June–1 July 2016; pp. 1–6.
13. Phan, V.-T.; Lee, H.-H. Performance enhancement of stand-alone DFIG systems with control of rotor and load side converters using resonant controllers. *IEEE Trans. Ind. Appl.* **2012**, *48*, 199–210. [CrossRef]
14. Xiao, S.; Geng, H.; Zhou, H.; Yang, G. Analysis of the control limit for rotor-side converter of doubly fed induction generator-based wind energy conversion system under various voltage dips. *IET Renew. Power Gener.* **2013**, *7*, 71–81. [CrossRef]

15. Peng, L.; Colas, F.; Francois, B.; Li, Y. A modified vector control strategy for DFIG based wind turbines to ride-through voltage dips. In Proceedings of the 2009 13th European Conference on Power Electronics and Applications, Barcelona, Spain, 8–10 September 2009; pp. 1–10.

16. Zhou, Y.; Bauer, P.; Ferreira, J.A.; Pierik, J. Operation of grid-connected DFIG under unbalanced grid voltage condition. *IEEE Trans. Energy Convers.* **2009**, *24*, 240–246. [CrossRef]

17. Tapia, A.; Tapia, G.; Ostolaza, J.X.; Saenz, J.R. Modeling and control of a wind turbine driven doubly fed induction generator. *IEEE Trans. Energy Convers.* **2003**, *18*, 194–204. [CrossRef]

18. Cardenas, R.; Pena, R.; Alepuz, S.; Asher, G. Overview of Control Systems for the Operation of DFIGs in Wind Energy Applications. *IEEE Trans. Ind. Electron.* **2013**, *60*, 2776–2798. [CrossRef]

19. Leonhard, W. *Control of Electrical Drives*; Springer-Verlag: New York, NY, USA, 2001.

20. Quang, N.P.; Districh, J.A. *Vector Control of Three-Phase AC Machines*; Springer: Berlin, Germany, 2008.

21. Thang, N.T.; Ban, N.T.; Hai, N.T. A novel method for excitation control of DFIG connected to the grid on the basis of similar signals from rotor. *Appl. Mech. Mater.* **2013**, *336*, 1153–1160. [CrossRef]

22. Trong, T.N.; Tien, B.N.; Thanh, H.N. Excitation control system of DFIG connected to the grid on the basis of similar signals from rotor. In Proceedings of the 2013 IEEE International Conference on Mechatronics and Automation, Takamatsu, Japan, 4–7 August 2013; pp. 738–742.

 *processes*

Article

Mold Level Predict of Continuous Casting Using Hybrid EMD-SVR-GA Algorithm

Zhufeng Lei * and Wenbin Su

School of Mechanical Engineering, Xi'an Jiaotong University, 28 West Xianning Road, Xi'an 710049, China;
wbsu@mail.xjtu.edu.cn
* Correspondence: leizhufeng@stu.xjtu.edu.cn

Received: 22 February 2019; Accepted: 22 March 2019; Published: 26 March 2019

Abstract: The prediction of mold level is a basic and key problem of continuous casting production control. Many current techniques fail to predict the mold level because of mold level is non-linear, non-stationary and does not have a normal distribution. A hybrid model, based on empirical mode decomposition (EMD) and support vector regression (SVR), is proposed to solve the mold level in this paper. Firstly, the EMD algorithm, with adaptive decomposition, is used to decompose the original mold level signal to many intrinsic mode functions (IMFs). Then, the SVR model optimized by genetic algorithm (GA) is used to predict the IMFs and residual sequences. Finally, the equalization of the predict results is reconstructed to obtain the predict result. Several hybrid predicting methods such as EMD and autoregressive moving average model (ARMA), EMD and SVR, wavelet transform (WT) and ARMA, WT and SVR are discussed and compared in this paper. These methods are applied to mold level prediction, the experimental results show that the proposed hybrid method based on EMD and SVR is a powerful tool for solving complex time series prediction. In view of the excellent generalization ability of the EMD, it is believed that the hybrid algorithm of EMD and SVR is the best model for mold level predict among the six methods, providing a new idea for guiding continuous casting process improvement.

Keywords: empirical mode decomposition; support vector regression; genetic algorithm; mold level; continuous cast

---

1. Introduction

In the modern steel industry, high efficiency continuous casting technology has become the most internationally competitive core technology. The continuous casting process is a complex and continuous phase change process. There are many factors that affect the quality of slabs [1]. The research of the key technologies and cores in the high quality steel continuous casting process is mainly focused on mold level precision, the segment, and secondary cooling dynamic control [2].

Mold level is non-linear and non-stationary in terms of the time scale and does not satisfy Gaussian normal distribution. Therefore, the development and adoption of an effective signal processing method to predict the Mold level is hugely challenging. This brings great difficulties for the prediction of Mold level. The prediction of Mold level is crucial for improving the adaptive control of the continuous casting process. Therefore, an accurate prediction of mold level cannot only guarantee the quality of slab products, but also improve the automation level of the continuous casting manufacturing industry.

Precise mold level monitoring is regarded as the key to improving continuous casting production quality [3,4]. The mold level is an important reference for casting speed control, segment roll gap control, mold cooling water control, and stopper rod opening control. In the continuous casting production process, the fluctuation of mold levels will cause large amounts of slag in the mold to be involved in the molten steel, which will seriously affect the quality of the slab and may even lead to accidents in the casting process, such as slab breakout and steel overflow at the top of the mold.

Continuous high mold level operation will lead to overflow accidents, where the impurities float on the liquid surface, resulting in surface defects of strand and internal defect of the cast product, which in turn affects the surface and internal quality of the slab. Lowering the casting speed results in excessive fluctuations in the mold level, which affects the productivity and production rhythm. This impacts the quality of the slab and causes unplanned shutdown because of the stick and damage of the tundish slide-gate.

The mold level predict model can control and maintain the mold level according to the mold level historical data when the casting speed is disordered. As shown in Figure 1, after the transition of tundish, hot metal enters into mold through slide-gate, and the lower end of slide-gate is under the surface of mold. Hot metal transforms into solidified shell in mold and enters the root roll segment through the narrow surface of mold. When the casting speed is over-low, the hot metal level in mold is over-high, will cause overflow accident; when the casting speed is over-high, the hot metal level in mold is over-low, which leads to the bulging of root roll segments and cause results in a break-out accident. The stopper rod can control the flow rate of the hot metal when it flows into the mold.

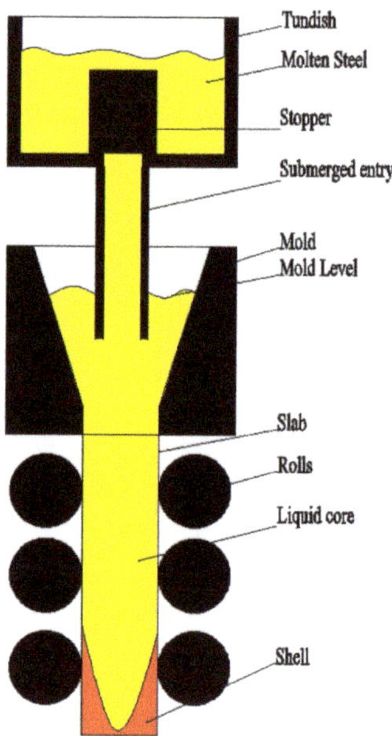

Figure 1. Mold level model.

Some researchers have adopted many methods for the prediction of time series. Numerical weather predict model is used for predicting future wind speed using mathematical models [5], multiple regression, exponential smoothing, autoregressive moving average model (ARMA), and many others for wind speed predict, power predict, and stock trend predict. Weron et al. [6] explained the complexities, strengths, and weaknesses of the available solutions for electricity price predicting, the opportunities and threats that predicting tools provide. However, the traditional time series predict methods, such as regression analysis and grey predict [7], have some shortcomings, where the prediction accuracy of signals with large fluctuations need to be improved [8].

In recent years, due to the rapid development of science and technology, artificial intelligence technology has been widely used and introduced into the prediction of time series [9]. Artificial neural networks (ANN) [10] and support vector regression (SVR) [11] methods are the main tools for dealing with non-linear, non-stationary time series. ANN is an artificial intelligence method developed in the 1940s that can simulate human brain biological processes [9]. Wang et al. [12] proposed a new back propagation neural network algorithm to apply to a semi-distributed model. Fei He et al. [13] gives an advance artificial intelligent technology based on the genetic algorithm (GA) and the back propagation (BP) neural networks. SVR is a small sample machine learning method, based on statistical learning theory, Vapnik-Chervonenkis (VC) dimension theory, and minimum structural risk principles [10]. Based on limited sample information, it seeks the best compromise between model complexity and learning ability to achieve the best promotion effect [14,15]. Y. Liu established a method for online predictions of the silicon content in blast furnace ironmaking processes [16]. Silvano Cincotti et al. [17] used the SVM model to forecast the electricity spot-prices of the Italian power exchange (IPEX), which provided a better prediction accuracy, closely followed by econometric technique. Existing studies have shown that the ANN method takes a long time to calculate and is prone to localized minimization [3,18–20], leading to overfitting and poor predict results. SVR is more adaptable to overfitting than ANN because the parameters of SVR can be improved by means of global optimization.

Many studies show that ANN or SVR methods were employed for the prediction of time series but only a few pieces of literature have combined the two methods [21]. For that reason, we apply the combination of empirical mode decomposition (EMD) and SVR for prediction of time series. The concept of hybrid predict appears in numerical weather prediction (NWP)-based predicts, such as NWP-ANN/ARMA for solar radiation predict [20]. Ye and Liu [22] used EMD-SVR for short-term wind power predict and achieved good predict results.

In this paper, we present a novel predict method for time series of mold level, based on the combination of EMD and SVR with global optimization. The results of simulation experiments display their effective and competitive advantages by using the proposed hybrid algorithm. First, the original mold level signal is decomposed by EMD into several intrinsic mode functions (IMFs). Then the improved SVR model is optimized by GA and used to predict the subsequences. Finally, the predict sequence is reconstructed to obtain the predict result. The rest of this paper is organized as follows. In Section 2, the basic algorithms EMD and SVR are introduced. In Section 3, we present a novel predict method for time series of mold level based on the combination of EMD and SVR with global optimization. Section 4 consists of experimental results and analysis. Finally, we conclude our work in Section 5.

2. Basic Algorithm Research

2.1. EMD Algorithm

EMD is an adaptive signal processing technique suitable for non-linear and non-stationary processes [23]. In 1998, Huang et al. [24] proposed the empirical mode decomposition technology, which has been widely used in biomedicine [25,26], speech recognition [27], system modeling [28–30], and process control [31,32]. Based on the time scales, EMD local features, such as local maxima, local minima, and zero-crossings, decompose the signal into several IMFs and a residual, the IMFs are orthogonal to each other. Modal decomposition is determined by the signal itself.

EMD satisfies the following basic assumptions:

(1)　In the entire data set, the number of extreme values and the number of zero crossings must be equal or at most have one point of difference.

(2)　At any point, the average is defined by the local maximum envelope, and the minimum envelope is zero.

EMD steps are shown in Figure 2.

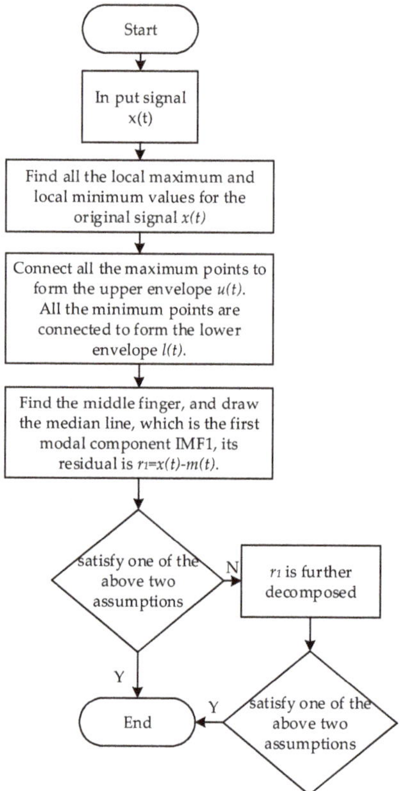

Figure 2. The empirical mode decomposition (EMD) algorithm.

Finally, the original signal is decomposed into:

$$x(t) = \sum_{i=1}^{N} c_i + r_N \tag{1}$$

where *x(t)* is the original signal, c_i is the IMF, N is the number of IMFs and r_N is the residual. $i = 1, 2, \ldots N$.

2.2. SVR Algorithm

SVR does not solve only the classification problem, but also solves the regression problem. The basic model is the largest linear classifier as defined in the feature space [33]. SVR aims to achieve a distinction between the samples by constructing a hyperplane for classification, so that the sorting interval between the samples is maximized and the sample to the hyperplane distance is minimized.

Set a training data set for a feature space:

$D = \{(x_1, y_1), (x_2, y_2) \ldots, (x_m, y_m)\},$

$x_i \in \chi = \Re^n, y_i \in y = \{+1, -1\}, i = 1, 2, \ldots, N$ where x_i is the *i*-th feature vector, y_i is the class

tag of x_i.

The corresponding equation of the classification hyperplane is as follows:

$$h(x) = \omega \cdot x + b \tag{2}$$

where x is the input vector, ω is the weight, b is the offset.

The classification decision function is as follows:

$$Sign(h(x))\tag{3}$$

$$\begin{cases} h(x) > 0, y_i = 1 \\ h(x) < 0, y_i = -1 \end{cases}\tag{4}$$

The support vector machine is implemented to find the ω and b when the interval between the separation hyperplane and the nearest sample point is maximized. When the training set is linearly separable, the sample points belonging to different classes can be separated by one or several straight lines with the largest interval. The maximum interval is solved by the following formula:

$$\max \gamma_i = y_i \left(\frac{\omega}{\|\omega\|} \cdot x_i + \frac{b}{\|\omega\|} \right)\tag{5}$$

$$s.t. y_i \left(\frac{\omega}{\|\omega\|} \cdot x_i + \frac{b}{\|\omega\|} \right) \geq \gamma, i = 1, 2, \ldots, N\tag{6}$$

Thus, we can obtain the linear separable support vector machine optimization problem.

$$\min_{\omega,b} \frac{1}{2} \|\omega\|^2\tag{7}$$

$$s.t. y_i (\omega \cdot x_i + b) - 1 \geq 0, i = 1, 2, \ldots, N\tag{8}$$

In the actual data set, there are many specific points, making the data set linear inseparable; in order to solve this problem, we introduce a slack variable for each sample point. $\xi_i \geq 0$ so that

$$y_i (\omega \cdot x_i + b) \geq 1 - \xi\tag{9}$$

For each slack variable ξ_i, pay a price ξ_i, and the optimization problem becomes:

$$\frac{1}{2} \|\omega\|^2 + C \sum_{i=1}^{N} \xi_i\tag{10}$$

where $C > 0$ is the penalty factor.

Most of the data in the actual data are linearly inseparable. Therefore, these data should be mapped to a high-dimensional feature space through non-linear mapping, and the non-linear problem is transformed into a linear problem. The linear indivisible problem is transformed into a linear separable problem.

The kernel functions are introduced as follows:

$$K(x_i, x_j) = \varphi(x_i) \cdot \varphi(x_j)\tag{11}$$

where the value of the kernel equals the inner product of two vectors, x_i and x_j.

At this point, we obtain:

$$W(\alpha) = \frac{1}{2} \sum_{i=1}^{N} \sum_{j=1}^{N} \alpha_i \alpha_j y_i y_j K(x_i, x_j) - \sum_{i=1}^{N} \alpha_i\tag{12}$$

where $\alpha_i > 0, i = 1, 2, \ldots, N$ is the lagrangian multiplier and N is the number of samples.

In this paper, the radial basis function (RBF) is chosen as the support vector machine kernel function, and the expression is as follows:

$$K(x_i, x) = \exp\left(\frac{-\|x_i - x\|^2}{2g^2}\right)$$ (13)

where g is the kernel function coefficient.

At this point, the classification function becomes:

$$f(x) = sign\left[\sum_{i=1}^{N} \alpha_i y_i \exp\left(\frac{-\|x_i - x\|^2}{2g^2}\right) + b\right]$$ (14)

3. Hybrid Algorithm Research

The accuracy of prediction for the mold level is influenced by many factors. In order to improve the accuracy of prediction for the mold level, a predicting model, based on a Hybrid predict algorithm for the mold level, is proposed.

First, the original signal is subjected to data pre-processing to remove singular points. Then all data is marked in the range of 0 to 1 to improve computational efficiency. Finally, the hybrid model is used for data predict.

3.1. Feasibility Analyses

EMD and SVR algorithm can be hybridized into an efficient hybrid algorithm. The reasons can be summarized as follows:

EMD is a decomposition algorithm that can decompose complex signals into simple signals. SVR is a regression algorithm. The fusion of EMD and SVR can helpful to predict complex signals, improve the ability of complex signal prediction, and it can obtain high-quality solutions. For this reason, the predict accuracy of hybrid algorithm is higher.

GA is a global optimization algorithm. The best model of SVR has a strong dependence on the kernel function parameters. GA can calculate the optimal SVR kernel function parameters faster and lay the foundation for the establishment of the best SVR model. Consequently, the efficiency of hybrid algorithm is higher.

EMD algorithm is an adaptive decomposition algorithm. There is no influence of human factors. The SVR model is more robust to nonlinear signals after optimization by the GA algorithm. Therefore, hybrid algorithm is more robust.

3.2. The Full Procedure of EMD-SVR$_{GA}$

EMD is an adaptive decomposition algorithm, based on the original signal, which can decompose complex signals into simple signals. Obviously, the prediction of simple signals is simpler than the prediction of complex signals, and the calculation cost is small. SVR is an excellent predictive algorithm, especially robust to nonlinear signals. GA algorithm is used to optimize parameters of the SVR kernel function, which further improves the SVR prediction accuracy. In order to make full use of the above algorithms, we present a hybrid algorithm, in which the GA is incorporated into the SVR. In EMD-SVR$_{GA}$, SVR is used to predict simple signals, which is decomposed by EMD, and the computational cost is greatly reduced. The GA algorithm optimizes the parameters of the SVR kernel function, improves the SVR calculation efficiency, and makes the hybrid algorithm achieve faster convergence speed. In addition, the adaptive characteristics of EMD and the robustness of SVR to nonlinear, non-stationary, non-Gaussian distributed signals further improves the generalization ability of the hybrid algorithm. The EMD-SVR$_{GA}$ algorithm can be described by the following steps as shown in Table 1, and the corresponding flowchart is shown in Figure 3.

Table 1. The EMD-SVR$_{\text{GA}}$ algorithm.

Step 1. Decompose the mold level signal data into several IMFs and one residual by the EMD.
Step 2. Global optimization of C and g in SVR is performed using GA to determine the SVR model.
Step 3. Predicted IMFs (PIMF) obtained by SVR for each IMF.
Step 4. Predicted residual sequence obtained by SVR for residual sequence.
Step 5. Sum all predicted IMFs and residual sequences to obtain the predicted signal.

The flow chart of EMD-SVR is shown in Figure 3.

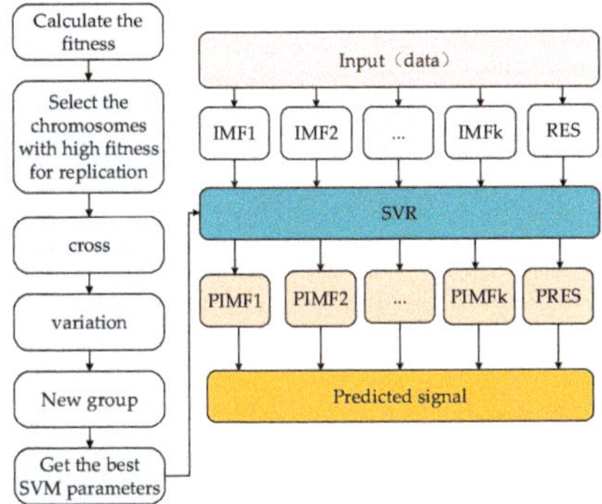

Figure 3. The flow chart of EMD-SVR. IMF: intrinsic mode functions; SVR: support vector regression; PIMF: predicted intrinsic mode functions.

From the above, we can see that GA gives the optimal parameters of the SVR kernel function, improves the convergence speed of SVR, EMD decomposes the original signal of mold level, reduces the complexity of mold level signal, and further predicted by SVR. The roles of the three algorithms are different. In a word, EMD performs signal decomposition, and the GA performs SVR kernel function parameter optimization to make the SVR obtain the prediction result precisely and faster.

4. Experiments Studies

4.1. Problem Prescription

In order to clearly express the applicability, superiority, and generalization capability of the model applications, the mold level data of actual process parameters are used in this paper. These were collected from the continuous casting machines developed by the China National Heavy Machinery Research Institute Co., Ltd., Xi'an, China. We used an eddy current sensor to collect the mold-level signal at a steady cast speed. The cast speed is 0.9 m/min, and the tundish temperature is 1562 °C. Most of the disturbances are non-linear and non-stationary, and the long-term predict model is difficult to establish. This paper presents mold level predict model is important for mold level control to propose new ideas to improve the continuous casting automatic control.

A continuous casting production process data acquisition graph is presented in Figure 4. The time interval $\Delta t = 1$ h, and the sampling frequency is 3 Hz.

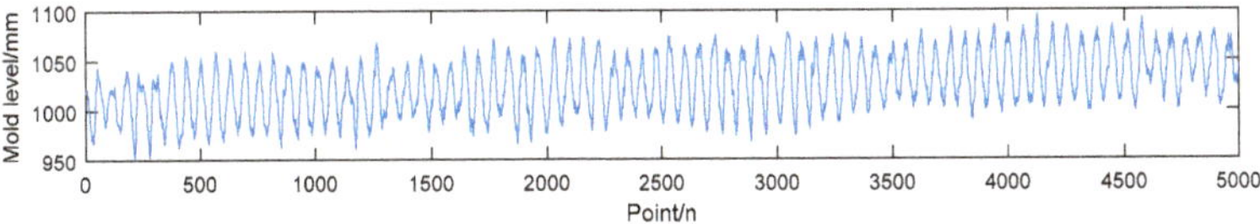

Figure 4. Mold level. The unit of mold level is mm, n is the number of point.

The main technical parameters of the continuous casting machine are shown in Table 2.

Table 2. Main technical parameters of the continuous casting machine.

Project	Specification
Continuous casting machine model	Curved continuous caster
Secondary cooling category	Aerosol cooling, dynamic water distribution
Gap control	Remote adjustment, dynamic soft reduction
Basic arc radius/mm	9500
Mold length/mm	900
Metallurgical length/mm	39,200
Mold vibration frequency/time/min	25–400
Mold vibration amplitude/mm	2–10
Slab width/mm	900–2150
Slab thickness/mm	230/250
Working speed/m/min	0.8–2.03

4.2. EMD-SVR

During the continuous casting production process, the data was intercepted for one hour, the singularity points were removed according to the Layda criteria, the data from the first 40 min was used as the training set, and the last 20 min of data was used as the test set to verify the validity of the model. Descriptive statistics of mold level data are also given in Table 3. Mold-level data EMD results is shown in Figure 5, there is a trend term in Figure 5, which clearly shows that the mold level data is non-stationary. As shown in Figure 6, C is 95.5729 and g is 0.39511, through the global optimization of the GA. Then the SVR model was determined, each IMF and residual sequence is predicted by the SVR model. The final predict signal, as shown in Figures 7 and 8.

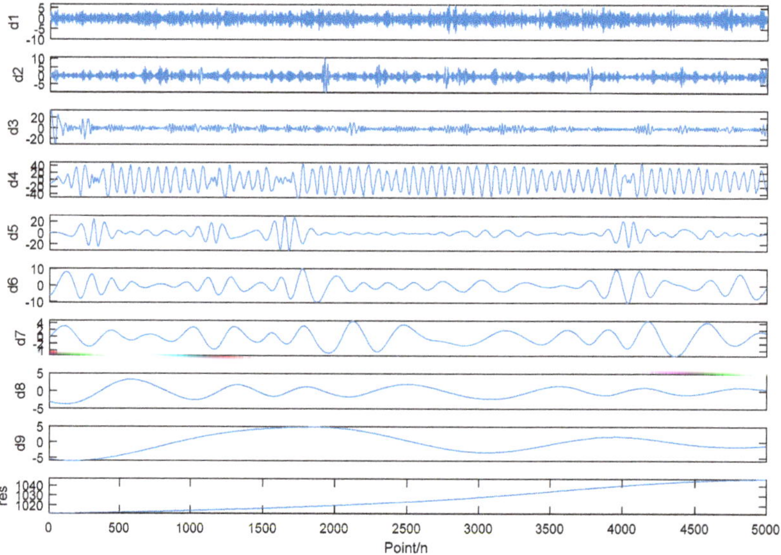

Figure 5. Mold-level data EMD results. d_i is the i-th IMF, the unit of d_i is mm, n is the number of Point, res is residual.

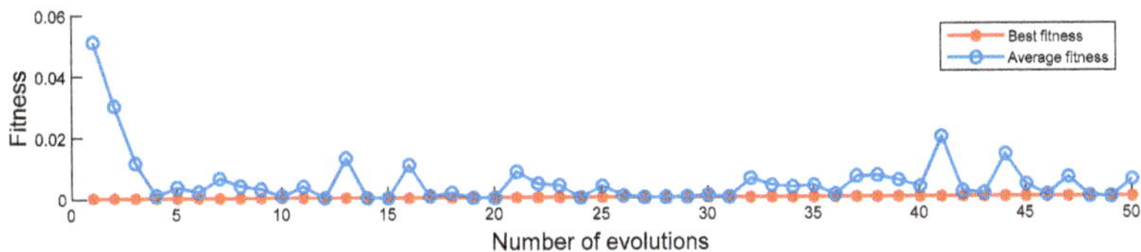

Figure 6. C and *g* optimization results.

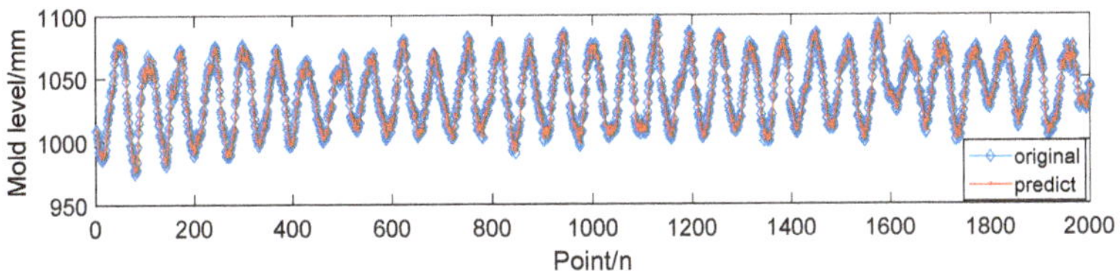

Figure 7. Comparison of EMD-SVR prediction results with original mold level data.

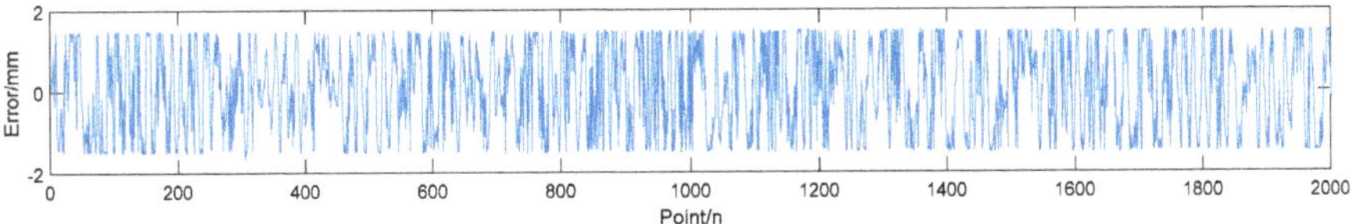

Figure 8. The EMD-SVR predict error.

Table 3. Descriptive statistics of mold level.

Project	Mold Level
Max-Min Values (mm)	951–1096
Mean (mm)	1026.7607
Standard Deviation (mm)	28.3495
Skewness	−0.02506
Kurtosis	2.1049

4.3. Experimental Results and Analyses

We made a comparison between our proposed EMD-SVRGA method and the five kinds of algorithms, i.e., the algorithms include WT-ARMA, WT-SVR, WT-SVR$_{GA}$, and EMD-ARMA, EMD-SVR$_{GA}$. The computational results of the above methods are listed in Table 3 in detail. The parameter values of the model are displayed in Table 4.

Table 4. Parameter values of the model. ACF is Autocorrelation coefficient, PACF is Partial autocorrelation coefficient.

	C	*g*	ACF	PACF
WT-ARMA	-	-	2	2
WT-SVR	100	1	-	-
WT-SVR$_{GA}$	16.3485	0.01773	-	-
EMD-ARMA	-	-	2	2
EMD-SVR	100	1	-	-
EMD-SVR$_{GA}$	95.5729	0.39511	-	-

The performances of the three hybrid methods models are verified by four statistical indicators in this paper, and the best hybrid predict model that is suitable for continuous casting process parameters is selected.

Correlations between the original data and the predict data, which is characterized by correlation coefficients (CC)

$$R = \frac{Cov(P, A)}{\sqrt{Var(P) \cdot Var(A)}} \tag{15}$$

CC is defined as a statistical indicator used to reflect the close relationship between variables, the larger the CC, the better the algorithm performance.

Root-mean-square Error (RMSE)

$$RMSE = \sqrt{\frac{\sum_{i=1}^{n} (P_i - A_i)^2}{n}} \tag{16}$$

RMSE is defined as reflect the degree of dispersion of a data set, measure the deviation between the observed value and the true value, the smaller the RMSE, the better the algorithm performance.

Mean Absolute Error (MAE)

$$MAE = \frac{\sum_{i=1}^{n} |P_i - A_i|}{n} \tag{17}$$

MAE is defined as average value of absolute error, reflect the actual situation of predict error better. The smaller the MAE, the better the algorithm performance.

Mean Absolute Percentage Error (MAPE)

$$MAPE = \frac{\sum_{i=1}^{n} \left| \frac{P_i - A_i}{A_i} \right|}{n} \times 100 \tag{18}$$

MAPE can be used to measure the outcome of a model predict. The smaller the MAPE, the better the algorithm performance.

Where P_i and A_i are the i-th predicted and actual values, respectively, and n is the total number of predict.

Predict model test results are shown in Table 5 and Figure 9.

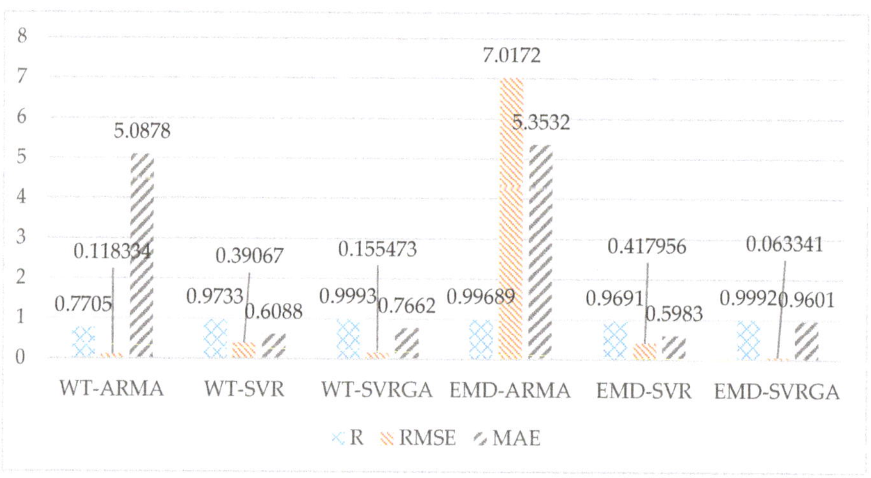

Figure 9. Comparison between EMD-SVR$_{GA}$ and other methods.

Table 5. Predict model test results.

	R	RMSE	MAE
WT-ARMA	0.7705	0.118334	5.0878
WT-SVR	0.9733	0.390670	0.6088
WT-SVR$_{GA}$	0.9993	0.155473	0.7662
EMD-ARMA	0.99689	7.0172	5.3532
EMD-SVR	0.9691	0.417956	0.5983
EMD-SVR$_{GA}$	0.9992	0.063341	0.9601

As shown in Table 5 and Figure 9, among the EMD based hybrid methods, the EMD-SVR had a better performance than EMD-ARMA, and EMD-SVR$_{GA}$ had better performance than EMD-SVR. Only the MAE is 0.9601, it is larger than the other methods. The other three indicators are all the best predictors of performance. Among the WT based hybrid methods, the WT-SVR had a better performance than WT-ARMA, and WT-SVR$_{GA}$ had better performance than WT-SVR, the performance of WT-SVR$_{GA}$ is the best, however, it is not stable. Only two indicators are better than the other two methods. RMSE is worse than WT-ARMA, and MAE is worse than WT-SVR.

The SVR based hybrid methods had better performance than ARMA based hybrid methods, due to the advantage of SVR as a non-linear data regression algorithm. The SVR based hybrid algorithm greatly improves the prediction accuracy compared to the traditional ARMA algorithm, and shows a strong generalization ability and robustness.

Among the SVR-based hybrid methods, R of the WT-SVR$_{GA}$ is close to the R of EMD-SVR$_{GA}$. Only MAE of WT-SVR$_{GA}$ is 0.7662, it is better than EMD-SVR$_{GA}$. Although the performance of WT-SVRGA predict is equivalent to that of EMD-SVR$_{GA}$, the wavelet transform is a kind of transcendental non-adaptive transformation. The transform effect depends on the selection of the basis function and the order of transformation. There are many human factors. For instance, the prediction effect is not stable, adaptability cannot be guaranteed to all data, and the generalization ability is far less than EMD-SVR$_{GA}$.

It is observed from the results that the R of SVR$_{GA}$-based hybrid method is improved by 0.2287 compared with the ARMA based hybrid method, RMSE is reduced by 6.9538. R of EMD hybrid method is reduced by 0.0001, RMSE is reduced by 0.054993. The EMD-SVR$_{GA}$ hybrid method greatly improves the prediction accuracy compared to the traditional ARMA algorithm, and shows a strong generalization ability and robust.

4.4. Predicting Accuracy Significance Tests

In order to verify the significant advantages of the proposed model in terms of prediction accuracy, some statistical tests are implemented in this section. Based on Dong et al. [34] and Fan et al. [35] suggested the Wilcoxon signed-rank test [36] and Friedman test [37], which are simultaneously applied in this paper.

The statistic W of the Wilcoxon signed-rank test:

$$W = \min\{r^+, r^-\} \tag{19}$$

The statistic F of the Friedman test:

$$F = \frac{12N}{k(k+1)}\left[\sum_{j=1}^{k} R_j^2 - \frac{k(k+1)^2}{4}\right] \tag{20}$$

where N is the total number of predict results; k is the number of compared models; R_j is the average rank sum obtained in each predict value for each compared model as shown in Equation (21):

$$R_j = \frac{1}{N} \sum_{i=1}^{N} r_i^j \qquad (21)$$

where r_i^j is the rank sum from 1 (the smallest predict error) to k (the worst predict error) for i-th predict result, for j-th compared model.

If the associated p-value of F meets the criterion of not acceptance, the null hypothesis, equal performance among all compared models, are also not held.

The results of Wilcoxon signed-rank test and Friedman test are listed in Table 6.

Table 6. Results of Wilcoxon signed-rank test and Friedman test.

| Compared Models | Wilcoxon Signed-Rank Test | | | | Friedman Test |
| | $\alpha = 0.025$ | | $\alpha = 0.05$ | | $\alpha = 0.05$ |
	h-Value	p-Value	h-Value	p-Value	
EMD-SVR$_{GA}$ vs. WT-ARMA	1	0	1	0	
EMD-SVR$_{GA}$ vs. WT-SVR	1	0	1	0	H_0: The results of the six algorithms are equal
EMD-SVR$_{GA}$ vs. WT-SVR$_{GA}$	1	0	1	0	$F = 35.8$
EMD-SVR$_{GA}$ vs. EMD-ARMA	1	0	1	0	$p = 0$ (Reject H_0)
EMD-SVR$_{GA}$ vs. EMD-SVR	1	0	1	0	

where α is the level of significance, p returns whether the population producing the two independent samples is the same significant probability, and h returns the result of the hypothesis test. If the overall difference between x and y is not significant, then h is zero. If the overall difference between x and y is significant, then h is 1. If p is close to zero, then the null hypothesis can be questioned. H_0 is the assumption of Friedman Test.

5. Conclusions

In this paper, a novel hybrid method, based on EMD-SVR$_{GA}$, is proposed. The proposed method is applied to predict the mold level. In this method, the original mold level signal decomposed to several IMFs by the EMD algorithm. SVR is optimized by GA, predicted IMFs and predicted residual sequences are obtained by optimized SVR. The predicted result is reconstructed by EMD. The use of EMD-SVR$_{GA}$ model can achieve accurate predictions of non-linear, non-stationary data. This model uses GA to improve the global search capability of the parameters in the SVR models and avoid falling into local optimization, thereby optimizing SVR algorithm for better accuracy. In this paper, six predict algorithms were calibrated by four statistical indicators. The experimental results demonstrate the reliability and validity of the proposed EMD-SVR$_{GA}$ model in predict Mold level. The model has a strong generalization, capability and robustness.

The precise prediction of mold level provides a new idea for the continuous casting process improvement. Short-term mold level prediction can effectively avoid confusion of production rhythm caused by crystallizer level fluctuations, and long-term predictions can effectively avoid accidents, such as slab breakout and steel overflow at the mold top. Accurate predictions of mold levels has important practical significance.

Author Contributions: W.S. conceived and designed the research; Z.L. performed the experiment and wrote the manuscript.

Funding: This work was financially supported by the National Natural Science Foundation of China (NO. 51575429).

Acknowledgments: Q.G., X.L., H.Z., B.H., and Y.Z. are acknowledged for their valuable technical support.

Nomenclature

EMD	Empirical mode decomposition
IMF	Intrinsic mode function
TS	Time series
ARMA	Autoregressive moving average
ACF	Autocorrelation function
PACF	Partial autocorrelation function
SVR	Support vector regression
GA	Genetic Algorithm
WT	Wavelet Transform
CC	Correlation Coefficient
RMSE	Root-mean-square error
MAE	Mean Absolute Error

References

1. Ataka, M. Rolling Technology and Theory for the Last 100 Years: The Contribution of Theory to Innovation in Strip Rolling Technology. *ISIJ Int.* **2015**, *55*, 89–102. [CrossRef]
2. Jin, X.; Chen, D.F.; Zhang, D.J.; Xie, X.; Bi, Y.Y. Water model study on fluid flow in slab continuous casting mould with solidified shell. *Ironmak. Steelmak.* **2011**, *38*, 155–159. [CrossRef]
3. Shen, B.Z.; Shen, H.F.; Liu, B.C. Water modelling of level fluctuation in thin slab continuous casting mould. *Ironmak. Steelmak.* **2009**, *36*, 33–38. [CrossRef]
4. Wang, X.; Zhang, S.; Yao, M.; Ma, H.; Zhang, X. Effect of casting process on mould friction during wide, thick slab continuous casting. *Ironmak. Steelmak.* **2014**, *41*, 464–473. [CrossRef]
5. Lynch, P. The origins of computer weather prediction and climate modeling. *J. Comput. Phys.* **2008**, *227*, 3431–3444. [CrossRef]
6. Weron, R. Electricity price predicting: A review of the state-of-the-art with a look into the future. *Int. J. Predict.* **2014**, *30*, 1030–1081. [CrossRef]
7. Lee, W.-J.; Hong, J. A hybrid dynamic and fuzzy time series model for mid-term power load predicting. *Int. J. Electr. Power Energy Syst.* **2015**, *64*, 1057–1062. [CrossRef]
8. Dai, S.; Niu, D.; Li, Y. Daily Peak Load Predicting Based on Complete Ensemble Empirical Mode Decomposition with Adaptive Noise and Support Vector Machine Optimized by Modified Grey Wolf Optimization Algorithm. *Energies* **2018**, *11*, 163. [CrossRef]
9. Wang, W.-C.; Chau, K.-W.; Cheng, C.-T.; Qiu, L. A comparison of performance of several artificial intelligence methods for predicting monthly discharge time series. *J. Hydrol.* **2009**, *374*, 294–306. [CrossRef]
10. Azad, H.B.; Mekhilef, S.; Ganapathy, V.G. Long-Term Wind Speed Predicting and General Pattern Recognition Using Neural Networks. *IEEE Trans. Sustain. Energy* **2014**, *5*, 546–553. [CrossRef]
11. Gaudioso, M.; Gorgone, E.; Labbe, M.; Rodriguez-Chia, A.M. Lagrangian relaxation for SVM feature selection. *Comput. Oper. Res.* **2017**, *87*, 137–145. [CrossRef]
12. Wang, J.; Shi, P.; Jiang, P.; Hu, J.; Qu, S.; Chen, X.; Chen, Y.; Dai, Y.; Xiao, Z. Application of BP Neural Network Algorithm in Traditional Hydrological Model for Flood Predicting. *Water* **2017**, *9*, 48. [CrossRef]
13. He, F.; Zhang, L. Mold breakout prediction in slab continuous casting based on combined method of GA-BP neural network and logic rules. *Int. J. Adv. Manuf. Technol.* **2018**, *95*, 4081–4089. [CrossRef]
14. Fan, G.-F.; Peng, L.-L.; Hong, W.-C.; Sun, F. Electric load predicting by the SVR model with differential empirical mode decomposition and auto regression. *Neurocomputing* **2016**, *173*, 958–970. [CrossRef]
15. Nie, H.; Liu, G.; Liu, X.; Wang, Y. Hybrid of ARIMA and SVMs for Short-Term Load Predicting. *Energy Procedia* **2012**, *16*, 1455–1460. [CrossRef]
16. Liu, Y.; Gao, Z. Enhanced just-in-time modelling for online quality prediction in BF ironmaking. *Ironmak. Steelmak.* **2015**, *42*, 321–330. [CrossRef]
17. Cincotti, S.; Gallo, G.; Ponta, L.; Raberto, M. Modeling and predicting of electricity spot-prices: Computational intelligence vs classical econometrics. *Ai Commun.* **2014**, *27*, 301–314. [CrossRef]

18. Ghosh, S.K.; Ganguly, S.; Chattopadhyay, P.P.; Datta, S. Effect of copper and microalloying (Ti, B) addition on tensile properties of HSLA steels predicted by ANN technique. *Ironmak. Steelmak.* **2009**, *36*, 125–132. [CrossRef]

19. Hong, W.-C. Chaotic particle swarm optimization algorithm in a support vector regression electric load predicting model. *Energy Convers. Manag.* **2009**, *50*, 105–117. [CrossRef]

20. Voyant, C.; Muselli, M.; Paoli, C.; Nivet, M.-L. Numerical weather prediction (NWP) and hybrid ARMA/ANN model to predict global radiation. *Energy* **2012**, *39*, 341–355. [CrossRef]

21. Ren, Y.; Suganthan, P.N.; Srikanth, N. A Comparative Study of Empirical Mode Decomposition-Based Short-Term Wind Speed Predicting Methods. *IEEE Trans. Sustain. Energy* **2015**, *6*, 236–244. [CrossRef]

22. Ye, L.; Liu, P. Combined Model Based on EMD-SVM for Short-term Wind Power Prediction. *Proc. CSEE* **2011**, *31*, 102–108.

23. Lei, Y.G.; Lin, J.; He, Z.J.; Zuo, M.J. A review on empirical mode decomposition in fault diagnosis of rotating machinery. *Mech. Syst. Signal Process.* **2013**, *35*, 108–126. [CrossRef]

24. Huang, N.E.; Shen, Z.; Long, S.R.; Wu, M.L.C.; Shih, H.H.; Zheng, Q.N.; Yen, N.C.; Tung, C.C.; Liu, H.H. The empirical mode decomposition and the Hilbert spectrum for nonlinear and non-stationary time series analysis. *Proc. R. Soc. A Math. Phys. Eng. Sci.* **1998**, *454*, 903–995. [CrossRef]

25. Bajaj, V.; Pachori, R.B. Classification of Seizure and Nonseizure EEG Signals Using Empirical Mode Decomposition. *IEEE Trans. Inf. Technol. Biomed.* **2012**, *16*, 1135–1142. [CrossRef] [PubMed]

26. Priya, A.; Yadav, P.; Jain, S.; Bajaj, V. Efficient method for classification of alcoholic and normal EEG signals using EMD. *J. Eng.* **2018**, *2018*, 166–172. [CrossRef]

27. Prasanna Kumar, M.K.; Kumaraswamy, R. Single-channel speech separation using combined EMD and speech-specific information. *Int. J. Speech Technol.* **2017**, *20*, 1037–1047. [CrossRef]

28. Guo, Z.; Zhao, W.; Lu, H.; Wang, J. Multi-step predicting for wind speed using a modified EMD-based artificial neural network model. *Renew. Energy* **2012**, *37*, 241–249. [CrossRef]

29. Tang, J.; Zhao, L.; Yue, H.; Yu, W.; Chai, T. Vibration Analysis Based on Empirical Mode Decomposition and Partial Least Square. *Procedia Energy* **2011**, *16*, 646–652. [CrossRef]

30. Xu, G.; Tian, W.; Qian, L. EMD- and SVM-based temperature drift modeling and compensation for a dynamically tuned gyroscope (DTG). *Mech. Syst. Signal Process.* **2007**, *21*, 3182–3188. [CrossRef]

31. Luo, L.; Yan, Y.; Xie, P.; Sun, J.; Xu, Y.; Yuan, J. Hilbert-Huang transform, Hurst and chaotic analysis based flow regime identification methods for an airlift reactor. *Chem. Eng. J.* **2012**, *181*, 570–580. [CrossRef]

32. Srinivasan, R.; Rengaswamy, R.; Miller, R. A modified empirical mode decomposition (EMD) process for oscillation characterization in control loops. *Control Eng. Pract.* **2007**, *15*, 1135–1148. [CrossRef]

33. Zhang, X.Y.; Zhou, J.Z. Multi-fault diagnosis for rolling element bearings based on ensemble empirical mode decomposition and optimized support vector machines. *Mech. Syst. Signal Process.* **2013**, *41*, 127–140. [CrossRef]

34. Dong, Y.; Zhang, Z.; Hong, W.-C. A Hybrid Seasonal Mechanism with a Chaotic Cuckoo Search Algorithm with a Support Vector Regression Model for Electric Load Predicting. *Energies* **2018**, *11*, 1009. [CrossRef]

35. Fan, G.-F.; Peng, L.-L.; Hong, W.-C. Short term load predicting based on phase space reconstruction algorithm and bi-square kernel regression model. *Appl. Energy* **2018**, *224*, 13–33. [CrossRef]

36. Wilcoxon, F. Individual comparisons by ranking methods. *Biom. Bull.* **1945**, *1*, 80–83. [CrossRef]

37. Friedman, M. A comparison of alternative tests of significance for the problem of m rankings. *Ann. Math. Stat.* **1940**, *11*, 86–92. [CrossRef]

 *processes*

Article

A Novel Robust Method for Solving CMB Receptor Model Based on Enhanced Sampling Monte Carlo Simulation

Wen Hou [1], **Yunlei Yang** [2], **Zheng Wang** [2], **Muzhou Hou** [2,*], **Qianhong Wu** [3] and **Xiaoliang Xie** [4]

[1] Lushan Binjiang Experimental School, Changsha 410013, China; hmzw@csu.edu.cn
[2] School of Mathematics and Statistics, Central South University, Changsha 410083, China;
    yunleiy@126.com (Y.Y.); zhengwang@csu.edu.cn (Z.W.)
[3] School of Geoscience and Info-Physics, Central South University, Changsha 410083, China;
    mzhoumcm@sina.com
[4] School of Mathematics and Statistics, and Mobile E-business Collaborative Innovation Center of Hunan
    Province, Hunan University of Commerce, Changsha 410205, China; hnucmath207@163.com
* Correspondence: houmuzhou@sina.com; Tel.: +86-137-8708-8322

Received: 15 February 2019; Accepted: 19 March 2019; Published: 23 March 2019

Abstract: The traditional effective variance weighted least squares algorithms for solving CMB (Chemical Mass Balance) models have the following drawbacks: When there is collinearity among the sources or the number of species is less than the number of sources, then some negative value of contribution will appear in the results of the source apportionment or the algorithm does not converge to calculation. In this paper, a novel robust algorithm based on enhanced sampling Monte Carlo simulation and effective variance weighted least squares (ESMC-CMB) is proposed, which overcomes the above weaknesses. In the following practical instances for source apportionment, when nine species and nine sources, with no collinearity among them, are selected, EPA-CMB8.2 (U.S. Environmental Protection Agency-CMB8.2), NKCMB1.0 (NanKai University, China-CMB1.0) and ESMC-CMB can obtain similar results. When the source raise dust is added to the source profiles, or nine sources and eight species are selected, EPA-CMB8.2 and NKCMB1.0 cannot solve the model, but the proposed ESMC-CMB algorithm can achieve satisfactory results that fully verify the robustness and effectiveness of ESMC-CMB.

Keywords: CMB receptor model; effective variance weighted least squares algorithm; enhanced sampling Monte Carlo simulation

---

1. Introduction

Atmospheric particulate matter (PM10 and PM2.5, with diameters less than 10 μm and 2.5 μm) is a mixture of solid or liquid particles suspended in the air, and is an important air pollutant in urban environments [1–3]. Epidemiological studies have shown that PM2.5/PM10 and an increase in respiratory symptoms, lung cancer mortality, and cardiovascular disease are closely related [4–10]. China is one of the countries with the most serious PM2.5 pollution in the world. In recent years, a total of 28 provinces and cities have reported heavy PM2.5 pollution phenomena; on average, each province has an annual total of nearly 20 days of heavy pollution.

At present, haze is frequent in China, affecting a wide range and having a long duration, which causes inconvenience to public life, threatens human health, and causes great concern for society and the government. Understanding and clarifying the potential sources and their contributions of PM2.5 is important [4]. The work of source apportionment of PM2.5 has become one of the core strategies in the prevention and control of atmospheric pollution.

The CMB (Chemical Mass Balance) air quality model [5,6] is the most important model of atmospheric particulate matter source apportionment technology [7], recommended by the United States' EPA (Environmental Protection Agency), mainly used to study the TSP (Total Suspended Particulate), PM2.5, PM10, and VOC (Volatile Organic Compounds) as well as other sources of pollutants and their contribution. CMB receptor models are established according to the principle of mass balance, and the chemical concentration of pollutants can be expressed by the sum of the product of the species richness and the source contribution.

The CMB receptor model [8,9] is composed of a set of linear equations, which indicates that the receptor concentration of each chemical element is equal to the linear sum of the product of the element content and the source contribution concentration. The basic principle of CMB model is mass conservation. It is assumed that there are several sources (J) that contribute to atmospheric particulates in the receptor, and that: (1) compositions of source emissions are constant over the period of ambient and source sampling; (2) the number of sources or source categories is less than or equal to the number of species; (3) the chemical composition of the particulate matter emitted by the various sources is significantly different; (4) the chemical composition of the particulate matter emitted by the source class is relatively stable; (5) all sources that make an obvious contribution to the receptors have their respective emission characteristics; (6) there is no interaction between the particles emitted by the source class, so the change in the process of transmission can be ignored; and (7) measurement uncertainties are random, uncorrelated, and normally distributed. Then the total substance concentration measured on the receptor is the linear sum of the contribution of each source.

The methods for solving CMB equations mainly include: (1) trace element method [10]; (2) linear programming solution [11]; (3) ordinary weighted least squares method [12]; (4) ridge regression weighted least squares [13]; (5) partial least squares [14]; (6) neural networks [15]; and (7) effective variance weighted least squares (EVWLS) with or without an intercept [16].

At present, the most commonly used algorithm for solving CMB model is the EVWLS method [17], which is derived by minimizing the weighted sums of the squares of the differences between the measured and the calculated values of C_i and F_{ij}, and is a practical method for calculating the contribution of the source S_j and the error σ_{S_j}:

$$\mathrm{min}\, m^2 = \sum_{i=1}^{I} \frac{\left(C_i - \sum_{j=1}^{J} F_{ij} \times S_j\right)^2}{V_{eff,i}}, \tag{1}$$

where the effective variance is $V_{eff,i} = \sigma_{C_i}^2 + \sum_{j=1}^{J} \sigma_{F_{ij}}^2 \times S_j^2$, σ_{S_j} (μgm^{-3} or g/g) is the uncertainty in source contribution S_j (μgm^{-3} or g/g), σ_{C_i} (μgm^{-3} or g/g) is the uncertainty in the ambient concentrations species i, and $\sigma_{F_{ij}}$ is the uncertainty in the fraction of species i in the source j profile.

The EVWLS method is actually an improvement over the ordinary weighted least squares method to minimize the sum of squares of the differences between the weighted chemical composition measurements and the calculated values.

However, there are some weaknesses to the above algorithms, such as near collinear sources resulting in incorrect source contributions, and the requirement that the number of chemical species be greater than or equal to the number of sources. At the same time, most of the above algorithms are finally transformed into optimization algorithms, which are mostly NP (Non-deterministic Polynomial) problems. So, in general, we get a locally optimal value or suboptimal value instead of a globally optimal value. So, instability is a fatal drawback to these algorithms, that is to say that different runs of the same input dataset at different times using the same algorithm may produce very different outputs or exhibit high variance with the same diagnostic criteria.

The Monte Carlo method [18], also known as stochastic simulation or statistical experiments, is based on statistical theory, according to the law of large numbers, using computer simulation technology [19] to solve some practical problem that is difficult to figure out directly with mathematical or other methods. The Monte Carlo method uses computer programs and mathematical models [20] to simulate practical random phenomena, through simulation experiments to get experimental data, and then infers from the analysis to get the law of certain phenomena. Monte Carlo simulation [19] is a method for exploring the solution and sensitivity of a complex system by varying the parameters within the statistical constraints. It is widely used in many fields such as engineering [21], environmental science [22], statistical physics [23], biophysics [24], materials science [25], and financial engineering [26]. Many practical problems are often accompanied by many random factors. If we take these factors into account, the model will become too complex to solve. However, we can utilize the Monte Carlo method to generate a random number to simulate these complicated phenomena, and then find out the operation law. The validity of the Monte Carlo method relies on the sampling process in simulation. However, the simple Monte Carlo algorithm converges too slowly, and it is easy to converge to local extreme points.

In this paper, we explore a novel robust method for solving CMB receptor model based on enhanced sampling Monte Carlo simulation, which overcomes the shortcomings of the above algorithms. In other words, when collinearity exists in the source profiles or the number of source profiles is greater than the number of species, the ESMC-CMB (Enhanced Sampling Monte Carlo CMB) algorithm can come to the correct results for source apportionment. In general, these enhanced sampling methods can be employed to help us quickly find an optimal stable solution when the model is complex, nonlinear, or involves more than just a couple uncertain parameters.

This paper is organized as follows. Section 2 provides a literature review about the CMB model and enhanced sampling Monte Carlo simulation. In Section 3, the proposed enhanced sampling Monte Carlo CMB algorithm (ESMC-CMB) is described. Section 4 presents the related numerical experiments and a comparison with various traditional algorithms. Finally, conclusions are given in Section 5.

2. CMB Model and Enhanced Sampling Monte Carlo Simulation

Methods commonly used for the particulate source apportionment include receptor model, source emission inventory, and source dispersion models. The source emission inventory method determines its contribution rate by investigating and accounting for emission factors and activity levels for different source categories. The source dispersion model is a combination of meteorological conditions, emission sources, and chemical processes to assess the distribution and contribution of different source classes in three dimensions [27]. The receptor model is a commonly used model in source apportionment.

In general, due to source j with constant emission rate E_j, the source contribution S_j present at a receptor during a sampling period of length T is

$$S_j = D_j \cdot E_j, \tag{2}$$

where:

$$D_j = \int_0^T d\left[\vec{u}(t), \sigma(t), \vec{x}_j\right] dt. \tag{3}$$

D_j is a dispersion factor depending on atmospheric stability (σ), wind velocity (u) and the location of source j with respect to the receptor (x_j). All parameters in Equation (2) vary with time, so the instantaneous D_j must be integrated over time period T [27].

The CMB receptor model consists of a solution of a linear equation that represents the chemical concentration of each receptor as the product of source profile abundance and source contribution. Resource profile abundances (i.e., mass fractions of certain chemicals or other properties emitted from each source) and receptor concentrations (estimated with appropriate uncertainties) are used as input data for CMB. In order to distinguish the contribution of source types, the measured chemical and physical properties must occur in different proportions of source emissions, and the changes of these proportions between source and recipient can be neglected or approximated. The CMB model calculates the contribution values of each source and the uncertainties of these values. The principle of the CMB receptor model is shown in Figure 1.

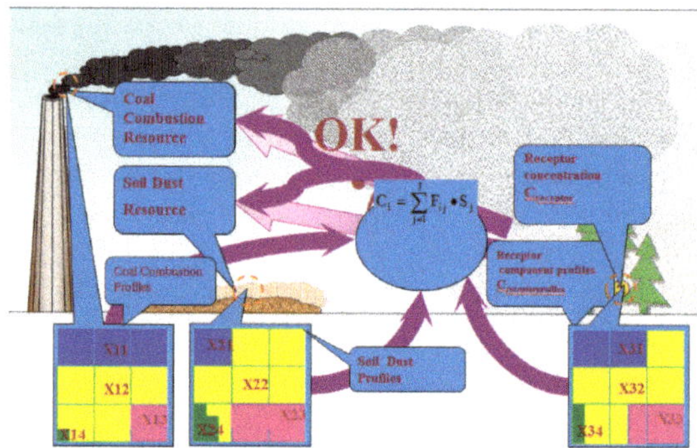

Figure 1. The principle of the Chemical Mass Balance (CMB) receptor model.

The receptor model was used to identify the source of the receptor and determine the quantitative contribution of various sources to the receptor by analyzing the chemical tracers of the source of the environmental samples and the emission sources. If there is no interaction between their emissions to cause mass removal, the total mass measured at the receptor C is a linear sum of the contributions of the individual sources S_j:

$$C = \sum_{j=1}^{J} D_j \cdot E_j = \sum_{j=1}^{J} S_j. \tag{4}$$

Similarly, the mass concentration of elemental component i, C_i, will be

$$C_i = \sum_{j=1}^{J} F_{ij} \cdot S_j i = 1, 2, \cdots, I, \tag{5}$$

where F_{ij} is the mass fraction of source contribution S_j composed of element i at the receptor. The number of chemical species (I) must be greater than or equal to the number of sources (J) for a unique solution to these equations.

Equations (4) and (5) are based on material immortality and mass conservation. In Equation (5), C_i and S_j are the inputs to the model, and F_{ij} is the source contribution we need to calculate.

There are several methods to solve the CMB receptor models: (1) the tracer element method [28]; (2) an ordinary weighted least squares solution [28]; (3) a linear programming solution [29], which maximizes the sum of the source contributions; (4) a ridge regression weighted least squares solution with or without an intercept [30] that is one approach for handling the multi-collinearity; (5) a neural networks solution; and (6) an EVWLS solution, which is the most common algorithm.

At present, the most commonly used algorithm to solve the CMB model is the effective variance least squares method, because this method is a practical method to calculate the error σ_{S_j} of source contribution S_j. The effective variance least squares method is actually an improvement on the ordinary weighted least squares method, which minimizes the sum of squares of the difference between the measured and calculated values of the weighted chemical components:

$$\min m^2 = \sum_{i=1}^{I} \frac{(C_i - \sum_{j=1}^{J} F_{ij} \times S_j)^2}{V_{eff,i}}, \tag{6}$$

where the effective variance is $V_{eff,i} = \sigma_{C_i}^2 + \sum_{j=1}^{J} \sigma_{F_{ij}}^2 \times S_j^2$, σ_{S_j} (μgm^{-3} or g/g) is the uncertainty in source contribution S_j (μgm^{-3} or g/g), σ_{C_i} (μgm^{-3} or g/g) is the uncertainty (i.e., measurement errors) in the ambient concentrations species i, and $\sigma_{F_{ij}}$ is the uncertainty (i.e., measurement errors) in the fraction of species i in the source j profile.

The matrix form of the CMB model is as follows:

$$C_{i \times 1} = F_{i \times j} S_{j \times 1}.$$ (7)

The steps of EVWLS iterative algorithm for solving the CMB model (Equation (7)) are as follows:

1. Set the initial estimate of the source contributions equal to zero:

$$S_j^{k=0} = 0 j = 1, 2, \cdots, J.$$ (8)

2. Calculate the diagonal components $V_{eff,i}$ of the effective variance matrix. All off-diagonal components of this matrix are equal to zero:

$$V_{eff,i}^k = \sigma_{C_i}^2 + \sum_{j=1}^{J} (S_j^k)^2 \times \sigma_{F_{ij}}^2.$$ (9)

3. Calculate the $K + 1$ value of S_j:

$$S_j^{k+1} = \left(F^T (V_e^k)^{-1} F \right)^{-1} F^T (V_e^k)^{-1} C.$$ (10)

4. If the result of Equation (10) is greater than 1%, the previous iteration is executed; if less than 1%, the iteration is terminated.

$$\text{If } \left| S_j^{k+1} - S_j^k \right| / S_j^{k+1} > 0.01, \text{ go to step 2. If } \left| S_j^{k+1} - S_j^k \right| / S_j^{k+1} \leq 0.01, \text{ go to step 5.}$$

5. Calculate the value of σ_{S_j} in the $K + 1$ step iteration, then

$$\sigma_{S_j} = \left[\left(F^T (V_e^{k+1})^{-1} F_{jj} \right)^{-1} \right]^{1/2} j = 1, 2, \cdots, J,$$ (11)

where $C = (C_1, \cdots, C_I)^T$ is a column vector with C_i as the ith component; $S = (S_1, \cdots, S_J)^T$ is a column vector with S_j as the jth component; F is an $I \times J$ matrix of F_{ij}, the source composition matrix; σ_{ci} is one standard deviation uncertainty of the C_i measurement; $\sigma_{F_{ij}}$ is one standard deviation uncertainty of the F_{ij} measurement; and V_e is diagonal matrix of effective variances.

The above algorithm shows that the input parameters of the model are: the measured values of the concentration spectrum of the chemical components of the receptor C_i and the standard deviation σ_{C_i} of C_i, the measured values F_{ij} of the source chemical composition spectrum and the standard deviation $\sigma_{F_{ij}}$ of F_{ij}. The output parameters of the model are: the calculated source contribution values of S_j and the standard deviation σ_{S_j} of S_j, the calculated source contribution values of chemical composition S_{ij}, and the standard deviation $\sigma_{S_{ij}}$ of S_{ij}.

In the actual work of source apportionment, there are two commonly used software tools, EPA-CMB8.2 (V8.2, EPA, Washington, USA, 2004) and NKCMB1.0 (V1.0, Nankai University, Tianjin, China, 2005), which are the concrete implementation of above effective variance least squares algorithm for solving the CMB model.

The CMB receptor model is one of the standard methods used by the U.S. Environmental Protection Agency (EPA) to assess air quality. The practical tool software EPA-CMB8.2 based on the CMB model and the effective variance least squares algorithm is recommended by the EPA. NKCMB1.0 is a practical software tool for PM2.5 source apportionment, developed by the Key Laboratory of Urban Air Particulate Pollution Prevention and Control, Nankai University, Tianjin China, based on the CMB receptor mathematical model and the corresponding effective variance least squares algorithm. NKCMB1.0 is more suitable for source analysis and calculation in China's more complex air quality environment.

As a stochastic method, Monte Carlo modeling can be used to describe and analyze complex problems by computer simulation sampling based on probability theory combined with certain statistical methods. Although the method emerged in the 1940s, it was limited to defense-related nuclear technology because it required sufficient computing resources to analyze the neutron behavior in matter [20]. With the rapid development of high-speed computers, the Monte Carlo simulation method is more and more widely used [19,20].

The basic idea of the Monte Carlo method is to establish an appropriate probability model or stochastic process so that its parameters (such as the probability of events, the mathematical expectation of random variables) are equal to the solution of the problem. Then repeated random sampling test of the model or process are carried out. With the statistical analysis to the results, the final calculation of the parameters, the approximate solution is obtained.

For example, in a Monte Carlo Simulation problem we represent the quantity we want to know as the expected value of a random variable Y, such as $\mu = E(Y)$. Then we generate values $Y_1, \cdots, Y_n$ randomly and independently from the distribution of Y and get their average:

$$\hat{\mu} = \frac{1}{n}\sum_{i=1}^{n} Y_i, \tag{12}$$

as the estimate of μ.

However, the convergence speed of the above simple sampling Monte Carlo method is too slow; for a large dimension sampling space, the time to complete the sampling calculation is intolerable.

This paper will explore a new, enhanced sampling method to accelerate the convergence of the algorithm from the following aspects.

Firstly, in the process of solving the receptor CMB model, if the diagnostic indicator $PM = \sum_{j=1}^{J} \eta_j = \sum_{j=1}^{J} S_j/C < \lambda$, the results did not meet the requirements. So we could sample in the following space $PM = \sum_{j=1}^{J} \eta_j = \sum_{j=1}^{J} S_j/C \geq \lambda$, for which the dimensions of the sample space will be reduced to some extent, and in the following experiment, $\lambda = 0.7$ will be selected. In the new sampling space, the Gibbs sampling method will be used.

Gibbs sampling [31–33] or a Gibbs sampler is a MCMC (Markov chain Monte Carlo) algorithm for obtaining a sequence of observations that are approximated from a specified multivariate probability distribution. Like other MCMC algorithms, Gibbs sampling from Markov chain can be regarded as a special case of the Metropolis-Hastings algorithm; its sampling distribution can be deduced from the properties of the Markov chain and probability transition matrix, and it finally converges to joint distribution. The name of the algorithm originated from Josiah Willard Gibbs and was proposed by brothers Stewart and Donald Gemman in 1984 [31–33]. Gibbs sampling is suitable for multivariate distribution, where conditional distribution is easier to sample than edge distribution. At the same time, in order to accelerate the convergence speed of the simulation process, in this paper we adopt the enhanced Gibbs sampling algorithm from [34], called the enhanced sampling algorithm for short.

In order to overcome the shortcomings of the effective variance algorithm for solving the CMB model, in this paper, the EVWLS (effective variance weighted least square) algorithm will be combined with the Monte Carlo simulation algorithm of enhanced sampling to obtain a novel robust ESMC-CMB algorithm for solving the CMB receptor model. The algorithm is programmed by using MATLAB (V8.5, Mathworks, Natick USA, 2015) and implemented through numerical experiments with a real background. By comparing with the results of EPA-CMB 8.2 and NKCMB 1.0, the accuracy, robustness, and superiority of ESMC-CMB algorithm are fully verified.

3. Solving CMB Model Based on Enhanced Sampling Monte Carlo Simulation

For the CMB model with consideration of random error:

$$C_i = \sum_{j=1}^{J} F_{ij} \cdot S_j + \varepsilon_i \, i = 1, 2, \cdots, I, \tag{13}$$

where C_i is the ambient concentration of species i, S_j is the source contribution of source j, F_{ij} is the fraction of species i in source j, ε_i is for errors. The number of chemical species (I) must be equal to or greater than the number of sources (J) for a unique solution to these equations. Equation (13) is solved by an effective variance weighted least squares approach: minimizing χ^2, where

$$\chi^2 = \sum_{i=1}^{I} \left[\frac{\left(C_i - \sum_{j=1}^{J} F_{ij} S_j \right)^2}{\sigma_{C_i}^2 + \sum_{j=1}^{J} \alpha_{F_{ij}}^2 S_j^2} \right]. \tag{14}$$

In the CMB model, uncertainties in the source contribution are estimated as

$$\sigma_{S_j} = \left(\sum_{i=1}^{I} \frac{F_{ij}^2}{\sigma_{C_i}^2 + \sum_{j=1}^{J} \alpha_{F_{ij}}^2 S_j^2} \right)^{-1/2}, \tag{15}$$

where σ_{S_j} ($\mu g m^{-3}$ or g/g) is the uncertainty in source contribution S_j ($\mu g m^{-3}$ or g/g), σ_{C_i} ($\mu g m^{-3}$ or g/g) is the uncertainty in the ambient concentrations species i, and $\sigma_{F_{ij}}$ is the uncertainty in the fraction of species i in the source j profile. Uncertainties in input variables are propagated by inversely weighting the EV (effective variance).

In this paper a new method for solving CMB receptor model based on the enhanced sampling Monte Carlo simulation was proposed as follows:

$$\begin{cases} \min \chi^2 = \sum_{i=1}^{I} \left[\dfrac{(C_i - \sum_{j=1}^{J} F_{ij} S_j)^2}{\sigma_{C_i}^2 + \sum_{j=1}^{J} \alpha_{F_{ij}}^2 S_j^2} \right] \\[4pt] st. \begin{cases} \text{Generate random inputs}: S_j \\ \text{with Enhanced Gibbs sampler} \\ \sum_{j=1}^{J} S_j \leq C \\ S_j \geq 0 \\ PM = \sum_{j=1}^{J} \eta_j = \sum_{j=1}^{J} S_j / C \geq \lambda \\ i = 1, 2, \cdots, I j = 1, 2, \cdots, J \end{cases} \\[4pt] \sigma_{S_j} = \left(\sum_{i=1}^{I} \dfrac{F_{ij}^2}{\sigma_{C_i}^2 + \sum_{j=1}^{J} \alpha_{F_{ij}}^2 S_j^2} \right)^{-1/2} \end{cases} \tag{16}$$

Then we can get the following ESMC-CMB algorithm:

Algorithm ESMC-CMB: Given the initial receptor and source profile data C_i, σ_{C_i}, F_{ij}, $\sigma_{F_{ij}}$, $i = 1, 2, \cdots, I j = 1, 2, \cdots, J$, the number of source and receptor components I, the number of source J, $obj = 10^{100}$, the number of simulation times N, $n = 0$.

Step 1: Generate random variables with the enhanced sampling Monte Carlo method proposed in this paper: $S_j \geq 0\,j = 1, 2, \cdots, J$.

Step 2: If $\sum_{j=1}^{J} S_j \geq C$, go to step 1.

Step 3: $n = n + 1$, Calculate χ^2, if $\chi^2 < obj$, then $obj = \chi^2\ objS_j = S_j$.

Step 4: if $n < N$ then step 1.

Step 5: Calculate χ^2, $\eta_j = \frac{objS_j}{C}$, σ_{S_j}.

4. Application to a Realistic Case

This realistic case focuses on the dataset from a city in China. The profiles of the receptor and source component are shown in Tables 1 and 2.

Table 1. Receptor component profiles.

Ele.	Conc.	STDE	Ele.	Conc.	STDE
TOT	111.8677	54.19443	Co	0.000505	0.000458
Na	0.381248	0.149582	Ni	0.006908	0.00752
Mg	0.201556	0.094942	Cu	0.055663	0.076044
Al	2.647172	2.03143	Zn	0.237994	0.184731
Si	2.435858	1.56244	Pb	0.111147	0.091934
P	0.061124	0.039434	OC	20.2725	12.6826
K	1.372987	0.862706	EC	3.855547	2.132063
Ca	2.912185	1.292981	Cl	0.26934	0.560002
Ti	0.014792	0.00704	NO_3	4.703921	5.350789
Cr	0.018382	0.012077	SO_4	17.27229	7.314421
Mn	0.041736	0.035401	NH_4	9.960722	5.706486
Fe	4.122549	6.704566			

Note: Ele. = Elements, Conc. = Concentration ($\mu g/m^3$), STDE = Standard Deviation.

EPA-CMB8.2 and NKCMB1.0 software can be used to solve the CMB model when the number of sources or source categories is less than or equal to the number of species. So, firstly, we select nine sources (Soil Dust, Construction Dust, Coal Combustion, Cooking Smoke, Biomass Burning, Industrial Processes, $NO_3{}^-$, $SO_4{}^{2-}$, Vehicular Emissions) and nine components (Al, Si, K, Ca, Fe, OC, EC, NO_3, SO_4), and use EPA-CMB8.2 and NKCMB1.0 to calculate source apportionment with the data in Tables 1 and 2; the results are shown in Figures 2 and 3.

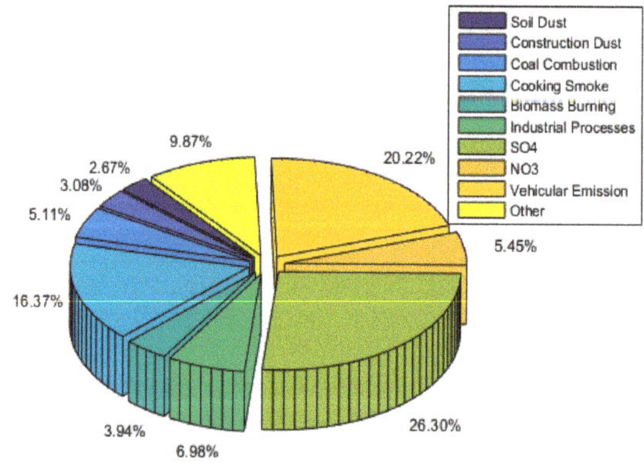

Figure 2. The result using EPA-CMB8.2 with nine sources and nine species.

Table 2. Source component profiles.

Ele.	Raise Dust		Soil Dust		Construction Dust		Coal Combustion		Cooking Smoke		Biomass Burning		Industrial Processes		NO_3^-		SO_4^{2-}		Vehicular Emission	
	Conc.	STDE	Conc.	STDE	Conc.	STDE	Conc.	STDE	Conc.	STDE	Conc.	STDE	Conc.	STDE	Conc.	STDE	Conc.	STDE	Conc.	STDE
Na	0.004722	0.001744	0.007309	0.004183	0.002478	0.000735	0.006365	0.004774	0.008617	0.005742	0.002959	0.002666	0.008600	0.000100	0	0.000001	0	0.000001	0.009363	0.005199
Mg	0.007276	0.001945	0.014675	0.006636	0.008546	0.002448	0.011922	0.008209	0.017405	0.012709	0.004460	0.003831	0.015600	0.000400	0	0.000001	0	0.000001	0.010941	0.006701
Al	0.088236	0.011353	0.118910	0.038482	0.069371	0.031392	0.239006	0.182281	0.012367	0.007308	0.031969	0.026745	0.005300	0.000100	0	0.000001	0	0.000001	0.010639	0.007285
Si	0.137211	0.033887	0.232882	0.076667	0.098363	0.022080	0.081033	0.081762	0.013954	0.014697	0.051409	0.047943	0.013100	0.001300	0	0.000001	0	0.000001	0.012261	0.006434
P	0.00081	0.000228	0.000874	0.000383	0.000264	0.000115	0.000311	0.000262	0.000321	0.000191	0.000126	0.000092	0.000000	0.000001	0	0.000001	0	0.000001	0.002077	0.000802
K	0.013932	0.003267	0.018596	0.006236	0.017332	0.002317	0.008941	0.007336	0.013851	0.015402	0.104925	0.065980	0.033000	0.000700	0	0.000001	0	0.000001	0.012172	0.005053
Ca	0.108035	0.028816	0.125479	0.085679	0.274893	0.043775	0.056683	0.086205	0.012212	0.007046	0.008350	0.007519	0.092000	0.001900	0	0.000001	0	0.000001	0.013024	0.006604
Ti	0.002224	0.000548	0.003509	0.001219	0.002643	0.001152	0.045129	0.030700	0.007512	0.006108	0.001460	0.001747	0.000400	0.000040	0	0.000001	0	0.000001	0.006982	0.003594
Cr	0.000138	0.000050	0.000322	0.000198	0.000084	0.000024	0.000795	0.000887	0.000449	0.000214	0.000872	0.001734	0.000300	0.000030	0	0.000001	0	0.000001	0.001887	0.002440
Mn	0.000501	0.000169	0.000722	0.000303	0.000322	0.000132	0.000193	0.000169	0.000187	0.000162	0.000079	0.000116	0.009800	0.000100	0	0.000001	0	0.000001	0.000901	0.001107
Fe	0.030867	0.009165	0.038558	0.016161	0.013179	0.007315	0.053666	0.032867	0.019319	0.014766	0.010202	0.008648	0.367000	0.000200	0	0.000001	0	0.000001	0.034335	0.022235
Co	0.000011	0.000003	0.000026	0.000012	0.000006	0.000004	0.000013	0.000017	0.000003	0.000003	0.000006	0.000013	0.000300	0.000030	0	0.000001	0	0.000001	0.000028	0.000033
Ni	0.000046	0.000019	0.000135	0.000082	0.000032	0.000004	0.000568	0.000988	0.000211	0.000131	0.000278	0.000550	0.000100	0.000100	0	0.000001	0	0.000001	0.001459	0.001336
Cu	0.000123	0.000048	0.000249	0.000149	0.000070	0.000019	0.000294	0.000233	0.000407	0.000332	0.000178	0.000147	0.000400	0.000100	0	0.000001	0	0.000001	0.001014	0.001473
Zn	0.000579	0.000181	0.000838	0.000476	0.000155	0.000050	0.000649	0.000530	0.001357	0.000918	0.000543	0.000496	0.009800	0.000900	0	0.000001	0	0.000001	0.000952	0.000729
Pb	0.000225	0.000127	0.000121	0.000065	0.000035	0.000006	0.000117	0.000088	0.000115	0.000085	0.000051	0.000036	0.003200	0.000300	0	0.000001	0	0.000001	0.000207	0.000262
OC	0.040941	0.007951	0.024068	0.005945	0.040341	0.007325	0.121996	0.115939	0.642280	0.409048	0.397684	0.092230	0.008200	0.000800	0	0.000001	0	0.000001	0.345982	0.172765
EC	0.006195	0.000620	0.000056	0.000006	0.001326	0.000133	0.013801	0.001380	0.018633	0.001863	0.042355	0.004236	0.004800	0.000480	0	0.000001	0	0.000001	0.147948	0.079996
Cl	0.002261	0.001599	0.004345	0.007539	0.001112	0.001547	0.005714	0.004869	0.008058	0.004448	0.169234	0.084954	0.007200	0.002300	0	0.000001	0	0.000001	0.004652	0.003824
NO$_3$	0.006385	0.001638	0.006381	0.004647	0.001362	0.000347	0.006629	0.006358	0.010071	0.008292	0.004203	0.004376	0.000000	0.000001	0.794872	0.079487	0	0.000001	0.009028	0.004308
SO$_4$	0.045996	0.014860	0.015446	0.006891	0.024699	0.004715	0.052591	0.047759	0.026123	0.020926	0.021658	0.015497	0.016600	0.002300	0	0.000001	0.727273	0.072727	0.013842	0.010331
NH$_4$	0.001191	0.000957	0.001302	0.000726	0.000190	0.000137	0.013185	0.014081	0.010247	0.014778	0.084053	0.051916	0.000000	0.000001	0.205128	0.020513	0.272727	0.027273	0.009194	0.007708

Note: Ele. = Elements, Conc. = Concentration (%), STDE = Standard Deviation.

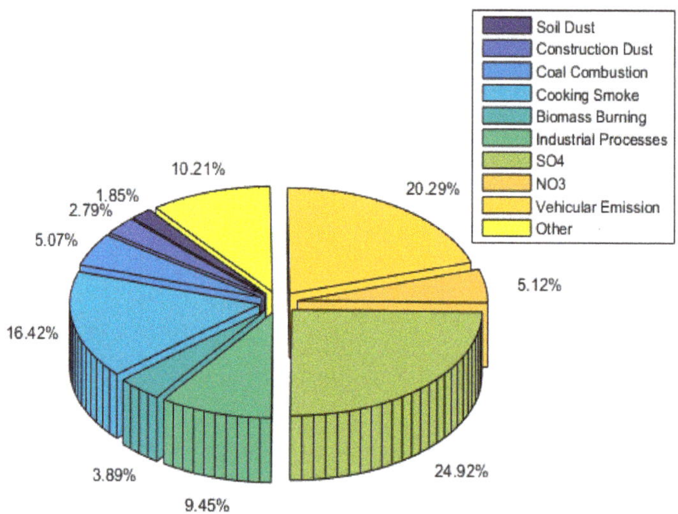

Figure 3. The result using NKCMB1.0 with nine sources and nine species.

With the same selection of the source profiles and receptor components and the same dataset, we use our proposed ESMC-CMB algorithm to calculate the source apportionment, and the results are shown in Figure 4. Table 3 shows the numerical comparison of the contribution rates of the above three algorithms.

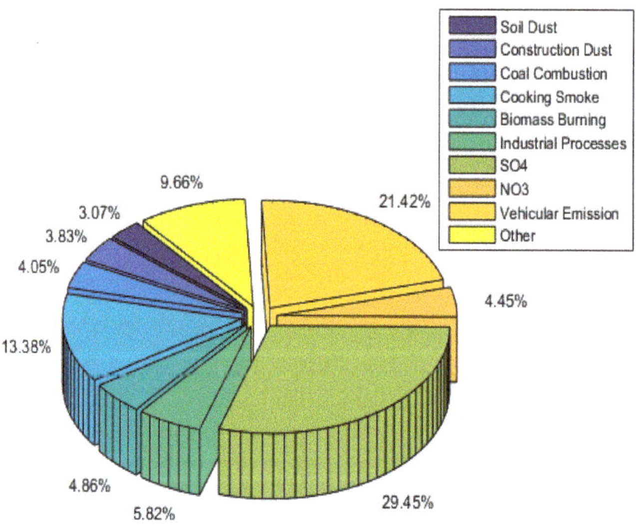

Figure 4. The result using proposed ESMC-CMB algorithm (with nine sources and nine species.

Table 3. A numerical comparison of EPA-CMB8.2, NKCMB1.0, and ESMC-CMB.

Algorithms Source Contribution	EPA-CMB8.2	NKCMB1.0	ESMC-CMB
Soil Dust	0.026698964	0.018464739	0.030721789
Construction Dust	0.030752527	0.027916899	0.038269989
Coal Combustion	0.051126227	0.050733701	0.04052869
Cooking Smoke	0.163715722	0.164181048	0.13384989
Biomass Burning	0.039397644	0.038893606	0.048647818
Industrial Processes	0.069811171	0.094471832	0.058153522
SO_4^{2-}	0.263039684	0.249173003	0.294499399
NO_3^{-}	0.054531719	0.051178676	0.04448467
Vehicular Emissions	0.202244424	0.202902245	0.214208926
Other	0.098681918	0.102084251	0.096635306

From the results of Figures 2–4 and Table 3, we can see that the results of source apportionment calculated with the above three algorithms are very close, and the correctness of the ESMC-CMB algorithm is verified.

If eight species such as Al, Si, K, Ca, Fe, OC, EC, and NO_3^{-} are selected, then the software EPA-CMB8.2 and NKCMB1.0 cannot be used because the number of species is less than the number of sources, but the proposed algorithm ESMC-CBM can calculate the results in Figure 5.

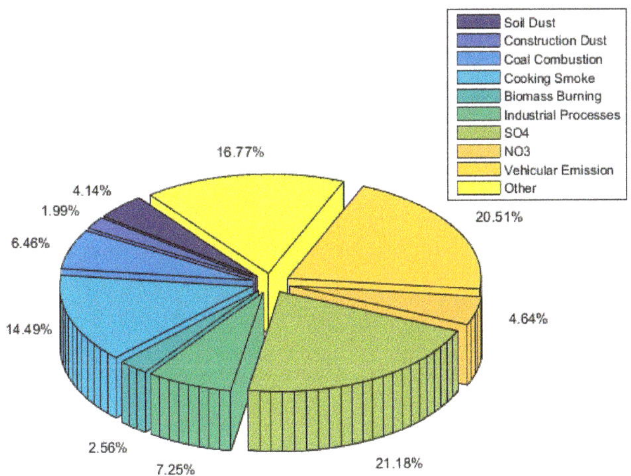

Figure 5. The result using ESMC-CMB with nine sources and eight species.

As there is strong collinearity between the sources Raise Dust (RD) and Soil Dust, if RD is added to the source profiles (Soil Dust, Construction Dust, Coal Combustion, Cooking Smoke, Biomass Burning, Industrial Processes, NO_3^{-}, SO_4^{2-}, Vehicular Emissions) to participate in the calculation using EPA-CMB8.2 and NKCMB1.0, some values of source contribution will be negative, so correct results cannot be obtained. However, using our proposed ESMC-CMB algorithm, we can get the correct value of the source apportionment as shown in Figure 6.

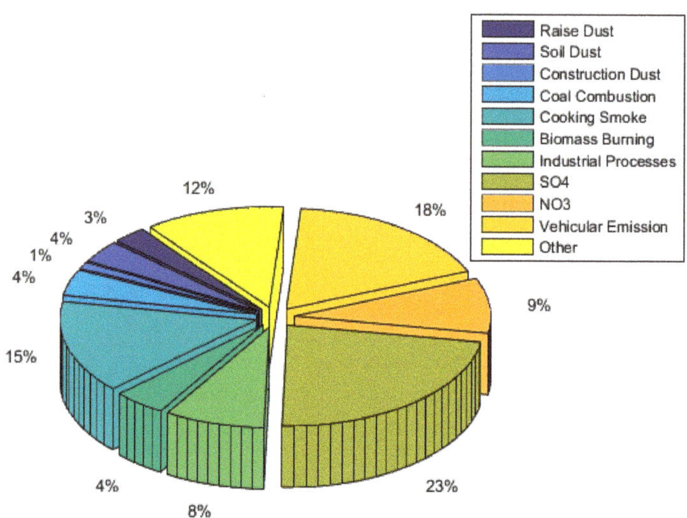

Figure 6. The result using ESMC-CMB with 10 sources and nine species including RD (Raise Dust) collinear with Soil Dust.

A comparison of the above results is given in Table 4. As can be seen clearly from Table 4, in the practical instances for source apportionment, when nine species and nine sources, with no collinearity among them, are selected, EPA-CMB8.2, NKCMB1.0, and ESMC-CMB can obtain similar results. However, because there is strong collinearity between source Raise Dust (RD) and Soil Dust, when the source Raise Dust is added to the source profiles, or nine sources and eight species are selected, EPA-CMB8.2 and NKCMB1.0 cannot solve the model, but the proposed ESMC-CMB algorithm can come to a satisfactory results, which fully verify the robustness and effectiveness of ESMC-CMB.

Table 4. A comparison of NKCMB1.0 and MC-CMB.

Conditions / Algorithms	EPA-CMB8.2	NKCMB1.0	ESMC-CMB
Number of sources ≤ number of species and existing no collinearity	Having results	Having results	Having results
Number of sources > number of species	No results	No results	Having results
The collinearity exist in sources	No results	No results	Having results

5. Conclusions

In this paper, a new robust algorithm for a CMB receptor model based on enhanced sampling Monte Carlo simulation and the effective variance weighted least squares is proposed. Because of the weaknesses of the traditional algorithms and software for CMB receptor source apportionment model such as collinearity and the requirement that the number of chemical species be greater than or equal to the number of sources, in many cases, software such as EPA-CMB8.2 and NKCMB1.0 cannot obtain results for the source apportionment or some values of the source contribution are negative. However, the proposed robust novel ESMC-CMB algorithm can overcome the above weaknesses and achieve satisfactory results. In the realistic source apportionment experiments, firstly, we selected nine sources (Soil Dust, Construction Dust, Coal Combustion, Cooking Smoke, Biomass Burning, Industrial Processes, NO_3^-, SO_4^{2-}, Vehicular Emissions) with no collinearity among them and nine species (Al, Si, K, Ca, Fe, OC, EC, NO_3, SO_4), and used the EPA-CMB8.2, NKCMB1.0, and ESMC-CMB algorithms to calculate source contributions, and got similar results, but when we selected eight species and nine sources or added Raise Dust to the source profiles, because of the collinearity with Soil Dust, EPA-CMB8.2 and NKCMB1.0 could not obtain correct results; however, the proposed ESMC-CMB algorithm can calculate the right results for source apportionment. This has fully demonstrated the robustness and effectiveness of the ESMC-CMB algorithm.

Although the ESMC-CMB algorithm has many advantages, there is often missing data in the actual problem. How to further improve the ESMC-CMB algorithm in the case of missing data is the next area of research to tackle.

Due to the limitations of the CMB model, in the realistic study of air pollution, the results of source analysis from the ESMC-CMB algorithm should be referred to the calculation results of other models, such as PMF (Positive Matrix Factorization) and CMAQ (Community Multiscale Air Quality), to obtain more reasonable results.

Author Contributions: Conceptualization, M.H.; Data curation, W.H.; Formal analysis, M.H., W.H., and Y.Y.; Methodology, W.H. and Z.W.; Writing, Q.W. and X.X.

Funding: This study was funded by the Natural Science Foundation of China under Grants 61375063, 61271355, 11301549, and 11271378.

Conflicts of Interest: The authors declare no conflict of interest.

References

1. Shen, G.; Wang, W.; Yang, Y.; Zhu, C.; Min, Y.; Xue, M.; Ding, J.; Li, W.; Wang, B.; Shen, H.; et al. Emission factors and particulate matter size distribution of polycyclic aromatic hydrocarbons from residential coal combustions in rural Northern China. *Atmos. Environ.* **2010**, *44*, 5237–5243. [CrossRef]
2. Kong, S.; Ji, Y.; Lu, B.; Chen, L.; Han, B.; Li, Z.; Bai, Z. Characterization of PM10 source profiles for fugitive dust in Fushun-a city famous for coal. *Atmos. Environ.* **2011**, *45*, 5351–5365. [CrossRef]
3. Zheng, J.; Che, W.; Zheng, Z.; Chen, L.; Zhong, L. Analysis of Spatial and Temporal Variability of PM10 Concentrations Using MODIS Aerosol Optical Thickness in the Pearl River Delta Region, China. *Aerosol Air Qual. Res.* **2013**, *13*, 862–876. [CrossRef]
4. Zheng, M.; Salmon, L.G.; Schauer, J.J.; Zeng, L.; Kiang, C.S.; Zhang, Y.; Cass, G.R. Seasonal trends in PM2.5 source contributions in Beijing, China. *Atmos. Environ.* **2005**, *39*, 3967–3976. [CrossRef]
5. Friedlander, S.K. Chemical element balances and identification of air pollution sources. *Environ. Sci. Technol.* **1973**, *7*, 235–240. [CrossRef] [PubMed]
6. Cooper, J.A.; Watson, J.G., Jr. Receptor oriented methods of air particulate source apportionment. *J. Air Pollut. Control Assoc.* **1980**, *30*, 1116–1125. [CrossRef]
7. Gordon, G.E. Receptor models. *Environ. Sci. Technol.* **1988**, *22*, 1132–1142. [CrossRef] [PubMed]
8. Watson, J.G. Overview of receptor model principles. *J. Air Pollut. Control Assoc.* **1984**, *34*, 619–623. [CrossRef]
9. Hidy, G.M.; Venkataraman, C. The chemical mass balance method for estimating atmospheric particle sources in Southern California. *Chem. Eng. Commun.* **1996**, *151*, 187–209. [CrossRef]
10. Miller, M.; Friedlander, S.; Hidy, G. A chemical element balance for the Pasadena aerosol. *J. Colloid Interface Sci.* **1972**, *39*, 165–176. [CrossRef]
11. Hougland, E. Chemical element balance by linear programming. In Proceedings of the 73rd Annual Meeting of the Air Pollution Control Association, Atlanta, GA, USA, 19–24 June 1983.
12. Gartrell, G.; Friedlander, S. Relating particulate pollution to sources: The 1972 California aerosol characterization study. *Atmos. Environ.* **1975**, *9*, 279–299. [CrossRef]
13. Watson, J.G.; Robinson, N.F.; Chow, J.C.; Henry, R.C.; Kim, B.; Nguyen, Q.T.; Meyer, E.L.; Pace, T.G. *Receptor Model Technical Series, Vol. III (1989 Revision) CMB7 User's Manual*; US Environmental Protection Agency: Washington, DC, USA, 1990.
14. Geladi, P.; Kowalski, B.R. Partial least-squares regression: A tutorial. *Anal. Chim. Acta* **1986**, *185*, 1–17. [CrossRef]
15. Song, X.-H.; Hopke, P.K. Solving the chemical mass balance problem using an artificial neural network. *Environ. Sci. Technol.* **1996**, *30*, 531–535. [CrossRef]
16. Watson, J.G.; Cooper, J.A.; Huntzicker, J.J. The effective variance weighting for least squares calculations applied to the mass balance receptor model. *Atmos. Environ.* **1984**, *18*, 1347–1355. [CrossRef]

17. Shi, G.L.; Zeng, F.; Li, X.; Feng, Y.C.; Wang, Y.Q.; Liu, G.X.; Zhu, T. Estimated contributions and uncertainties of PCA/MLR–CMB results: Source apportionment for synthetic and ambient datasets. *Atmos. Environ.* **2011**, *45*, 2811–2819. [CrossRef]

18. Mahadevan, S. Monte carlo simulation. In *Mechanical Engineering-New York And Basel-Marcel Dekker*; Marcel Dekker Inc.: New York, NY, USA, 1997; pp. 123–146.

19. Brémaud, P. *Markov Chains: Gibbs Fields, Monte Carlo Simulation, and Queues*; Springer Science & Business Media: Berlin, Germany, 2013; Volume 31.

20. Sobol, I.M. Global sensitivity indices for nonlinear mathematical models and their Monte Carlo estimates. *Math. Comput. Simul.* **2001**, *55*, 271–280. [CrossRef]

21. Smith, A. *Sequential Monte Carlo Methods in Practice*; Springer Science & Business Media: Berlin, Germany, 2013.

22. Hanna, S.R.; Chang, J.C.; Fernau, M.E. Monte Carlo estimates of uncertainties in predictions by a photochemical grid model (UAM-IV) due to uncertainties in input variables. *Atmos. Environ.* **1998**, *32*, 3619–3628. [CrossRef]

23. Landau, D.P.; Binder, K. *A Guide to Monte Carlo Simulations in Statistical Physics*; Cambridge University Press: Cambridge, UK, 2014.

24. Friedland, W.; Dingfelder, M.; Kundrát, P.; Jacob, P. Track structures, DNA targets and radiation effects in the biophysical Monte Carlo simulation code PARTRAC. *Mutat. Res./Fund. Mol. Mech. Mutagen.* **2011**, *711*, 28–40. [CrossRef] [PubMed]

25. Ohno, K.; Esfarjani, K.; Kawazoe, Y. *Computational Materials Science: From AB Initio to Monte Carlo Methods*; Springer Science & Business Media: Berlin, Germany, 2012; Volume 129.

26. Glasserman, P. *Monte Carlo Methods in Financial Engineering*; Springer Science & Business Media: Berlin, Germany, 2003; Volume 53.

27. Watson, J.G. Chemical Element Balance Receptor Model Methodology for Assessing the Sources of Fine and Total Suspended Particulate Matter in Portland, Oregon. Ph.D. Thesis, Department of Environmental Science, Oregon Graduate Center, Oregon City, OR, USA, 1979.

28. Christensen, W.F.; Gunst, R.F. Measurement error models in chemical mass balance analysis of air quality data. *Atmos. Environ.* **2004**, *38*, 733–744. [CrossRef]

29. Cheng, M.; Hopke, P.K. *Linear Programming Procedure and Regression Diagnostics for least-Squares Solution Using CMB Receptor Model, in Receptor Methods for Source Apportionment—Real World Issues and Applications*; Air Pollution Control Association: Pittsburgh, PA, USA, 1986.

30. Gleser, L.J. Some thoughts on chemical mass balance models. *Chemom. Intell. Lab. Syst.* **1997**, *37*, 15–22. [CrossRef]

31. Yue, K.; Wu, H.; Liu, W.; Zhu, Y. Representing and processing lineages over uncertain data based on the Bayesian network. *Appl. Soft Comput.* **2015**, *37*, 345–362. [CrossRef]

32. Kozumi, H.; Kobayashi, G. Gibbs sampling methods for Bayesian quantile regression. *J. Stat. Comput. Simul.* **2011**, *81*, 1565–1578. [CrossRef]

33. Gilks, W.R.; Wild, P. Adaptive rejection sampling for Gibbs sampling. *Appl. Stat.* **1992**, *41*, 337–348. [CrossRef]

34. Arroyo, D.; Emery, X.; Peláez, M. An enhanced Gibbs sampler algorithm for non-conditional simulation of Gaussian random vectors. *Comput. Geosci.* **2012**, *46*, 138–148. [CrossRef]

Article

Numerical Investigation of SCR Mixer Design Optimization for Improved Performance

Ghazanfar Mehdi [1,*], Song Zhou [1,*], Yuanqing Zhu [1], Ahmer Hussain Shah [2] and Kishore Chand [3]

[1] College of Energy and Power Engineering, Harbin Engineering University, Harbin 150001, China; zhuyuanqing@hrbeu.edu.cn
[2] Department of Textile Engineering, Balochistan University of Information Technology, Engineering and Management Sciences, Quetta 87300, Pakistan; ahmer.shah@buitms.edu.pk
[3] College of Material Science and Chemical Engineering, Harbin Engineering University, Harbin 150001, China; kishor.vallasai@gmail.com
* Correspondence: ghazanfarmehdi22@gmail.com (G.M.); songzhou@hrbeu.edu.cn (S.Z.); Tel.: +86-130-3004-8065 (G.M.); +86-138-4506-3167 (S.Z.)

Received: 1 February 2019; Accepted: 18 March 2019; Published: 22 March 2019

Abstract: The continuous increase in the number of stringent exhaust emission legislations of marine Diesel engines had led to a decrease in NO_x emissions at the required level. Selective catalyst reduction (SCR) is the most prominent and mature technology used to reduce NO_x emissions. However, to obtain maximum NO_x removal with minimum ammonia slip remains a challenge. Therefore, new mixers are designed in order to obtain the maximum SCR efficiency. This paper reports performance parameters such as uniformity of velocity, ammonia uniformity distribution, and temperature distribution. Also, a numerical model is developed to investigate the interaction of urea droplet with exhaust gas and its effects by using line (LM) and swirl (SM) type mixers alone and in combination (LSM). The urea droplet residence time and its interaction in straight pipe are also investigated. Model calculations proved the improvement in velocity uniformity, distribution of ammonia uniformity, and temperature distribution for LSM. Prominent enhancement in the evaporation rate was also achieved by using LSM, which may be due to the breaking of urea droplets into droplets of smaller diameter. Therefore, the SCR system accomplished higher urea conversion efficiency by using LSM. Lastly, the ISO 8178 standard engine test cycle E3 was used to verify the simulation results. It has been observed that the average weighted value of NO_x emission obtained at SCR outlet using LSM was 2.44 g/kWh, which strongly meets International Maritime Organization (IMO) Tier III NO_x (3.4 g/kWh) emission regulations.

Keywords: selective catalyst reduction system; emission control; marine Diesel engine; urea; ammonia

---

1. Introduction

Environmental safety is one of the hottest research areas nowadays due to increased public awareness. To achieve this goal, efforts are continuously made to reduce pollution and develop green processes. Exhaust emissions from marine Diesel engine are responsible for producing severe environmental pollution, especially nitrogen oxide (NO_x) emissions [1,2]. The automobile Diesel engine produces exhaust emissions species such as carbon monoxide, carbon dioxide, and hydrocarbon in abundance compared to NO_x. On the other hand, marine Diesel engine produces more detrimental NO_x emissions [3]. Hence, in order to reduce exhaust emissions from ships, many national and international organizations have promulgated regulations and also enforced strict requirements on NO_x emissions in Emission Control Areas [4]. In 2016, International Maritime organization (IMO) Tier

III regulations on NO_x emissions have already been enforced in North America emission control areas, including the East and West Coast of USA and Caribbean. As reported, it will also be enforced in North Sea and Baltic Sea in future [5]. Both high pressure fuel injection and exhaust gas recirculation systems have potential to reduce NO_x emissions, but due to poor engine performance results and continuous increase in engine emission legislations, more primitive and improved processes are needed to overcome the said issues [6]. The most convenient and easy option is the treatment of exhaust gas. Selective Catalyst Reduction (SCR) is technically mature, and is the most prominent after-treatment technology used to meet the latest NO_x emission regulations due to its high NO_x removal efficiency, cost-effectiveness, and good fuel economy [7]. However, challenges related with SCR system include improper mixture of urea water solution (UWS) with exhaust gas and ammonia leakage [8]. UWS (32.5% urea) is injected into exhaust gas [9,10]. Urea decomposition occurs in three steps: firstly, water is evaporated from urea water solution; secondly, the urea pyrolysis reaction occurs, which results in the decomposition of urea into isocyanic acid (HNCO) and ammonia (NH_3); and lastly, hydrolysis of HNCO occurs which produces NH_3 and CO_2 [11,12]. Ammonia is used as a reducing agent and cannot be used directly due to poisonous nature, storing and handling difficulties [13]. It has been proved that NO_x is mainly 90% composition of NO at the exhaust of the marine Diesel engine. The key reactions involved in SCR system are described as follows [14–17].

$CO(NH_2)_2 \rightarrow NH_3 + HNCO$	(Urea pyrolysis reaction)
$HNCO + H_2O \rightarrow NH_3 + CO_2$	(HNCO hydrolysis)
$4NH_3 + 4NO + O_2 \rightarrow 4N_2 + 6\,H_2O$	(Standard SCR reaction)
$4NH_3 + 2NO + 2NO2 \rightarrow 4N2 + 6\,H_2O$	(Fast-speed SCR reaction)
$8NH_3 + 6NO_2 \rightarrow 7N_2 + 12H_2O$	(Slow SCR Reaction)

A static mixer is commonly adopted to generate uniform distribution of ammonia at the inlet of SCR catalyst [18]. CFD code (Fire 8.3, 2004) has been used to optimize the design of SCR system. The authors have done many studies about the decomposition and evaporation of urea water solution without using static mixer. Birkhold [19] studied the urea droplet's evaporation and decomposition at different temperatures of exhaust gas. Strom [20] studied the effects of turbulent velocity and different forces on distribution and movements of urea water droplet in the exhaust species. In addition, other authors have established various studies to evaluate the effect of static mixer on the SCR performance. Sivanandi Rajadurai [21] studied the distribution of ammonia by using wire mesh mixer. Zhang [22] investigated the uniformity index of ammonia by adopting delta wing mixer. Shazam Williams and Ming Chen studied the velocity and ammonia distribution in the straight pipe together with static mixer having two rows of tabs [23]. Azael and Ibarra investigated the interaction of fluid with a double vortex mixer to improve the performance of SCR. The evaporation effect and droplet crushing was significantly improved. The distribution of NH_3 was also studied, revealing a smooth and uniform distribution using a double vortex mixer; however, the velocity uniformity distribution, reaction temperature, and droplet distribution time was not studied [24]. One of the disadvantages of the SCR system is that it occupies the additional space on vessels. Mostly, authors are working to simplify the SCR system according to the space requirements. One of the authors used an 18 mm distance between the two mixers without considering the impact of velocity on SCR system [18]. Park et al. reported that if the continuous uneven impact of velocity occurs, which causes excessive aerodynamic velocity and temperature, it will lead to the thermal fatigue failure, which reduces the service life [25]. The first mixer can create turbulence intensity and sudden impact on velocity, which directly affects the second mixer. By considering both of the references, it can be perceived that the optimum condition which can satisfy both of the conclusions from different studies can lie within limits. It therefore looked suitable to trial for 0.2 m distance. One of the study used series of only swirl mixer (SM) near the injection point and in line [18]. However, use of a line mixer (LM) can result in increasing the mixing flow (exhaust gas plus urea droplets) in the blind corner near the pipe wall, while SM results in increasing the mixing flow in the center [25]. Some studies investigated LM with different blade angles [25,26], and others used SM with different blade angles [18,27]. In this study LM was used in front of SM.

Initially, LM and SM were used separately and then analyzed for the combined effect of both (LSM). In this work, the blade angle was 45° for both mixers because it has been proven that static mixers consisting of bitched blades with an angle of inclination angle of 45° can generate higher turbulence intensity and a swirling flow in the pipe [28]. Furthermore, many authors have used a greater number of blades than this study. One author used 36 blades in only LM [25]. In this study, 18 (LM:12, SM:6) blades were used and distributed in two difference places in the pipe. In addition, if a single mixer has many blades at a certain location, it tends to decrease the wall temperature, which ultimately results in deposit formation [29]. Hence it is necessary to distribute number of blades in two mixers with different locations to increase the mixing performance and prevents deposit formation. Generally, SCR performance depends upon the velocity uniformity and ammonia uniformity near the inlet of the SCR catalyst. Therefore, a suitable design of mixers for increasing the efficiency of SCR is needed. Furthermore, a numerical model has been developed, which describes the impact of mixer on the ammonia and velocity uniformity, droplet residence time, and temperature distribution in the pipe. In addition, it is also necessary to consider reaction temperature and wall temperature, as it affects the catalyst performance and deposit formation [1]. In this paper, two different types of mixers—line (LM) and swirl (SM) type mixers alone and in combination (LSM)—were used to indicate the effects on the performance of SCR. The main objective of this investigation is to create the optimum SCR design, to achieve higher uniformity index for velocity and ammonia distribution, better evaporation rate, and droplet distribution by consideration of the reaction temperature and wall temperature distribution based on CFD. Furthermore, for the verification of the simulated results of ISO 8178 a standard marine Diesel engine test cycle E3 was used [30].

2. Computational Model Formulation

2.1. Geometric Model

The complications in the design and arrangements of SCR system leads to difficulties in the uniformity of flow field and ammonia in the exhaust flow by using only urea water solution (UWS) and species diffusion. It is therefore necessary to use static mixer to optimize the mixing of exhaust gas flow with UWS. It has been proved by previous studies that uniformity of flow velocity and ammonia distribution can be increased with a suitable mixer, which directly enhances the performance of SCR catalyst. The optimized mixers have ability of complete rapid mixing of UWS droplet with exhaust gas which accelerates the pyrolysis reaction. Therefore, the catalyst conversion efficiency was improved [31].

In this study, two different types of mixers are used—line type mixer (LM) and swirl type (SM). In the first part of study, the SCR performance was analyzed by LM and SM separately. The second part of study focused on the combined effects of LM and SM. The distance between two mixers was kept 0.2 m. The pipe diameter was 100 mm. Injection point was located 0.1 m away from the LM and 0.3 m away from the SM, when both mixers were used together. For the single mixer, the position was taken as 0.1 m away from the injection point. The complete systematic design of SCR system is represented in Figure 1.

The design of LM includes 4 × 4 rows of blades and the angle of blades with horizontal plane was 45°. This design represents the improved heat transfer efficiency and abolishes the blind corners. Four blades were used at the center rows while two at the corners. One column of blades shows a downward direction and the other shows an upward direction. The design construction of SM was adopted in order to enhance the breakup effect of droplets and increase the turbulence intensity which ultimately results rapid mixing and pyrolysis of urea. The SM mixer possess six number of blades tilted at an angle of 45° with the horizontal. By analyzing the result of individual mixers on SCR performance, a system was designed using both LM and SM in combination to achieve an improvement in performance. Considering the complexity of SCR system, the mesh grid is generated by tetrahedral method for adequate adaptation as shown in Figure 2. Away from the solid wall boundary, the grid was stretched which was the good comprise between the mesh points and computational costs. Furthermore, mesh independence test has been performed to calculate the ammonia uniformity without using any mixer, as shown in Figure 3. After mesh independence analysis, the total number of cells in the mesh system was 1.8 million; because a further increased in mesh cells has no significant effect on the uniformity index.

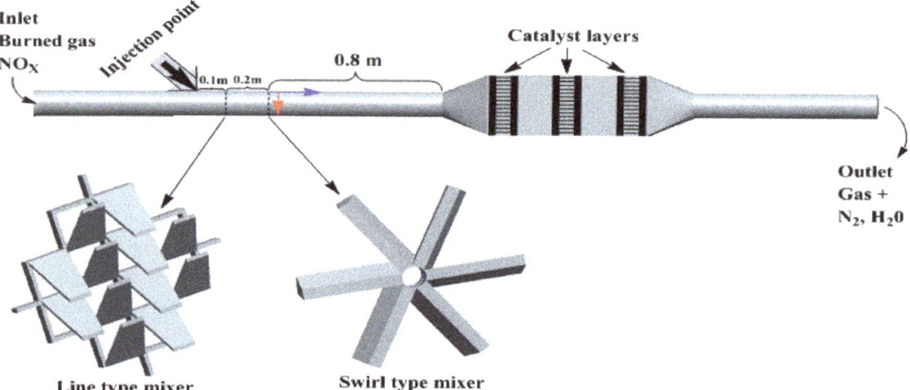

Figure 1. Systematic structure of the selective catalyst reduction (SCR) system.

Figure 2. Mesh grids of computational domain.

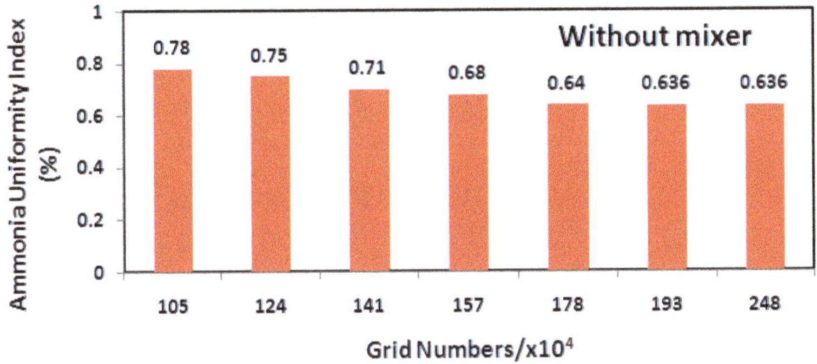

Figure 3. Mesh grid independence analysis.

2.2. Numerical Procedure

Numerical studies have been done for the optimum design of SCR system in the form of flow uniformity and the reduction of NO_X emissions [22,32]. In this study, the flow field distribution of velocity was taken uniform and thermal expansion in the catalyst was negligible. Exhaust gases from the engine were adopted as the ideal gas mixture in the CFD. Incompressible steady flow was used. The flow was simulated by using the standard K-epsilon model (K-ε) in combination with wall treatment function to evaluate the turbulent flow velocity in the exhaust system because the Reynolds number is usually very high in an exhaust pipe. It is a two equation model that gives a general description of turbulence by means of two transport partial differential equations. The first transported variable is the turbulence kinetic energy (K) and the second transported variable is the rate of dissipation of turbulence energy (ε) [25,33]. The injection and decomposition process of UWS were simulated by transport model. Evaporation and decomposition performances were described by using a mixture model of multicomponents. The concentration of 32.5 wt% UWS was injected at the temperature window of 313 K. Pressure swirl atomizer was used for injection of UWS. KH-RT model was used for breakup of droplet. Firstly, water was evaporated from urea droplets during decomposition. The rate of water evaporation was predicted by Raouli's law of multi component mixture. Raoult's law is based on the assumptions that the vapor phase behaves as an ideal gas and the liquid phase is an ideal solution in mixing of the flow. The rate of water evaporation can be obtained by Equation (1).

$$\frac{dm}{dt} = MW_{water} k_c \left(\frac{x_{water} P_{sat,water}}{R T_P} - C_{water} \right) \tag{1}$$

If the fraction of water in UWS is going beyond the threshold value (0.01), then the UWS droplets are considered as the particles of urea and it can be easily calculated by the decomposition reactions of urea particles. As the urea particles react with exhaust gas species, it results in NH_3 and HNCO. The mass change rate can be obtained by Equation (2).

$$\frac{dm_P}{dt} = MW_i \sum_{r=1}^{N_R} R_{i,r} \tag{2}$$

The species transport and continuity equation are used to define the variation of gases produced from decomposition reactions such as NH_3, CO_2, and HNCO (Equations (3) and (4)).

$$\frac{\partial(\rho Y_i)}{\partial t} + \nabla \cdot (\rho \vec{V} Y_i) = -\nabla \cdot \vec{J_i} + R_i + S_i \tag{3}$$

$$\frac{\partial \rho}{\partial t} + \nabla \cdot (\rho \vec{V}) = S_m \tag{4}$$

By concentration gradients, the mass diffusion in laminar flows used and it was modeled by Fick's law, as shown in Equation (5).

$$\vec{j}_i = -\rho D_{i,m} \nabla Y_i - D_{T,i} \nabla T/T \tag{5}$$

The turbulence due to turbulent diffusion is directly responsible for the quick mixing and species transport in turbulent flows therefore the mass diffusion can be obtained by Equation (6).

$$\vec{j}_i = -(\rho D_{i,m} + \mu/Sc_t) \nabla Y_i - D_{T,i} \nabla T/T \tag{6}$$

The diffusive process is modeled by using turbulent diffusivity (D_T) in the K-ε model. The D_T can be obtained from Equation (7) [34].

$$D_T = v_T/Sc_t \tag{7}$$

Schmidt turbulence number (Sc_t) is usually taken as 0.7. In the turbulence model, the turbulent diffusivity D_T directly depends upon the use of turbulent viscosity V_T. Turbulent viscosity is obtained from given formula in the K-ε model Equation (8).

$$v_T = C_u k^2/\varepsilon \tag{8}$$

In the K-ε model, closure coefficient (C_u) is taken as 0.09.

Energy dissipation rate and turbulent kinetic energy are obtained from a standard two-layer K-ε turbulent model as showed in Equations (9) and (10), respectively.

$$\frac{\partial k}{\partial t} + \frac{\partial u_i k}{\partial x_i} = \frac{\partial}{\partial x_i}\left(\left(v + \frac{v_t}{\sigma_k}\right) + \frac{\partial k}{\partial x_i}\right) + G - \varepsilon \tag{9}$$

$$\frac{\partial \varepsilon}{\partial t} + \frac{\partial u_i \varepsilon}{\partial x_i} = \frac{\partial}{\partial x_i}\left(\left(v + \frac{v_t}{\sigma_\varepsilon}\right) + \frac{\partial \varepsilon}{\partial x_i}\right) + \frac{\varepsilon}{K}(C_{\varepsilon 1} G - C_{\varepsilon 2}\varepsilon) \tag{10}$$

where,

$$G = -\overline{u_i u_i}\frac{\partial u_i}{\partial x_i} = v_t\left(\frac{\partial u_i}{\partial x_i} + \frac{\partial u_i}{\partial x_i}\right)\frac{\partial u_i}{\partial x_i}, \text{ and } v_t = C_\mu \frac{K^2}{\varepsilon}$$

$$C_{\varepsilon 1} = 1.44,\ C_{\varepsilon 2} = 1.92,\ C_\mu = 0.09, \sigma_K = 1.0, \sigma_\varepsilon = 1.3$$

For describing the conservation equation of gas–liquid phase and spray phenomenon, the discrete phase model was used. By the discrete phase model under the Lagrangian coordinate system, it could be constructed a governing equation to analyze the motion law of droplet and motion of liquid droplet which can be given by Equation (11).

$$\frac{d\vec{u}_p}{dt} = \frac{\vec{u} - \vec{u}_p}{\tau_r} + \vec{g}\frac{(\rho_p - \rho)}{\rho_p} + \vec{F} \tag{11}$$

Gradient diffusion is responsible for the rate of vaporization at low vaporization rates [35]. Transportation of urea droplet vapor was made at the bulk of gas and droplet surface because of the concentration gradient. At high vaporization rates, the fluctuation of urea water droplets into exhaust gases is directly associated with the effects of evaporating species under convective flow [36]. UWS droplets include the mixtures of two parts; therefore, the evaporation of UWS is separated into two different periods. During the first stage of decomposition, contents of water evaporate quickly from UWS droplets. In the later stage, molten urea evaporates [37,38]. In this study, UWS droplets were taken as multicomponent, therefore the total vaporization rate of UWS is the sum of vaporization rates for all UWS components. The evaporation of urea solution is the heat and mass transfers between the urea droplet and exhaust gases. Diffusion is controlled by concentration gradients described by

Fick's law which was used as diffusion also occurs in multispecies systems that are experiencing the Stefan flow also and the model included the internal recirculation and Stefan flow effects [39].Thus the evaporation rate can be described by Equation (12).

$$m_p c_p \frac{dT_P}{dt} = hA_P(T_\infty - T_P) - \frac{dm_p}{dt} h_{fg} + \varepsilon_p + A_p \sigma (\theta_R^4 - T_p^4) \tag{12}$$

Radiation properties are very complex to be described by the flow and have less importance. It was therefore neglected in our study. Moreover, some studies also have neglected this effect due to the complexity of system [25,26].

A honeycomb-type catalyst is often used in the SCR system, and its single channel is very small to the outline size of the catalyst [32]. In this study, the catalytic reaction section was taken as a multiple holes section, and the porous medium model was adopted for evaluation. In the porous medium model, the momentum source term was included to the momentum equation and showed in Equation (13).

$$S_i = -\left(\sum_{j=1}^{3} D_{ij} \mu v_j + \sum_{j=1}^{3} C_{ij} \frac{1}{2} \rho |v_j| v_j \right) \tag{13}$$

If each position of catalyst has the same nature in gas convection and diffusion in the channel are ignored, the above equation can be simplified as Equation (14).

$$\nabla p = -\frac{\mu}{\alpha} v \tag{14}$$

2.3. Boundary Conditions

CFD boundary conditions were selected according to the ISO standard 8178 marine Diesel engine. The initial temperature and exhaust flow rate before the catalyst reactor weretaken as 327 °C and 1.6 kg/s, respectively. The quantity of NO_x emission was 256 ppm. The inlet and outlet diameter of catalyst reactor were 100 mm. The boundary conditions at the inlet of catalytic reactor was set to the mass flow rate, turbulent kinetic energy was set as 6% of the average velocity, and the characteristic length was 10% of the inlet diameter. The boundary condition at the outlet of catalytic reactor was defined as outlet pressure and the outlet pressure was same as atmospheric pressure. The boundary conditions at the wall of catalytic reactor were set as nonslip velocity and frictionless. The material is defined as iron.

The working parameters of the SCR system are shown in Tables 1 and 2. The working process of the SCR system was studied and simulated in this paper. Flow rate, temperature at the diesel exhaust, exhaust pressure, and urea injection rate increased with respect to engine load. Furthermore, urea water solution was injected by using air assisted injector at the upstream of mixer with the ten numbers of streams. Pressure atomizer injector was used with the injection hole diameter of 0.0007 m. The spray angle was kept 60° with spray pressure of 6 bar. The rate of multitude hole was 0.89 in the porous medium model. The thermal conductivity of solid phase coefficient was taken as 1.7 W/(m·k) and the specific heat capacity was 1016 J/(kg·k).

Table 1. Exhaust conditions of marine Diesel engine.

Parameters	25%	50%	75%	100%
Flow (kg/s)	0.516	1.2	1.63	1.94
Pressure (bar)	1.4	2.17	3.1	3.8
Temperature (°C)	254	290	327	395
Urea Injection (kg/s)	0.0091	0.013	0.0161	0.0192

Table 2. Injection system and catalyst conditions.

	Parameters	Value
Injection Conditions	Injection type model	Pressure swirl atomizer
	Injection inner hole diameter	0.0007 m
	Urea temperature	313 k
	Number of streams	10
Catalyst Conditions	Inverse absolute Permeability (m^{-2})	1.85
	Inertia resistance (m^{-1})	85
	Porosity	0.89
	Surface to volume ratio (1/m)	1275

3. Results and Discussion

3.1. Uniformity

Catalyst efficiency depends upon the two important parameters: flow velocity uniformity and uniform distribution of ammonia. Long-time irregular scattering and non uniformity of flow velocity will cause the excessive aerodynamic velocity and temperature, which directly impacts the structure and performance of catalyst. As the result, thermal stress can be developed, which leads to the fatigue failure and effects the service life of catalyst [25]. It is therefore recommended that the velocity of UWS must be well distributed as much as possible, at the front of the SCR catalyst reactor inlet. Equation (15) can be used to calculate the uniformity index of flow velocity and ammonia distribution.

$$UI_{flow} = 1 - \frac{1}{2}\sum_{i=1}^{n}\frac{|V_i - V_{mean}|A_i}{AV_{mean}} \tag{15}$$

3.1.1. Velocity Uniformity

The distribution of velocity field of LM, SM and the combination of both (LSM) is represented in Figure 4. The exhaust gases flow with LM was initially very uneven; downstream of mixer at the center of pipe. Furthermore, a baffle gap was also generated at the downstream but the velocity distribution inside the baffle seems smooth. As the distance was increased to the reactor inlet, uniformity was also increased, as shown in Figure 4a. SM effectively created the rotational air flow generating swirling mixing flow at the pipe wall. Swirl flow shows well mixing of exhaust gases and urea solution. Moreover, a high baffle gap was produced at the downstream of mixer, but the velocity distribution inside the baffle was highly uneven. However, flow velocity of mixture at the middle seems comparatively low as in Figure 4b.

Figure 4c showed the combined effects of LM and SM. In comparison with pure rotating blade and line type blade structures of SM and LM, respectively, the velocity uniformity was enhanced when LSM was employed. The arrangement of LSM not only generated a strong swirl flow beside the mixer wall, but also increased the velocity near the wall of pipe.

Catalyst effectiveness and utilization rate directly depends on the velocity uniformity index of UWS and exhaust gas mixture along the axial direction at the catalyst reactor inlet. The relation of the velocity uniformity index from the mixer downstream to catalyst reactor inlet is shown in Figure 5. Uniformity index of flow velocity of LM, SM, and LSM mixers were calculated as 0.93, 0.86, and 0.95 at the upstream of reactor inlet (0.8 m), respectively. Uniformity index of LSM at the downstream of mixers (0.1 m) is relatively low because of continuous and sudden impacts of two mixers on the flow velocity. However, with the increase of distance the uniformity of velocity for LSM was greatly improved due to the turbulence outcome produced from the sudden impact effect of two mixers. Therefore, the overall uniformity index of flow velocity of LSM before the reactor inlet was relatively high as compared to LM and SM.

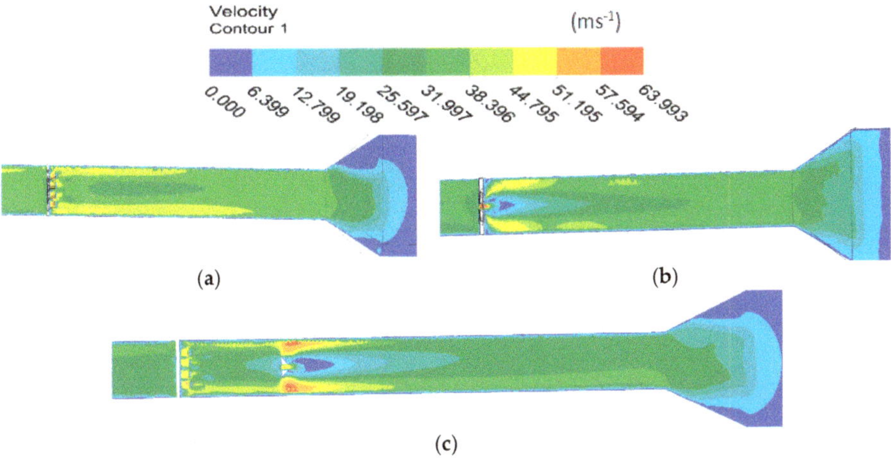

Figure 4. Velocity contour maps. (a) Line (LM) type mixer, (b) swirl (SM) type mixer, and (c) two mixers (LSM).

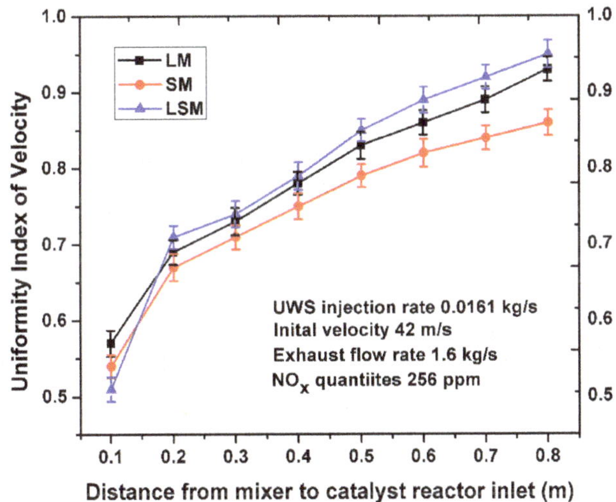

Figure 5. Distribution of velocity uniformity.

To illuminate the SCR reactor internal stability and uniformity index of flow velocity, the distribution at the reactor entrance was extracted and is shown in Figure 6. The main objective of displaying the velocity uniformity at the reactor inlet was to remove the turbulence effect produced by mixers and to show actual velocity distribution more clearly. Figure 6a,c shows that the exhaust gases gained high turbulent kinetic energy, resulting in the better velocity uniformity. However, SM (Figure 6b) shows that the radial velocity gradient was high and resulted in poor uniformity, but the velocity flow was slightly decreased with the increase of distance.

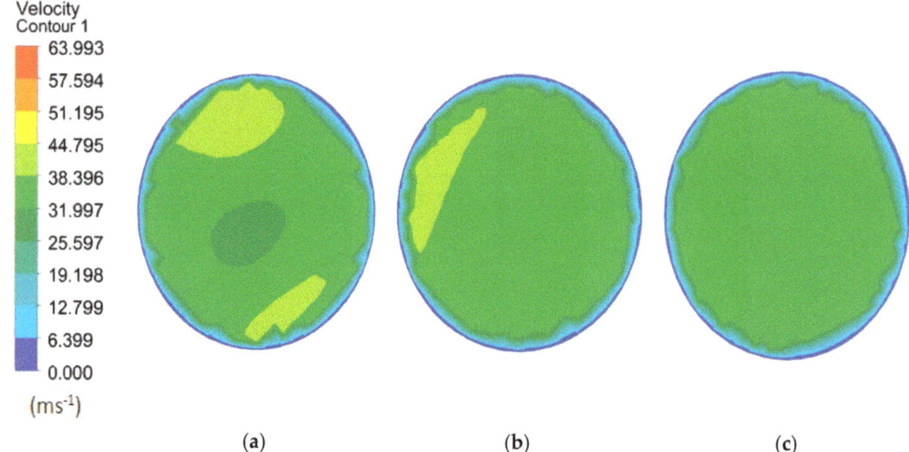

(a)   (b)   (c)

Figure 6. Velocity contour maps at the inlet of catalyst inlet. (**a**) Line (LM) mixer, (**b**) swirl (SM) mixer, (**c**) and two mixers (LSM).

3.1.2. Ammonia Uniformity

The irregular and insufficient ammonia distribution will result in NO_X escape [40,41]. The NH_3 uniformity distribution is an important index of the SCR system for estimating performance [42–44]. The water from the urea droplets evaporates entirely and converts urea in NH_3 gas when it reaches to reactor inlet. Reductant uniformity index is the most significant factor used to evaluate, whether the catalyst will achieve the highest denitrification rate and lowest ammonia leakage. The plane was created in CFD at a certain distance/area and then divided the plane into the number of points and finds the mass fraction of ammonia on each single point. After that, the standard deviation and average value of all the points were calculated. Finally, the formula for the uniformity index was used to find ammonia uniformity at certain distances. The ammonia uniformity index from the mixer downstream to catalyst reactor inlet in the axial direction was calculated and is shown in Figure 7. The NH_3 uniformity distribution of LM, SM, and LSM were calculated as 0.87, 0.94, and 0.96 at the upstream of reactor inlet, respectively.

Ammonia uniformity index of LSM at the downstream of mixers is relatively low due to uninterruptedly and sudden impacts of two mixers continuously. However, the uniformity index of ammonia of two mixers was greatly enhanced with the increase of distance due to high turbulence outcome produced from the impact effect of two mixers. Hence, generally, the ammonia uniformity index is the combination of two (line and swirl type) mixers before the catalyst reactor inlet is relatively high, as compared with the separately use of line and swirl type mixers.

Uniformity index of ammonia distribution at the outlet of reactor inlet was extracted to clarify the result as much as possible and showed in Figure 8. At first, the plane was created in CFD at the catalyst inlet, plane was divided into number of cells and value of molar concentration of ammonia at each single point was calculated. Next, the standard deviation and average value of ammonia concentration at the catalyst inlet were calculated. Finally, the uniformity index was used to calculate the ammonia uniformity at catalyst inlet. In LM, a high concentration of ammonia distribution is located at the upper side of the pipe wall, and some places at the mid of pipe also show that ammonia distribution was not smooth (Figure 8a), resulting in poor catalyst performance. In SM, ammonia seems well distributed as a whole but the small area at the upper side of pipe shows the high concentration (Figure 8b). Distribution of ammonia was greatly improved when using LSM and much less ammonia was deposited at sides and center (Figure 8c).

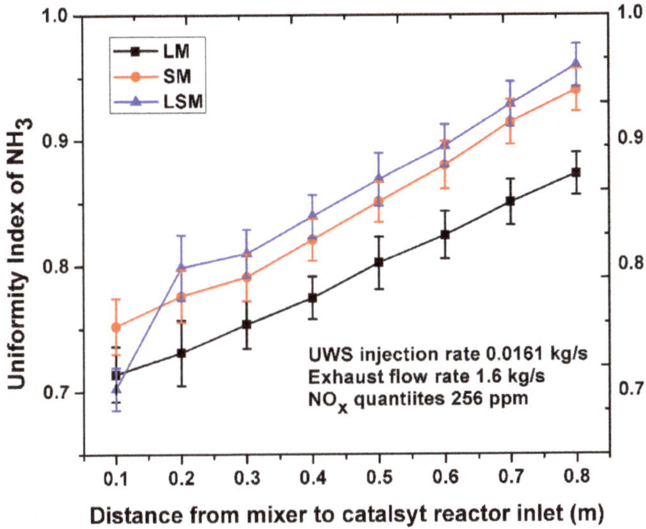

Figure 7. Distribution of ammonia uniformity.

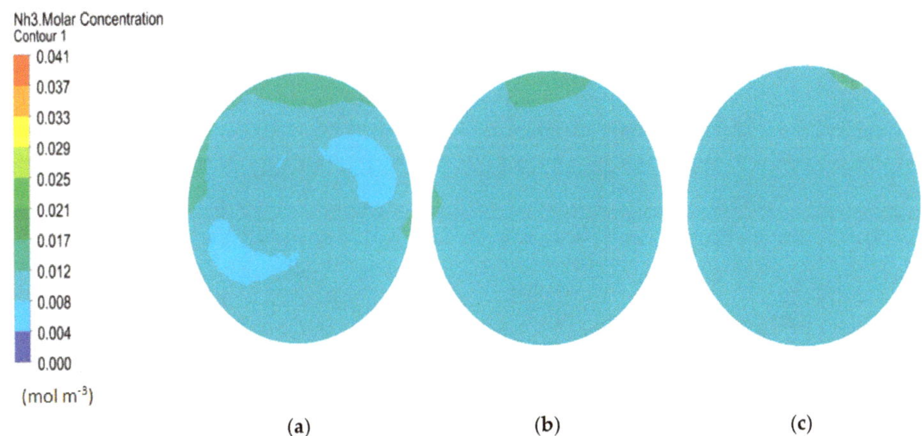

(a)   **(b)**   **(c)**

Figure 8. Velocity contour maps at the inlet of catalyst inlet. (**a**) Line (LM) mixer, (**b**) swirl (SM) mixer, and (**c**) two mixers (LSM).

The values of indexes of the velocity and ammonia uniformity of LM, SM, and LSM from the location of mixer downstream to catalyst reactor inlet are tabulated in Table 3. It is very clear that LM reveals a good uniformity index for velocity but, on the contrary, it exhibits a poor uniformity index for ammonia. SM behaves opposite to the LM: it possesses good ammonia uniformity but shows poor velocity distribution.

Table 3. Uniformity index at the inlet of the catalyst reactor.

Uniformity Index	LM	SM	LSM
Velocity Uniformity	0.93	0.86	0.95
Ammonia Uniformity	0.87	0.94	0.96

It is difficult to obtain the good uniformity of the both important indicators at the same time while using only single mixer. In comparison, LSM can be categorized with the good uniformity for both, the velocity and ammonia distribution due to high turbulent intensity. The values calculated for uniformity were 0.95 and 0.96 for the velocity and ammonia, respectively. For the purpose of mass transfer and homogenization, when the mixture species passes through the second mixer of LSM, it creates swirl and endorse the mass transfer effect near the pipe wall. The shear and swirl flows with the pipe wall produce molecular diffusion and eddy diffusion under forced convection. Hence, the mixing of ammonia with exhaust gases is comparatively sufficient which results the effective improvement in the uniformity of velocity and ammonia, simultaneously.

3.2. Droplet Residence Time

UWS injection in the exhaust gases lowers the temperature of the gas phase due to heat transfer phenomena. Figure 9 represents the UWS droplets residence time for diameter ranges from 0.007 mm to 0.07 mm. The plane was located at the downstream of injector. By counting the frequency of droplets with different diameters to pass through the plane for calculation purpose, the plane was divided into number of sections with each section having a fixed width. Each section was analyzed for the droplet, which was assumed to be spherical. After that, each droplet diameter was calculated from droplet volume obtained by taking into account the volume fraction occupying the computational meshes. Also, the cumulative probability distribution function was used to describe the probability of finding the diameter of the secondary droplets in a sample of splashing drops. The mass of splashing droplets from the wall depends upon the splashing energy of the droplets [45]. In the system of LM, SM, and LSM, while the UWS droplets were injected into the gas phase, the UWS droplets residence time ranged from 0.004 s to 0.121 s, 0.004 s to 0.091 s, and 0.004 s to 0.064 s, respectively. From this analysis, the LSM-installed system showed obvious differences of 47% and 29% decreases of residence time compared to LM and SM, respectively.

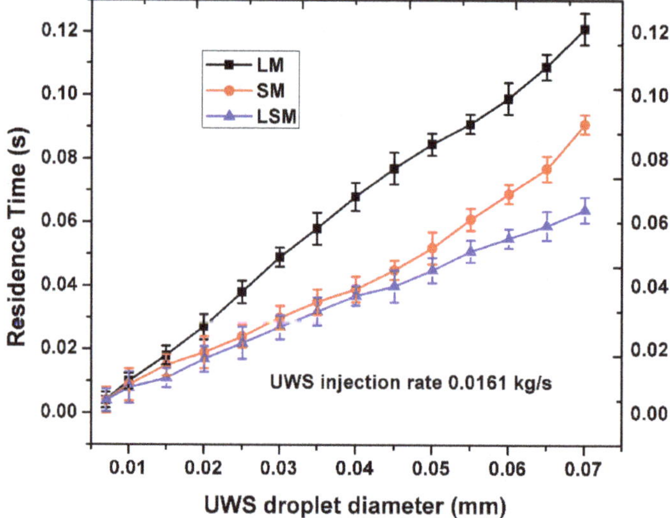

Figure 9. Residence time of urea droplets with the use of different mixers.

The angle of injection of UWS was 30, and the injection was done at the center of pipe; therefore, it is difficult for droplets to move far in radial directions but as an alternative of it, droplets will try to moves toward center. The droplets with bigger diameter will not follow the flow but start to move in radial directions. Small droplets evaporate faster than the bigger diameter and can easily be blown to the center of pipe. However, with the use of LSM, the mixing ability of UWS droplets and exhaust gas enhance sufficiently, which resist the gathering of larger UWS droplets and avert the temperature drop at the center of pipe. Moreover, the combination of two mixers results to increase the turbulence intensity which produces high evaporation rate. Overall, in the system of LSM, the residence time of UWS droplets decreases not only due to the improvement in the distribution of droplets and but also due of high turbulence impact of two mixers.

3.3. Urea Conversion

Better conversion or evaporation of UWS results the enough production of ammonia for SCR reaction. With the use of mixers, primary UWS droplets can be distributed and broken up into secondary droplets resulting in quick evaporation of UWS. Conversion of urea from the mixer downstream to the catalyst inlet (Figure 1 blue arrow) for LM, SM, and LSM are shown in Figure 10. It has been observed that there is deficient conversion of urea with the use of single mixer. As the decomposition distance increases, the conversion of urea also increases gradually. For the system of LM and SM, the conversion of urea is 76.1% and 83.2%, respectively. However, for the LSM system, the conversion increases up to 95.4%. Urea conversion is directly related with the mixing of flow distribution and droplet residence time in the SCR system.

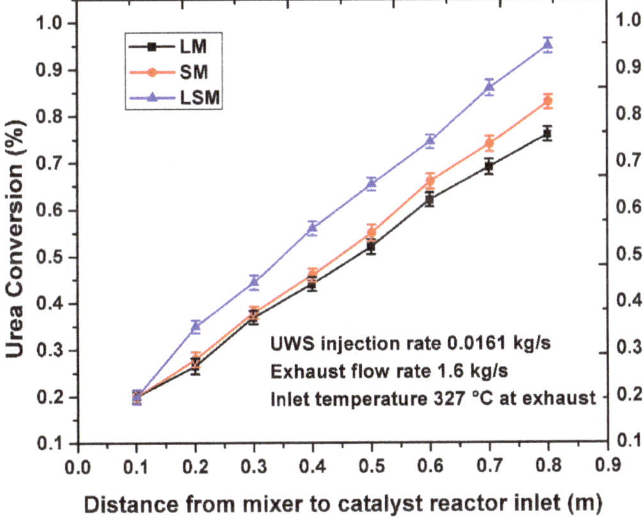

Figure 10. Urea conversion from mixer downstream to catalyst inlet.

Conversion of urea takes place in two stages, one is water vaporization stage, while the other urea vaporization stage as shown in Figure 11. Initially, the evaporation of pure water occurs from UWS droplets; therefore it becomes unstable in the exhaust gas in the later stage. As a result, the decomposition of urea produces NH_3 and HCNO. The droplets with bigger diameter can only undergo water vaporization stage and hence does not produces sufficient ammonia in long distances. With the use of LSM, bigger droplets are break up into small droplets and need short time to evaporate which ultimately results in increased evaporation rate.

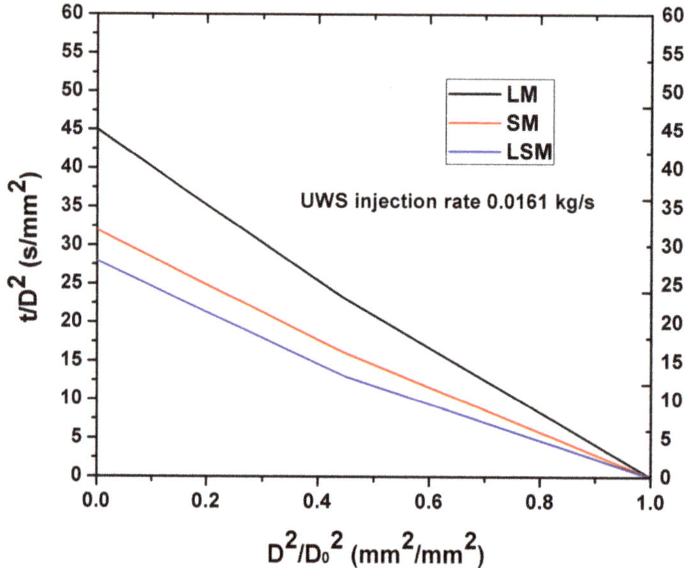

Figure 11. Urea water solution (UWS) droplet at two vaporization stages.

3.4. Relation of Temperature with SCR Performance

Two important factors relating to the SCR system temperature include deposit formation and catalyst reaction efficiency. SCR is a means of converting NO_x emissions into N_2 and H_2O by using the catalyst and O_2 in the temperature window of 280 to 420 °C [11]. The rate of reaction will be slow and unwanted reactions will occur, if the temperature goes below 280 °C, resulting in poor SCR catalyst performance. Ammonia will start to burn without reacting with the NO_x emissions if the temperature goes above 420 °C. It is therefore recommended to control the reaction temperature of SCR system [46].

The wall temperature is the prominent source for finding the deposit formation. Urea starts to decompose rapidly, associated with the secondary reactions, if the temperature is more than 163 °C. If the temperature is in between the range of 133 to 163 °C, pyrolysis of urea occurs slowly. Urea crystals are formed certainly as temperature is reduced below 133 °C. Once the temperature goes down, urea crystals are produced and the exhaust pipe will be blocked, which is responsible for decreasing the mixing performance and increasing the back pressure [18]. Concurrently, the overall catalytic reactor efficiency is decreased considerably. Once the urea injected into the system, a number of small droplets produce collisions with the pipe wall and mixer. The increase in collision strength of droplets with pipe wall decreased the temperature. Under the Leidensforst temperature, droplets of urea are separated into four boiling phenomenon as maintained by three values of temperature: 140, 190, and 300 °C. Heat transfer to the liquid film from pipe wall increased at ~180 °C [9]. If the wall temperature is lower than the boiling point temperature (T_b) of the UWS droplet, the droplets will stick on the wall and, if the temperature is higher than T_b, the droplets will rebound after striking with the wall [45]. Therefore, it is important to obtain maximum liquid film temperature for improved performance of urea decomposition and decrease the deposit formation.

Temperature Distribution along the Mixer Downstream

The temperature distribution along the radial direction at the mixer downstream (Figure 1 red arrow) as represented in Figure 12. The pipe diameter is 10 cm; therefore, the initial position (0 cm) stands for the upper edge of the pipe and 10 cm shows the position of lower edge. A plane was created in CFD at a certain distance and then divided into number of points. The value of temperature on each single point was found. Next, the standard deviation and average value of all the points were calculated. The minimum temperature of LM at the top edge is 274 °C, which is 32 °C less than SM and LSM.As it has been proved that after the urea injection process a large number of droplets collide with the mixer and the pipe wall: the wall temperature decreases with the increase in collision intensity [29]. In all types of mixers arrangement, the temperature difference at the center of the pipe is not obvious. However, the temperatures of LM and SM at the bottom edge of pipe were 276 °C and 281 °C, respectively. The temperatures at top edge and bottom edge of pipe are 304 °C and 301 °C respectively. Generally, the combination of two mixers (LSM) have good temperature distribution for both upper edge as well as lower edge, which is beneficial for catalyst reaction performance and also very suitable to prevent the deposit formation.

The axial wall temperature distribution from the mixer downstream to the catalyst inlet (blue arrow in Figure 1) as represented in Figure 13. Wall temperature distributions of LM and SM at the catalyst inlet were289 °C and 294 °C, respectively, but the combination of two mixers (LSM) was 300 °C due to better heat transfer effect. A continuous decrease in the temperature was observed from 0.3 to 0.5 m distance; with minimal temperature of 270 °C in SM. Low temperature region produces direct effect on the wall of pipe and reaction performance, without creates disturbance for the mixer.

Generally, temperature plays an important role in the reaction performance and deposit formation. With the use of LSM, the temperature remains above 280 °C in axial and radial directions, which is very helpful for preventing the unwanted reactions and deposit formations.

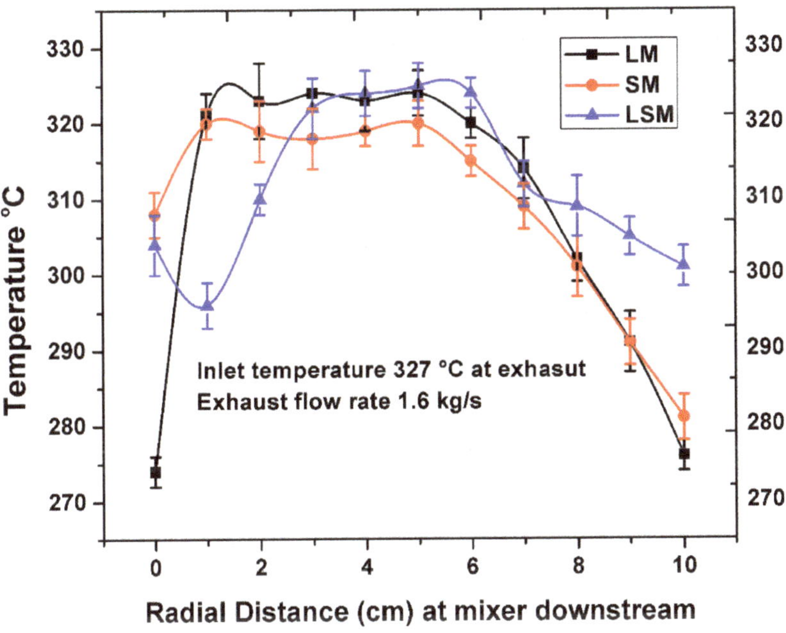

Figure 12. Radial distribution of temperature in the mixer downstream.

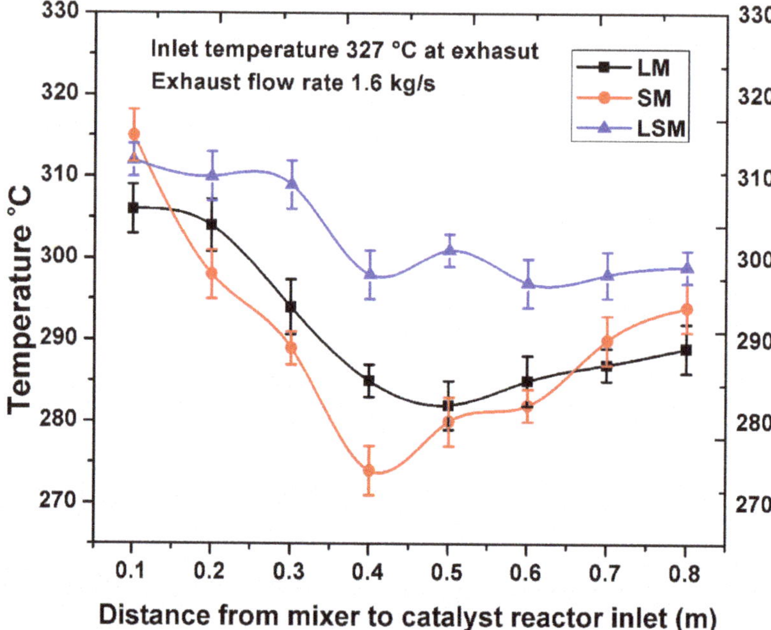

Figure 13. Axial distribution of temperature from mixer downstream to catalyst inlet in pipe wall.

3.5. Working Performance of LSM-Based SCR System

The standard ISO 8178 is the international marine Diesel engine test cycle used to measure the exhaust emissions from ships. By following the control requirements of exhaust emission as per IMO Tier III, standard marine Diesel engine test cycle has been divided into two parts one is ISO 8178 D2 test cycle for marine Diesel engine operated with constant speed and other is ISO 8178 E3 test cycle with propelling character for marine Diesel engine. In this study marine Diesel engine with propelling character was studied. Hence, the ISO 8178 E3 test cycle was used to measure the exhaust emissions. The weightage average value of NO_x exhaust emission was calculated by ISO 8178 marine Diesel engine test cycle E3 as shown in Table 4 [30].

Table 4. ISO 8178 marine Diesel engine test cycle E3.

Type ISO 8178 E3 Mode	1	2	3	4
Load (%)	25	50	75	100
Speed (%)	63	80	91	100
Weightage factor	0.15	0.15	0.5	0.2

The overall weighted NO_x exhaust emission level in g/kWh can be calculated using Equation (16) [47].

$$EF_x = \frac{\sum\limits_{i=1}^{n} m_i WF_i}{\sum\limits_{i=1}^{n} p_i WF_i} \tag{16}$$

The main purpose of SCR system is to decrease the NO_x emissions and to prevent the ammonia leakages responsible for air pollution. Figure 14 represents the NO_x removal efficiency under different loading conditions. It was observed that, there is a small difference in between theoretical value (standard) and calculated value (simulation). At low load, the NO_x removal efficiency is low, because under lower loading conditions, the exhaust temperature of Diesel engine is comparatively low, which directly affects the catalyst performance; as a result, the catalyst efficiency decreases. Figure 15 shows the NH_3 escaping rate at different engine loading conditions. The rate of NH_3 escaping decreases with the increase of engine load. NH_3 escaping rate is relatively high at low load due to lower temperature of the exhaust gas, resulting in incomplete catalyst reaction. The reaction of NH_3 oxidization is neglected for design of SCR model parameters in this study; therefore, more NH_3 slipping occurs at low loading conditions. However, at higher loading conditions (75% and 100%), the NH_3 escaping rate is less than 10 ppm, which meets the design requirement of the SCR system.

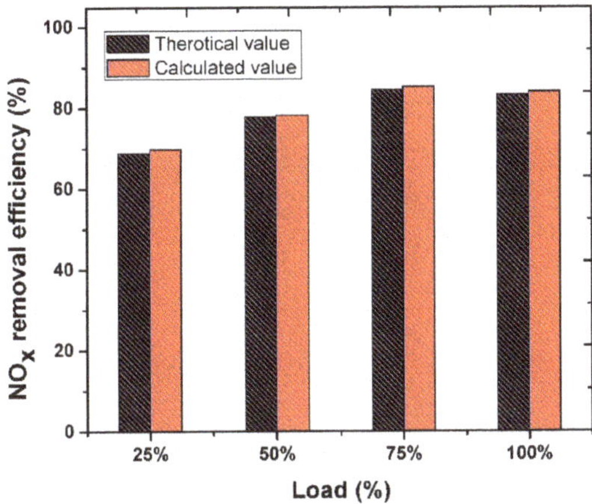

Figure 14. NO_x removal efficiency under different engine loading conditions.

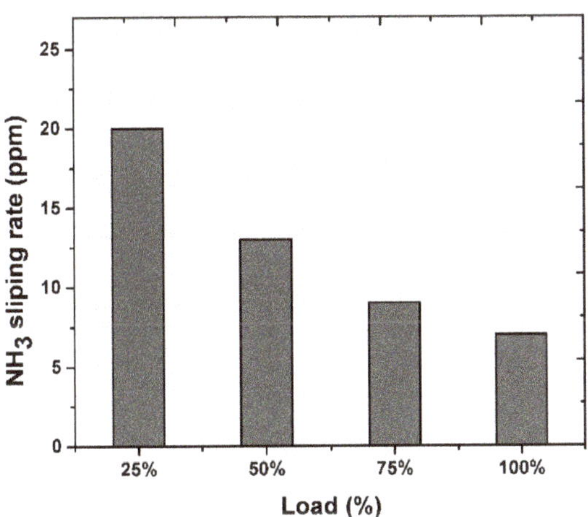

Figure 15. NH_3 escaping rate under different engine loading conditions.

Verification of simulated results was confirmed by using ISO 8178 standard marine Diesel engine test cycle E3. According to emission regulations of IMO Tier III, the value of NO_x emission should be less than 3.4 g/kWh under all loading conditions. Figure 16 shows a NO_x emission value of the LSM-based SCR system at different loading conditions. It was observed that NO_x emission decreases as the load increases. The average weighted value of NO_x emission was 2.44 g/kWh for four different loads at the downstream of SCR catalyst. Hence, the system based on using LSM strongly meets the standard of IMO Tier III NO_x emission regulations effectively.

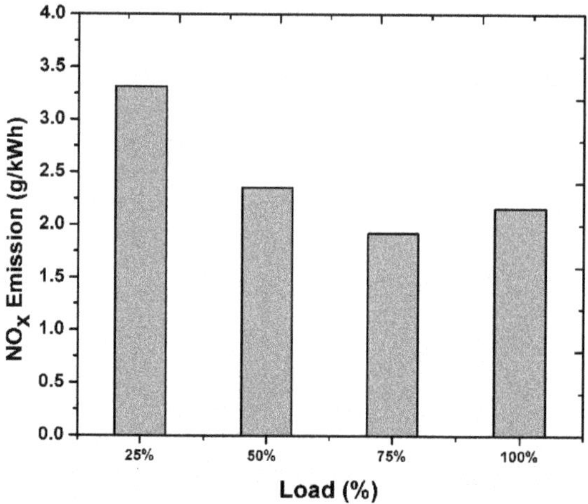

Figure 16. NO_x emission under different engine loading conditions.

4. Conclusions

In this work, numerical methods were used to analyze the mixing performance, effects of mixers on the evaporation rate of urea, residence time of urea droplet in the pipe, and temperature distribution for the catalyst reaction and deposit formation. The prime results are shown below.

- For the in-line type mixer (LM), the uniformity index of velocity was good (0.93) but the uniformity of ammonia was poor (0.87). In contrary to LM, the swirl type mixer (SM) has good ammonia uniformity (0.94) but poor uniformity index of velocity (0.86). However, better values were observed by using combination of two mixers (LSM). The uniformity index of velocity and ammonia uniformity achieved the values of 0.95 and 0.96, respectively, for LSM-based SCR system.
- The residence time UWS was studied. The results show that the residence time of urea droplets for LSM-based SCR system was 0.064 s, which represents47% and 29% decreases compared to LM and SM, respectively. Furthermore, the conversion of urea into ammonia is highly related with the residence time of urea droplets in the pipe. Hence, urea conversion achieves the value of 95.4% by using LSM, which is 19.3% and 12.2% higher than the value of LM and SM, respectively.
- It was also observed that the combination of two mixers (LSM) have good temperature distribution than the LM and SM for radial and axial directions, the temperature at catalyst inlet in axial direction was 300 °C for LSM-based SCR system which is suitable for the catalyst reaction performance and prevents the deposit formation.
- Finally, the simulated results of the model parameters were compared and verified by using ISO 8178 standard marine Diesel engine test cycle E3. The average weighted value of NO_x emission was calculated as 2.44 g/kWh for four different loads. Hence, it is concluded that the system based on using LSM strongly meets the standard of IMO Tier III NO_x emission regulations effectively.

Author Contributions: Conceptualization, Methodology, Software, Data Curation, Writing—Original Draft Preparation and Validation, G.M. and S.Z.; Formal Analysis and Investigation Y.Z.; Supervision, Funding Acquisition, and Project Administration S.Z.; Writing—Review and Editing, A.H.S. and K.C.

Funding: We gratefully acknowledge the financial support of the National Key and Research and Development Program of China (No: 2016YFCO205400) and the provincial funding of the National Projects of Heilongjiang Province in China (No: GX17A020).

Conflicts of Interest: The authors declare no conflict of interest.

Abbreviations

CFD	Computational Fluid Dynamics
ECA	Emission Control Areas
IMO	International Maritime Organization
LM	Line Mixer
LSM	Line Swirl Mixer
SM	Swirl Mixer
SCR	Selective Catalyst Reduction
UWS	Urea Water Solution

Nomenclature

Symbol	Name (Unit)
R	Gas constant ($J\,kg^{-1}\,K^{-1}$)
P	Pressure (Pa)
T	Reaction Temperature (K)
U	Fluid velocity (m/s)
P	Droplet density (m^2/s^3)
K	Turbulent kinetic energy (m^2/s^2)
V_t	Velocity with time (m/s)
C_μ	Closure coefficient
C_s	Volume concentration (m^3)
$\vec{j_i}$	Diffusion flux
D_S	Component diffusion coefficient
S_m	Chemical reaction component mass
$\vec{V}$	Gas velocity vector
R_i	Net production rate of species
S_i	Rate of creation from dispersed phase to user defined phase
Y_i	Species destruction molar rate
D_T	Turbulent diffusivity (m^2/s)
$D_{T,i}$	Thermal diffusion coefficient (m^2/s)
$D_{T,m}$	Mass diffusion coefficient
C_{water}	Molar concentration of water
K_c	Mass transfer coefficient (m/s)
T_p	Droplet temperature
$R_{i,r}$	Species destruction molar rate
m_p	Droplet mass (kg)
u_p	Velocity of liquid droplet
ρ_p	Density of liquid droplet
F	Force except drag force
C_p	Specific heat of liquid droplet (J/K)
A_p	Droplet surface area (m_2)
$T\infty$	Droplet environment temperature
h	Convective heat transfer coefficient ($W/(m^2K)$)
h_{fg}	latent heat of vaporization (KJ/kg)
V_i	Carrier Nominal velocity (m/s)
V_{mean}	Average velocity (m/s)
A_i	Cell area (m^2)
A	Sectional area of the plane (m^2)
EF_x	Weighted emission level (g/kWh)
m_i	Mass emission rate (g/h)
WF_i	Weighting factor
P_i	Engine load
Ar	Pre-exponential factor
Ea	Activation energy (J/kmol)
MW	Species molecular weight

Greek symbols

E	turbulent dissipation (m^2/s^3)
A	viscosity coefficient
ε_p	radiant heat transfer rate of liquid droplet
μ_t	turbulent viscosity (m^2/s)
$P_{sat,\,water}$	Vapour pressure of water

Dimensionless numbers

Pr	Prandtl number
S_{ct}	Turbulent Schmidt number
S_c	Schmidt number

References

1. Zhanga, C.; Suna, C.; Wua, M.; Lub, K. Optimisation design of SCR mixer for improving deposit performance at low temperatures. *Fuel* **2019**, *237*, 465–474. [CrossRef]
2. Chen, W.; Fali, H.; Qin, L.; Jun, H.; Zhao, B.; Yangzhe, T.; Fei, Y. Mechanism and Performance of the SCR of NO with NH3 over Sulfated Sintered Ore Catalyst. *Catalysts* **2019**, *9*, 90. [CrossRef]
3. Hu, N.; Zhou, P.; Yang, J. Reducing emissions by optimising the fuel injector match with the combustion chamber geometry for a marine medium-speed diesel engine. *Transp. Res. Part D Transp. Environ.* **2017**, *53*, 1–16. [CrossRef]
4. International Maritime Organization. *Annex VI of MARPOL 73/78, Regulations for the Prevention of 418 Air Pollution from Ships and NOx Technical Code*; International Maritime Organization: London, UK, 2008; pp. 15–19.
5. Yewen, G.; Stein, W.W. A potentially overestimated compliance method for the Emission Control Areas. *Transp. Res. Part D* **2017**, *55*, 51–66.
6. Zamboni, G.; Moggia, S.; Capobianco, M. Hybrid EGR and turbocharging systems control for low NO_x, and fuel consumption in an automotive diesel engine. *Appl. Energy* **2016**, *165*, 839–848. [CrossRef]
7. Nova, I.; Tronconi, E. *Urea-SCR Technology for deNO$_x$ after Treatment of Diesel Exhausts*; Springer: New York, NY, USA, 2014; pp. 12–18.
8. Koebel, M.; Elsener, M.; Madia, G. Reaction pathways in the selective catalytic reduction process with NO and NO_2 at low temperatures. *Ind. Eng. Chem. Res.* **2001**, *40*, 52–59. [CrossRef]
9. Sadashiva, P.S. Review on Selective Catalytic Reduction (SCR)-A Promising Technology to mitigate NO_x of Modern Automobiles. *Int. J. Appl. Eng. Res.* **2018**, *13*, 5–9.
10. Shahariarm, G.M.H.; Lim, O.T. A Study on Urea-Water Solution Spray-Wall Impingement Process and Solid Deposit Formation in Urea-SCR de-NOx System. *Energies* **2018**, *12*, 125. [CrossRef]
11. Koebel, M.; Elsener, M.; Kleemann, M. Urea-SCR: A promising technique to reduce NO_x emissions from automotive diesel engines. *Catal. Today* **2000**, *59*, 335–345. [CrossRef]
12. Fang, H.L.; DaCosta, H.F.M. Urea thermolysis and NOx reduction with and without SCR catalysts. *Appl. Catal. B Environ.* **2003**, *46*, 14–34. [CrossRef]
13. Jeong, S.J.; Sang, J.L.; Kim, W.S. Numerical Study on the Optimum Injection of Urea-Water Solution for SCR DeNO$_x$ System of a Heavy-Duty Diesel Engine to Improve DeNO$_x$ Performance and Reduce NH3 Slip. *Environ. Eng. Sci.* **2008**, *25*, 1017–1036. [CrossRef]
14. Varna, A.; Spiteri, A.C.; Wright, Y.M. Experimental and numerical assessment of impingement and mixing of urea-water sprays for nitric oxide reduction in diesel exhaust. *Appl. Energy* **2015**, *157*, 824–837. [CrossRef]
15. Grout, S.; Blaisot, J.B.; Pajot, K. Experimental investigation on the injection of an urea-watersolution in hot air stream for the SCR application: Evaporation and spray/wall interaction. *Fuel* **2013**, *106*, 166–177. [CrossRef]
16. Choi, C.; Sung, Y.; Choi, G.M.; Kim, D.J. Numerical Analysis of Urea Decomposition with Static Mixers in Marine SCR System. *J. Clean Energy Technol.* **2015**, *3*, 39–42. [CrossRef]
17. Haitao, D.; Dawei, M.; Renbin, Z.; Bowen, S.; Jun, H. Impact of Control Measures on Nitrogen Oxides, Sulfur Dioxide and Particulate Matter Emissions from Coal-Fired Power Plants in Anhui Province, China. *Atmosphere* **2019**, *10*, 35.

18. Tana, L.; Fenga, P.; Yangb, S.; Guoa, Y.; Liua, S.Z.L. CFD studies on effects of SCR mixers on the performance of urea conversion and mixing of the reducing agent. *Chem. Eng. Process. Process Intensif.* **2018**, *123*, 82–88. [CrossRef]

19. Birkhold, F.; Meingast, U.; Wassermann, P. Modeling and simulation of the injection of urea-water-solution for automotive SCR DeNOx-systems. *Appl. Catal. B Environ.* **2007**, *70*, 119–127. [CrossRef]

20. Ström, H.; Lundström, A.; Andersson, B. Choice of urea-spray models in CFD simulations of urea-SCR systems. *Chem. Eng. J.* **2009**, *150*, 69–82. [CrossRef]

21. Rajadurai, S. *Improved NOx Reduction Using Wiremesh Thermolysis Mixer in Urea SCR System*; SAE Technical Paper 2008; SAE: Zürich, Switzerland, 2008.

22. Zhang, X.; Romzek, M.; Morgan, C. *3-D Numerical Study of Mixing Characteristics of NH₃ in Front of SCR*; SAE Technical Paper; SAE: Zürich, Switzerland, 2006.

23. Chen, M.; Williams, S. *Modeling and Optimization of SCR-Exhaust Aftertreatment Systems*; SAE, World Congress and Exhibition; SAE: Zürich, Switzerland, 2005.

24. Capetillo, A.; Ibarra, F. Multiphase injector modeling for automotive SCR systems: A full factorial design of experiment and optimization. *Comput. Math. Appl.* **2017**, *74*, 89–97. [CrossRef]

25. Park, T.; Kim, Y.S.; Lee, I.; Choi, G.; Kim, D. Effect of static mixer geometry on flow mixing and pressure drop in marine SCR applications. *Int. J. Nav. Arch. Ocean Eng.* **2014**, *6*, 27–38. [CrossRef]

26. Yuanqing, Z.; Zhang, R.; Zhou, S.; Huang, C.; Feng, Y.; Shreka, M.; Zhang, C. Performance Optimization of High-Pressure SCR System in a Marine Diesel Engine. Part I: Flow Optimization and Analysis. *Top. Catal.* **2019**. [CrossRef]

27. Olivier, F.; Toshiyuki, Y. Development of Static Mixer Device for Heavy Duty Diesel Engine SCR After-treatment System. *Calsonic Kansei Tech. Rev.* **2012**, *9*, 50–55.

28. Hyman, D. Mixing and agitation. *Adv. Chem. Eng.* **1999**, *3*, 119–202.

29. Kuhnke, D. Spray Wall Interaction Modeling by Dimensionless Data Analysis. Ph.D. Thesis, Universitat Darmstadt, Darmstadt, Germany, 2004.

30. Emission Test Cycles. Available online: www.dieselnet.com/standards/cycles/iso8178.php (accessed on 20 December 2018).

31. Yong, Y. *Development of a 3D Numerical Model for Predicting Spray, Urea Decomposition and Mixing in SCR Systems*; SAE Technical Papers; SAE: Zürich, Switzerland, 2007.

32. Jeong, S.; Lee, S.; Kim, W.; Lee, C. *Simulation on the Optimum Shape and Location of Urea Injector for Urea-SCR System of Heavy-Duty Diesel Engine to Prevent NH3 Slip*; SAE Technical Paper; SAE: Zürich, Switzerland, 2005.

33. Yakhot, V.; Orszag, S.; Thangam, S.; Gatski, T.; Speziale, C. Development of turbulence models for shear flows by a double expansion technique. *Phys. Fluids A* **1992**, *4*, 1510–1520. [CrossRef]

34. Andersson, B.; Andersson, R.; Håkansson, L. *Computational Fluid Dynamics for Engineers*; Cambridge University Press: Cambridge, UK, 2011; pp. 56–66.

35. Wang, H.; Fischman, G.S. Role of liquid droplet surface diffusionin the vapor-liquidsolid whisker growth mechanism. *J. Appl. Phys.* **1994**, *76*, 1557–1562. [CrossRef]

36. Berlemont, A.; Grancher, M.S.; Gouesbet, G. Heat and mass transfer coupling between vaporizing droplets and turbulence using a Lagrangian approach. *Int. J. Heat Mass Transf.* **1995**, *38*, 3023–3034. [CrossRef]

37. Schaber, P.M.; Colson, J.; Higgins, S. Study of the urea thermal decomposition (pyrolysis) reaction and importance to cyanuric acid production. *Am. Lab* **1999**, *31*, 13–21.

38. Schaber, P.M.; Colson, J.; Higgins, S. Thermal decomposition (pyrolysis) of urea in an open reaction vessel. *Thermochim. Acta* **2004**, *424*, 131–142. [CrossRef]

39. Law, C.K. Recent advances in droplet vaporization and combustion. *Prog. Energy Combust. Sci.* **1982**, *8*, 171–201. [CrossRef]

40. Spiteri, A.; Eggenschwiler, P.D.; Liao, Y.; Wigley, G.; Michalow-Mauke, K.A.; Elsener, M. Comparative analysis on the performance of pressure and air assisted urea injection for selective catalytic reduction of NOₓ. *Fuel* **2015**, *161*, 269–277. [CrossRef]

41. Shahariar, G.M.H.; Lim, O.T. Investigation of urea-water solution spray impingement on the hot surface of automotive SCR system. *J. Mech. Sci. Technol.* **2018**, *32*, 2935–2946. [CrossRef]

42. Xu, Z.; Liu, J.; Fu, J. Experimental investigation on the urea injection and mixing module for improving the performance of Urea-SCR in diesel engines. *Can. J. Chem. Eng.* **2018**, *96*, 1417–1429. [CrossRef]

43. Paramadayalan, T.; Pant, A. Selective catalytic reduction converter design: The effect of ammonia nonuniformity at inlet. *Korean J. Chem. Eng.* **2013**, *30*, 2170–2177. [CrossRef]

44. Fischer, S.; Fischer, R.B.; Lauer, T. Impact of the turbulence model and numerical approach on the prediction of the ammonia homogenization in an automotive SCR system. *SAE Int. J. Eng.* **2012**, *5*, 1443–1458. [CrossRef]

45. Rourke, P.J.O.; Amsden, A.A. *A Spray/Wall Interaction Submodel for the KIVA-3 Wall Flm Model*; SAE Technical Paper; SAE: Zürich, Switzerland, 2000.

46. Marin, P.; Fissore, D.; Barresi, A.A. Simulation of an industrial-scale process for the SCR of NO_x based on the loop reactor concept. *Chemical Eng. Process. Process Intensif.* **2009**, *48*, 311–320. [CrossRef]

47. Gysel, N.R.; Robert, L.; Russell; Welch, W.A.; Cocker, I.D.R. Impact of Aftertreatment Technologies on the In-Use Gaseous and Particulate Matter Emissions from a Tugboat. *Energy Fuels* **2016**, *30*, 684–689. [CrossRef]

Article

Implementation of Maximum Power Point Tracking Based on Variable Speed Forecasting for Wind Energy Systems

Yujia Zhang [1], Lei Zhang [2,*] and Yongwen Liu [3]

[1]   School of Automation, Northwestern Polytechnical University, Xi'an 710072, China; zyjagllfb@163.com
[2]   College of Information Science and Engineering, Henan University of Technology, Zhengzhou 450001, China
[3]   Software Engineering College, Zhengzhou University of Light Industry, Zhengzhou 450000, China; yongwen.liu@zzuli.edu.cn
*   Correspondence: zhanglei1003261964@163.com; Tel.: +86-150-0382-0682

Received: 30 January 2019; Accepted: 8 March 2019; Published: 15 March 2019

Abstract: In order to precisely control the wind power generation systems under nonlinear variable wind velocity, this paper proposes a novel maximum power tracking (MPPT) strategy for wind turbine systems based on a hybrid wind velocity forecasting algorithm. The proposed algorithm adapts the bat algorithm and improved extreme learning machine (BA-ELM) for forecasting wind speed to alleviate the slow response of anemometers and sensors, considering that the change of wind speed requires a very short response time. In the controlling strategy, to optimize the output power, a state feedback control technique is proposed to achieve the rotor flux and rotor speed tracking purpose based on MPPT algorithm. This method could decouple the current and voltage of induction generator to track the reference of stator current and flux linkage. By adjusting the wind turbine mechanical speed, the wind energy system could operate at the optimal rotational speed and achieve the maximal power. Simulation results verified the effectiveness of the proposed technique.

Keywords: maximum power tracking (MPPT); wind speed forecasting; wind energy system (WES); state feedback controller

1. Introduction

Wind, used as widely distributed huge reserves of green energy [1], has dramatic increased in grid-connected power these years. Thus, to improve the efficiency of wind power system becomes an essential part, both to reduce the costs of wind power generation systems, and to increase the proportion of renewable energy in the national power grid. High efficiency, good robustness and low costs have become the research focus of wind energy harnessing.

Currently, the literature extensively investigates modeling effective wind turbine systems to optimize and effectively utilize the turbine power output through the maximum power point tracking (MPPT) technique. By the adoption of variable speed wind turbine (VSWT), adjusting the rotation speed of wind turbine rapidly according to the variable wind speed can achieve for high efficiency to harness wind source [2]. These technique adopted by MPPT controller mainly can be categorized into four types: by controlling of Tip Speed Ratio (TSR), adopting Power Signal Feedback (PSF) control, Perturb and Observe (P&O) method, and Optimal Torque Control (OTC) method. TSR is a constant value which is dependent of wind velocity. It is the only parameter that can be set to provide the maximum power output from wind which is related to rotor radius and the blades rotational speed [3]. Such method needs two sensors to measure the wind and turbine speeds by an anemometer and a tachometer respectively for that both wind speed and rotational speed are feedback signals [4]. Compared to TSR (nomenclature can be seen in Table 1), PSF only uses one sensor to measure rotational speed. Error between measured turbine power and reference power is delivered to the controller, then the output power is adjusted to the reference value. The efficiency is good and the reliability is better than TSR [5]. While the P&O method does not necessarily need any sensor beside the electrical measurement devices, its reliability is strong but efficiency is not good. The Optimal Torque Control (OTC) extracts optimum torque by measuring angular velocity [6], which is similar to PSF control yet adopts mechanical torque equation.

Table 1. Acronyms and nomenclature.

BA	bat algorithm
MPPT	maximum power point tracking
VSWT	variable speed wind turbine
PSF	adopting power signal feedback
TSR	tip speed ratio
OTC	optimal torque control
FIS	fuzzy inference system
ELM	extreme learning machine
SLFNN	single-hidden layer feed-forward neural networks
MAE	mean absolute error
MSE	mean square error
VOC	voltage oriented control
P&O	perturb and observe

However, these conventional techniques fail to consider that the wind speed is a discrete nonlinear parameter set which is not compliant to a certain law of variation. That requires the anemometers to measure wind velocity in time and the controller should also respond quickly to the wind speed fluctuation, then drives the mechanical rotor to rotate with the optimized direction and speed. Usually the measurements and controlling process should not exceed one second to harness the wind energy in highest efficiency [7]. Yet in large-scale wind turbines, anemometers and controllers have large volume and great inertia, leading to a slow response [8]. Therefore, typically when a rotor speed instruction has not completed, the controller should execute in an opposite direction. In this way, the wind turbine is more likely to have mechanical fatigue. Besides, because of the hysteresis effect, the current optimal tracking point is not the current maximum power point.

Such problems are noticed by scholars in recent years, with the development of computational intelligence and numerical optimization, some novel MPPT frameworks are also proposed to tackle this issue. These framework often involves: the fuzzy inference system (FIS), taking multi-objectives into account [9]; nonlinear control [10], with Boukhezzar et al. adapts a two-mass model with a wind speed estimator for variable-speed wind turbine control [11]; robust control, via controlling the rotor angular speed to control the tip-speed ratio [12]; adaptive control [13] and the like. Some scholars even proposed hybrid models with the combination of artificial intelligence algorithm and conventional control methods to achieve a higher efficiency [14]. Though it can effectively deal with high non-linearity of wind turbines, the training process is indeed time-costing and introduces a huge amount of iteration parameters like weights and bias into systems [15]. In order to avoid generating these parameters and to save hardware, here we consider extreme learning machine (ELM) to forecasting wind velocity, which do not need to use back-propagation method to updates weights and thresholds and only the linear least square solution is needed [15]. With the prediction scheme, a novel controlling strategy to intervene the hysteresis effect in both wind speed measurement and controlling process is proposed in this paper.

The proposed controlling strategy can be concluded into three steps, first is to optimize the weights and thresholds of extreme learning machine (ELM) by bat algorithm (BA) to forecast the short–term wind speed, since the random distributed weights and bias of traditional ELM algorithm are likely to be not convergent [16]. The second is that we adapt the forecasting values to calculate the optimal reference rotational speed based on MPPT algorithm. Meanwhile the anemometer would measure the current wind velocity thus to feed the input of BA-ELM. Finally the state feedback control and optimal control technique will be implemented in wind energy conversion system, to achieve maximal power point thus the current asynchronous machine torque can achieve high efficiency [17].

The outline of rest is organized as follows. Section 2 describes the wind energy conversion system. Section 3 introduces the detailed design procedure of speed forecasting method with BA-ELM for wind energy conversion system. Section 4 presents the state feedback controller for induction machine with the MPPT algorithm. The simulation results are presented in Section 5. Finally, some comments conclude this work in Section 6.

2. Wind Energy Conversion System

The wind energy conversion system is a high nonlinear and complex coupled system. The system comprises of the wind turbine, transmission device, induction machine and converters [18]. The basic components and general scheme of a wind energy system are shown in Figure 1, which contain the following main parts:

1. Wind turbine, which is a installation capturing wind energy by blades and transferring the wind kinetic power to mechanical torque;
2. Induction machine, which can convert the power from the mechanical side into the electrical side.
3. Gearbox and shaft, which is a transmission device to adapt the rotation speed for the generator;
4. Power converters, it is composed of the grid side inverter and the machine side rectifier, connected by a DC-bus.

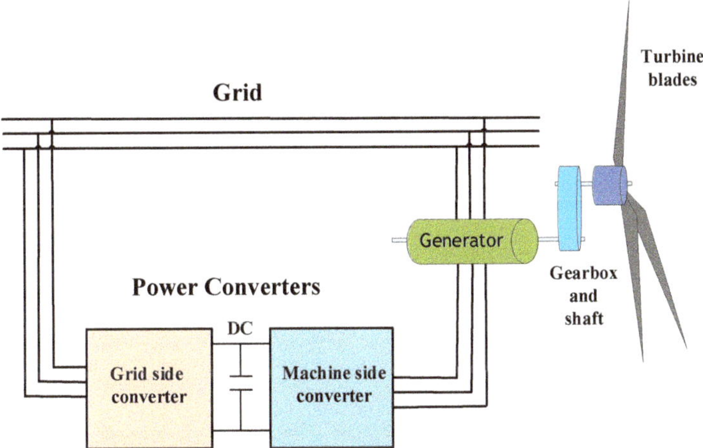

Figure 1. Block diagram of the wind turbine.

The following subsections will focus on the wind turbine model and induction machine model.

2.1. Wind Turbine Model

The wind turbine extracts power by the wind blades in the turbine nacelle, then converts it into mechanical power. The wind kinetic power can be formulated as follows:

$$P_w = \frac{1}{2}\rho\pi R^2 V_w^3 \tag{1}$$

where P_w represents the power input of wind turbine, R determines the radius of blades, V_w indicates the wind speed.

The tip speed ratio λ can be expressed by

$$\lambda = \frac{R\omega_r}{V_w} \tag{2}$$

where ω_r represents the rotor speed.

The power extracted from the wind is

$$P_m = \frac{1}{2}C_p(\lambda, \beta)\rho\pi R^2 V_w^3 \tag{3}$$

where P_m represents the mechanical power, $C_p(\lambda, \beta)$ is a non-linear power coefficient depending on the design of turbine [19], which is:

$$C_p(\lambda, \beta) = 0.5176(\frac{116}{\lambda_i} - 0.4\beta - 5)e^{-\frac{21}{\lambda_i}} + 0.0068\lambda \tag{4}$$

with

$$\frac{1}{\lambda_i} = \frac{1}{\lambda + 0.08\beta} - \frac{0.035}{\beta^3 + 1} \tag{5}$$

where β is the blade pitch angle. The power coefficient is a nonlinear function of tip speed ratio λ and the blade pitch angle β, and its curve is plotted in Figure 2.

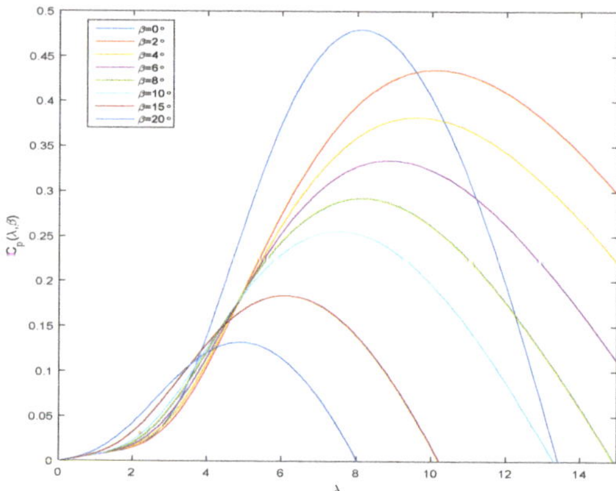

Figure 2. Power coefficient $C_p(\lambda, \beta)$ versus tip-speed ratio in different pitch angle.

It can be seen that two points can be concluded to extract more power in the wind energy system [20], which are:

(a) When the blade pitch angle β does not change, the peak values of power coefficiency $C_p(\lambda, \beta)$ corresponds to a unique tip speed ratio λ, where the conversion of wind energy is expected.

(b) As the blade pitch angle β increases, the wind energy use coefficient $C_p(\lambda, \beta)$ decreases obviously. Thus for tracking more wind power, β should be set into a small value.

The torque T_L caused by the wind turbine can be computed as

$$T_L = \frac{P_{\mathrm{m}}}{\omega_r} = \frac{1}{2}\rho \frac{C_p(\lambda, \beta)}{\lambda}\pi R^3 V_w^2 \tag{6}$$

From Equation (6), it can be noticed that the turbine T_L is related to the wind speed V_w and the characteristics of the turbine, that is the power coefficiency $C_p(\lambda, \beta)$.

2.2. Induction Machine Model

2.2.1. Mechanical Equations

In the block wind turbine blades, the aerodynamic torque model can be used to describe dynamic relationship between the high-speed rotor shaft and the low-speed axial-flow fan of the wind turbine, which is composed of a spring and a damper. Formulations for the system can be established as follows:

$$J_m \dot{\omega}_r = T_e - f_r \omega_r - T_L \tag{7}$$

where J_m represents the rotary inertia of wind turbine pales.

2.2.2. State Space Equation of the Induction Machine Motor

The state space equation of induction machine in the well-known inductor part flux reference frame (α, β) can be expressed as follows [21]:

$$\dot{i}_{s\alpha} = -c_1 i_{s\alpha} + c_2 c_3 \psi_{r\alpha} + c_3 p \omega_r \psi_{r\beta} + \frac{u_{s\alpha}}{c_4} \tag{8}$$

$$\dot{i}_{s\beta} = -c_1 i_{s\beta} + c_2 c_3 \psi_{r\beta} - c_3 p \omega_r \psi_{r\alpha} + \frac{u_{s\beta}}{c_4} \tag{9}$$

$$\dot{\psi}_{r\alpha} = c_5 i_{s\alpha} - c_6 \psi_{r\alpha} - p \omega_r \psi_{r\beta} \tag{10}$$

$$\dot{\psi}_{r\beta} = c_5 i_{s\beta} - c_6 \psi_{r\beta} + p \omega_r \psi_{r\alpha} \tag{11}$$

$$T_e = \frac{p M_{sr}}{L_r}(i_{s\beta} \psi_{r\alpha} - i_{s\alpha} \psi_{r\beta}) \tag{12}$$

where $u_{s\alpha}$ and $u_{s\beta}$ are the stator voltages, $i_{s\alpha}$ and $i_{s\beta}$ are the stator currents, $\psi_{r\alpha}$ and $\psi_{r\beta}$ are the rotor fluxes, p is the number of pole pairs, M_{sr} is the mutual inductance, L_r is the rotor inductance, T_e is the electromagnetic torque. Moreover, the variables $c_1, c_2, c_3, c_4, c_5, c_6$ are defined as follows

$$c_1 = \frac{R_s + R_r \frac{M_{sr}^2}{L_r^2}}{\sigma L_s}, c_2 = \frac{K}{T_r}, c_3 = K, c_4 = \sigma L_s, c_5 = \frac{M_{sr}}{T_r}, c_6 = \frac{1}{T_r}$$

with the related parameters T_r, σ, K are

$$T_r = \frac{L_r}{R_r}, \sigma = 1 - \frac{M_{sr}^2}{L_s L_r}, K = \frac{M_{sr}}{\sigma L_r}$$

where R_s is stator resistance, R_r is rotor resistance, L_s is stator inductance.

Substitute Equation (12) into Equation (7), the dynamic equation can be expressed by the following form

$$\dot{\omega}_r = c_7(\psi_{r\alpha} i_{s\beta} - \psi_{r\beta} i_{s\alpha}) - \frac{f_r}{J_m}\omega_r - \frac{T_L}{J_m} \tag{13}$$

with $c_7 = \frac{pM_{sr}}{J_m L_r}$ is a known constant.

2.2.3. Current Flux Model

In the induction machine model, the stator voltage and stator current can be measured, while the secondary flux can not be measured. Here, the current flux model is proposed to estimate the value of secondary flux. From voltage equation and flux equation of induction machine, the flux current model of induction machine is given as follows

$$\frac{d\psi_r}{dt} = -\frac{1}{T_r}\psi_r + \frac{M_{sr}}{T_r}i_s - j\left(\omega_{mr} - p\omega_r\right)\psi_r \tag{14}$$

where ψ_r is the induced-part flux, i_s is inductor current, ω_{mr} is the induced-part flux vector rotational speed, j is the imaginary unit. This equation represents the so-called "current model" of the induction machine.

3. Wind Speed Forecasting

As an anticipatory control strategy, predict wind velocity is more conducive to smooth the output power of wind turbines, since the power generation process is a complex nonlinear process, which will be affected by the wind speed and torque of the power generator. Besides, wind speed forecasting can estimate daily output of wind turbines in advance, and improve the planning ability of wind farms for electric power transmission and distribution. In addition, the maximum power point tracking can be achieved precisely by wind forecasting even without anemometers. Considering that there is a significant time-lag in the wind speed measurement of wind turbines for the anemometers' inertia is great, suitable prediction model should be chosen from the model base according to certain principles. Here we build a hybrid prediction model of wind speed on the basis of applying the modified bat algorithm (BA) to optimize the initial weights in the layers of extreme learning machine (ELM). The prediction model can predict the future state of wind velocity on the basis of the estimated mechanical torque from the output of WES model, so as to settle down the problem of system lag from measurement of the anemometer. Thus to keep correspondence with current wind speed by adapting the mechanical speed ω_r mentioned above from blade shafts to predict the one-step wind speed, current maximum power can be tracking in higher accuracy which eliminates the effect of the system lag.

3.1. BA Algorithm

The bat algorithm (BA) is a novel metaheuristic algorithm adapted in prediction model is to optimize the input weights between layers of the studied training network [22]. Given a data set of of historical wind speed input to the network, the short term wind speed forecasting will perform as the output of prediction system.

The BA rules can be concluded as the real-time dynamic adjustment of the location, loudness and pulse emission of the virtual BA bats when hunting and foraging, bats change the frequency, loudness and pulse emissivity, and choose the best solution until the end specified iteration loop or achieve specified accuracy. Here we denote y^t to be the bat positions, with the parameter t is the current iteration number.

f_i as the pulse frequency in a range $[f_{min}, f_{max}]$, and v_i is the velocities of bats. Initially each bat is assigned with a random frequency. An iterative loop is presented as follows:

$$f_i = f_{min} + (f_{max} - f_{min})\beta \ , i = 0, 1, \cdots, N-1 \tag{15}$$

$$v_i^t = v_i^{t-1} + (y_i^t - y_*)f_i \ , i = 0, 1, \cdots, N-1 \tag{16}$$

$$y_i^t = y_i^{t-1} + v_i^t \ , i = 0, 1, \cdots, N-1 \tag{17}$$

In the above formulation, $\beta \in [0, 1]$, parameter N denotes specified loop accounts which end the iteration loop. Here the newest obtained position y_N^t will be evaluated with the fitness function to determine whether the solution exhibits the best current performance.

3.2. BA-ELM Network

Extreme Learning Machine (ELM) is a supervised learning algorithm originated from single-hidden layer feed-forward neural networks (SLFNN) proposed by Guangbin Huang [15], and it achieves high precision in the performance of classification and forecasting. Unlike other gradient-based learning algorithms, the main idea is that the weights between the input layer and the hidden layer, the bias of the hidden layer of ELM do not need to be adjusted. The solution is very efficient in that only the least norm and the least square solution are needed (ultimately resolved into Moore-Penrose inverse problem). Therefore, the algorithm has the advantages of using very few training parameters and achieving extremely fast speed [23].

Here for standard SLFNN, the constructed model is formulated as following [24].

$$\sum_{i=1}^{\tilde{N}} g\left(w_i \cdot x_j + b_i\right)\beta_i = t_j . j = 1, 2, ..., N \tag{18}$$

$$\hat{t}_j(w, \beta, b) = [\hat{t}_{1j} \cdots \hat{t}_{mj}]_{m \times 1}^T$$

$$= [\sum_{i=1}^{l} \beta_{i1} g(w_1 X_j + b_i) \cdots \sum_{i=1}^{l} \beta_{im} g(w_1 X_j + b_i)]_{m \times 1}^T, \tag{19}$$

Which:

$$X_j = \begin{bmatrix} X_{1j} & X_{2j} & X_{3j} & \cdots & X_{n-1j} & X_{nj} \end{bmatrix}^T \tag{20}$$

In the above formulation, w_i denotes the weights between the input and the hidden layer. X_j denotes the inputs historical wind speed data. b_i denotes the bias. In addition, $g(x)$ denotes the activate function. β_i which equals to $(\beta_{ij})_{\hat{N} \times M}$, represents the weight matrix between the hidden layer and the output layer [4]. Here the above formulation can also be denoted as:

$$H\beta = T \tag{21}$$

H is the hidden layer output matrix of neural network. The equation has a unique solution when H is reversible (that means the number of hidden layers equals to the number of input data). While in most cases, the number of hidden layers is far less than the number of inputs. Common solutions to solve this problem include the gradient descend method to iterate parameters. However, it easily falls into the problem of over-trained and time-costing. To overcome its weakness, the literature points out that a lot of experiments show that H do not need to be adjusted, considering that adjusting input weights and biases of hidden layer will bring no possible gains [25]. Thus in ELM, when the input weights and biases are set once for all, the formulation (21) equals to find the linear least square solution for formulation (22).

$$\left\| H\hat{\beta} - T \right\| = \min_{\beta} \left\| H\beta - T \right\| \tag{22}$$

The detailed ELM training process can be found in [15]. Since the hidden layer nodes are pre-allocated and input weights and biases remain unchanged, some initial weights and biases are likely to remain non-optimized values [26]. For example, randomly allocated input weights might be zero, thus some hidden layer nodes would fail, leading that the model cannot converge or a slow convergence. To optimize these parameters, BA is introduced to optimize the weights and thresholds of ELM.

The BA-ELM training process can be concluded into two steps. First is to optimize the initial weights (w,β), and biases of the hidden layer b with the usage of bat algorithm. Second is to train the constructed ELM with these parameters and find the linear least square solution. The structure of BA-ELM is shown in Figure 3.

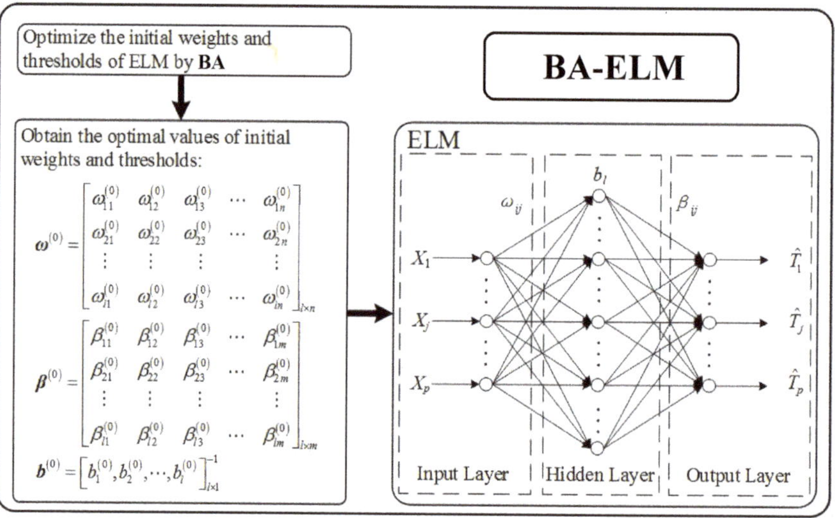

Figure 3. Block of multi-step forecasting.

When running an iteration loop of BA, the fitness function will be invoked to update for the current best weights and thresholds, which aims to reduce the error between the predicted value and the actual wind speed value. Thus the current best of these parameters will be updated and substituted into ELM when an iteration of BA finished. Then the iteration loop of BA will stop when satisfy the objective tolerance of BA.

To evaluate the errors between the actual wind velocity and predicted ones, here two percentage error indexes are employed: mean absolute error (MAE), and the mean square error (MSE). In addition, the proposed BA-ELM are compared with traditional ELM, the promoting percentages are defined as follows:

$$\varsigma_{MAE} = \left| \frac{MAE_1 - MAE_2}{MAE_1} \right| \times 100\% \tag{23}$$

$$\xi_{MSE} = \left| \frac{MSE_1 - MSE_2}{MSE_1} \right| \times 100\% \tag{24}$$

Therefore we can output the forecasting wind speed at the end of our training process. The maximum rotor speed $w_{r,opt}$, can be deduced by applying Equation (5), which is a control signal to obtain the optimal tip speed ratio.

For the effective reduction of prediction error, the short-term wind speed prediction technique is applied in our model. Due to the uncertainty of system disturbance, it is necessary to improve the accuracy of the model and performance index in real time. The implementation of the ideology could be realized by solving the linear least square problem of β and T. In each step, the real time output values of the system are detected and compared with the predicted values to correct the prediction error. When the system is influenced by such factors as the non-linearity, interference, adaptation of constructed model and the like, feedback compensation will correct the prediction output in time to make the optimization.

To verify the prediction accuracy of proposed BA-ELM network, we choose two indicators to test the precision of multi-step forecasting: the mean absolute error (MAE) and the mean square error (MSE). Training and testing data collected from three sites of different wind farms.

The speed forecasting results with BA-ELM algorithm is shown in Table 2. Here two points can be concluded:

(1) Single step wind speed forecasting of ELM and BA-ELM is more accurately compared to two-step forecasting, which indicates BA-ELM achieves high precision in shorter term forecasting.

(2) Both networks can get good training and testing accuracy in forecasting. With bat algorithm to optimize the input weights and thresholds, BA-ELM is more inclined to obtain higher precision according to the MAE and MSE indexes, which illustrates that BA could improve the forecasting performance of ELM.

Table 2. Forecasting Performance of Modified ELM.

		ELM		BA-ELM		Percentage Improvement		
		1-Step	2-Step	1-Step	2-Step		1-Step	2-Step
Site 1	MAE	0.138	0.233	0.109	0.193	ξ_{MAE}	21.014	17.167
	MSE	0.055	0.062	0.037	0.060	ξ_{MSE}	32.727	3.226
Site 2	MAE	0.159	0.245	0.127	0.236	ξ_{MAE}	20.126	3.673
	MSE	0.047	0.079	0.039	0.069	ξ_{MSE}	17.021	12.658
Site 3	MAE	0.175	0.244	0.132	0.40	ξ_{MAE}	24.751	1.639
	MSE	0.026	0.108	0.025	0.086	ξ_{MSE}	3.846	20.370

4. Controller Design

In order to achieve the MPPT control, the state feedback controller is designed by measuring the stator current and flux linkage compared with the desired current and flux reference. Considering the time lag in the turbine's wind speed measurement, we adopt the turbine mechanical speed ω_r from blade shafts to predict the one-step wind speed V_w [27]. The rotor flux is estimated with current flux model. For gird side, the voltage oriented control (VOC) is proposed. The overall control scheme for wind turbine system is shown in Figure 4. This figure contain two main parts: Grid Side and Machine Side. The Grid Side was designed with Voltage Oriented Control strategy. The Machine Side was designed by combined BA-ELM algorithm, MPPT and State Feedback Control strategy together.

4.1. MPPT Control Objective

From Equation (4), we can deduce that the MPPT algorithm should be applied to extract the maximum power from the wind turbine when the rotor speed is below the rated speed [28]. Moreover, a rotor power control mechanism should be activated when the rotor speed exceeds the speed of rated power. On account that different wind speed corresponds to unique optimal turbine speed, and the maximum power point is expected so that the power can be extracted as much as possible.

From Equation (2), the reference of generator speed is calculated by the optimal tip-speed ratio λ_{opt}, which is

$$\omega_{r,opt} = \frac{\lambda_{opt} V_w}{R} \tag{25}$$

Currently, the wind speed used in the control system is usually measured with an anemometer. While in practical, the wind speed is changeable parameters [29]. In this paper, to record its real time value and to optimize the power output, the estimated effective wind speed $\hat{V}_w$ is obtained with wind speed forecasting method mentioned above.

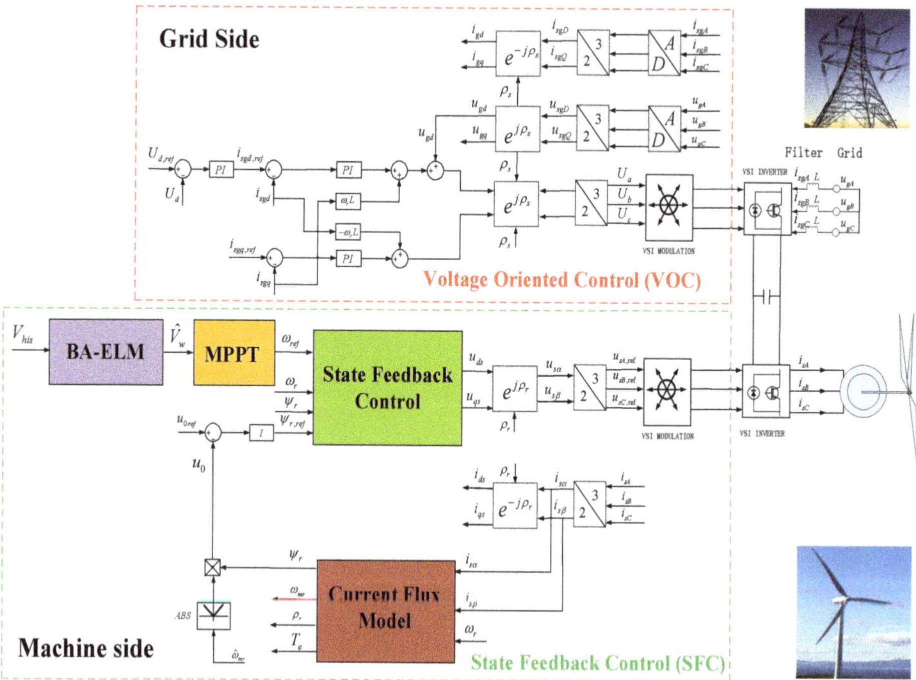

Figure 4. Block diagram of the wind energy conversion scheme.

The block diagram of the MPPT technique is presented in Figure 5. It can be seen that the low pass filter was added to give a clear reference value of generator speed, which can avoid the turbulence of turbine mechanics.

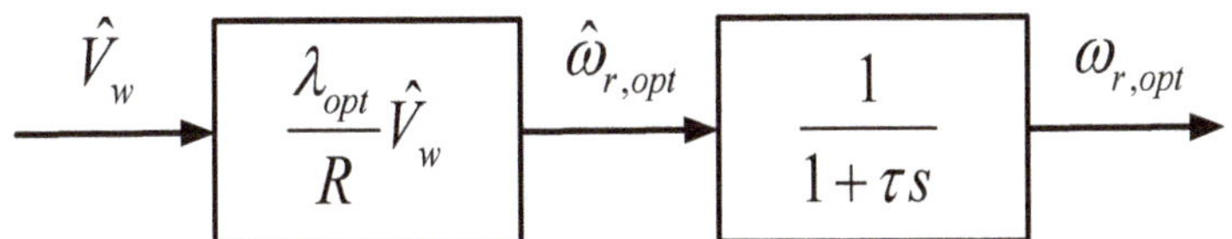

Figure 5. Block diagram of the MPPT technique.

4.2. Control System Design for Machine Side

4.2.1. Internal Loop Design

To represent the internal loop, rewrite the model from the model (α, β) stationary reference frame (8)–(12) to (d, q) rotary reference frame is given as follows:

$$\dot{i}_{ds} = -c_1 i_{ds} + \omega_{mr} i_{qs} + c_2 c_3 \psi_{dr} + c_3 p \omega_r \psi_{qr} + \frac{u_{ds}}{c_4} \tag{26}$$

$$\dot{i}_{qs} = -c_1 i_{qs} - \omega_{mr} i_{ds} + c_2 c_3 \psi_{qr} - c_3 p \omega_r \psi_{dr} + \frac{u_{qs}}{c_4} \tag{27}$$

$$\dot{\psi}_{dr} = c_5 i_{ds} - c_6 \psi_{dr} + (\omega_{mr} - p\omega_r) \psi_{qr} \tag{28}$$

$$\dot{\psi}_{qr} = c_5 i_{qs} - c_6 \psi_{qr} - (\omega_{mr} - p\omega_r) \psi_{dr} \tag{29}$$

$$\dot{\omega}_r = c_7 \left(\psi_{dr} i_{qs} - \psi_{qr} i_{ds} \right) - \frac{f_r}{J_m} \omega_r - \frac{T_L}{J_m} \tag{30}$$

where ω_{mr} is the rotating speed of the reference frame, which can be chosen arbitrarily, if we choose:

$$\frac{d\rho_r}{dt} = \omega_{mr} = p\omega_r + c_5 \frac{i_{qs}}{\psi_{dr}} = p\omega_r + \frac{M_{sr}}{T_r} \frac{i_{qs}}{\psi_{dr}} \tag{31}$$

Then Equation (29) will become

$$\dot{\psi}_{qr} = -\frac{1}{T_r} \psi_{qr} \tag{32}$$

It is obvious that $\dot{\psi}_{qr}$ converge to zero exponentially, which means that $\psi_r = \psi_{dr}$, then the state equations of LIM will become:

$$\dot{i}_{ds} = -c_1 i_{ds} + p\omega_r i_{qs} + c_5 \frac{i_{qs}^2}{\psi_r} + c_2 c_3 \psi_r + \frac{u_{ds}}{c_4} \tag{33}$$

$$\dot{i}_{qs} = -c_1 i_{qs} - p\omega_r i_{ds} - c_5 \frac{i_{ds} i_{qs}}{\psi_r} - c_3 p\omega_r \psi_r + \frac{u_{qs}}{c_4} \tag{34}$$

$$\dot{\psi}_r = c_5 i_{ds} - c_6 \psi_r \tag{35}$$

$$\dot{\rho}_r = p\omega_r + c_5 \frac{i_{qs}}{\psi_r} \tag{36}$$

$$\dot{\omega}_r = c_7 \psi_r i_{qs} - \frac{f_r}{J_m} \omega_r - \frac{T_L}{J_m} \tag{37}$$

Now, the two control inputs u_{ds} and u_{qs} are designed through a state feedback as follows:

$$u_{ds} = c_4 \left[-p\omega_r i_{qs} - c_5 \frac{i_{qs}^2}{\psi_r} - c_2 c_3 \psi_r + v_{ds} \right] \tag{38}$$

$$u_{qs} = c_4 \left[p\omega_r i_{ds} + c_5 \frac{i_{ds} i_{qs}}{\psi_r} + c_3 p\omega_r \psi_r + v_{qs} \right] \tag{39}$$

where v_{ds} and v_{qs} are additional control inputs that will be designed in the following stages.

Subtract Equations (38) and (39) in the induction machine model (33)–(37), we obtain the following equations:

$$\dot{i}_{ds} = -c_1 i_{ds} + v_{ds} \tag{40}$$

$$\dot{i}_{qs} = -c_1 i_{qs} + v_{qs} \tag{41}$$

$$\dot{\psi}_r = c_5 i_{ds} - c_6 \psi_r \tag{42}$$

$$\dot{\rho}_r = p\omega_r + c_5 \frac{i_{qs}}{\psi_r} \tag{43}$$

$$\dot{\omega}_r = c_7 \psi_r i_{qs} - \frac{f_r}{J_m} \omega_r - \frac{T_L}{J_m} \tag{44}$$

Using PI algorithm to design the control inputs v_{ds} and v_{qs}, we obtain:

$$v_{ds} = -k_{pd} \left(i_{ds} - i_{ds,ref} \right) - k_{id} \int_0^t \left(i_{ds} - i_{ds,ref} \right) d\tau \tag{45}$$

$$v_{qs} = -k_{pq} \left(i_{qs} - i_{qs,ref} \right) - k_{iq} \int_0^t \left(i_{qs} - i_{qs,ref} \right) d\tau \tag{46}$$

In this subsection, the state feedback terms are used to decoupling the system. Then, two PI controllers are proposed to achieve current tracking. The two desired currents $i_{ds,ref}$ and $i_{qs,ref}$ will be designed in the next subsection.

4.2.2. External Loop Design

In this section, we design the desired current $i_{ds,ref}$ and $i_{qs,ref}$. In the LIM, the flux reference $\psi_{r,ref}$ is set equal to constant. From Equation (32) we know ψ_{qr} convergence to zero exponentially, which means $\psi_{r,ref} = \psi_{dr,ref}$. One can determine from this equation that the desired current $i_{ds,ref}$ can be expressed by:

$$i_{ds,ref} = \frac{c_6}{c_5} \psi_{r,ref} \tag{47}$$

By using PI controller above, we can ensure that the current i_{ds} converges to $i_{ds,ref}$, which means that the flux ψ_{dr} converges to $\psi_{dr,ref}$.

After that, we use the $i_{qs,ref}$ and $\psi_{r,ref}$ to replace i_{qs} and ψ_{dr}, rewrite Equation (44) as follows:

$$\dot{\omega}_r = c_7 \psi_{r,ref} i_{qs,ref} - \frac{f_r}{J_m} \omega_r - \frac{T_L}{J_m} \tag{48}$$

Using PI controller to design the control input $i_{qs,ref}$, we obtain:

$$i_{qs,ref} = -k_{pv}\left(\omega_r - \omega_{r,ref}\right) - k_{iv}\int_0^t \left(\omega_r - \omega_{r,ref}\right)d\tau \tag{49}$$

Here, the desired rotor speed $\omega_{r,ref}$ is given with MPPT algorithm, which can be obtained from Equation (25), which is:

$$\omega_{r,ref} = \hat{\omega}_{r,opt} = \frac{\lambda_{opt}\hat{V}_w}{R} \tag{50}$$

where $\hat{V}_w$ is the wind speed forecast value with MBA-ELM algorithm.

4.3. Control System Design for Grid Side

In this part, the grid-side converter control has been adopted on the basis of an effective method: voltage oriented control (VOC), as shown in Figure 4. This method is based on the idea that the injected currents can be decoupled into the direct d and quadrature q components. For the reason that the aim is to directly control the dc-link voltage, the control scheme is provided with a further control loop, which output the direct reference current. To make the reactive power flow with the grid can be remaining to zero, the quadrature current reference is set to zero.

5. Simulation Results

To verify the proposed algorithm, simulation is operated on MATLAB/Simulink R2014a (Matlab 2014a, The MathWorks, Natick, Apple Hill Campus, MA, USA, 2014). Considering the randomness and volatility of wind speed, in order to avoid frequent switch between forecasting model and wind turbine model, the average wind speed is taken as the basis of model switching. In the process of switching, try to diminish the disturbance between models. The values of related parameters [30] are given in Table 3.

Table 3. System Specifications.

Symbol	Parameter	Value
λ_{opt}	Optimal tip speed ratio	7
β_{opt}	Optimal blade pitch angle	$0°$
R	Blade radium of turbine blades (m)	2.5
$C_{p,max}$	Power coefficient	0.45
P_{pal}	Generator rated power (kW)	5.5
V_{wpal}	Generator rated speed (rpm)	1500
P_{rated}	Rated power (kW)	2.2
U_{rated}	Rated voltage (V)	220
p	Number of pole pairs	2
R_r	Rotor resistance (Ω)	1.52
R_s	Stator resistance (Ω)	2.9
L_r	Rotor inductance (H)	0.229
L_s	Stator inductance (H)	0.223
M_{sr}	Mutual inductance (H)	0.217
J_m	Moment of inertia (kg·m^2)	0.0048
f_r	Viscous friction coefficient (Nm·s/rad)	8.29×10^{-5}

In this paper, the blade pitch β is set as optimal zero, which means that $\beta_{opt} = 0°$. In consideration of tracking the maximum power, we control the wind wheel torque through dominating rotational speed. The controller parameters are given as $k_{pd} = 10, k_{id} = 200; k_{pq} = 10, k_{iq} = 200; k_{pv} = 2, k_{iv} = 20$. The parameters of wind energy system is given in Table 3. To verify the performance of designed controller, based on MPPT algorithm, two typical wind speed signals are tested in the Matlab/Simulink environment. The maximum value of $C_p(\lambda, \beta)$ ($C_{p,\max} = 0.45$) is obtained when $\beta = 0^o$ and $\lambda = 7$, as shown in Figure 2. In the simulation, the reference rotor flux is set as $\psi_{r,ref} = 0.7Wb$.

5.1. Constant Wind Speed Signal Tracking Performance

In this subsection, the wind speed signal V_w is presented as step function, which is

$$V_w = \begin{cases} 10\,\text{m/s}, & 0\,\text{s} \leq t < 6\,\text{s} \\ 20\,\text{m/s}, & 6\,\text{s} \leq t < 12\,\text{s} \\ 15\,\text{m/s}, & 12\,\text{s} \leq t < 18\,\text{s} \end{cases} \tag{51}$$

Figure 6a shows that the pattern of the flux ψ_r follows up the desired reference flux linkage $\psi_{r,ref}$ with the steady state error almost zero. It also indicates the suitable adjusting time is achieved. The rotor speed has the great tracking performance as exhibited in Figure 6b with the step change of reference speed. The profile of current tracking performance is totally in accordance with the theoretical value, the induction generator current adjust itself quickly to follow its reference, with the tracking error converges to zero in real time on both d, q axis, as displayed in Figure 6c,d. Correspondingly, Figure 6e presents the profile of the control inputs u_{ds} and u_{qs}. The three-phase primary voltage U_a, U_b, U_c is presented in Figure 6f. It can be seen that the variation of the wind velocity plays an essential faction in the change of V/I frequency. The proposed technique, suitable for both the variable power and constant power working regions, has been verified also on a real wind speed profile.

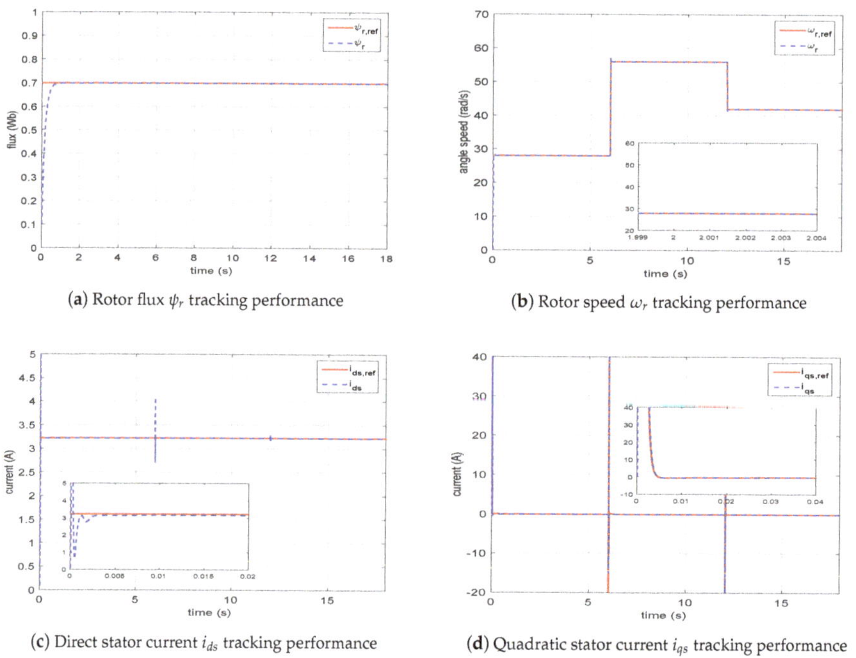

(**a**) Rotor flux ψ_r tracking performance

(**b**) Rotor speed ω_r tracking performance

(**c**) Direct stator current i_{ds} tracking performance

(**d**) Quadratic stator current i_{qs} tracking performance

Figure 6. *Cont.*

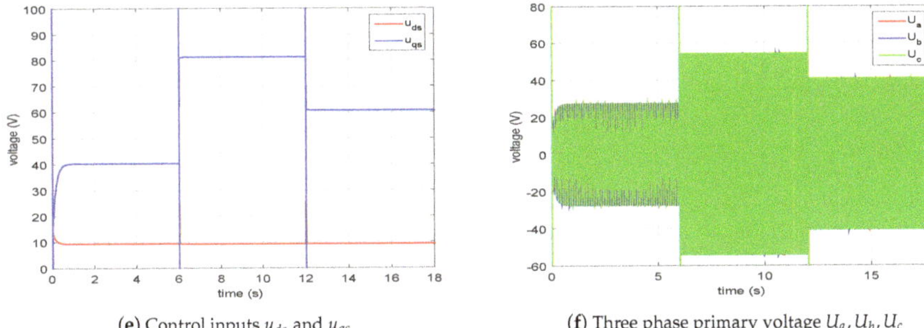

(**e**) Control inputs u_{ds} and u_{qs}

(**f**) Three phase primary voltage U_a, U_b, U_c

Figure 6. Constant wind speed signal test performance.

5.2. Various Wind Speed Signal Tracking Performance

In this subsection, the emulator of wind turbine uses the equation to approximate the wind speed V_w, which is:

$$V_w = V_{w,av}\left(1 - 0.18\cos\left(2\pi t\right) - 0.18\cos\left(2\pi t/60\right)\right) \tag{52}$$

where $V_{w,av}=10$ m/s is the average wind speed.

Figure 7a shows the rotor flux tracking performance. It can be seen tracking error rapidly converges to zero and there is no overshoot. Moreover, the adjustment time is very small. The rotor speed has the same characteristics as exhibited in Figure 7b with the various wind speed signal. The direct and quadratic stator currents tracking performance are presented in Figure 7c,d. It can be seen that the value of i_{qs} varies with different wind speed while the value of i_{qs} keeps constant. Correspondingly, the control inputs u_{ds} and u_{qs} are shown in Figure 7e, and the three-phase primary voltage U_a, U_b, U_c is given in Figure 7f. It is clearly that control inputs are smooth and continuous.

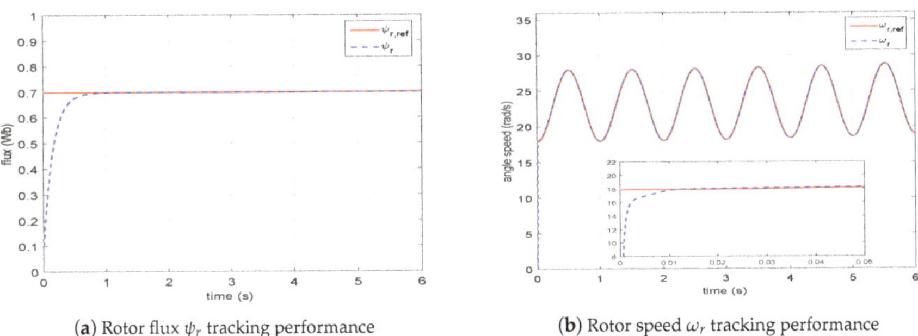

(**a**) Rotor flux ψ_r tracking performance

(**b**) Rotor speed ω_r tracking performance

Figure 7. *Cont.*

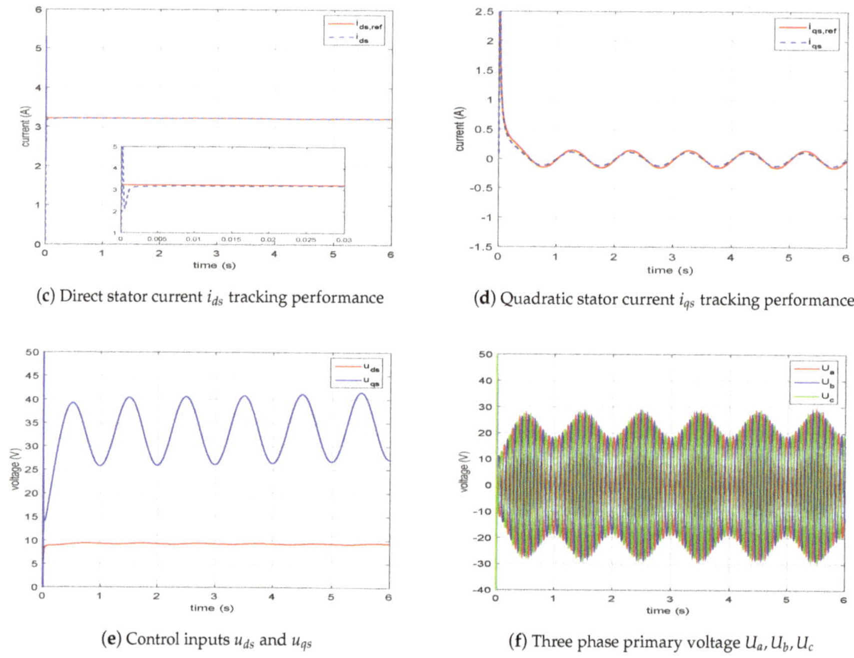

(**c**) Direct stator current i_{ds} tracking performance

(**d**) Quadratic stator current i_{qs} tracking performance

(**e**) Control inputs u_{ds} and u_{qs}

(**f**) Three phase primary voltage U_a, U_b, U_c

Figure 7. Various wind speed signal test performance.

The correct behavior of the system has been verified also on a real wind speed profile on a daily scale. Results show a good behavior of the system, capable of extracting the maximum generable power at low wind speeds and the rated power at high wind speed by properly driving the blade pitch actuators.

6. Conclusions

This paper investigates a complete modeling of a maximum power point tracking (MPPT) wind energy system with BA-ELM prediction model proposed to tackle the lag of wind speed measurement in turbines and eliminate the discontinuity of wind speed sequence. The state feedback control technique is adapted and combined with speed forecasting model to tracking the maximum power in induction generator, in which the turbine torque has been compensated based on the control law. Here the PI controller is also applied to enhance the controlling performance and robustness. Simulation results show that the flux linkage and turbine rotational speed tracking the reference value almost without oscillation and back to stable state, which indicates its highly acceptable tracking performance, considering the quick reaction and following-up time. Thus, the maximal power point in WES can be obtained with the tracking characteristics allow for the industrial variable-speed-tracking application. It can be seen that the proposed technique is a great method and can be adopted in the industrial applications.

Author Contributions: Literature search, Y.Z. and L.Z. and Y.L.; figures, Y.Z. and Y.L.; study design, Y.Z., L.Z. and Y.L.; data collection, Y.L.; data analysis, L.Z.; writing, Y.Z., L.Z. and Y.L.

Funding: This research received no external funding.

References

1. Yang, B.; Zhang, X.; Yu, T.; Shu, H.; Fang, Z. Grouped grey wolf optimizer for maximum power point tracking of doubly-fed induction generator based wind turbine. *Energy Convers. Manag.* **2017**, *133*, 427–443. [CrossRef]
2. Muyeen, S.M.; Tamura, J.; Murata, T. *Stability Augmentation of a Grid-connected Wind Farm*; Green Energy & Technology; Springer: London, UK, 2008.
3. Fathabadi, H. Maximum mechanical power extraction from wind turbines using novel proposed high accuracy single-sensor-based maximum power point tracking technique. *Energy* **2016**, *113*, 1219–1230. [CrossRef]
4. Delfino, F.; Pampararo, F.; Procopio, R.; Rossi, M. A Feedback Linearization Control Scheme for the Integration of Wind Energy Conversion Systems Into Distribution Grids. *IEEE Syst. J.* **2012**, *6*, 85–93. [CrossRef]
5. Ardjal, A.; Mansouri, R.; Bettayeb, M. Nonlinear synergetic control of wind turbine for maximum power point tracking. In Proceedings of the International Conference on Electrical Engineering, Boumerdes, Algeria, 29–31 October 2017; pp. 1–5.
6. Oh, K.Y.; Park, J.Y.; Lee, J.S.; Lee, J.K. Implementation of a torque and a collective pitch controller in a wind turbine simulator to characterize the dynamics at three control regions. *Renew. Energy* **2015**, *79*, 150–160. [CrossRef]
7. Nguyen, N.T. A novel wind sensor concept based on thermal image measurement using a temperature sensor array. *Sens. Actuators A Phys.* **2004**, *110*, 323–327. [CrossRef]
8. Pao, L.Y.; Johnson, K.E. Control of Wind Turbines. *Control Syst. IEEE* **2011**, *31*, 44–62.
9. Liu, J.; Meng, H.; Hu, Y.; Lin, Z.; Wang, W. A novel MPPT method for enhancing energy conversion efficiency taking power smoothing into account. *Energy Convers. Manag.* **2015**, *101*, 738–748. [CrossRef]
10. Tang, C.Y.; Guo, Y.; Jiang, J.N. Nonlinear Dual-Mode Control of Variable-Speed Wind Turbines With Doubly Fed Induction Generators. *IEEE Trans. Control Syst. Technol.* **2011**, *19*, 744–756. [CrossRef]
11. Boukhezzar, B.; Siguerdidjane, H. Nonlinear Control of a Variable-Speed Wind Turbine Using a Two-Mass Model. *IEEE Trans. Energy Convers.* **2011**, *26*, 149–162. [CrossRef]
12. Iyasere, E.; Salah, M.H.; Dawson, D.M.; Wagner, J.R.; Tatlicioglu, E. Robust nonlinear control strategy to maximize energy capture in a variable speed wind turbine with an internal induction generator. *Control Theory Technol.* **2012**, *10*, 184–194. [CrossRef]
13. Wei, C.; Zhang, Z.; Qiao, W.; Qu, L. An Adaptive Network-Based Reinforcement Learning Method for MPPT Control of PMSG Wind Energy Conversion Systems. *IEEE Trans. Power Electron.* **2016**, *31*, 7837–7848. [CrossRef]
14. Bekakra, Y.; Attous, D.B. Optimal tuning of PI controller using PSO optimization for indirect power control for DFIG based wind turbine with MPPT. *Int. J. Syst. Assur. Eng. Manag.* **2014**, *5*, 219–229. [CrossRef]
15. Huang, G.B.; Zhu, Q.Y.; Siew, C.K. Extreme learning machine: A new learning scheme of feedforward neural networks. In Proceedings of the IEEE International Joint Conference on Neural Networks, Budapest, Hungary, 25–29 July 2004; pp. 985–990.
16. Saraswathi, S.; Sundaram, S.; Sundararajan, N.; Zimmermann, M.; Nilsenhamilton, M. ICGA-PSO-ELM Approach for Accurate Multiclass Cancer Classification Resulting in Reduced Gene Sets in Which Genes Encoding Secreted Proteins Are Highly Represented. *IEEE/ACM Trans. Comput. Biol. Bioinform.* **2011**, *8*, 452–463. [CrossRef] [PubMed]
17. Kooning, J.D.M.D.; Vandoorn, T.L.; Vyver, J.V.D.; Meersman, B.; Vandevelde, L. Displacement of the maximum power point caused by losses in wind turbine systems. *Renew. Energy* **2016**, *85*, 273–280. [CrossRef]
18. Yaramasu, V.; Wu, B. Predictive Control of Three-Level Boost Converter and NPC Inverter for High Power PMSG-Based Medium Voltage Wind Energy Conversion Systems. *IEEE Trans. Power Electron.* **2014**, *29*, 5308–5322. [CrossRef]
19. Yamakura, S.; Kesamaru, K. Dynamic simulation of PMSG small wind turbine generation system with HCS-MPPT control. In Proceedings of the International Conference on Electrical Machines and Systems, Sapporo, Japan, 21–24 October 2012; pp. 1–4.
20. Li, S.; Li, J. Output Predictor based Active Disturbance Rejection Control for a Wind Energy Conversion System with PMSG. *IEEE Access* **2017**, *1*. [CrossRef]

21. Merabet, A.; Tanvir, A.; Beddek, K. Speed Control of Sensorless Induction Generator by Artificial Neural Network in Wind Energy Conversion System. *IET Renew. Power Gen.* **2017**, *10*, 1597–1606. [CrossRef]

22. Yang, X. Bat algorithm for multi-objective optimisation. *Int. J. Bio-Inspir. Comput.* **2012**, *3*, 267–274. [CrossRef]

23. Huang, G.B.; Wang, D.H.; Lan, Y. Extreme learning machines: A survey. *Int. J. Mach. Learn. Cybern.* **2011**, *2*, 107–122. [CrossRef]

24. Huang, G.B.; Chen, Y.Q.; Babri, H.A. Classification ability of single hidden layer feedforward neural networks. *IEEE Trans. Neural Netw.* **2000**, *11*, 799–801. [CrossRef]

25. Huang, G.B.; Chen, L.; Siew, C.K. Universal approximation using incremental constructive feedforward networks with random hidden nodes. *IEEE Trans. Neural Netw.* **2006**, *17*, 879–892. [CrossRef]

26. Bartlett, P.L. The sample complexity of pattern classification with neural networks: The size of the weights is more important than the size of the network. *IEEE Trans. Inf. Theory* **1998**, *44*, 525–536. [CrossRef]

27. Cirrincione, M.; Pucci, M.; Vitale, G. Growing Neural Gas-Based MPPT of Variable Pitch Wind Generators With Induction Machines. *IEEE Trans. Ind. Appl.* **2010**, *48*, 1006–1016. [CrossRef]

28. Vermillion, C.; Grunnagle, T.; Lim, R.; Kolmanovsky, I. Model-Based Plant Design and Hierarchical Control of a Prototype Lighter-Than-Air Wind Energy System, With Experimental Flight Test Results. *IEEE Trans. Control Syst. Technol.* **2014**, *22*, 531–542. [CrossRef]

29. Heier, S. *Grid Integration of Wind Energy*; Wiley Online Library: Hoboken, NJ, USA, 2014.

30. Cirrincione, M.; Pucci, M.; Vitale, G. Growing neural gas (GNG)-based maximum power point tracking for high-performance wind generator with an induction machine. *IEEE Trans. Ind. Appl.* **2011**, *47*, 861–872. [CrossRef]

 *processes*

Article

A Flexible Responsive Load Economic Model for Industrial Demands

Reza Sharifi [1], Amjad Anvari-Moghaddam [2,*], S. Hamid Fathi [1] and Vahid Vahidinasab [3]

[1] Electrical Engineering Department, Amirkabir University of Technology, Tehran 15916-34311, Iran; reza.sharifi@aut.ac.ir (R.S.); fathi@aut.ac.ir (S.H.F.)

[2] Department of Energy Technology, Aalborg University, 9220 Aalborg East, Denmark

[3] Department of Electrical Engineering, Shahid Beheshti University, Tehran 19839-69411, Iran; v_vahidinasab@sbu.ac.ir

* Correspondence: aam@et.aau.dk; Tel.: +45-9356-2062

Received: 18 February 2019; Accepted: 4 March 2019; Published: 8 March 2019

Abstract: The best pricing method for any company in a perfectly competitive market is the pricing scheme with regards to the marginal cost. In contrast to this environment, there is a market with imperfect competition. In this market, the price can be affected by some players in the generation/demand side (i.e., suppliers and/or buyers). In the economic literature, "market power" refers to a company that has the power to affect prices. In fact, market power is often defined as the ability to divert prices from competitive levels. In the electricity market, especially because of the integration of intermittent renewable energy resources (RESs) along with the inflexibility of demand, there are levels of market power on the supply side. Hence, implementation of demand response (DR) programs is necessary to increase the flexibility of the demand side to deal with the intermittency of renewable generations and at the same time tackle the market power of the supply side. This paper uses economic theories and mathematical formulations to develop a flexible responsive load economic model (FRLEM) based on real-time pricing (RTP) to show modification of the load profile and mitigation of the energy costs for an industrial zone. This model was developed based on constant elasticity of the substitution utility function, known as one of the most popular utility functions in microeconomics.

Keywords: demand-side management; economic demand response model; consumer utility function; electricity market restructuring

---

1. Introduction

The objective of demand-side management (DSM) in the industrial sector is to improve the profile of electrical loads by two means: (1) energy efficiency solutions; and (2) demand response (DR) programs. An energy efficiency solution reduces the electricity consumed to provide a certain service, with the primary goal of reducing electricity costs and protecting the environment. These programs reduce the total electricity consumption and peak electricity load with the help of energy-efficient equipment and other efficiency improvement means. These programs consist of activities such as installing thermal insulators, low-power equipment, and so on [1].

The DR program refers to a set of measures aimed at encouraging a voluntary change in the consumers' electricity usage pattern in response to changes in electricity prices or grid reliability conditions. The increase in the demand-side capacity following the participation in such programs could effectively reduce the electricity costs [2] and improve the robust operation of energy systems [3], as well as enhance system reliability [4,5]. The US Department of Energy has defined DR as changes in electric usage by end-use customers from their normal consumption patterns in response to changes in the price of electricity over time (passive participation), or to incentive payments (active participation) designed to induce lower electricity use at times of high wholesale market prices or when system reliability is jeopardized [6]. This DR definition is focused on the price of the wholesale market and the occurrence of power crises. However, what is important for consumers is the retailer's price, which includes the cost of energy transmission, distribution, and peripheral services. The Nordic Electricity Market has provided a more precise definition of DR as a voluntary temporary adjustment of electricity demand as a response to a price signal or a reliability-based action [7].

The DR programs are classified into two major categories and several subcategories, as shown in Figure 1 [8].

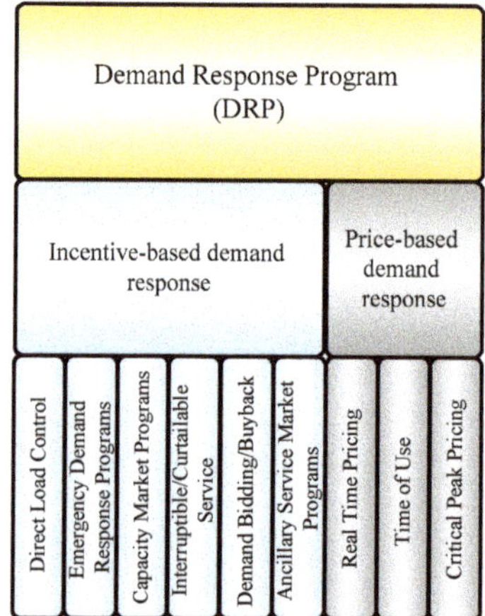

Figure 1. Classification of demand response (DR) programs.

In the industrial sector, electrical loads can be categorized into several groups in terms of their response to DR programs. In [9], these loads are divided into three categories:

- The loads related to production machinery and equipment that apply a variable force on a raw material over a defined cycle time (e.g., mechanical and hydraulic equipment, presses, welding equipment, etc.). These loads cannot be modified, but can be switched on/off when necessary. Therefore, these electrical loads can respond well to direct load curtailment commands while supporting demand-shifting-type DR programs.
- The loads related to production machinery and equipment that apply a steady continuous force to move fluids (e.g., pumps, fans, blowers, air compressors, etc.). The power consumption of these loads can be adjusted as needed, so they respond to all DR programs.
- The loads related to production machinery and equipment that change the phase, composition, or chemical properties of a raw material and run continuously unless stopped for maintenance. Any interruption or change in the power supply of such machinery may result in the failure of the production process or poor product quality. These loads never respond to DR programs.

According to the above-mentioned classification of electrical loads in the industrial sector, we can conclude that some of these electrical loads (known as flexible demand) may respond well to DR programs [9] However, there are several technical/operational challenges and obstacles ahead of the application of such programs in the industrial sector. These mainly include [10–12]:

- Absence of proper responsive load economic models,
- Poor understanding of the technical potentials of DR programs in the industrial sector,
- Insufficient economic returns of DR programs,
- Inadequate time to adjust production or take action,

- Absence of smart platforms for reciprocal communication between retailers and end-use customers.

In general, access to a flexible responsive load economic model (FRLEM) greatly contributes to our knowledge about the impact of consumer participation in DR programs and their influence on the load profile [13–15]. Therefore, many such models have been developed for DR programs of different types, such as time of use (TOU), critical peak pricing (CPP), and real-time pricing (RTP)-based programs [16–20]. However, most of these models utilize the concept of price elasticity, which assumes the point elasticity as the price elasticity of demand at a particular point on the demand curve. In other words, these models linearize the demand curve at a specific operating point instead of considering the entire demand. This means that they consider the price elasticity of demand as a pre-known fixed value at a point on the demand curve. Since the price elasticity varies along the demand curve, this assumption creates a discontinuity in the decision-making process. Therefore, these models will not perform well for the industrial sector, where a given DR model is supposed to consider countless consumers with different loads [21,22].

In References [23,24], economic theories and mathematical formulations are used to present a novel model for TOU-based DR programs with effective market-oriented models for residential consumers to change their time of consumption (i.e., to achieve adjustability in the DR model). Application of DR programs for the operation management of energy hubs is investigated in Reference [25], where responsive thermal and electrical loads are presented with related temporal behavior. Authors in [26] presented a day-ahead multi-objective optimization model for a building energy management system under TOU price-based DR programs, which merges building-integrated photovoltaic with other generations to optimize the economy and occupants' comfort by the synergetic dispatch of source–load–storage. In [27], an incentive DR program is proposed to assist customers to participate in the program. The proposed plan provides tools that can help the customers' premises to take part in DR programs. Likewise, a multi-level demand charge, along with an RTP program and incentivized signals is proposed in [28] for the participation of thermal energy storage (TES)-integrated commercial buildings in DR programs. The findings demonstrate that the existence of TES in a commercial building could result in higher flexibility for joining DR programs, and so to further decrease energy costs while maintaining the level of comfort for residents.

In this paper, we use economic theories and mathematical formulations to develop an RLEM for RTP-based programs with 24-h time intervals. Our aim in this effort was to achieve a model with two primary features: adaptability and adjustability. Here, adaptability refers to the applicability of the proposed model to all types of consumers regardless of their response to the DR program. Moreover, since the operating plans of industrial consumers may vary depending on their policies, extensive disparity in the hours of power usage in the industrial sector might be needed. Because of this, the model to be developed for this sector has to be sufficiently adjustable, which is also considered in this work. In light of the reviewed literature, the contributions of this work can be summarized as:

- A novel model is presented based on economic theories and mathematical formulations to enable industrial customer response to RTP-based DR programs for cost minimization,
- An efficient framework is proposed for industrial load management by considering different levels of participation in DR programs,
- A working platform is introduced to support key functionalities of an industrial DR program known as adaptability and adjustability.

The rest of this paper is given as follows: Section 2 explains theory of consumer choice. Simulation results together with model validation under different test scenarios are presented in Section 3. Finally, Section 4 concludes the paper by summarizing the main results.

2. Theory of Consumer Choice

As mentioned in the introductory part of the paper, a model to be used for an RTP-based DR program should be able to account for a wide range of customers by featuring both adaptability and adjustability. Many of the existing models are based on the concept of price elasticity of demand, which cannot meet these features in a continuous process, as required for RTP-based programs.

The theory of consumer choice is a well-known microeconomics theory that explores how consumers spend their economic resources according to their preferences and subject to their budget constraints. Two important instruments of this theory are the utility function and the budget constraint. The interactions of these functions and constraints determine how consumers make their spending decisions [29,30].

The utility function is an economic concept that represents the interest in gaining greater returns. There is no specific method for formulating this function, and it is often derived through empirical methods. One of the most popular standard utility functions developed for microeconomic analyses is the constant elasticity of substitution (CES).

This function is widely popular among economists working on microeconomic problems, and is normally applied where there are several different commodities available for consumption. The CES utility function for n commodities is [31]:

$$U(X_1, X_2, X_3, \ldots, X_n) = \left(\sum_{i=1}^{n} \alpha_i^{1-\rho} X_i^{\rho} \right)^{\frac{1}{\rho}} \quad 0 \neq \rho \prec 1 \quad \sum_{i=1}^{n} \alpha_i = 1; \ \alpha_i > 0 \tag{1}$$

In the above formulation, $(\rho - 1)^{-1}$ is the elasticity of substitution and α_i denotes the share factors.

To expand this function to the electricity market, we assume that electrical energy with a specific price P_1 is considered as a commodity (X_1), and electrical energy at a specific price of P_2 is considered as a commodity (X_2). Thus, assuming n electricity price levels for a 24-h period, we will have n commodities.

Parameter ρ determines the consumer's desire to participate in DR program. The greater the value of ρ, the more willing the customer will be to participate in DR program. This parameter also provides the first feature of a worthy DR model, known as adaptability. This means that by changing/fine-tuning this parameter, a broad range of customers in the electricity market can be considered. In the same way, parameter α_i (or share factors) accounts for adjustment of consumption ratio over the time in accordance with the customer's request. That is, changing the α_i parameter provides a new arrangement of the consumption in each hour. This enables the model to account for adjusting the amount of consumption in every time interval, which is the second feature of a worthy DR model, known as adjustability of consumption levels. From now on, parameter ρ is named as the adaptability parameter while α_i is named as the adjustability parameter.

In the theory of consumer choice, consumer's spending decisions are subject to budget constraints. This theory considers the consumer behavior as a maximization problem where the goal is to obtain the highest profit from limited resources. Since a consumer's desire for making more profit is unlimited, the only factor that limits consumption is the limited budget. Therefore, the utility function subject to the budget constraint is:

$$\begin{aligned} U(C_1, C_2, C_3, \ldots, C_{24}) \\ \text{s.t.} \quad B = \sum_{i=1}^{24} (C_i \cdot P_i) \end{aligned} \tag{2}$$

Here, it is assumed that the consumer adjusts their flexible power consumptions (C_i) according to the pricing scheme. Therefore, the consumer aims to maximize the following utility function:

$$Max\left\{ U(C_1, C_2, C_3, \ldots, C_{24}) = \left(\sum_{i=1}^{24} \alpha_i^{1-\rho} C_i^{\rho}\right)^{\frac{1}{\rho}} \right\}; \quad 0 \neq \rho \prec 1; \quad \sum_{i=1}^{n} \alpha_i = 1; \quad \alpha_i > 0 \tag{3}$$

$$\text{s.t.} \quad B = \sum_{i=1}^{24} (C_i \cdot P_i)$$

According to the method of Lagrange multipliers [32]:

$$L = \left(\sum_{i=1}^{24} \alpha_i^{1-\rho} C_i^{\rho}\right)^{\frac{1}{\rho}} + \lambda \left[B - \sum_{i=1}^{24} C_i \cdot P_i\right] \tag{4}$$

The partial derivatives of Equation (4) with respect to any C_i and λ are given by:

$$\frac{dL}{dC_i} = \alpha_i^{1-\rho} C_i^{\rho-1} \cdot \left(\sum_{j=1}^{24} \alpha_j^{1-\rho} C_j^{\rho}\right)^{\frac{1-\rho}{\rho}} - \lambda P_i = 0 \quad \Rightarrow \lambda P_i = \alpha_i^{1-\rho} C_i^{\rho-1} \cdot \left(\sum_{j=1}^{24} \alpha_j^{1-\rho} C_j^{\rho}\right)^{\frac{1-\rho}{\rho}} \tag{5}$$

$$\frac{dL}{d\lambda} = B - \sum_{i=1}^{24} C_i \cdot P_i = 0 \tag{6}$$

Taking (5) and (6) into account, one can easily conclude that:

$$C_2 = \frac{\alpha_2}{\alpha_1}\left(\frac{P_2}{P_1}\right)^{\frac{1}{\rho-1}} \cdot C_1$$

$$C_3 = \frac{\alpha_3}{\alpha_1}\left(\frac{P_3}{P_1}\right)^{\frac{1}{\rho-1}} \cdot C_1$$

$$\vdots \tag{7}$$

$$C_{24} = \frac{\alpha_{24}}{\alpha_1}\left(\frac{P_{24}}{P_1}\right)^{\frac{1}{\rho-1}} \cdot C_1$$

By substituting the relations extracted from (7) into (6), we arrive at:

$$B = C_1 \cdot P_1 + \frac{\alpha_2}{\alpha_1}\left(\frac{P_2}{P_1}\right)^{\frac{1}{\rho-1}} \cdot C_1 \cdot P_2 + \frac{\alpha_3}{\alpha_1}\left(\frac{P_3}{P_1}\right)^{\frac{1}{\rho-1}} \cdot C_1 \cdot P_3 + \ldots\ldots + \frac{\alpha_{24}}{\alpha_1}\left(\frac{P_{24}}{P_1}\right)^{\frac{1}{\rho-1}} \cdot C_1 \cdot P_{24};$$

$$\Rightarrow C_1 = \frac{B}{P_1 + \frac{\alpha_2}{\alpha_1}\left(\frac{P_2}{P_1}\right)^{\frac{1}{\rho-1}} \cdot P_2 + \ldots + \frac{\alpha_{24}}{\alpha_1}\left(\frac{P_{24}}{P_1}\right)^{\frac{1}{\rho-1}} \cdot P_{24}} \tag{8}$$

Assuming Ω as the denominator of the above formulation, we have:

$$\Omega = P_1 + \frac{\alpha_2}{\alpha_1}\left(\frac{P_2}{P_1}\right)^{\frac{1}{\rho-1}} \cdot P_2 + \frac{\alpha_3}{\alpha_1}\left(\frac{P_3}{P_1}\right)^{\frac{1}{\rho-1}} \cdot P_3 + \ldots\ldots + \frac{\alpha_{24}}{\alpha_1}\left(\frac{P_{24}}{P_1}\right)^{\frac{1}{\rho-1}} \cdot P_{24} \tag{9}$$

Therefore, the load model for a 24-h daily period in the presence of an RTP-based DR program is:

$$\left.\begin{aligned} C_1 &= \frac{B}{\Omega} \\ C_2 &= \frac{\alpha_2}{\alpha_1}\left(\frac{P_2}{P_1}\right)^{\frac{1}{\rho-1}} \cdot \frac{B}{\Omega} \\ &\vdots \\ C_{24} &= \frac{\alpha_{24}}{\alpha_1}\left(\frac{P_{24}}{P_1}\right)^{\frac{1}{\rho-1}} \cdot \frac{B}{\Omega} \end{aligned}\right\} \Rightarrow \left(\begin{aligned} C_h &= \frac{\alpha_h}{\alpha_1}\left(\frac{P_h}{P_1}\right)^{\frac{1}{\rho-1}} \cdot \frac{B}{\Omega} \\ h &= 1,2,\ldots,24 \end{aligned}\right) \tag{10}$$

Since the industrial production in any industrial zone is directly related to their electricity usage, there is no difference between the total electricity consumed over a 24-h period before and after the implementation of the DR program. In other words, the change will be in the pattern (hours) of consumption, not in the total power consumed, which means:

$$C_{\text{primary}} = \sum_{h=1}^{24} C_h = \sum_{h=1}^{24} \frac{\alpha_h}{\alpha_1} \left(\frac{P_h}{P_1} \right)^{\frac{1}{\rho-1}} \cdot \frac{B}{\Omega} \tag{11}$$

where C_{primary} is total electricity consumed before the DR program.

$$B = \frac{\Omega . C_{\text{primary}}}{\sum\limits_{h=1}^{24} \frac{\alpha_h}{\alpha_1} \left(\frac{P_h}{P_1} \right)^{\frac{1}{\rho-1}}} \tag{12}$$

By substituting (12) into (10), it can be deduced that:

$$C_h = \frac{\alpha_h}{\alpha_1} \left(\frac{P_h}{P_1} \right)^{\frac{1}{\rho-1}} \cdot \frac{C_{\text{primary}}}{\sum\limits_{h=1}^{24} \frac{\alpha_h}{\alpha_1} \left(\frac{P_h}{P_1} \right)^{\frac{1}{\rho-1}}} \tag{13}$$

As shown in (13), the demand for each hour C_h is a function of budget B, price P_h, adjustability parameters α_h, and adaptability parameter ρ. The consumers' budget is limited and prices are determined by market mechanism and retailers. Thus, the parameters related to the customer are ρ and α_h. These parameters decide the extent of consumer participation in the DR program, and can be used to estimate how the consumer responds to such programs.

3. Performance Evaluation

In this section, we examine the performance of the proposed model. The load profile used for this purpose (Figure 2) belongs to an industrial zone that was adopted from [33].

In this industrial zone, as shown in Table 1, operation of industrial units is continuous and are scheduled in three shifts per day [34]. Also, the electric energy demand is divided into non-flexible and flexible parts. In each work shift, it is assumed that the minimum demand relates to the non-flexible part while the rest of the electrical demand relates to the flexible loads. Hence, we are only able to manage the flexible part of the demand to reduce the cost of purchasing electrical energy.

Table 1. Work shifts.

Day Shift	Evening Shift	Night Shift
08:00–15:00	16:00–23:00	00:00–07:00

Note that the work shift times can affect the cost of a company. In general, employers pay a premium to the night-shift workers. Hence, there is less willingness to engage workers in the night shift from the company owner's perspective. Since the cost of paying for night-shift workers may be higher than the cost of electricity, there should be a trade-off between these two. The proposed DR model in this paper could also address this issue with the help of the adjustability parameters α. Wages are assumed to be constant for any time within a given work shift.

Therefore, the values of adjustability parameters α_h are considered different among different work shifts while remaining unchanged within a work shift, that is:

$$\begin{cases} \alpha_{day} = \alpha_{h_1} & h \in 08:00 - 15:00 \\ \alpha_{evening} = \alpha_{h_2} & h \in 16:00 - 23:00 \\ \alpha_{night} = \alpha_{h_3} & h \in 00:00 - 07:00 \end{cases} \tag{14}$$

As an example, the condition $\alpha_{day} > \alpha_{night}$ denotes that the company owner's desire to use electrical energy for the day shift is higher than for the night shift. On the other hand, when $\alpha_{day} = \alpha_{evening} = \alpha_{night}$, the industrial DR participant demonstrates a homogeneous behavior along the day, meaning that there is no difference among the work shifts from the company owner's viewpoint.

In the following sections, the effectiveness of this model is validated under two case studies.

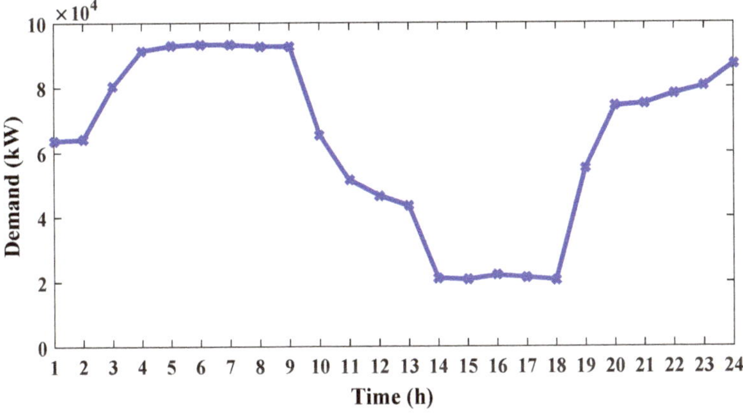

Figure 2. Load profile used in the evaluation.

3.1. Case I

In the first case, we simulated four scenarios, illustrated in Table 2. In the first scenario simulation, customers (e.g., industrial plants) showed no response to price changes, that is, there was no participation on the part of consumers. In the next three scenarios, consumers participated by the amounts shown in the third column of Table 2. As stated earlier, the higher the ρ value, the greater the participation of customers in the DR program. The following assumptions were also made for the aforementioned scenarios:

(a) The price signal for the examined industrial zone was provided in two ways (Note: in both pricing methods, the average price was the same for over 24 h):

    (1)   RTP pricing (as shown in Figure 3a, belongs to the date 28 August 2017, available in [35]);

    (2)   TOU pricing according to work shifts (as shown in Figure 3b);

(b) The adjustability parameters values were the same for all work shifts, that is, $\alpha_{day} = \alpha_{evening} = \alpha_{night}$.

Table 2. Performance evaluation of the proposed model in different scenarios. RTP: real-time pricing; TOU: time of use.

No.	Program	Partnership Level	Energy Consumption (MW)	Budget (Euro/kW)	Budget Change (%)
1	TOU	–	1528	69,064	base
	RTP		1528	71,508	base
2	TOU	0.3	1528	63,906	−8.07
	RTP		1528	63,273	−13.01
3	TOU	0.5	1528	62,769	−10.02
	RTP		1528	62,045	−15.25
4	TOU	0.8	1528	59,254	−16.55
	RTP		1528	58,433	−22.37

The results obtained by modelling of the scenarios are presented in Table 2. As the fourth column of this table shows, all scenarios had the same total energy consumption. This is consistent with our second assumption, which states that industrial units do not reduce their total energy usage but shift the usage from high-price hours to low-price hours to minimize their electricity bill without altering their output.

According to the budget column of the first scenario in Table 2, the RTP program had greater cost implications for customers than TOU program. This is a realistic result, as implementation of RTP program is more likely to transfer the risk of the wholesale market from retailers to consumers.

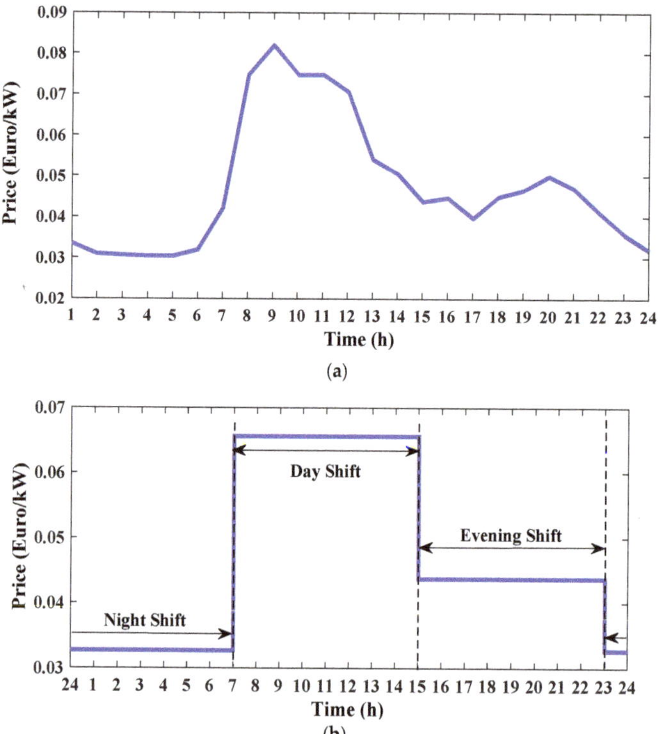

Figure 3. Electricity prices according to (**a**) real-time pricing (RTP) pricing; (**b**) time of use (TOU) pricing.

In scenario No. 2, although both RTP and TOU programs were modelled with the same participation level ($\rho = 0.3$), the customer's electricity budget in RTP was less than in TOU.

This was also true for the reduction in electricity budget, meaning that even a low participation in the RTP program resulted in a lower electricity budget compared to the base scenario and TOU.

An examination of Table 2 also reveals similar results for scenarios No. 3 and 4. However, as the participation rates assumed in these scenarios were greater than before, the consequent reductions in electricity budget were more significant, which implies that industrial customers can reduce their electricity bill without reducing their consumption.

To get a better insight into the level of commitment in industrial DR programs over the work shifts, the load profiles are plotted in Figures 4 and 5 for the different scenarios in Table 2. It can be observed that with the increase of ρ, the consumer's wish to take part in DR programs, especially during the night shift, increased, as the price of electric energy is low. This effect is clearly illustrated in Figure 5 for each time interval.

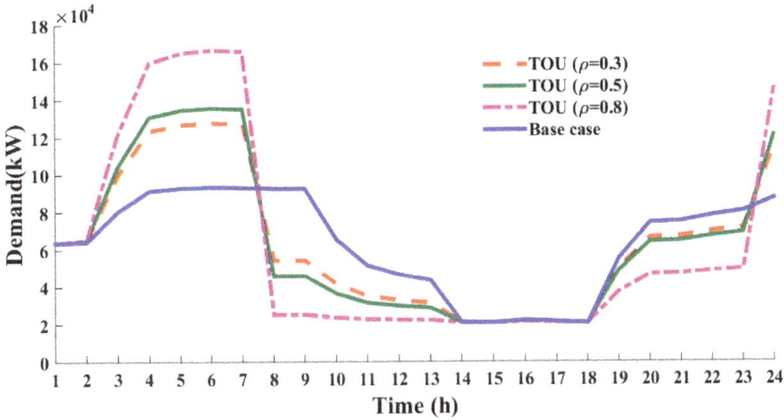

Figure 4. The impact of TOU programs on load profile.

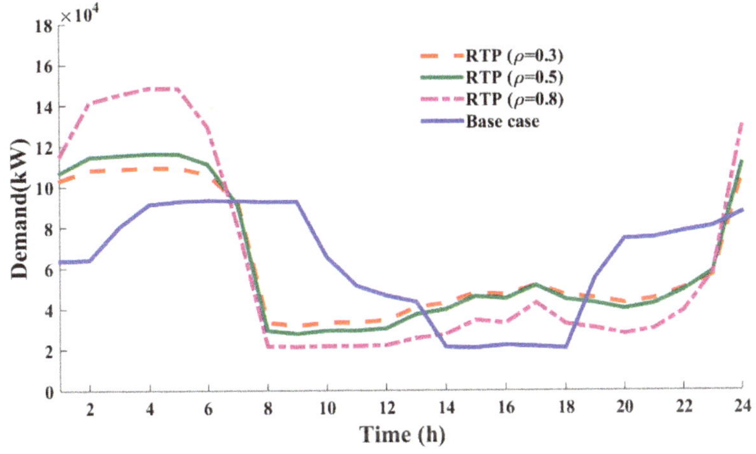

Figure 5. The impact of RTP programs on load profile.

3.2. Case II

Table 3 shows the changes in the budget and demand for adaptability parameter ρ and adjustability parameters $\alpha = (\alpha_{day}, \alpha_{evening}, \alpha_{night})$.

In Table 3 for a given ρ, when $\alpha_{day} = \alpha_{evening} = \alpha_{night}$, as mentioned before, there was no difference between the work shifts. Because of this, consumption shifted to a night shift when the price of electrical energy was low. On the other hand, different behavioral patterns during a given working day could be emulated by considering different adjustability values. As an example, in row 3 of Table 3, $\alpha_{day} = 1.1 \times \alpha_{evening}$ and $\alpha_{day} = 1.3 \times \alpha_{night}$, which denotes that the company owner's desire to use electrical energy for the day shift was higher than the evening shift and much higher than the night shift. Since the price of electrical energy for the day shift is higher than other shifts, the amount of required budget is increased relative to the previous state.

In Table 3 for a given α, an increase in ρ (and consequently consumers' participation) led to a decrease in the budget, which reflects the consumer's tendency to move towards more participation in the DR program. As mentioned and shown in Table 3, the total daily power consumption in the absence and presence of DR action remained the same and only shifted among different working hours.

As shown in Figures 6 and 7, the required budget and day shift consumption was reduced by increasing the level of participation ρ. On the other hand, with the increase of the adjustability parameters α_{day} for the day shift (which means an increase in the inclination of electricity consumption for day shift with high energy prices), the amount of the budget and day-shift consumption also increased.

The results obtained for DR model during the night shift are shown in Figure 8. As can be seen, with the increase of ρ, there was a growing inclination in the power consumption of this shift due to the lower electricity prices. In a like manner, with the increase of the adjustability parameters α_{night}, the consumption tendency increased at the night shift.

Table 3. Changes in the budget and demand.

Scenario	Budget (Euro) and Demand (kW)		
	$\rho = 0.3$	$\rho = 0.5$	$\rho = 0.8$
base case—without DR		71,508 and 1528	
$\alpha_{day} = \alpha_{evening} = \alpha_{night}$	63,273 and 1528	62,045 and 1528	58,433 and 1528
$\alpha_{day} = 1.1 \times \alpha_{evening}$ and $\alpha_{day} = 1.3 \times \alpha_{night}$	64,046 and 1528	62,727 and 1528	58,741 and 1528

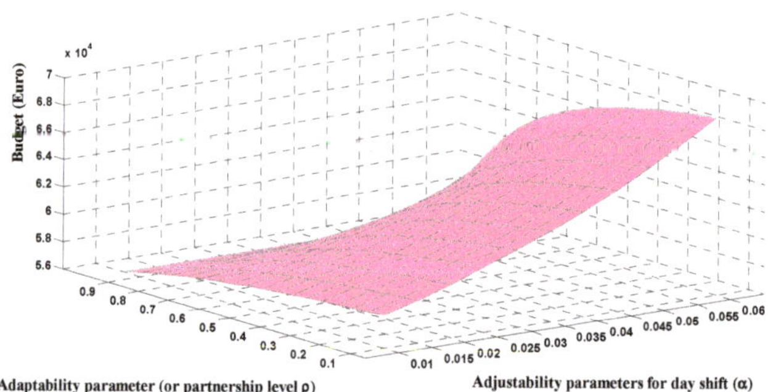

Figure 6. Changes in the budget for different values of adaptability and adjustability parameters.

As can be seen from the simulation results, the introduction of the two mentioned control parameters in an industrial DR action creates a strong tool to adjust the amount of participation and consumption according to customer preferences. So, the proposed DR model is able to reconcile to different customers with different flexibilities against prices, and allows the consumption levels to be adjusted over different work shifts.

This section may be divided by subheadings. It should provide a concise and precise description of the experimental results, their interpretation, as well as the experimental conclusions that can be drawn.

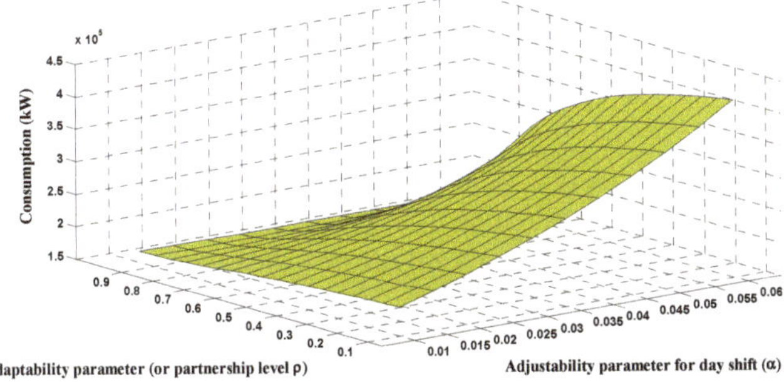

Figure 7. Changes in the day-shift consumption for different values of adaptability and adjustability parameters.

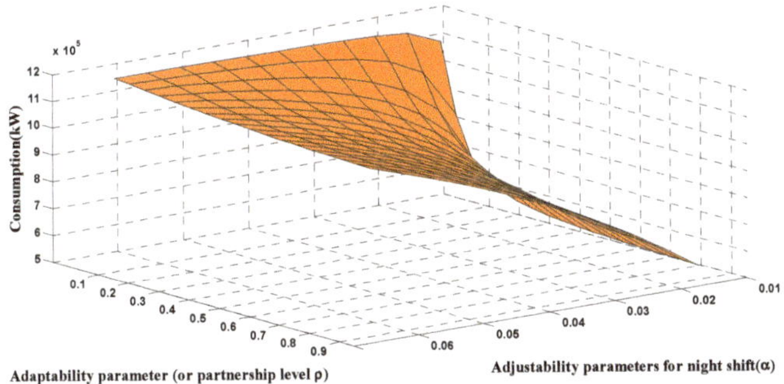

Figure 8. Changes in the night-shift consumption for different values of adaptability and adjustability parameters.

4. Conclusions

The development of flexible responsive load economic models is an important requirement for the proper evaluation of the consumer participation impact on load profiles. In this paper, economic models and mathematical formulation were used to develop a new economic model for RTP-based DR programs. Compared to the existing models reported in the same area, the model presented in this paper was able to integrate consumers' tendency to modify the load profile and reallocate the consumption at any time. This feature enabled the DR model to be customized for every type of consumer with different preferences. Unlike the previous DR models, which are mainly based on the concept of price elasticity at a specific operating point, the proposed model considered the whole demand curve modelling for DR actions. Additionally, the presented model demonstrated that the classical economic decision-making process for each consumer can be applied. In this way, the decision maker can identify and assess all possible options and consequences of DR implementation and select the most logical way of doing so. For example, retailers can use this model to build optimal bidding curves or help network operators get the information they need about responding to the DR program in order to determine network tariffs that are appropriate for managing congestions. It was also demonstrated that the proposed model enables system operators to investigate the impact of different participation levels on the aggregated load profile.

Author Contributions: Conceptualization, R.S.; methodology, R.S., A.A.-M.; software, R.S.; resources, A.A.-M., S.H.F., and V.V.; writing—original draft preparation, R.S.; writing—review and editing, A.A.-M., V.V.; supervision, A.A.-M., S.H.F., and V.V.

Funding: This research received no external funding.

Conflicts of Interest: The authors declare no conflict of interest.

References

1. Sharifi, R.; Fathi, S.H.; Vahidinasab, V. A review on demand-side tools in electricity market. *Renew. Sustain. Energy Rev.* **2017**, *72*, 1–7. [CrossRef]
2. Wang, Y.; Wang, Y.; Yujing, H.; Yu, H.; Du, R.; Zhang, F.; Zhang, F.; Zhu, J. Optimal Scheduling of the Regional Integrated Energy System based on energy price Demand Response. *IEEE Trans. Sustain. Energy* **2018**. [CrossRef]
3. Majidi, M.; Mohammadi-Ivatloo, B.; Anvari-Moghaddam, A. Optimal robust operation of combined heat and power systems with demand response programs. *Appl. Therm. Eng.* **2019**, *149*, 1359–1369. [CrossRef]
4. Vahedipour-Dahraie, M.; Anvari-Moghaddam, A.; Guerrero, J.M. Evaluation of reliability in risk-constrained scheduling of autonomous microgrids with demand response and renewable resources. *IET Renew. Power Gener.* **2018**, *12*, 12657–12667. [CrossRef]
5. Amini, M.; Almassalkhi, M. Investigating delays in frequency-dependent load control. In Proceedings of the 2016 IEEE Innovative Smart Grid Technologies—Asia (ISGT-Asia), Melbourne, Australia, 28 November–1 December 2016; pp. 448–453.
6. Safdarian, A.; Fotuhi-Firuzabad, M.; Lehtonen, M. Impacts of time-varying electricity rates on forward contract scheduling of DisCos. *IEEE Trans. Power Deliv.* **2014**, *29*, 733–741. [CrossRef]
7. THEMA Consulting Group: "Demand Response in the Nordic Electricity Market—Input to Strategy on Demand Flexibility. *TemaNord* **2014**, *553*. Available online: http://www.nordicenergy.org/wp-content/uploads/2014/10/Demand-response-in-the-Nordic-electricity-market.pdf (accessed on 21 March 2016). [CrossRef]
8. Willis, H.L.; Powell, R.W.; Tram, H.N. Computerized methods for analysis of impact of demand side management on distribution systems. *IEEE Trans. Power Deliv.* **1987**, *2*, 1236–1243. [CrossRef]
9. Starke, M.; Alkadi, N.; Ma, O. Assessment of Industrial Load for Demand Response Across US Regions of the Western Interconnect. No. ORNL/TM-2013/407; ORNL/DOE, September 2013. Available online: https://info.ornl.gov/sites/publications/Files/Pub45942.pdf (accessed on 7 November 2016).
10. Sharifi, R.; Fathi, S.H.; Vahidinasab, V. Customer baseline load models for residential sector in a smart-grid environment. *Energy Rep.* **2016**, *2*, 74–81. [CrossRef]
11. Bhattarai, B.P.; Lévesque, M.; Bak-Jensen, B.; Pillai, J.R.; Maier, M.; Tipper, D.; Myers, K.S. Design and cosimulation of hierarchical architecture for demand response control and coordination. *IEEE Trans. Ind. Inform.* **2017**, *13*, 1806–1816. [CrossRef]
12. Pinson, P.; Madsen, H. Benefits and challenges of electrical demand response: A critical review. *Renew. Sustain. Energy Rev.* **2014**, *39*, 686–699.
13. Aalami, H.A.; Parsa-Moghaddam, M.; Yousefi, G.R. Modelling and prioritizing demand response programs in power markets. *Electr. Power Syst. Res.* **2010**, *80*, 426–435. [CrossRef]
14. Ashot, M.; Kennedy, S.W. A novel demand response model with an application for a virtual power plant. *IEEE Trans. Smart Grid.* **2015**, *6*, 230–237.
15. Deng, R.; Yang, Z.; Chow, M.Y.; Chen, J. A survey on demand response in smart grids: Mathematical models and approaches. *IEEE Trans. Ind. Inform.* **2015**, *11*, 570–582. [CrossRef]
16. Kai, M.; Hu, G.; Spanos, C.J. A cooperative demand response scheme using punishment mechanism and application to industrial refrigerated warehouses. *IEEE Trans. Ind. Inform.* **2015**, *11*, 1520–1531.
17. Sharifi, R.; Anvari-Moghaddam, A.; Fathi, S.H.; Guerrero, J.M.; Vahidinasab, V. Dynamic pricing: An efficient solution for true demand response enabling. *J. Renew. Sustain. Energy* **2017**, *9*, 065502. [CrossRef]

18. Vahedipour-Dahraie, M.; Rashidizadeh-Kermani, H.; Najafi, H.R.; Anvari-Moghaddam, A.; Guerrero, J.M. Stochastic security and risk-constrained scheduling for microgrid with demand response and renewable energy resources. *IET Gener. Transm. Distrib.* **2017**, *11*, 1812–1821. [CrossRef]

19. Xu, F.Y.; Loi, L.L. Novel active time-based demand response for industrial consumers in smart grid. *IEEE Trans. Ind. Inform.* **2015**, *11*, 1564–1573. [CrossRef]

20. Huang, X.; Hong, S.H.; Li, Y. Hour-ahead price based energy management scheme for industrial facilities. *IEEE Trans. Ind. Inform.* **2017**, 132886–132898. [CrossRef]

21. Mohajeryami, S.; Moghaddam, I.N.; Doostan, M.; Vatani, B.; Schwarz, P. A novel economic model for price-based demand response. *Electr. Power Syst. Res.* **2016**, *135*, 1–9. [CrossRef]

22. Vahedipour-Dahraie, M.; Najafi, H.; Anvari-Moghaddam, A.; Guerrero, J. Study of the Effect of Time-Based Rate Demand Response Programs on Stochastic Day-Ahead Energy and Reserve Scheduling in Islanded Residential Microgrids. *Appl. Sci.* **2017**, *7*, 378. [CrossRef]

23. Sharifi, R.; Anvari-Moghaddam, A.; Fathi, S.H.; Guerrero, J.M.; Vahidinasab, V. Economic demand response model in liberalised electricity markets with respect to flexibility of consumers. *IET Gener. Transm. Distrib.* **2017**, *11*, 4291–4298. [CrossRef]

24. Sharifi, R.; Anvari-Moghaddam, A.; Fathi, S.H.; Guerrero, J.M.; Vahidinasab, V. An optimal market-oriented demand response model for price-responsive residential consumers. *Energy Effic.* **2018**, 1–3. [CrossRef]

25. Alipour, M.; Zare, K.; Abapour, M. MINLP probabilistic scheduling model for demand response programs integrated energy hubs. *IEEE Trans. Ind. Inform.* **2018**, *14*, 79–88. [CrossRef]

26. Wang, F.; Zhou, L.; Ren, H.; Liu, X.; Talari, S.; Shafie-khah, M.; Catalão, J.P. Multi-objective optimization model of source-load-storage synergetic dispatch for building energy system based on TOU price demand response. *IEEE Trans. Ind. Appl.* **2018**, *54*, 1017–1028. [CrossRef]

27. Eissa, M.M. Developing incentive demand response with commercial energy management system (CEMS) based on diffusion model, smart meters and new communication protocol. *Appl. Energy* **2019**, *236*, 273–292. [CrossRef]

28. Cao, Y.; Du, J.; Soleymanzadeh, E. Model predictive control of commercial buildings in demand response programs in the presence of thermal storage. *J. Clean. Prod.* **2019**, *218*, 315–327. [CrossRef]

29. Thaler, R. Toward a positive theory of consumer choice. *J. Econ. Behav. Organ.* **1980**, *1*, 39–60. [CrossRef]

30. Lancaster, K.J. A new approach to consumer theory. *J. Political Econ.* **1966**, *72*, 132–157. [CrossRef]

31. Uzawa, H. Production functions with constant elasticities of substitution. *Rev. Econ. Stud.* **1962**, *29*, 291–299. [CrossRef]

32. Bellman, R. Dynamic programming and Lagrange multipliers. *Proc. Natl. Acad. Sci. USA* **1956**, *42*, 767–769. [CrossRef]

33. Tao, H.; Pinson, P.; Fan, S. Global energy forecasting competition 2012. *Int. J. Forecast.* **2014**, *30*, 357–363.

34. Geliebter, A.; Gluck, M.E.; Tanowitz, M.; Aronoff, N.J.; Zammit, G.K. Work-shift period and weight change. *Nutrition* **2000**, *16*, 27–29. [CrossRef]

35. NordPool. Price Calculation. Available online: http://nordpoolspot.com/Market-data1/Dayahead/Area-Prices/ALL1/H (accessed on 13 October 2016).

 *processes*

Article

An Efficient Energy Management in Office Using Bio-Inspired Energy Optimization Algorithms

Ibrar Ullah [1,*], Zar Khitab [2], Muhammad Naeem Khan [1] and Sajjad Hussain [3]

[1]   Department of Electrical Engineering, Capital University of Science and Technology,
     Islamabad 44000, Pakistan; naeembannu@uetpeshawar.edu.pk
[2]   Army Public College of Management & Sciences (APCOMS), Rawalpindi 46000, Pakistan;
     zar.khitab@apcoms.edu.pk
[3]   School of Engineering, University of Glasgow, Glasgow G12 8QQ, UK; Sajjad.Hussain@glasgow.ac.uk
*   Correspondence: ibrarullah@uetpeshawar.edu.pk; Tel.: +92-333-905-1548

Received: 29 January 2019; Accepted: 1 March 2019; Published: 7 March 2019

Abstract: Energy is one of the valuable resources in this biosphere. However, with the rapid increase of the population and increasing dependency on the daily use of energy due to smart technologies and the Internet of Things (IoT), the existing resources are becoming scarce. Therefore, to have an optimum usage of the existing energy resources on the consumer side, new techniques and algorithms are being discovered and used in the energy optimization process in the smart grid (SG). In SG, because of the possibility of bi-directional power flow and communication between the utility and consumers, an active and optimized energy scheduling technique is essential, which minimizes the end-user electricity bill, reduces the peak-to-average power ratio (PAR) and reduces the frequency of interruptions. Because of the varying nature of the power consumption patterns of consumers, optimized scheduling of energy consumption is a challenging task. For the maximum benefit of both the utility and consumers, to decide whether to store, buy or sale extra energy, such active environmental features must also be taken into consideration. This paper presents two bio-inspired energy optimization techniques; the grasshopper optimization algorithm (GOA) and bacterial foraging algorithm (BFA), for power scheduling in a single office. It is clear from the simulation results that the consumer electricity bill can be reduced by more than 34.69% and 37.47%, while PAR has a reduction of 56.20% and 20.87% with GOA and BFA scheduling, respectively, as compared to unscheduled energy consumption with the day-ahead pricing (DAP) scheme.

Keywords: appliance scheduling techniques; bacterial foraging algorithm (BFA); energy management system; energy optimization algorithms; grasshopper optimization algorithm (GOA); smart grid

---

1. Introduction

With the increased use of modern technologies and smart appliances in every field of life, energy consumption is rapidly increasing. The rising electricity demand cannot be fulfilled by the traditional electric power grid. That is why the smart grid is becoming more popular to fulfil daily electricity demand. The smart grid (SG) is supposed to be the incorporation of information technologies (IT) in the existing power grids to increase their robustness and consistency. Smart meters (SM) are used for communication and energy monitoring purposes in SG. To schedule smart appliances in residential, commercial and industrial sectors, an energy management controller (EMC) is installed at the consumer premises. Demand side management (DSM) has many strategies that help to solve the energy optimization problem by peak clipping, load shifting, strategic conservation, flexible load shifting, strategic load growth and valley filling. By using these strategies, the load is shifted from high demand timings to low demand timings [1]. The two main functionalities of DSM are proper management of the load and demand response (DR) [2]. Consumer load management is also known as DSM. It is the process of shifting electricity demand from high-demand (on-peak) hours to low-demand(off-peak) hours to decrease the energy cost. DR is the consumer's response to variable pricing signals. There are two shapes of DR: in the form of energy price reduction or some incentives to consumers [3,4].

The main objectives of the energy management system (EMS) are the reduction of the energy bill, PAR and consumer discomfort. Many algorithms have been deigned to accomplish the aforementioned objectives. For cost and energy consumption minimization, mixed integer linear programming (MILP), mixed integer nonlinear programming (MINLP), non-integer linear programming (NILP) and convex programming were used in [5–8]. However, these techniques are used for fewer appliances and have a large convergence time. In order to overcome these deficiencies, researchers use meta-heuristic techniques to resolve the issue of energy optimization. For cost minimization, the genetic algorithm (GA) was proposed by the authors in [9,10]. For cost minimization and aggregated power consumption, differential evolution (DE) and ant colony optimization (ACO) were used in [11,12].

In this research work, we use GOA and BFA techniques for a single office using the DAP pricing signal. The simulation is performed in MATLAB, and we obtained the results of PAR, cost and average waiting time. The rest of the paper is divided into the following sections: Related work is illustrated in Section 2. Section 3 discusses the problem statement and approach. Section 4 depicts the system model and problem formulation. The proposed schemes are described in Section 5. Simulation results are illustrated in Section 6 to demonstrate some of the achievements. The paper is concluded in Section 7.

2. Related Work

In SG, numerous algorithms have been proposed by researches, for energy-efficient optimization in residential, commercial and industrial areas, for the benefits of both consumers and the utility. The main targets of researchers have been balancing the load and decreasing electricity cost. Different parameters such as pricing mechanisms, types of appliances and different user demands are considered.

Hybrid bacterial foraging and genetic (HBG) algorithm-based DSM for smart homes was proposed by the authors in [13]. They focused on peak load reduction, cost minimization, user comfort maximization and load shifting. Through HBG cost, PAR and waiting time were reduced compared to GA and BFA. A smart community-based energy optimization technique was discussed in [14]. The authors focused on the end-user's high comfort level and less energy usage with integration of renewable energy sources using particle swarm optimization (PSO). A time-constrained nature-inspired algorithm-based home energy management (HEM) system was proposed by the authors in [15]. GA, moth-flame optimization algorithm (MFO) and their hybridization were proposed for energy bill reduction and achieving end-users' high comfort level. A HEM system using cuckoo search was proposed in [16]. The performance of GA and the cuckoo search algorithm was compared with respect to the reduction of energy cost, PAR and user discomfort by using the DAP signal. Cuckoo search incorporation with levy flights of some kind of birds and fruit-flies were considered for the breading strategy in [17]. In many optimization problems, because of its generic and robust nature, the cuckoo search is superior to GA and PSO. The authors used GA, TLBO (teacher learning-based algorithm), LP (linear programming) and TLGO (teacher learning genetic optimization) algorithms for appliances scheduling in [18]. They categorized flexible appliances as "time flexible" and "power flexible" for proficient energy consumption of consumers in SG. This approach enables energy consumers to schedule their appliances to get optimized energy consumption. This approach also maximizes the comfort level of customers with restricted total energy consumption. In [19], the authors discussed optimal operation methods for a micro-grid. They used the improved adaptive evolutionary algorithm and swarm optimization algorithm. In [20], the authors presented an optimal scheduling scheme for residential appliances, using the DAP signal in smart homes. This algorithm reduced peak cost to 22.6% and normal price to 11.7%. This approach does not rely on the energy optimization approach. In [21], the authors discussed load shifting, cost minimization and the energy storage system

(ESS). They proposed a system that enables the user to buy energy during low demand timings and sale their stored energy to the utility during on-peak hours.

In [22], the gradient-based particle swarm optimization (GPSO) technique for demand response (DR) in smart homes, considering the load and energy price uncertainties, was discussed. Having an optimal scheduling of power, the heuristic-based genetic algorithm (GA) was used for demand response (DR) in HEM systems in [23]. In this paper, the authors used GA, TLBO (teaching learning-based optimization), EDE (enhanced differential evolution) and proposed EDTLA (enhanced differential teaching learning algorithm) for minimization of the residential total energy cost and end-user discomfort level. The authors in [24] discussed the cooperative multi-swarm particle swarm optimization (PSO) technique for achieving their goals of cost minimization; however, they did not considered PAR. In [25], the authors used GA with the DAP scheme for optimally scheduling the load demand. In [26], the authors introduced a load balancing mechanism in commercial, residential and industrial areas. They compared the usage of electricity with GA and without GA in DSM. By using GA-based DSM, they reduced the electricity usage during peak hours. However, PAR and end-user discomfort were not discussed. In [27], the authors used PSO for scheduling of smart electric appliances for electricity cost minimization. They took different cases of changing the renewable energy consumption rate and user comfort level and applied it in a smart community as a case study. The aforementioned optimization techniques achieved the energy cost minimization and PAR reduction by losing the end-user comfort. Therefore, in this work, we have explored and analyzed two bio-inspired algorithms for the energy optimization problem in the residential sector: GOA and BFA. This is because the algorithms were developed based on the perfect optimization behavior of naturally-available organisms. Through simulations, we have shown that using bio-inspired optimization algorithms, the energy cost and PAR can be reduced compared to the unscheduled load.

3. Problem Statement and Approach

Traditional electric power grids are unable to fulfill today's electricity demand. This deficiency has raised the demand for an energy management system. Through different techniques and algorithms, we can solve the energy optimization problem. Researchers have applied different bio-inspired algorithms, however, they have not considered the end-user comfort by reducing their waiting time along with the reduction of energy cost and PAR. Therefore, in this work, we use the GOA and BFA techniques for the office energy management system (OEMS), using DAP. Eight appliances have been considered, named automatically operating appliances (AOAs). We have divided 12-h office-timings into 60 time slots of 12 min in duration. The simulation results show that by using GOA and BFA in OEMS, we can reduce the total cost to 34.69% and 37.47% and PAR to 56.20% and 20.87%, respectively.

4. System Model and Problem Formulation

4.1. Model Architecture

The efficient utilization of the existing energy resources is necessary in our daily life. The proposed system model architecture is depicted in Figure 1. It consists of a smart meter (SM), the energy management controller (EMC), automatically-operated appliances (AOAs) and advance distribution and communication systems. EMC receives the required energy consumption outline from all connected appliances, which schedule the energy consumption pattern according to the pricing signal. The utility sends the pricing signal to the smart meter, which is then forwarded to the EMC. At the same time, the SM receives the consumed electricity reading from EMC and transmits it to the utility. Through a wireless communication network, i.e., Wi-MAX, ZigBee, Bluetooth, WiFi, GSM or GPRS or using PLC (power line communication), the utility and SM communicate with each other. In this work, we considered only a single office with eight appliances. In our case, the decision would be made after every 12 min, not 1 h, because we have divided 24 h into 120 time slots, each equal to

12 min. Here, we considered 120/2 slots for offices because offices only consume energy during the daytime, which is denoted by symbol s.

$$s \in S = \{1, 2, 3...60\} \tag{1}$$

The scheduled vector of office energy consumption (OEC) for a single appliance is:

$$OEC_{ac}^{s} = \{OEC_{ac}^{1}, OEC_{ac}^{2}, OEC_{ac}^{3}...OEC_{ac}^{60}\} \tag{2}$$

Total energy consumption OEC_T is calculated as:

$$OEC_T = \sum_{s=1}^{60} (\sum_{i=1}^{12} OEC_i^s) \tag{3}$$

Table 1 gives the specifications of different appliances in an office.

Table 1. Specifications of office automatically operating appliances (AOAs).

S. No.	AOAs	LOT	Power Rating (kW)	OTIs
1	Air conditioner	30	4.00	1–60
2	Computer	40	0.25	5–55
3	Electric kettle	2	3.00	1–55
4	Coffee maker	3	2.00	10–45
5	Water dispenser	45	2.5	1–60
6	Oven	5	5.00	10–50
7	Fan	25	3.5	1–60
8	Light	35	2	1–60

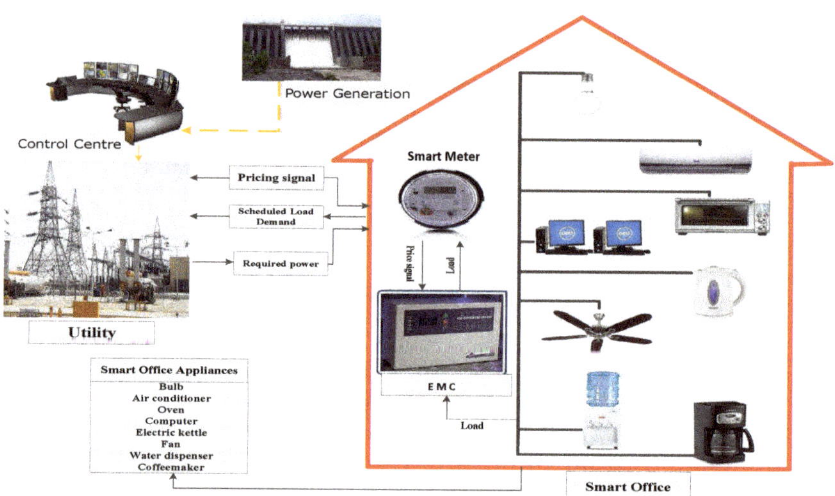

Figure 1. The proposed system model architecture.

4.2. Problem Formulation

In the proposed work, we formulate our problems of: (a) end-user high comfort level, (b) consumers' electricity bill minimization and (c) minimization of PAR by optimization of the energy consumption profiles of office appliances, using the multiple knapsack problem (MKP) scheduling technique. MKP is a capacity (resources) allotment problem. It consists of N number of capacities and Q number of objects [28].

MKP is a combinatorial problem. In MKP, the stuff quantity, having different weights and values, can be kept into a knapsack of a certain capacity, such that the worth of the knapsack should be maximum, as shown in Figure 2.

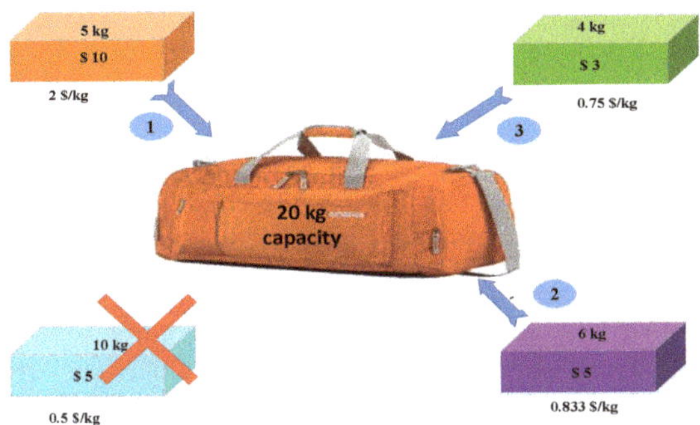

Figure 2. The multiple knapsack problem (MKP) formulation.

We consider U number of knapsacks and use MKP as the scheduling mechanism to map our problem as follows:

- Power capacities of consumers in every time interval are mapped as U number of knapsacks;
- Appliances in an office are mapped as "Q" number of objects;
- The weight of every object in MKP is mapped as appliances' consumed energy in every time interval. This is assumed to be time invariant;
- In MKP, the worth of each object in a particular time interval is mapped as the cost of appliances' consumed energy in that interval of time [29].

If OEC_T is the maximum energy capacity in every time slot, then the end-user electricity cost along with PAR can be minimized, keeping aggregated energy consumption of the cumulative office appliances within the maximum threshold limit of C_T.

Mathematically, this constraint can be shown as follows:

$$OEC_T \leq OEC_{max} \tag{4}$$

Here, OEC_T is the cumulative energy demand of the end-user and OEC_{max} is the maximum energy capacity in a particular interval of time available from the utility grid. MKP scheduling tells us to keep the total energy demand of the end-user less than or equal to this maximum energy capacity threshold.

4.3. The Electricity Cost

In order to calculate total energy cost, we use the following equation:

$$C = \sum_{h=1}^{60} (E_{rate} \times P_{rate}) \tag{5}$$

C is the total cost in 60 time slots; E_{rate} is the energy cost per hour; P_{rate} is the connected appliances' power rating.

4.4. The Power Consumption

The power consumption of each appliance is calculated by the following equation:

$$Load = \sum_{h=1}^{60} P_{rate} \times X \tag{6}$$

P_{rate} is the power rating, and X is the ON/OFF status of an appliance.

4.5. PAR

For RAR, the following equation is used.

$$PAR = \frac{max(load)}{Avg(load)} \tag{7}$$

4.6. Waiting Time

This is that time interval when a consumer wants to switch-ON an appliance. However, due to the scheduling of appliances, the consumer has to wait for a certain amount of time. Figure 3 shows that α is the appliance starting time, but actually, the appliance will start its operation at η. mathematically, the waiting time is given as:

$$\tau_w = \eta - \alpha \tag{8}$$

The normalized waiting time is given by:

$$\tau_w = \frac{\eta - \alpha}{(\beta - LOT) - \alpha} \tag{9}$$

where $(\beta - LOT)$ is the last starting time of an appliance, so that it will complete its operation.

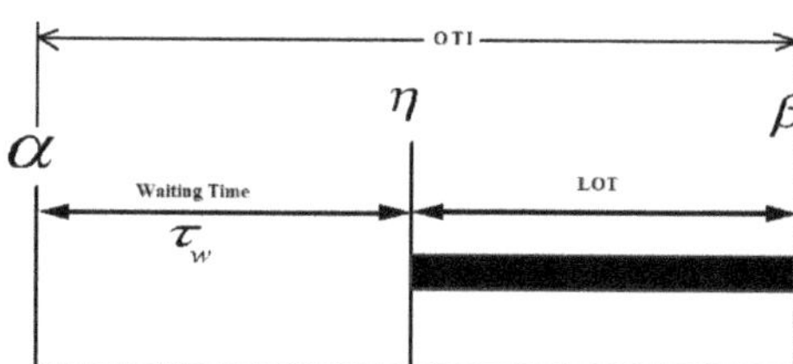

Figure 3. Appliances starting time, operation ending time and waiting time.

4.7. Objective Function

Our objective function can mathematically be expressed as follows:

$$min\left(\sum_{m=1}^{60} [w_1 \times \sum_{n=1}^{N} (OEC_{T,n} \times E_{rate}) + (w_2 \times \tau_w)] \right) \tag{10}$$

where E_{rate} is the energy cost in every interval of time. The aim of our objective function is to minimize electricity cost, while keeping a higher consumer comfort level by the reduction of waiting time. w_1 and w_2 are weighting factors of the two portions of our objective function. Their values can be either "0" or "1", so that $(w_1 + w_2) = 1$ [9]. This reveals that either w_1 or w_2 could be zero or one. This means that, if a consumer does not want to schedule his/her appliances, then these weighting factors will be $w_1 = 1$ and $w_2 = 0$ in the objective function.

5. Scheduling Algorithms

To solve the appliances optimal scheduling problem, in order to achieve the lowest energy cost, lower PAR and less user discomfort, different scheduling algorithms have been proposed in the literature. In this paper, we have proposed GOA and BFA. A brief description of both of these algorithms is given below.

5.1. Grasshopper Optimization Algorithm

A grasshopper is a kind of destructive insect, which is known as a pest because it damages crops. There are eleven thousand species of grasshopper [30]. It is generally considered as a flying animal. As a grasshopper reaches its adult stage, it passes through the stages of eggs, nymph and adult, as shown in the Figure 4.

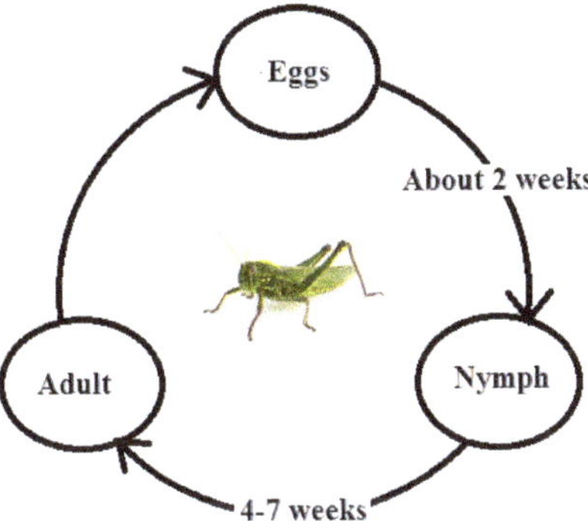

Figure 4. The lifecycle of a grasshopper.

Usually, grasshoppers can be seen in the form of a swarm in nature. They attack agricultural lands and become a nightmare for formers. The swarming behaviour is found in both adult and nymph grasshoppers [31,32]. Generally, a locust swarm contains five billion of grasshoppers and spread over an area of 60 square miles. Nature-inspired algorithms have a unique search mechanism for food. This consists of two techniques: exploration and exploitation.

(a) Exploration: The process in which the algorithm finds a new solution from the current solutions in the search space.

(b) Exploitation: The process in which algorithm searches the surrounding search space.

The mathematical model of GOA, which carries the swarming behaviour of a grasshopper, is given as in [33].

$$P_i = S_i + V_i + A_i \tag{11}$$

In the above equation, P_i shows the ith position of a grasshopper, S_i gives the social collaboration, V_i gives the gravitational force over a grasshopper and A_i is the air-advection. To produce randomness in the above equation, it becomes:

$$P_i = (x_1 \times S_i) + (x_2 \times V_i) + (x_3 \times A_i) \tag{12}$$

where x_1, x_2 and x_3 are random numbers between zero and one. S_i is modelled as:

$$S_i = \sum_{i=1}^{N} s(d_{ij})\hat{d}_{ij}\ldots\ldots\ldots\ldots j \neq i \tag{13}$$

where N is the number of search agents, s defines the social interaction between two grasshoppers i and j, d_{ij} is the respective distance between i^{th} and j^{th} grasshoppers and is given by:

$$d_{ij} = |P_j - P_i| \tag{14}$$

and:

$$\hat{d}_{ij} = \frac{(P_j - P_i)}{d_{ij}} \tag{15}$$

is the unit vector from the ith grasshopper to the jth grasshopper. Mathematically, the social force is given as follows:

$$s(d) = Fe^{\frac{-d}{l}} - e^{-r} \tag{16}$$

where F is the attractive force, d shows the distance and l is the measure of attraction.

The V component in Equation (1) is given as:

$$V_i = -v\hat{e}_v \tag{17}$$

where v is the gravitational force, and the negative sign shows its direction towards the centre of the Earth, while $\hat{e}_v$ is the unit vector in the direction of the Earth.

Now, the A component in Equation (1) is given as:

$$A_i = v\hat{e}_w \tag{18}$$

where v is the constant drift when there is a wind and $\hat{e}_w$ shows the wind directional unit vector. By putting the values of S, G and A in Equation (1), we get:

$$S_i = \sum_{i=1}^{N} s(|P_j - P_i|)\frac{(P_j - P_i)}{d_{ij}} - g\hat{e}_g + v\hat{e}_w \tag{19}$$

We utilize the above equation for the swarm in free space and use it in simulation to describe the interaction between the grasshoppers in a swarm. The steps involved in the GOA algorithm are given in Algorithm 1 and are depicted in Figure 5.

Algorithm 1: GOA algorithm.

---

1 **Initialization:** Generation of price signal according to the scheme used
2 LOTs' specification of appliances
3 power ratings of appliances
4 **Input:** variables u_b, l_b, dim, N
5 Initialize position of grasshopper
6 **for** $h = 1 \ to \ H$ **do**
7   Find electricity cost
8   Find cost of all appliances' LOTs
9   Find F_{best}, L_{best} and G_{best};
10   **for** $It = 1 \ to \ It_{Max}$ **do**
11   **for** $i = 1 \ to \ N_{VAR}$ **do**
12   **end**
13   Find the best position
14   **end**
15   Update the LOTs of appliances
16 **end**
17 **Output:** $OEC_T, load, PAR.$

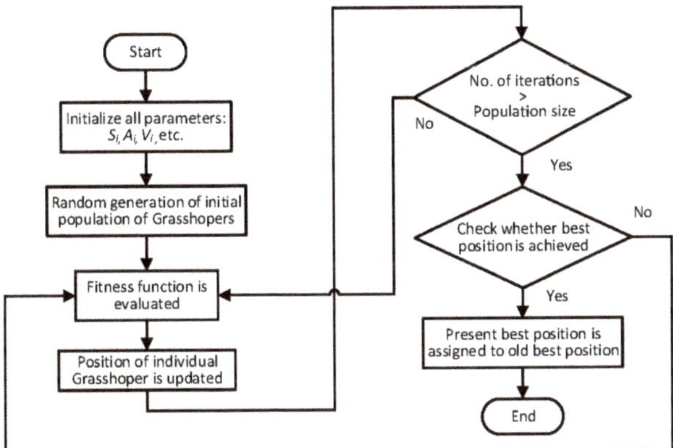

Figure 5. Flowchart of grasshopper optimization algorithm (GOA).

5.2. Bacterial Foraging Algorithm

BFA was proposed by Kevin Passino in 2002 [34]. In this algorithm, the group foraging strategy of a swarm of *Escherichia coli* (*E. coli*) bacteria is the key point. Bacteria forage for food and nutrients, to maximize their energy per unit time. By sending a signal, bacteria also communicate with each other. Because of predators, the prey may be mobile. Therefore, it is chased by the bacterium in an optimal way. When a bacterium maximizes its energy by getting sufficient food, then it does other activities like sheltering, mating, fighting, etc.

The following steps are needed in order to explain BFA.

(a)   Chemotaxis
(b)   Swarming
(c)   Reproduction
(d)   Elimination

5.2.1. Chemotaxis

In the chemotaxis step, the *E. coli* move from one place to another through flagella. According to the biological point of view, its motion is observed in two different ways: it may either swim or tumble.

To consider the chemotaxis movement of bacteria, we have the following equation:

$$\delta^j(i+1,k,l) = \delta^j(i,k,l) + Q(j)\frac{\Delta(j)}{\sqrt{\Delta^T(j)\Delta(j)}} \tag{20}$$

In the above equation, $\delta^j(i,k,l)$ shows the position of the jth bacterium at the ith chemotactic, the kth reproductive and the lth elimination-dispersal step. $Q(i)$ represents the size of the step taken by the bacterium in a random direction when it tumbles. Δ shows the vector in random direction $[-1, 1]$.

5.2.2. Swarming

The *E. coli* bacterium is blessed with swarming behaviour. In this step, bacteria cells form a ring-shaped structure and move in search of nutrients. A high level of succinate usage stimulates the cells, due to which attractant-aspartate is released by the cells, which helps them to bind in groups.

5.2.3. Reproduction

When a bacterium is in a feasible and nutritious environment, it reproduces, splits into two bacteria and keeps the number of cells in a swarm fixed.

5.2.4. Elimination and Dispersal

The scarcity of nutrients kills the bacterium or disperses them into another environment. They are also killed due to high temperature. If there is a poor condition in the environment, the bacteria may place themselves near a good food source, hence assisting chemotaxis.

To calculate the fitness of each bacterium, the following equation can be used:

$$F_j[i,k,l] = F_j[i,k,l] + F_{cc}(\delta_j[i,k,l], P[i,k,l]) \tag{21}$$

In the above equation, F_j shows the fitness of the bacterium and δ_j is the position of the bacterium.

$$F_{cc} = \sum_{d=1}^{d-1} (100 \times (\delta(j,d+1) - (\delta(j,d))^2)^2 + (\delta(j,d) - 1)^2) \tag{22}$$

In order to achieve the time-varying objective, we must put the objective function J_{cc} into the actual objective function F_j. The steps involved in the BFA algorithm are given in Algorithm 2 and are depicted in Figure 6.

Algorithm 2: Bacterial foraging algorithm.

---

1. **Initialization:** Generation of the price signal according to the scheme used, LOTs' specification of appliances, power ratings of appliances
2. **Input:** Give initial values to variables; $pop, N_p, N_e, N_c, N_r, N_s, D, C.$
3. Evaluate fitness for each bacterium (J_{last}).
4. **for** $l = 1$ to N_e **do**
5.   **for** $k = 1$ to N_r **do**
6.   **for** $j = 1$ to N_c **do**
7.   **for** $i = 1$ to N_p **do**
8.   Find new position of the bacterium
9.   Find the fitness
10.   **for** $s = 1$ to N_s **do**
11.   **end**
12.   **if** $J_i < J_{last}$ **then**
13.   Replace the previous position of the bacterium with the new position
14.   Go back to line 10
15.   **else**
16.   Assign a random direction
17.   Evaluate the fitness
18.   Go back to line no. 10
19.   **end**
20.   **end**
21.   **end**
22.   Evaluate the fitness of the bacterium
23.   Select the best one Random elimination and dispersal
24.   **end**
25.   **if** $1 < N_e$ **then**
26.   **else**
27.   Go back to the initial elimination step
28.   **end**
29. **end**
30. **Output:** $OEC_T, load, PAR.$

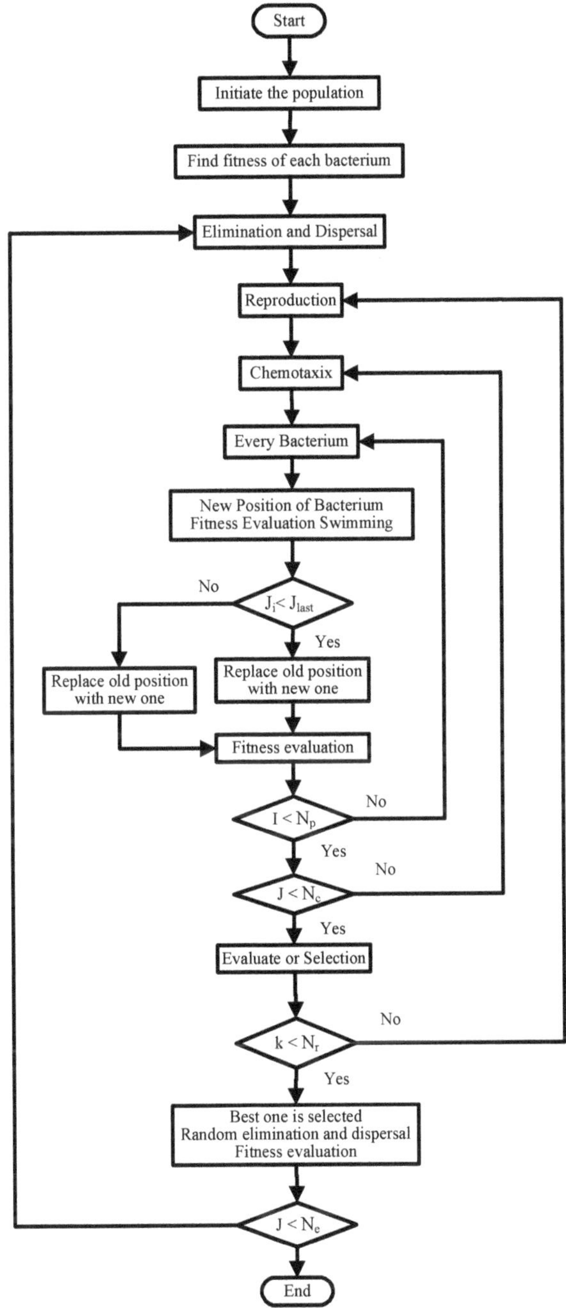

Figure 6. Flowchart of bacterial foraging algorithm (BFA).

6. Simulation Results

A substantial simulation was performed to show the performance of different algorithms in terms of minimization of electricity cost by shifting appliances from on-peak hours to off-peak hours, minimization of PAR and minimization of end-user discomfort due to waiting time. In this paper, eight appliances were selected, as shown in the Table 1. For comparison purpose, firefly algorithm (FA), cuckoo search algorithm (CSA) and ant colony optimization algorthm (ACO) are considered in the same scenario for thirty-days load scheduling. Figure 7 gives the day-ahead pricing (DAP) signal, taken from the daily report of the New York Independent System Operator (NYISO) [35]. The total time of 24 h was divided into 24 time slots. For an office, usually 8–12 h was used, so time was taken from 8:00–20:00. Figure 8 shows the daily unscheduled load and scheduled load with the GOA and BFA algorithms. The figure shows that GOA outperformed by eliminating the peak in the unscheduled load. Figure 9 shows the hourly unscheduled (Un-sch) and scheduled load with GOA and BFA cost. It is clear that the hourly cost is averaged compared to the unscheduled cost, especially high cost in the on-peak hours due to shifting of the load from on-peak hours to off-peak hours. Figure 10 shows that the office monthly load was equal for all algorithms, as each algorithm had to reschedule the appliances only. Figure 11 depicts the total monthly cost in dollars. In the unscheduled case, we had a maximum cost of 267.45 $; when scheduled by GOA, it became 174.67 $ (34.69% reduction); and in the case of BFA, it became 161.23 $ (37.47% reduction). The comparison of these proposed algorithms with state-of-the-art algorithms for the same scenario is depicted in Table 2. Figure 12 depicts the daily PAR. It is clear from the figure that our proposed schemes minimized the PAR. Before scheduling, the PAR value was 7.81, and after scheduling with GOA and BFA, the PAR values became 3.42 (56.20% reduction) and 6.18 (20.87% reduction), respectively. Figure 13 depicts the average waiting time. The waiting time in the case of GOA was 1.28 h, and BFA was 1.32 h. This shows that the waiting time of BFA was greater than GOA because it had reduced the total cost more than that reduced by BFA. Table 2 shows that, there is always a trade-off between energy cost and waiting time.

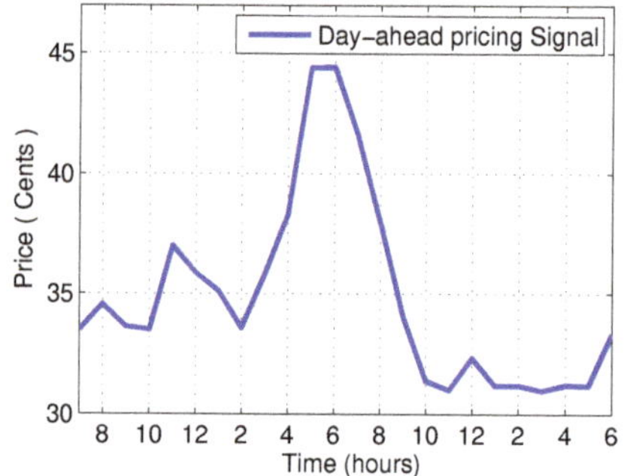

Figure 7. Day-ahead pricing signal [35].

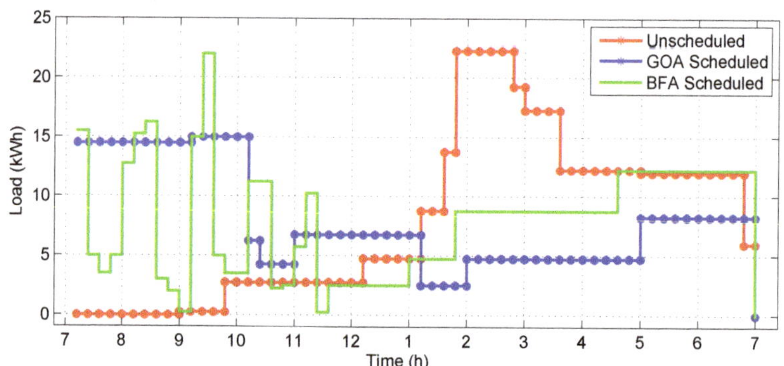

Figure 8. Hourly load for unscheduled load and scheduled load with the GOA and BFA algorithms.

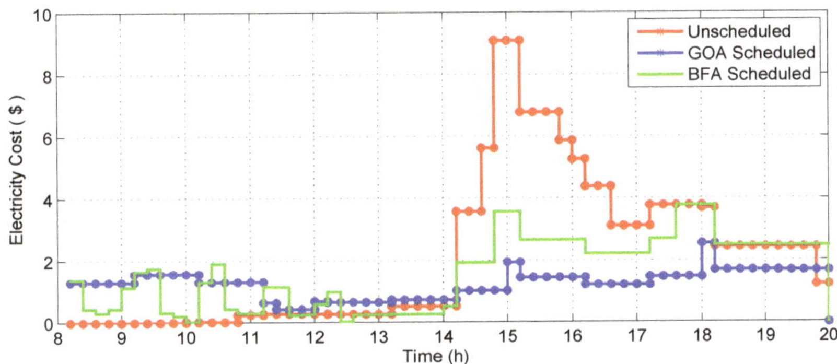

Figure 9. Hourly cost for unscheduled load and scheduled load with the GOA and BFA algorithms.

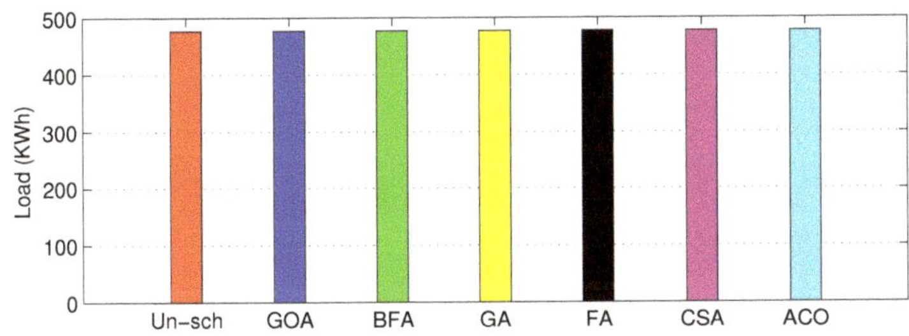

Figure 10. Total monthly load.

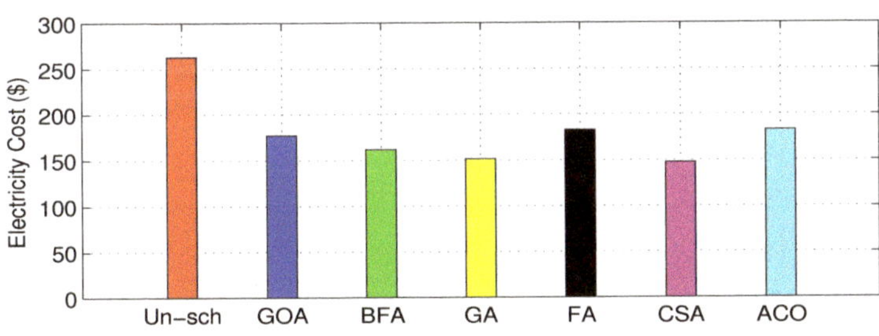

Figure 11. Total cost.

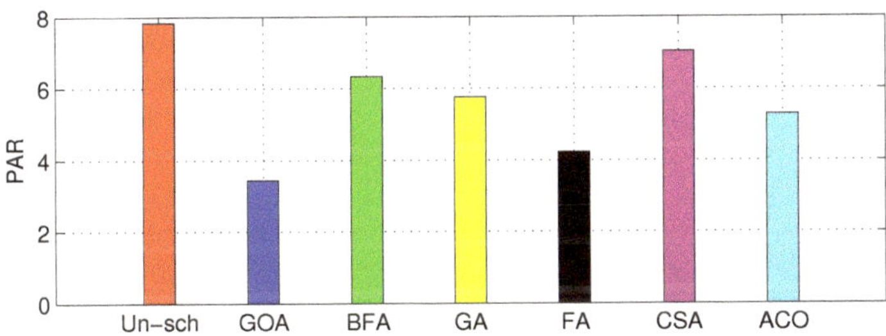

Figure 12. The peak-to-average power ratio (PAR).

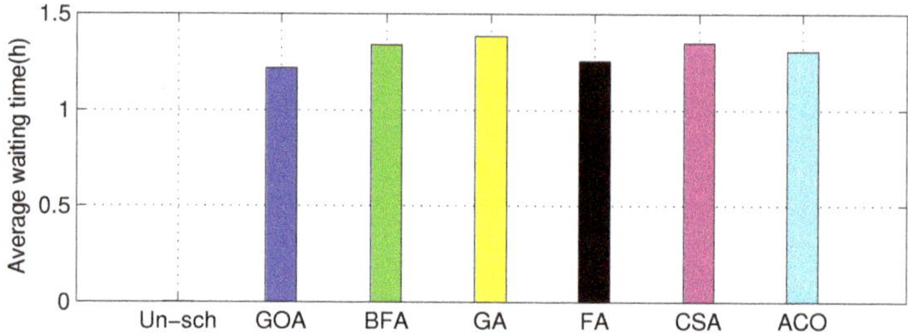

Figure 13. Average waiting time.

Table 2 compares the performance of the proposed algorithms with unscheduled load and state-of-the-art algorithms like firefly algorithm (FA), cuckoo search algorithm (CSA) and ant colony optimization (ACO) with respect to three parameters; energy cost, waiting time and PAR.

Table 2. Comparison of the unscheduled load and scheduled load with the GOA, BFA, GA, FA, CSA and ACO algorithms.

Techniques	Days	Cost ($)	Cost Reduction	Waiting Time (h)	PAR	PAR Change
Unschedule	30 days	267.45	–	–	7.81	–
GOA-scheduled	30 days	174.67	34.69%	1.28	3.42	56.20%
BFA-scheduled	30 days	161.23	37.47%	1.32	6.18	20.87%
GA-scheduled	30 days	150.07	43.89%	1.39	5.84	25.22%
FA-scheduled	30 days	177.39	33.68%	1.25	4.17	46.60%
CSA-scheduled	30 days	147.68	44.79%	1.38	7.11	08.96%
ACO-scheduled	30 days	176.83	33.89%	1.27	5.34	31.62%

Table 3 shows the run-time of the proposed algorithms using an Intel (R) Core (TM) i5 processor, with 4.00 GB of installed memory (RAM) and the 32-bit Windows 7 Operating system.

Table 3. Run-time of the proposed algorithms for 30 days of load scheduling.

Proposed Algorithm	No. of Days	Run-Time (s)
GOA	30 days	11.695
BFA	30 days	13.171

7. Conclusions

In this paper, we have proposed a novel technique of appliances' scheduling in an office. We used two nature-inspired optimization algorithms, GOA and BFA, to achieve our objective functions of end-user electricity bill minimization along with a reduction of PAR and user discomfort due to appliance scheduling. We considered only eight appliances to check our proposed algorithms' performance. We compared our results with a few state-of-the-art nature-inspired algorithms in the literature like GA, FA, CSA and ACO for the three mentioned fitness functions, i.e., minimization of the electricity bill, PAR and waiting time. Indeed, numerous countries in the world can fulfil electricity demand. However, keeping in view the minimization of the electricity bill, the reliability of the existing system and improvements towards smart grids to facilitate the customers, with increased dependency on electricity with automation, energy optimization is a big issue throughout the world. Furthermore, with increased electricity generation, carbon emission increases due to the use of different types of fuels, which pollute this biosphere day by day. Therefore, the advantage of these algorithms for energy optimization is not only to save money, but to reduce pollution, as well. The simulation results show that our proposed energy optimization schemes performed well in the case of minimization of PAR and cost. However, when energy cost is minimized, user waiting time will increase as a penalty. In the future, multi-objective algorithms will be designed to minimize the energy cost and PAR while keeping in view the high comfort level of consumers. Furthermore, the proposed algorithms will be applied to residential, commercial and industrial areas for the greater benefit of both the utility and consumers. For this purpose, more nature-inspired algorithms will be used and analyzed.

Author Contributions: I.U. proposed the idea and did the writing, review and editing; Z.K. did the formal analysis and comparisons; M.N.K. proposed the topology and system design; S.H. provided technical supervision, insights and additional ideas on the presentation. All authors equally contributed, revised and approved the manuscript.

Funding: This research received no external funding.

Conflicts of Interest: The authors declare no conflict of interest.

Abbreviations

The following abbreviations are used in this manuscript:

OEC	Office energy consumption
LOT	Length of operational time
OTI	Operational time interval
AOAs	Automatically operating appliances
PAR	Peak-to-average power ratio
Un-sch	Un-scheduled load
FA	Firefly algorithm
CSA	Cuckoo search algorithm
ACO	Ant colony optimization
s	Each time slot
C	The total electricity cost in sixty time slots
P_{rate}	Power rating of connected appliances
$Load$	Power consumption of each appliance
τ_w	Waiting time for an appliance
S	Set of 60 time slots
E_{rate}	Energy cost per hour
X	ON-OFF states of an appliance
α	Starting time of an appliance
η	Operational starting time of an appliance
β	Ending time of an appliance

References

1. Hashmi, M.; Hnninen, S.; Mki, K. Survey of smart grid concepts, architectures, and technological demonstrations world-wide. In Proceedings of the IEEE PES Conference on Innovative Smart Grid Technologies (ISGT Latin America), Medellin, Colombia, 19–21 October 2011; pp. 1–7.
2. Rahimi, F.; Ipakchi, A. Demand response as a market resource under the smart grid paradigm. *IEEE Trans. Smart Grid* **2010**, *1*, 82–88. [CrossRef]
3. Ahmed, A.; Manzoor, A.; Khan, A.; Zeb, A.; Ahmad, H. Performance Measurement of Energy Management Controller Using Heuristic Techniques. In Proceedings of the Conference on Complex, Intelligent, and Software Intensive Systems, Turin, Italy, 10–13 July 2017; pp. 181–188.
4. Ozturk, Y.; Senthilkumar, D.; Kumar, S.; Lee, G. An intelligent home energy management system to improve demand response. *IEEE Trans. Smart Grid* **2013**, *4*, 694–701. [CrossRef]
5. Mavrotas, G.; Karmellos, M. Multi-objective optimization and comparison framework for the design of Distributed Energy Systems. *Energy Convers. Manag.* **2019**, *180*, 473–495.
6. Sousa, T.; Morais, H.; Vale, Z.; Faria, P.; Soares, J. Intelligent energy resource management considering vehicle-to-grid: A simulated annealing approach. *IEEE Trans. Smart Grid* **2012**, *3*, 535–542. [CrossRef]

7.   Soares, J.; Sousa, T.; Morais, H.; Vale, Z.; Faria, P. An optimal scheduling problem in distribution networks considering V2G. In Proceedings of the IEEE Symposium on Computational Intelligence Applications in Smart Grid (CIASG), Paris, France, 11–15 April 2011; pp. 1–8.

8.   Tsui, K.M.; Chan, S.C. Demand response optimization for smart home scheduling under real-time pricing. *IEEE Trans. Smart Grid* **2012**, *3*, 1812–1821. [CrossRef]

9.   Zhao, Z.; Lee, W.C.; Shin, Y.; Song, K.B. An optimal power scheduling method for demand response in home energy management system. *IEEE Trans. Smart Grid* **2013**, *4*, 1391–1400. [CrossRef]

10. Arabali, A.; Ghofrani, M.; Etezadi-Amoli, M.; Fadali, M.S.; Baghzouz, Y. Genetic-algorithm-based optimization approach for energy management. *IEEE Trans. Power Deliv.* **2013**, *28*, 162–170. [CrossRef]

11. Tang, L.; Zhao, Y.; Liu, J. An improved differential evolution algorithm for practical dynamic scheduling in steelmaking-continuous casting production. *IEEE Trans. Evol. Comput.* **2014**, *18*, 209–225. [CrossRef]

12. Liu, B.; Kang, J.; Jiang, N.; Jing, Y. Cost control of the transmission congestion management in electricity systems based on ant colony algorithm. *Energy Power Eng.* **2011**, *3*, 17. [CrossRef]

13. Khalid, A.; Javaid, N.; Mateen, A.; Khalid, B.; Khan, Z.A.; Qasim, U. Demand Side Management using Hybrid Bacterial Foraging and Genetic Algorithm Optimization Techniques. In Proceedings of the 10th International Conference on Complex, Intelligent, and Software Intensive Systems (CISIS), Fukuoka, Japan, 6–8 July 2016; pp. 494–502.

14. Wang, K.; Li, H.; Maharjan, S.; Zhang, Y.; Guo, S. Green Energy Scheduling for Demand Side Management in the Smart Grid. *IEEE Trans. Green Commun. Netw.* **2018**, *2*, 596–611. [CrossRef]

15. Ullah, I.; Hussain, S. Time-Constrained Nature-Inspired Optimization Algorithms for an Efficient Energy Management System in Smart Homes and Buildings. *Appl. Sci.* **2019**, *9*, 792. [CrossRef]

16. Aslam, S.; Bukhsh, R.; Khalid, A.; Javaid, N.; Ullah, I.; Fatima, I. An Efficient Home Energy Management Scheme Using Cuckoo Search. In Proceedings of the International Conference on P2P, Parallel, Grid, Cloud and Internet Computing, Taichung, Taiwan, 27–29 October 2018; pp. 167–178.

17. Yang, X.-S.; Suash, D. Cuckoo search via Lévy flights. In Proceedings of the 2009 World Congress on Nature & Biologically Inspired Computing (NaBIC), Coimbatore, India, 9–11 December 2009; pp. 210–214.

18. Manzoor, A.; Javaid, N.; Ullah, I.; Abdul, W.; Almogren, A.; Alamri, A. An Intelligent Hybrid Heuristic Scheme for Smart Metering based Demand Side Management in Smart Homes. *Energies* **2017**, *10*, 1258. [CrossRef]

19. Gao, R.; Wu, J.; Hu, W.; Zhang, Y. An improved ABC algorithm for energy management of microgrid. *Int. J. Comput. Commun. Control* **2018**, *13*, 477–491. [CrossRef]

20. Shirazi, E.; Jadid, S. Optimal residential appliance scheduling under dynamic pricing scheme via HEMDAS. *Energy Build.* **2015**, *93*, 40–49. [CrossRef]

21. Adika, C.O.; Wang, L. Smart charging and appliance scheduling approaches to DSM. *Int. J. Electr. Power Energy Syst.* **2014**, *57*, 232–240. [CrossRef]

22. Huang, Y.; Wang, L.; Guo, W.; Kang, Q.; Wu, Q. Chance Constrained Optimization in a Home Energy Management System. *IEEE Trans. Smart Grid* **2018**, *9*, 252–260. [CrossRef]

23. Javaid, N.; Hussain, S.M.; Ullah, I.; Noor, M.A.; Abdul, W. Demand Side Management in Nearly Zero Energy Buildings Using Heuristic Optimizations. *Energies* **2017**, *10*, 1131–1159. [CrossRef]

24. Ma, K.; Hu, S.; Yang, J.; Xu, X.; Guan, X. Appliances scheduling via cooperative multi-swarm PSO under day-ahead prices and photovoltaic generation. *Appl. Soft Comput.* **2018**, *62*, 504–513. [CrossRef]

25. Asgher, U.; Rasheed, B.; Saad, A.; Rahman, A.; Alamri, A. Smart Energy Optimization Using Heuristic Algorithm in Smart Grid with Integration of Solar Energy Sources. *Energies* **2018**, *11*, 3494. [CrossRef]

26. Bharathi, C.; Rekha, D.; Vijayakumar, V. Genetic Algorithm Based Demand Side Management for Smart Grid. *Wirel. Pers. Commun.* **2017**, *93*, 481–502. [CrossRef]

27. Shi, K.; Li, D.; Gong, T.; Dong, M.; Gong, F.; Sun, Y. Smart Community Energy Cost Optimization Taking User Comfort Level and Renewable Energy Consumption Rate into Consideration. *Processes* **2019**, *7*, 63. [CrossRef]

28. Liu, Y.; Yuen, C.; Huang, S. Peak-to-Average Ratio Constrained Demand-Side Management with Consumer's Preference in Residential Smart Grid. *IEEE J. Sel. Top. Signal Process.* **2014**, *8*, 1084–1097. [CrossRef]

29. Yang, X.S. Firefly algorithms for multimodal optimization, in Stochastic Algorithms: Foundations and Applications. *Lect. Notes Comput. Sci.* **2009**, *5792*, 169–178.

30. Saremi, S.; Mirjalili, S.; Lewis, A. Grasshopper optimization algorithm: Theory and application. *Adv. Eng. Softw.* **2017**, *105*, 30–47. [CrossRef]

31. Simpson, S.J.; McCaffery, A.; HAeGELE, B.F. A behavioural analysis of phase change in the desert locust. *Biol. Rev.* **1999**, *74*, 461–480. [CrossRef]

32. Rogers, S.M.; Matheson, T.; Despland, E.; Dodgson, T.; Burrows, M.; Simpson, S.J. Mechanistically-induced behavioural gregarization in the desert locust *Schistocerca gregaria*. *J. Exp. Biol.* **2003**, *206*, 3991–4002. [CrossRef] [PubMed]

33. Topaz, C.M.; Bernoff, A.J.; Logan, S.; Toolson, W. A model for rolling swarms of locusts. *Eur. Phys. J. Spec. Top.* **2008**, *157*, 93–109. [CrossRef]

34. Passino, K.M. Biomimicry of bacterial foraging for distributed optimization and control. *IEEE Control Syst.* **2002**, *22*, 52–67.

35. Day-Ahead Pricing (DAP). NYISO (New York Independent System Operator). Available online: http://www.energyonline.com/Data/GenericData.aspx?DataId=11&NYISO-Day-Ahead-Energy-Price (accessed on 25 January 2019).

Article

An Integrated Delphi-AHP and Fuzzy TOPSIS Approach toward Ranking and Selection of Renewable Energy Resources in Pakistan

Yasir Ahmed Solangi [1,*], Qingmei Tan [1], Nayyar Hussain Mirjat [2], Gordhan Das Valasai [3], Muhammad Waris Ali Khan [4] and Muhammad Ikram [1]

[1] College of Economics and Management, Nanjing University of Aeronautics and Astronautics, Nanjing 211106, China; tanchangchina@sina.com (Q.T.); mikram@nuaa.edu.cn (M.I.)

[2] Department of Electrical Engineering, Mehran University of Engineering and Technology, Jamshoro 76062, Pakistan; nayyar.hussain@faculty.muet.edu.pk

[3] Department of Mechanical Engineering, Quaid-e-Awam University of Engineering, Science and Technology, Nawabshah 67480, Pakistan; valasai@gmail.com

[4] Faculty of Industrial Management, University Malaysia, Pahang 26300, Malaysia; waris@ump.edu.my

* Correspondence: yasir.solangi86@hotmail.com; Tel.: +86-186-5185-2672

Received: 31 December 2018; Accepted: 15 February 2019; Published: 25 February 2019

Abstract: Pakistan has long relied on fossil fuels for electricity generation. This is despite the fact that the country is blessed with enormous renewable energy (RE) resources, which can significantly diversify the fuel mix for electricity generation. In this study, various renewable resources of Pakistan—solar, hydro, biomass, wind, and geothermal energy—are analyzed by using an integrated Delphi-analytical hierarchy process (AHP) and fuzzy technique for order of preference by similarity to ideal solution (F-TOPSIS)-based methodology. In the first phase, the Delphi method was employed to define and select the most important criteria for the selection of RE resources. This process identified four main criteria, i.e., economic, environmental, technical, and socio-political aspects, which are further supplemented by 20 sub-criteria. AHP is later used to obtain the weights of each criterion and the sub-criteria of the decision model. The results of this study reveal wind energy as the most feasible RE resource for electricity generation followed by hydropower, solar, biomass, and geothermal energy. The sensitivity analysis of the decision model results shows that the results of this study are significant, reliable, and robust. The study provides important insights related to the prioritizing of RE resources for electricity generation and can be used to undertake policy decisions toward sustainable energy planning in Pakistan.

Keywords: Delphi; analytical hierarchy process; fuzzy technique for order of preference by similarity to ideal solution techniques; renewable energy (RE) resources; sustainable energy planning

---

1. Introduction

Energy is one of the key drivers for sustainable growth and economy of any country. In fact, in this era, the measure of the development of any economy is synonymous with the level of energy consumption. Energy, which is crucial in economic, environmental, technical, and socio-political aspects, has become one of the most discussed issues in recent times. Industrialization and technological developments have created a higher energy need worldwide [1]. However, the quantity of reserves of fossil fuels differs from one country to another. This situation has, therefore, resulted in an unavoidable economic dependency, major environmental concerns, technological issues, important social consequences, and serious political conflicts [2]. The existing situation and the future estimations for energy requirements make it crucial to explore alternate energy resources. Further, the current and future possible economic, environmental, technical, and socio-political consequences also push countries toward renewable energy (RE) resources. In this scenario, RE has become the answer to sustainable energy planning.

Pakistan is a developing country with striving economic progress and thus essentially requires a sufficient amount of energy for meeting the growth targets and attaining sustainable development. However, in the last two decades, the country has been coping with its worst ever energy crises, with regular power interruptions that have deteriorated the economy, thus adversely impacting the livelihood of people [3]. The ongoing energy crises in the country have paralyzed the economy, and the circular debt of the power sector alone has crossed over Pak Rupees 922 billion [4]. It is also unfortunate that only around 50% of the country population has access to on-grid electricity. Load shedding of 16–18 h in the rural areas is a common phenomenon while in the urban areas, the electricity is inaccessible for 10–12 h a day [5]. This huge shortfall of electricity is mainly owing to Pakistan's reliance on fossil fuels, which are a key source of huge import bills and expensive electricity. Apart from these issues, there are also various governance-related problems and hurdles behind these crises. Amongst these all, a lack of focus toward harnessing the indigenous RE resources is a noteworthy shortcoming that has not received any major attention in the planning and development processes in Pakistan. In order to eliminate energy shortfalls, it is required that abundantly available renewable resources be harnessed effectively [6]. Pakistan is fortunate to have various RE resources that include solar, hydro, biomass, wind, and geothermal energy [7]. All these RE resources have huge potential to generate electricity as well as help to eradicate energy deficits and enhance sustainable development of the country.

The total electricity generation capacity of Pakistan in 2017 has been reached to 29,944 MW [8]. Shares in electricity generation are comprised of natural gas, 33.6%, oil, 32.1%, coal, 0.2%, hydropower, 26.1%, nuclear, 5.7%, and renewable energy, only 2.2%. In the total energy mix of Pakistan, the share of non-renewable is, as such, higher and needs to be reduced in order to ensure long-term sustainability and energy security. In the wake of the global focus on reducing greenhouse gas (GHG) emissions and thus enhanced effort for the deployment of RE-based projects for electricity generation, Pakistan also needs to undertake essential measures toward development and completion of RE-based power projects. However, so far, no significant accomplishment toward harnessing RE-based projects has been witnessed in the country. In this context, serious and extensive energy planning and decision-making efforts are required to exploit RE resources for electricity generation.

RE resources are not only capable of meeting the ever-increasing demand for electricity, but they are also environment-friendly. These facts regarding RE resources are recognized globally, but Pakistan, though blessed with enormous RE potential, is making extremely slow progress in realizing the true potential of RE-based projects. In this context, the Alternative Energy Development Board (AEDB) and the Pakistan Council of Renewable Energy Technologies (PCRET) are two key organizations of the government undertaking RE projects and technology development activities, respectively [9]. However, the progress of these organizations is very slow and only some small RE-based projects have been installed in the country. The poor level of commitment from government, the overlapping management functions of the energy sector, and the lack of financial capacity and technical awareness are key barriers toward developing RE-based projects in Pakistan. Likewise, Usama et al. [10] have identified various key barriers that obstruct successful implementation of social sustainability practices in manufacturing firms using interpretive structural modeling (ISM).

It is, therefore, important that, for sustainable development, short-term, middle-term, and long-term energy planning consider various energy resource alternatives. As such, the various RE resources need to be evaluated in a systemic way, and they must be considered from the techno-economic and socio-political point of view. In this context, the aim of this study is to systematically prioritize the RE resources of Pakistan for sustainable energy planning. According to the authors, this is the very first attempt to propose and develop an integrated Delphi-analytical hierarchy process (AHP) and fuzzy technique for order of preference by similarity to ideal solution

(F-TOPSIS)-based methodology to undertake systematic prioritization of RE resources of Pakistan. This effort is expected to inspire policy and decision makers to consider a systematic planning process for resource selection to expedite the development of RE-based projects.

The rest of the paper is as follows: Section 2 presents related literature applied in the energy sector, while Section 3 provides a detailed analysis of various RE resources of Pakistan. The analysis-based proposed integrated decision framework is shown in Section 4. The results and relevant discussion are contained in Section 5, and Section 6 provides conclusions and recommendations derived from this study.

2. Related Literature

There are various energy planning-related studies where Delphi and Multi-Criteria Decision Making (MCDM) approaches have been comprehensively used with varying aims, objective, and specific criteria. These applications have considered energy planning and policymaking at different levels, risks assessments of long-term energy plans, the selection of the best RE technologies, energy scenario analysis, environmental concerns related to the energy sector, energy management problems, and the selection of power plants. The Delphi and F-TOPSIS methodology has been used to rank flood vulnerability and vulnerability characteristics in the South Han River basin in South Korea [11]. Some authors [12–14] have stated that MCDM approaches are well-suited to address strategic decision-making problems. MCDM methods provide a systemic and transparent way to enclose multiple conflicting objectives. MCDM based on multi-attribute value functions is often employed to support energy planning and policy, and to select and prioritize suitable alternatives. Wang et al. [15] in a detailed review provided various applications of MCDM methods in sustainable energy planning and concluded that AHP is the most prevailing and popular method.

It is apparent from the literature that MCDM methods are often used and are popular for decision making in sustainable energy planning, and they greatly help in addressing important criteria [15]. Amer and Daim [16] used the AHP method to suggest the optimal RE resource in Pakistan. In this study, authors ranked biomass as the best alternative for electricity generation; however, biomass only has a 5000 MW potential, which is not sufficient to meet the increasing energy demand of the country [17]. Another limitation of this study is that the authors used AHP to obtain the weights of the RE alternatives; however, literature has suggested that the combination or integration of various MCDM methods for one goal provide more refined and better results. Furthermore, Table 1 illustrates the summary of various energy-related studies from multi-criteria approaches.

Table 1. The summary of various energy related studies with multi-criteria perspectives.

No.	Focus	Method	Year	Reference
1	Multi-criteria decision making for plant location selection	Delphi-AHP and Preference Ranking Organization Method for Enrichment of Evaluations (PROMETHEE)	2013	[18]
2	Developing offshore wind farm siting criteria	Delphi	2018	[19]
3	Portfolio of renewable energy sources for achieving the 3-E policy goals in Taiwan	AHP	2011	[20]
4	An analysis on barriers to renewable energy development in Nepal	AHP	2018	[21]
5	Potential survey of photovoltaic power plants	AHP	2017	[22]
6	Supplier evaluation and selection in the gas and oil industry	Supply Chain Operations Reference (SCOR) metrics-AHP and TOPSIS	2018	[23]
7	Selection of wind power project location in the southeastern part of Pakistan	Factor analysis-AHP and Fuzzy TOPSIS	2018	[24]
8	Dam site selection in Iran	AHP-TOPSIS	2018	[25]
9	Assessing energy management performance in Turkey	AHP-TOPSIS and VlseKriterijuska Optimizacija I Komoromisno Resenje (VIKOR)	2018	[26]

Table 1. *Cont.*

No.	Focus	Method	Year	Reference
10	Considering the public opinion and geospatial conditions to distinguish energy alternatives for energy investment planning	Goal programming, AHP and F-TOPSIS	2018	[27]
11	Risk assessment and mitigation model for overseas steel-plant project investment	AHP—Fuzzy Inference System	2018	[28]
12	Evaluating water resource management strategies	Multiple Attribute Utility Theory (MAUT)-AHP- Elimination Et Choice Translating Reality (ELECTRE) and TOPSIS	2018	[29]
13	Sustainable assessment of economy-based and community-based urban regeneration	AHP	2018	[30]
14	The policy scenario analysis for accomplishing renewable energy sources targets	Fuzzy TOPSIS	2017	[31]
15	Identifying the most significant low-emission energy technologies development in Poland	Fuzzy AHP-Fuzzy TOPSIS	2018	[32]
16	Assessing the energy planning in Turkey	Analytic Network Process (ANP)-Fuzzy TOPSIS	2018	[33]
17	Strategic selection of renewable energy source for Turkey	Hesitant Fuzzy Linguistic (HFL)- Simple Additive Weighting (SAW) and HFL-TOPSIS	2018	[34]
18	Evaluation and prioritization of renewable energy alternatives	HFL-TOPSIS and Interval type-2 Fuzzy AHP	2017	[35]
19	Comparative analysis for optimum paper shredder selection	AHP– Graph Theory and Matrix Approach (GTMA) and AHP–TOPSIS	2018	[36]
20	A SWOT framework for analyzing the electricity supply chain	Strengths, Weaknesses, Opportunities and Threats (SWOT)-AHP and Fuzzy TOPSIS	2015	[37]
21	Selection of the best energy technology alternative in energy planning	Modified Fuzzy TOPSIS	2011	[38]
22	Identifying the barriers to renewable energy development in Pakistan	----	2009	[39]
23	Sustainable development through energy management	----	2014	[40]

In light of the above, this study attempts to contribute to the contemporary literature by proposing an integrated Delphi-AHP and F-TOPSIS methodology for RE resource prioritization. The Delphi approach translates Delphi qualitative assessments about the importance of the criteria into constraints on the weights that are exploited through the AHP approach. Further, the F-TOPSIS method has been employed to finally rank the RE alternatives. The following section described the detailed analysis of various RE resources of Pakistan.

3. Renewable Energy Potential of Pakistan

Pakistan is blessed with various RE resources such as solar, hydro, biomass, wind, and geothermal energy, with significant potential to produce electricity. However, there has been extremely slow growth in harnessing these resources, so they form a mere share in the overall energy mix of the country. The RE policy of 2006 aims at adding 10,000 MW of electricity from RE resources by 2030 [41]. However, the estimated potential of RE is far greater than 10,000 MW. With growing demand, the maximum potential for RE needs to be tapped. Table 2 provides a summary of the estimated potential of each RE resource and installed electricity generation capacity from these resources in Pakistan.

It is, therefore, required that government undertake extraordinary measures to explore, develop, and establish sustainable energy sources at the regional and national level to overcome the current energy crisis. In the following sub-sections, detailed analysis of five RE resources of Pakistan for electricity generation is provided.

Table 2. Estimated renewable energy potential of various source and installed capacity in Pakistan [42–44]. RE: renewable energy.

RE Resource	Potential (MW)	Installed (MW)
Solar	2,900,000	200
Hydro	60,000	6556
Biomass	5000	35
Wind	346,000	308
Geothermal	100,000	0

3.1. Solar Energy

Pakistan has plentiful solar energy throughout the year and across the country [9]. The solar energy potential of Pakistan is estimated to be 2,900,000 MW, which can be exploited extensively to meet the energy demand. Geographically, Pakistan receives the highest solar radiation in the region with more than 300 sunlight days with around 1800–2200 kWh/m^2 of annual radiation at a 26–28 °C average annual temperature, which can produce an electricity of 5.5–6 kWh/m^2/day [45]. As such, the exploitable solar resources are estimated to be greater than 50,000 MW, with more than 2500 h of sunlight in a year. There is an excellent potential for deploying solar energy projects in Baluchistan and Sindh, where the sun shines for 7–8 h a day, approximately 2300–2700 h/annum [9]. A solar map of Pakistan for direct normal radiation is depicted in Figure 1.

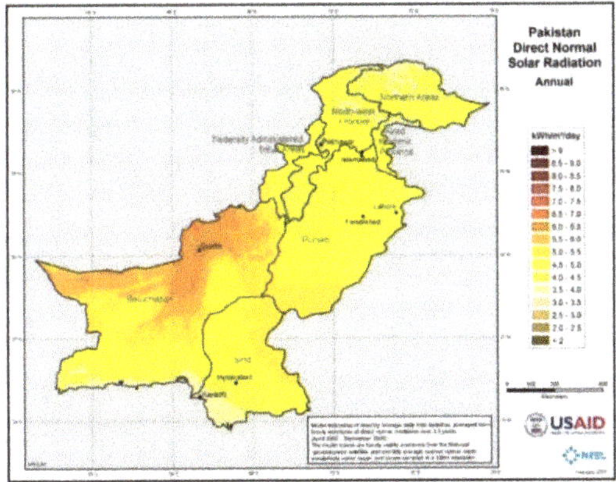

Figure 1. Solar map of Pakistan [46].

Unfortunately, due to a lack of interest and commitment of concerned government authorities, the development of RE resources including solar energy for generating electricity is at very early stages despite outstanding geographical conditions. In the meantime, with the penetration of solar-based technology in the market, various electricity consumers in both rural and urban areas have installed standalone photovoltaic units of 100–500 W for power generation. However, these individual efforts can be short-lived with an operational maintenance requirement, the availability of spares, and other challenges. With the potential of solar energy duly taken into consideration, and with an annual mean sunlight duration of 8–8.5 h a day, it is projected that around 40,000 villages in the country can be provided with electricity [47].

The only significant effort by government related to the harnessing of solar energy is the development of the Quaid-e-Azam Solar Power Park underway in the Bahawalpur district of Punjab. At completion, the total installed capacity of this solar project will be 1000 MW [48]. It is evident from this analysis that the share of solar energy for electricity generation is negligible and requires colossal efforts for the development of the solar energy sector to ensure sustainable supplies to address the demand–supply gap and ensure energy security.

3.2. Hydropower

Pakistan has an enormous resource potential of hydropower, with a suitable amount of water at appropriate terrains to generate electricity [49]. The northern parts of the country are rich with significant hydropower resources, while few resources are also identified in the southern part of the country. In all, it is estimated that these resources have a potential of 60,000 MW of electricity generation [50]. As such, Pakistan can sufficiently produce electricity from hydro resources, with only a mere potential exploited so far. About 89% of the potential is yet to be harnessed. The current total installed capacity of the hydropower units of Pakistan is only 6556 MW [42].

The key hydropower resources identified in the terrain of Hindukush, Himalayas, and the Karakoram ranges include flows from various rivers, namely Indus with a potential of 66% of electricity generation followed by Jhelum, 9%, Swat, 3%, Kunhar, 3%, Kandiah, 2%, Punch, 1%, and others, 16% [51]. Figure 2 illustrates the total estimated shares of identified hydro resources of major rivers of Pakistan.

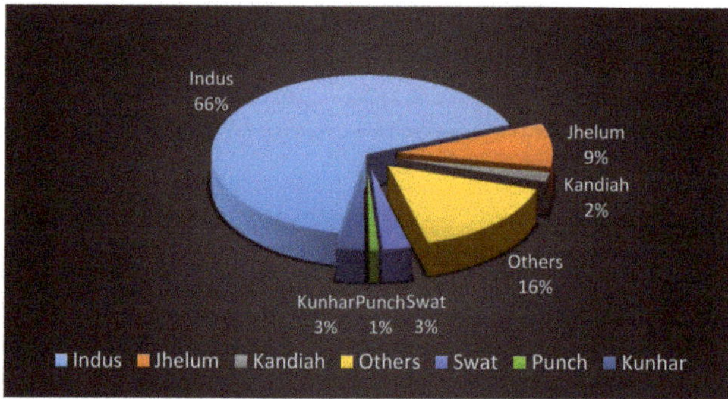

Figure 2. Identified hydro resources of Pakistan [51].

At present, hydropower is the most economically viable, environmentally friendly, and cheapest source of electricity in Pakistan. It contributes 26% of the total power generation in Pakistan. The Government of Pakistan (GoP) plans to add 13 hydro projects with a capacity of 20,733 MW by 2023, whereas six projects sites have been identified with the capacity and feasibility of 8650 MW of electricity [50]. Hydropower is one of the oldest and most mature RE resources in the world. Therefore, considering the huge hydropower potential, the GoP needs to give it top priority and undertake the development of hydro-based projects in the country.

3.3. Biomass Energy

Biomass is a sustainable RE resource and is widely available [52]. Pakistan is an agricultural country and is the 5th largest sugarcane producer in the world [48]. The average 50 million tons of sugar yields provides an estimated 10 million tons of bagasse annually from sugar mills. The AEDB has identified the bagasse potential of 1800 MW and the waste-to-power potential of 500 MW with the help of Germany, USA, and Denmark [49,53]. As such, the AEDB has begun to install biomass-to-energy power plants of 12 MW, 11 MW, and 9 MW in Punjab and Sindh. However, these initiatives are not sufficient compared to the significant potential of biomass energy of the country.

With around 62% of the population of country living in rural areas, it is estimated that, overall, including bagasse, 5000 MW potential of biomass is available in the country. Both agriculture and animal wastes are readily available in rural areas and are a cheap source that can be used for cooking and heating. Biomass energy potential is a promising source of energy seeking the greatest ever attention. Despite this huge potential of biomass in the country, no notable grid-connected power generation project has been developed yet. It is therefore high time that government takes care of significant biomass resource toward electricity generation.

3.4. Wind Energy

Electricity generation from wind energy has grown remarkably over the past decade; as such, wind energy is now the second leading source of RE for providing electricity globally. With its promising potential, energy experts believe that wind energy will provide one-third of total global electricity supplies by the year 2050. Countries like Germany, India, and Brazil have been leading wind energy development over the last few years [54]. China is, however, far ahead in the global wind power market with a total cumulative installed capacity of 188,232 MW followed by USA and Germany with 89,077 and 56,132 MW, respectively [55]. Figure 3 highlights the cumulative wind power installed capacity of the top 10 countries and the rest of the world in 2017.

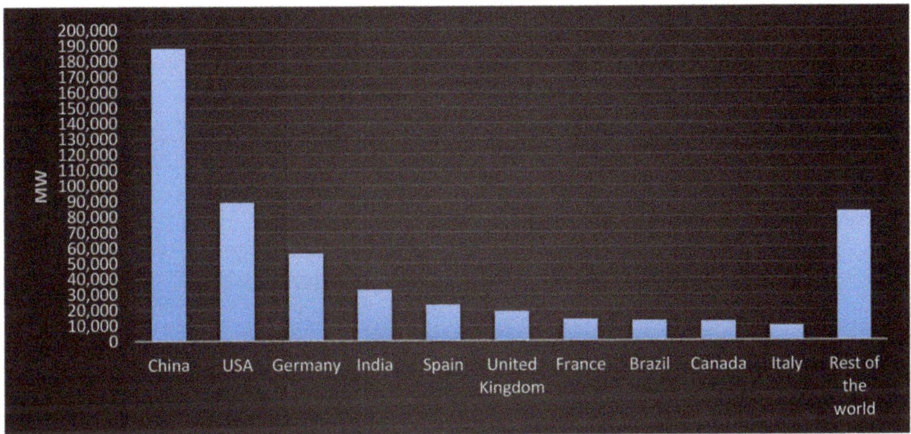

Figure 3. Global wind power capacity in MW [55].

3.4.1. Wind Energy Resources in Pakistan

Pakistan also has abundant wind energy potential for electricity generation. Wind projects, mainly comprised of 50–100 MW units in a wind corridor in the Sindh province, with a cumulative 500 W capacity were installed by 2003 [56]. PCRET also installed 26 micro units of wind energy, each of 500 W, in the village of Gul Muhammad, and it is stated to be the first wind energy-electrified village of Pakistan. It is estimated that Pakistan is capable of producing about 346,000 MW of wind power [24]. However, the first major wind power project of Pakistan only became functional in 2013. According to a survey, 50,000 MW of wind power potential has been identified in the southern regions of the country alone, i.e., in Sindh and Baluchistan, whereas a wind power capacity of 1000–1500 MW has been estimated in the Punjab province. The Pakistan Meteorological Department (PMD) surveyed the wind speed in the coastal areas of Sindh and Baluchistan and found a persistent wind speed of 5–7 m/s [57]. The PMD has also measured wind in two regions in Sindh, namely Gharo and Keti Bandar. Following a year-round survey and data collection, these sites were found to be ideal for wind power projects. The annual mean wind speed in Gharo was estimated to be 6.86 m/s from the above-ground level of 50 m, while 408.6 W/m^2 is the annual power density of the area. These figures show that this area has good wind potential and is economically feasible for large wind farms. The wind power potential of Pakistan is shown in Figure 4.

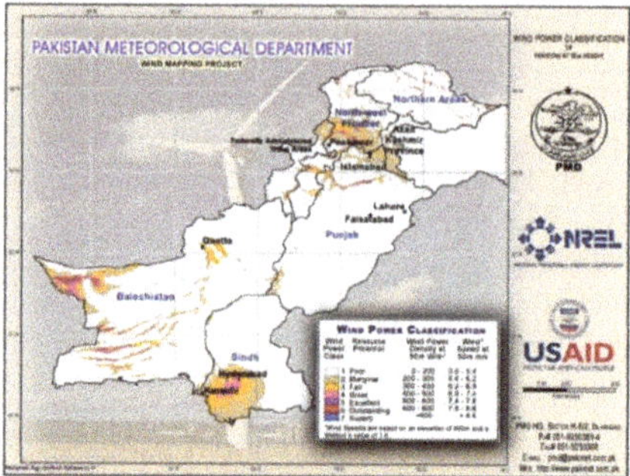

Figure 4. Wind resource map of Pakistan [46].

3.5. Geothermal Energy

Pakistan has an enormous potential of geothermal energy, especially in KPK, Baluchistan, Kashmir, and Himalayas, with temperatures estimated around 30–170 °C [45]. Geothermal energy is a type of heat energy that is present inside the surface of the earth in the form of volcanoes, hot springs, and hot water. Further, it is identified that Pakistan can produce up to 100,000 MW of electricity from geothermal resources [44]. However, no effort has been made by government to utilize geothermal energy, so substantial investment and planning is required for implementation of geothermal energy projects in Pakistan. This renewable resource can also be useful for space heating and cooling in buildings, greenhouses, hot water supply, fish farming, and establishing small industries requiring heat. A geothermal resource potential map is presented in Figure 5.

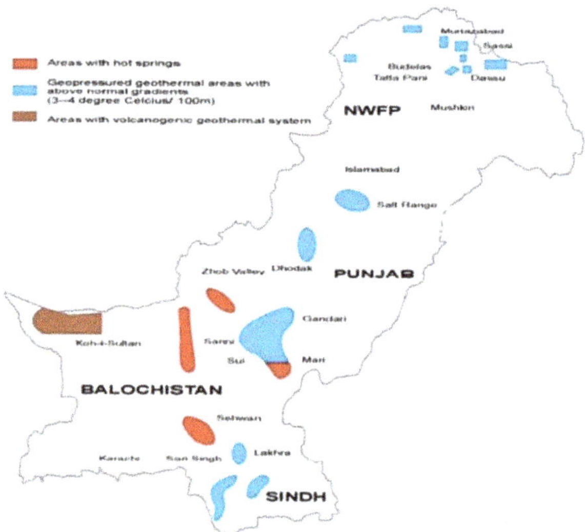

Figure 5. Geothermal resource map of Pakistan [58].

It is evident from the above analysis that Pakistan essentially needs to seriously plan and make the required policy decisions for the development of indigenous RE resources. However, it is understood that evaluation, selection, and prioritization of these resources is a complex process with divided opinions among policy-makers, and stakeholders may have a different view as well. This is common in most developing countries, where limited financial resources are major constraints in developing RE resources. As such, for the case of Pakistan we propose and develop a multi-criteria decision model taking into account various criteria, as well as experts' weights of these criteria, and subsequently prioritize the RE resources of the country.

4. An Integrated Decision Framework

The proposed integrated decision framework of the present study is provided in Figure 6. In the first instant, using the Delphi method, key criteria are identified; these criteria and sub-criteria are then determined using AHP methodology. Finally, the F-TOPSIS method is employed to prioritize the alternatives (i.e., renewable energy resources).

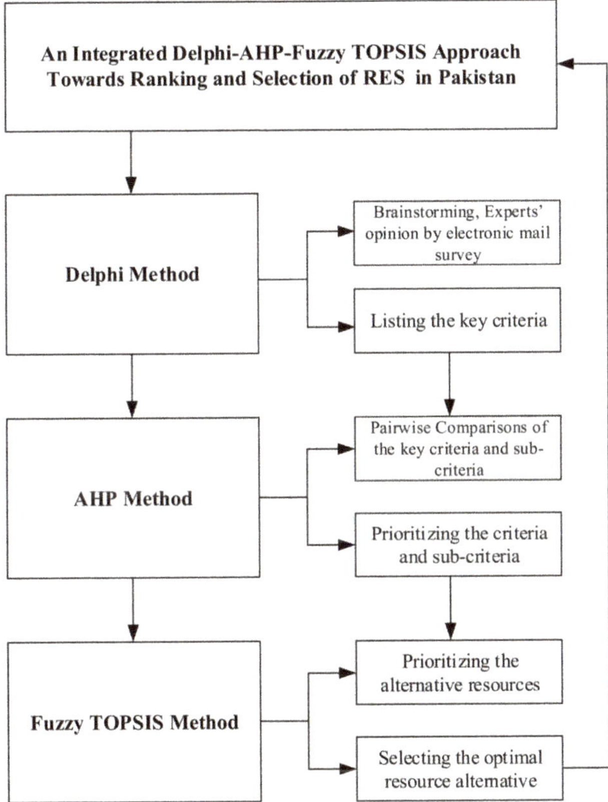

Figure 6. Decision model of the study.

The detailed elaboration of each module of the proposed decision model is provided in the following sub-sections.

4.1. The Delphi Method

The Delphi methodology was originally developed by the RAND Corporation in the 1950s for eliciting expert opinion [59]. This research method is a systematic and interactive method to collect and analyze experts' judgment or opinion by collecting data, brainstorming for problems, prioritizing the issues, forecasting, and decision-making [60]. It is a very useful method when there is a lack of clarity with decision makers and stakeholders regarding certain decisions.

In the precedent literature, the experts' group size seems to be different in each study, but a panel of experts between 9 and 18 participants is suggested in order to obtain pertinent results and avoid any disagreement among experts [61]. Skulmoski et al. [62] recommended that the participants possess knowledge of different aspects of decision problems in order to be designated for the Delphi survey panel. Relevant experience and knowledge about the survey, time capacity, willingness to participate, and strong communication skills are also desired. The Delphi approach is at the first stage of our decision framework, which includes review of the literature and a survey from experts to finalize appropriate criteria toward addressing the decision problem of prioritizing RE resources. The first round is exploratory in nature. Following a comprehensive literature review, key criteria influencing RE resource prioritization such as economic, environmental, technical, and socio-political considerations have been explored. This followed the questionnaire survey process, which was conducted through electronic mail sent to academia, stakeholders, industry, and government energy experts, in order to save expenses and time. In the study, we evaluated the coefficient of variation (CV) and content validity ratio (CVR) of the survey. When the CV values were less than 0.50, a further round for evaluating criteria was not undertaken [63]. Further, the CVR proposed by C.H. Lawshe [64] and its calculation suggested by Wilson et al. [65] were employed to measure agreement among experts, determining the importance of specific criteria. The CVR ranges from +1 to −1. A greater, positive value indicates that experts were in agreement that a criterion was important. A CVR higher than 0.29 can usually be regarded as a suitable assessment level. The CV is the ratio of the standard deviation to the mean. It is easy to compare the consistency of the overall obtained data using the CV, as presented in Equation (1).

$$CVR = \frac{NE - \frac{N}{2}}{\frac{N}{2}} \tag{1}$$

where NE is the number of experts representing assessed criteria, which is "important", and N = the total number of experts.

4.2. Analysis of RE Resources Using MCDM Techniques

Evaluation and selection decisions pertaining to prioritization of RE resources for long-term development is a complex process. This is mainly because the nature of decision problems are multi-faceted and owing to various constraints and limitations [66]. In this context, multi-criteria analysis of such complex problems is useful and appropriate for technology choice, considering long-term energy planning, resource potential, acquisition and deployment of renewables, and the uncertainty of future energy demand. Therefore, AHP and F-TOPSIS methodologies of MCDM are used in this study to analyze the decision problem with economic, environmental, technical, and socio-political aspects as main criteria, and solar, hydro, biomass, wind, and geothermal energy as alternatives in the decision model. It is anticipated that this approach will provide an improved mechanism for decision making in the RE sector compared to traditional assessment methods such as cost–benefit or techno-economic analysis. No study in the Pakistani context has evaluated RE resources based on an integrated AHP and F-TOPSIS framework. However, there are a few studies in the literature wherein decision evaluation has been undertaken using limited criteria and minimum sub-criteria.

4.2.1. Analytical Hierarchy Process

Various MCDM methods are often used and are popular in energy planning decision-making [67]. AHP is one of the most widely used MCDM methods in this context. It provides a means of decomposing a complex problem into a hierarchy of sub-problems that are further subjectively evaluated. The subjective assessments are later converted into numerical form and are arranged to rank each alternative on a numerical scale. Thomas L. Saaty developed the AHP methodology in the 1970s and is accomplished using the following steps [68].

Step 1. The decision problem is divided into different levels in a hierarchical manner comprised of goal, criteria, and sub-criteria. The elements at one level are related to those at other levels, and a hierarchical relationship is established between them.

Step 2. Corresponding to the hierarchical structure, data is collected from the decision makers based on a pairwise comparison of criteria on a numerical scale, which is illustrated in Figure 7.

Step 3. The various criteria are compared to determine the relative importance via the principal eigenvalue and the corresponding normalized eigenvector of the comparison matrix. The elements of the normalized eigenvector are then named as weights with respect to criteria and sub-criteria.

Step 4. The matrix consistency of order n is assessed. The matrix comparison undertaken in this method is subjective, and inconsistency is tolerated by AHP through redundancy in the approach. The consistency index (CI) must be within the required level if it fails, and the comparison may be repeated. The CI is calculated as

$$CI = \frac{(\lambda\max - n)}{(n-1)} \tag{2}$$

where λ_{max} is the maximum eigenvalue of the judgment matrix, and n is the number of elements in the judgement. The CI can be compared with a random consistency index (RI). The consistency ratio (CR) is calculated as

$$CR = \frac{CI}{RI} \tag{3}$$

where RI is the random consistency index. The average CI of a randomly generated pairwise comparison matrix of similar size is illustrated in Table 3. Saaty suggests that the value of CR should be less than 0.1, while meaningless results may be found for a value of more than 0.1 [69].

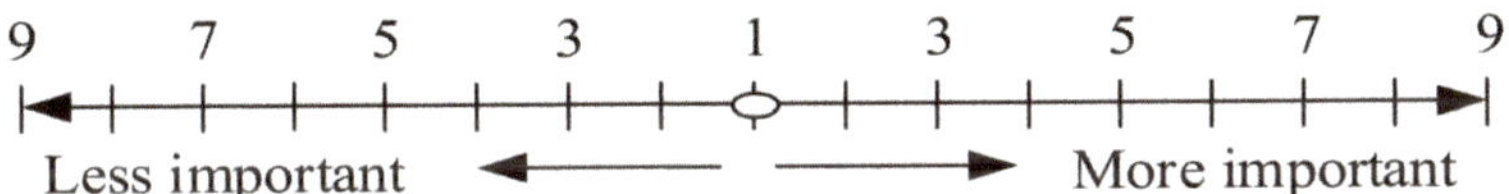

Figure 7. Pair-wise comparison matrix scale [70].

Table 3. Random index (RI) scale [71].

n	1	2	3	4	5	6	7	8	9	10
Random Index (RI)	0.00	0.00	0.058	0.90	1.12	1.24	1.32	1.41	1.45	1.49

Following implementation of the AHP methodology as per these steps, the F-TOPSIS method has been employed to finally evaluate and prioritize the best RE alternative.

4.2.2. F-TOPSIS

The TOPSIS technique was developed by Hwang and Yoon in 1981 [72]. It is based on the distance between positive and negative solutions i.e. the best alternative should be closest to the positive ideal solution, whereas the least favorable alternative should be farthest from the negative ideal solution. The fuzzy logic or fuzzy set theory is a powerful mathematical technique used to address uncertain and imprecise information in decision problems [73]. Fuzzy logic was developed by Lofti A. Zadeh in 1965 [74]. When fuzzy logic is combined with TOPSIS, it provides additional decision support, and this combined methodology is known as fuzzy TOPSIS (F-TOPSIS). This method can be based on linguistic variables with triangular fuzzy numbers (TFNs); thus, in this study, a TFN scale was utilized and is shown in Table 4.

Table 4. Linguistic scale for alternatives ranking [75].

Number	Linguistic Variable	TFNs
1	Very poor (VP)	(0,0,1)
2	Poor (P)	(0,1,3)
3	Rather Poor (RP)	(1,3,5)
4	Fair (F)	(3,5,7)
5	Rather Good (RG)	(5,7,9)
6	Good (G)	(7,9,10)
7	Very Good (VG)	(9,9,10)

The detailed F-TOPSIS methodology implementation steps are described as follows:

Step 1. Obtain the evaluation matrix of decision-makers.

Step 2. Define the fuzzy decision matrix $\widetilde{W}$.

$$\widetilde{W} = (w_{ij})_{m \times n} \tag{4}$$

where $w_{ij} = (w_{1ij}, w_{2ij}, w_{3ij})$.

Step 3. Compute the normalized fuzzy decision matrix, indicated by $\widetilde{W}$, is shown as

$$\widetilde{W} = [w_{ij}]_{m \times n} \tag{5}$$

where $i = 1, 2, 3, \ldots, m$ and $j = 1, 2, 3, \ldots, n$.

For the benefit criteria, conduct the normalization process using Equation (6).

$$w_{ij} = \left(\frac{w_{1ij}}{w_{3j}^*}, \frac{w_{2ij}}{w_{3j}^*}, \frac{w_{3ij}}{w_{3j}^*} \right). \tag{6}$$

For cost criteria, conduct the normalization process using Equation (7).

$$w_{ij} = \left(\frac{w_{1j}^-}{w_{3ij}}, \frac{w_{1j}^-}{w_{2ij}}, \frac{w_{1j}^-}{w_{1ij}} \right). \tag{7}$$

Step 5. Calculate the weighted normalized fuzzy decision matrix. The weighted normalized fuzzy decision matrix is presented using Equation (8).

$$\widetilde{V} = [v_{ij}]_{m \times n} \tag{8}$$

where $v_{ij} = w_{ij} \times w_j$

Step 6. Identify the distance between ideal positive solution (d_i^+) and negative ideal solution (d_i^-) using Equations (9) and (10).

$$d_i^+ = (v_1^*, v_2^*, v_3^*, \ldots, v_n^*) \tag{9}$$

where $V_j^+ = (1, 1, 1)\ j = 1, 2, 3, \ldots, n$.

$$d_i^- = (v_1^-, v_2^-, v_3^-, \ldots, v_n^-) \tag{10}$$

where $V_j^- = (0, 0, 0)\ j = 1, 2, 3, \ldots, n$.

Step 7. Calculate the closeness coefficient (CC_i) using Equation (11).

$$CC_i = \frac{d_i^-}{d_i^+ + d_i^-} \tag{11}$$

where $i = 1, 2, 3, \ldots, m$.

Step 8. Evaluate and prioritize the optimal alternative according to the CC_i value.

We have taken into account both cost and benefit criteria. Resource potential (ECA$_3$), job creation (SPA$_2$), and energy security (SPA$_3$) are considered as benefit criteria, while the rest are taken as cost criteria. All these criteria play a crucial role in assessing and prioritizing RE resources for electricity generation.

4.3. The Survey Respondents for the Delphi, AHP, and Fuzzy TOPSIS Study

It is critical to finalize and consult with specialized experts while implementing the Delphi, AHP, and F-TOPSIS methodologies since the understanding and relevancy of experts in assigning weights could be quite conflicting and uncertain [76]. In order to achieve the objective of this study, we approached 15 experts from academic institutions, government energy departments, stakeholders, and industries. However, out of these 15 only 10 experts agreed to participate in these study surveys. The demographic information of experts is given in Appendix A. YAAHP software (Version 10.5) was used to obtain the weights of main criteria and sub-criteria. Subsequently, the F-TOPSIS method was employed with Microsoft Excel to analyze and rank RE alternatives for electricity generation in Pakistan.

4.4. The Process of Delphi, AHP, and Fuzzy TOPSIS Methodology Implementation

Initially, the Delphi method was employed with the help of 10 experts' feedback to identify the main criteria and sub-criteria for evaluating RE resources of Pakistan. From the Delphi analysis, the authors shortlisted four main criteria and their 20 sub-criteria for further analysis. Secondly, AHP and F-TOPSIS methods of MCDM were employed to obtain the weights and rank the criteria, sub-criteria, and alternatives. The AHP method has the strength to analyze quantitative and qualitative factors altogether in one model. Therefore, expert assessment was employed at the AHP step to evaluate the four criteria and 20 sub-criteria, while the F-TOPSIS approach was employed to prioritize the five RE alternatives of the decision model.

4.4.1. Data Analysis of RE Resources

In this study, alongside wind, geothermal, and biomass resources, mini-hydropower and solar PV are considered as alternatives, while large hydropower and solar thermal are excluded from the study for being cost-intensive and technologically complex. Table 5 provides the key quantitative data related to RE resources that include average initial cost, O&M cost, efficiency, capacity factor, expected life of the RE plants, and CO$_2$ emissions avoided per year. Table 6 provides the information pertaining to the job creation, land requirements, and power generation cost of RE-based plants in the USA. It is pertinent to mention that the data of Table 6 is taken from a developed country where labor and other associated costs are quite expensive. It is, therefore, assumed that implementing RE-based plants in Pakistan will create more jobs than the USA. Moreover, for the land requirement, areas requiring a low amount of land are generally preferred for developing RE-based plants. It is also noted that in the case of wind energy plants, wind farms can also be used for other activities such as farming and cattle cropping. As such, wind farms can be far more useful compared to other RE resources where land cannot be utilized for other purposes.

Table 5. RE data of various RE resources of Pakistan [77,78].

RE Source	Initial Cost (m USD/kW)	O&M Cost (m USD/Year)	Efficiency (%)	Capacity Factor (%)	Expected Life of RE Plant	CO$_2$ (m Tons Avoided/Year)
Solar	570	57	80	25	25	0.16
Hydro	39,412	788	80	50	100	24.25
Biomass	3000	70	33	83	40	0.90
Wind	3650	7	96	34	20	0.30
Geothermal	2500	35	90	60	25	0.95

Table 6. Job creation, land requirements, and energy generation cost for RE technologies in USA [78,79].

RE Source	Job Creation Employees/500 MW	Land Requirement sKM^2/1000 MW	Energy Generation Cost ($/kWh)
Solar	5370	35	0.058
Hydro	2500	750	0.064
Biomass	36,055	5000	0.098
Wind	5635	100	0.044–0.20
Geothermal	27,050	18	0.04–0.14

Lastly, Figure 8 shows the public opinion regarding the acceptance of the implementation of RE-based plants in Portugal and Australia [80]. The public acceptance is one of the sub-criteria of the socio-political aspect in this study. The public opinion varies from country to country, but in general the public is favorable toward greener technologies for the electricity generation.

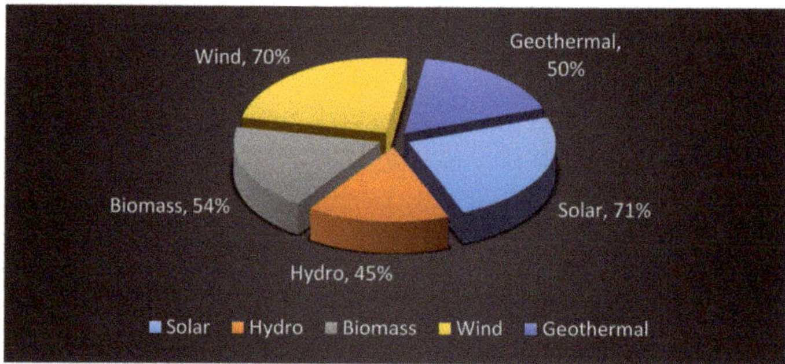

Figure 8. Public opinion for RE technologies implementation in Portugal and Australia.

5. Results and Discussion

Assessment and prioritization of the best RE resource is a difficult and complex decision problem. However, an attempt has been made in the context of Pakistan to address this decision problem by considering four main criteria and 20 sub-criteria to evaluate five RE resources for the electricity generation in Pakistan.

5.1. Delphi Results

In this phase, the experts were not only asked to assign the weights to the criteria but also to recommend additional criteria which they consider important for evaluating RE resources. The data analysis was undertaken using the CV and CVR values of each criterion. Each evaluation criterion met the required CV (less than 0.5) and CVR (greater than 0.29) levels. The detailed CV and CVR results of the various criteria and sub-criteria are given in Appendix A. Finally, the refined results regarding criteria and sub-criteria that would influence the assessment process of RE resources are shown in Table 7.

Table 7. Criteria refined by the experts.

Criteria	Reference
1. Economic Aspect (ECA)	
Initial cost (ECA$_1$)	[15,81,82]
Operation and Maintenance Cost (ECA$_2$)	[15,81,83]
Resource Potential (ECA$_3$)	[83,84]
Energy Generation Cost (ECA$_4$)	[82,85,86]
Expected life of RE plant (ECA$_5$)	[15,87]

Table 7. *Cont.*

Criteria	Reference
2. Environmental Aspect (ENA)	
CO_2 Emission reduction (ENA$_1$)	[83,86]
Impact on Environment (ENA$_2$)	[83,86]
Land Requirement (ENA$_3$)	[83,88,89]
Noise (ENA$_4$)	[90,91]
3. Technical Aspect (TA)	
Technology maturity (TA$_1$)	[81,83,86]
Efficiency (TA$_2$)	[83,85,92]
Capacity factor (TA$_3$)	[16,93]
Human Resource Expertise (TA$_4$)	[16,94]
Climate conditions (TA$_5$)	[90,95,96]
Reliability/Feasibility (TA$_6$)	[15,86,97]
4. Socio-Political Aspect (SPA)	
Public acceptance (SPA$_1$)	[81,83,86]
Job creation (SPA$_2$)	[81,83]
Energy security (SPA$_3$)	[16,82,98]
Institutional arrangement (SPA$_4$)	[99,100]
Regulatory mechanism (SPA$_5$)	Delphi consultation

The four main criteria and 20 sub-criteria were identified via the Delphi approach. These criteria are useful in assessing the potential of various RE resources of Pakistan. Later, each of the main criteria and sub-criteria are ranked via the AHP method. A brief description of each of these criteria is summarized as follows:

5.1.1. The economic aspect (ECA)

The economic aspect is significant for the selection and ranking of RE resources in Pakistan. The various sub-aspects (sub-criteria) have been identified from economic perspectives and are defined as follows:

Initial Cost (ECA$_1$)

Initial cost is defined as the total expenditure required for establishing a renewable power plant and comprises the labor, equipment, and infrastructure development costs as well. RE resources that require low initial cost are generally preferred.

Operation and Maintenance Cost (ECA$_2$)

O&M cost comprises the total cost of operating and undertaking the regular maintenance of the plant including the salaries of workers; the maintenance costs ensure system operation and reduce downtime. RE resources with a low O&M cost requirement are generally preferred.

Resource Potential (ECA$_3$)

Resource potential is the availability of renewable resources (solar, hydro, biomass, wind, and geothermal energy) in the region to produce energy. RE resources with more potential to generate electricity are preferable.

Energy Generation Cost (ECA$_4$)

Energy generation cost is defined as the cost of electricity generated from renewable power plants. RE resources with a lower electricity generation cost are preferable.

The Expected Life of RE Plant (ECA$_5$)

A renewable plant has a life expectancy. The expected life of the RE plant depends upon the raw resource used to generate electricity. Thus, this sub-criterion is very important when analyzing an RE plant. As such, RE resources with a greater plant life expectancy are preferable.

5.1.2. The environmental Aspect (ENA)

The environmental aspect plays a key role in the selection and ranking of various RE resources (i.e., solar, hydro, biomass, wind, and geothermal energy). These aspects include CO_2 emission reduction, impact on environment, land requirements, and noise, which are individually defined as follows:

CO_2 Emissions Reduction (ENA$_1$)

This criterion pertains to minimum CO_2 emissions from RE resources, including their production and transportation. RE resources that cause low CO_2 emissions are preferable.

Impact on Environment (ENA$_2$)

A RE power plant's impact on the environment and on its surroundings is important. As such, RE resources that cause low or no impact on the environment are preferable.

Land Requirement (ENA$_3$)

RE power plants require land for physical installation. RE resources that occupy a low amount of land and can be used for other purposes such as cultivation and farming are preferable.

Noise (ENA$_4$)

The probability of noise pollution due to the installation of RE power plants in the region is significant. As such, RE resources that have less or zero noise pollution are preferable.

5.1.3. The technical Aspect (TA)

The technical aspect is an important part of choosing an optimal RE resource for electricity generation. There are several key sub-aspects from the technical perspective and are defined as follows:

Technology Maturity (TA$_1$)

This indicates how technology is extensive at regional, national, and international levels. RE resources with a mature technology are generally suitable for electricity generation.

Efficiency (TA$_2$)

The efficiency of an electricity plant is denoted by the ratio of output energy to input energy. RE resources with greater efficiency are preferable.

Capacity Factor (TA$_3$)

The capacity factor indicates how useful and productive energy obtained from an RE source can be. RE resources with a high capacity factor are more useful for electricity generation.

Human Resource Expertise (TA$_4$)

Human resource (HR) experts and their availability in the country to operate and maintain the RE-based power plant is important. As such, RE resources that have more HR experts available are recommended.

Climate Conditions (TA$_5$)

Climate conditions affect generation from a power plant; the performance of RE-based plant depends on these conditions. RE resources that are favorable and useful in different climate conditions are generally preferred.

Reliability/Feasibility (TA$_6$)

Reliability is described as the ability of an electric power plant to perform essential functions under specified conditions. A constant and sufficient amount of electricity generation thus needs to be ensured. RE resources with higher reliability and feasibility are preferable.

5.1.4. The Socio-Political Aspect (SPA)

The socio-political aspect is crucial for the selection of RE resources in Pakistan. Similarly, this aspect has important sub-aspects (sub-criteria) and each of these has been described here:

Public Acceptance (SPA$_1$)

Public acceptance is defined as an acceptability and opinion about the utilization of RE plants in the region. RE resources with a more favorable opinion are preferable.

Job Creation (SPA$_2$)

RE plant installation promises job creation and opportunities for the locals in terms of both technical and non-technical positions. RE resources that create more job opportunities for the local people are preferable.

Energy Security (SPA$_3$)

Energy security will be strengthened by installing and utilizing RE resources in the country. RE resources that can generate more electricity and help the country to reduce its reliance on fossil fuels are preferable.

Institutional Arrangement (SPA$_4$)

Institutional arrangement is required for the development of RE resources. RE resources with an existing institutional arrangement are preferable.

The Regulatory Mechanism (SPA$_5$)

Mechanisms that support the promotion of tariffs, long-term contracts, or mandatory targets are crucial for the deployment of RE-based plants. RE resources with a suitable regulatory mechanism are preferable.

5.2. AHP Results

The AHP technique was employed to develop the pairwise comparison matrix of the identified criteria and sub-criteria. Each element of this matrix signifies a numerical importance with the other element in the matrix entry comparison. The calculations are partly based on actual objective data for priority weightage/ranking associated with sub-criteria.

The CR and RI were determined using Equations (2) and (3). The AHP model thus provides the results of various pairwise comparisons at different levels of the hierarchy. The detailed pairwise comparisons matrices of criteria and sub-criteria are provided in Appendix B. Further, Table 8 illustrates the local and global weight and overall ranking of each main criteria and sub-criteria. The results of AHP methodology in this study reveal that the economic aspect is considered the most important criterion, with a weight of 0.3695, followed by socio-political and technical aspect criteria, with comparative scores of 0.2959 and 0.1859, respectively. The environmental aspect secured the last place in the pairwise comparison of the main criteria. In the pairwise comparison of sub-criteria, the resource potential is ranked as the priority sub-criteria from economic perspective. These aspects are rightly identified as they pose serious challenges for the government to eradicate the on-going energy crisis and maximize the utilization of RE resources for sustainable development. Experts also ranked energy security as the second most important sub-criterion of the socio-political crtierion, followed by regulatory mechanism, job creation, institutional arrangement, and public acceptance. From the environmental perspective, CO_2 emission reduction and land requirements were considered the most important sub-criteria. Technology maturity was the most significant sub-criterion of the technical criterion. The preference of technology maturity over efficiency and reliability shows the lowered risk and sustainability of new renewable technologies.

Table 8. Overall priority weight and ranking of criteria and sub-criteria.

Criteria	Sub-Criteria	Local Weight	Global Weight	Overall Ranking
Economic Aspect (ECA)		**0.3695**		**1st (Criteria)**
	Initial cost (ECA_1)	0.1688	0.0624	5th
	Operation and Maintenance Cost (ECA_2)	0.1725	0.0637	4th
	Resource Potential (ECA_3)	0.3475	0.1284	1st
	Energy Generation Cost (ECA_4)	0.2164	0.0800	3rd
	Expected life of renewable energy plant (ECA_5)	0.0949	0.0350	13th
Environmental Aspect (ENA)		0.1487		4th (Criteria)
	CO_2 Emission reduction (ENA_1)	0.3695	0.0549	7th
	Impact on Environment (ENA_2)	0.1886	0.0281	15th
	Land Requirement (ENA_3)	0.3382	0.0503	9th
	Noise (ENA_4)	0.1037	0.0154	19th
Technical Aspect (TA)		0.1859		3rd (Criteria)
	Technology maturity (TA_1)	0.2804	0.0521	8th
	Efficiency (TA_2)	0.2203	0.0409	12th
	Capacity factor (TA_3)	0.1354	0.0252	17th
	Human Resource Expertise (TA_4)	0.0608	0.0113	20th
	Climate conditions (TA_5)	0.1415	0.0263	16th
	Reliability/Feasibility (TA_6)	0.1617	0.0301	14th
Socio-Political Aspect (SPA)		0.2959		2nd (Criteria)
	Public acceptance (SPA_1)	0.0771	0.0228	18th
	Job creation (SPA_2)	0.1563	0.0462	10th
	Energy security (SPA_3)	0.4275	0.1265	2nd
	Institutional arrangement (SPA_4)	0.1385	0.0410	11th
	Regulatory mechanism (SPA_5)	0.2006	0.0594	6th

It is evident from Table 9 that AHP methodology pairwise comparisons of criteria and sub-criteria have been accomplished.

Table 9. Final ranking of the renewable alternatives.

RE Source	DPIS (d_i^+)	DNIS (d_i^-)	CC_i	Rank
Solar	19.367	0.654	0.033	3
Hydro	19.352	0.672	0.034	2
Biomass	19.394	0.625	0.031	4
Wind	19.320	0.706	0.035	1
Geothermal	19.396	0.623	0.031	5

5.3. Fuzzy TOPSIS Results

Finally, F-TOPSIS methodology was employed to determine the optimal RE resource from the given five RE alternatives. Initially, we obtained the decision matrix by providing the TFNs of the alternatives, obtained the normalized decision matrix, and further weighted the normalized decision matrix, provided in detail in Appendix C. The results of F-TOPSIS indicate that the wind energy-based electricity generation is the most prioritized alternative since it has the least distance from the ideal solution. Table 9 provides a summary ranking of the five RE resources of Pakistan considered in this study. The closeness coefficient CC_i indicates that wind energy achieved the highest score, 0.035, followed by 0.034 for hydro, 0.033 for solar, 0.031 for biomass, and 0.031 for the geothermal alternative. Wind energy ranks first on the basis of the renewable potential in the country, a lower capital cost, the land's potential for cultivation and cropping, higher efficiency and public acceptance, good energy security, and the lower impact on the environment, with low or zero greenhouse gas emissions.

5.4. Sensitivity Analysis

A sensitivity analysis was undertaken to investigate any major or minor variation in experts' preferences that might change the results. As such, we examined the level of significance of criteria weights with the ranking of RE resources (i.e., alternative) one by one. Five scenarios were accordingly evaluated, revealing that the priority order of the alternatives does not vary and remains the same. Table 10 provides the criteria weights in the evaluated scenarios. Moreover, Table 11 shows the obtained results/rankings of the sensitivity analysis. Scenario 1 is the weights of this study, whereas the rest of the scenarios are considered for the sensitivity analysis. The results of Table 11 signify that results of Scenarios 2 and 3 have no impact on the ranking order of the alternatives; in Scenarios 4 and 5, the ranking order of alternatives (solar and hydro) has changed.

Table 10. Criteria weights employed for sensitivity analysis.

Criteria	Scenario 1 (Current Weight)	Scenario 2	Scenario 3	Scenario 4	Scenario 5
Economic (ECA)	0.3695	0.25	0.30	0.35	0.40
Environmental (ENA)	0.1859	0.25	0.20	0.15	0.10
Technical (TA)	0.1487	0.25	0.30	0.35	0.40
Socio-Political (SPA)	0.2959	0.25	0.20	0.15	0.10

Table 11. Results of the sensitivity analysis.

Alternative	Scenario 1	Scenario 2	Scenario 3	Scenario 4	Scenario 5
Solar	3	3	3	2	2
Hydro	2	2	2	3	3
Biomass	4	4	4	4	4
Wind	1	1	1	1	1
Geothermal	5	5	5	5	5

In summary, the sensitivity analysis reveals that there is no significant change in the main findings of the study; therefore, it appears insignificant to change the weights of the obtained results. Wind energy remained the highest ranked alternative, followed by hydro, solar, biomass, and geothermal. Thus, the study results are considered valid and robust.

6. Conclusions

Pakistan has not been able to diversify its electricity generation from fossil fuels to RE resources. Thus, the share of RE in the overall energy mix is negligible. The ongoing energy crises, the ever increasing circular debt of the power sector, and decreasing economic growth may cause further crises in the economy. It is also believed that energy demand will increase in the future. Given these facts, it is now impossible to neglect the indigenous RE resource development for electricity generation in Pakistan. As regards the decision problem pertaining to the prioritization of RE resources for harnessing, this study lays a foundation for planning and policy makers of the country to consider scientific decision aid methods.

As such, in this study, five majors RE resources of Pakistan, treated as alternatives, are analyzed and ranked on multiple decision criteria. In the AHP criteria weight analysis, the highest scores were attained for the economic aspect, followed by socio-political, technical, and environmental criteria. The quantitative and qualitative data was analyzed for overall synthesis. The AHP model results illustrated that resource potential and energy generation costs were the highest-ranked sub-criteria in the economic aspect. CO_2 emission reduction and land requirements were the top-ranked sub-criteria in the environmental aspect. Technology maturity and efficiency emerged as the highest-ranked sub-criteria in the technical aspect. In the socio-political aspect, energy security and regulatory mechanisms appeared as top-ranked sub-criteria. Therefore, all selected main criteria and the top-ranked sub-criteria should play a significant role in the development of RE resources in Pakistan. In terms of F-TOPSIS, alternative weights and thus priority ranking wind energy attained the highest score, followed by hydropower, solar, biomass, and geothermal resources. Furthermore, sensitivity analysis provided additional confidence regarding the results of this study. In general, this study highlights the importance of wind energy, which can be effectively utilized for electricity generation in Pakistan. The integrated Delphi, AHP, F-TOPSIS methodologies, in fact, enabled us to understand the decision variants more specifically with regard to common criteria, thereby synthesizing the qualitative criteria into numerical values and thus providing results with adequate clarity.

The top ranking of the wind energy resource for harnessing and exploitation for electricity generation is also in consonance with international and national level development in the wind energy sector. It is therefore recommended that the GoP not only adopt a scientific basis for resolving energy planning and policy decision problems but also consider results of this study, which emphasizes the exploitation and development of available RE resource potential for sustainable electricity generation. These key recommendations in terms of prioritizing wind energy on other RE resources is viable and should be adopted because wind energy is the most abundant, economical, and environment-friendly source of electricity generation. We thus propose the following policy recommendations to promote the development of RE resources in Pakistan:

1. It is recommended that government should utilize advanced innovation or technology for the development of wind, hydro, solar, biomass, and geothermal energy.
2. Pakistan can use a competitive advantage to promote the overall industrial development and then create job opportunities and a sizable domestic market for the development of RE resources.
3. RE enterprises should be promoted to engage in the green energy industry and increase research and development to help enterprises. The government should organize information sharing platforms and technical seminars, and combine promotion policies to achieve sustainable development.
4. The electricity generation cost from the RE power plants must be low for residential or commercial use because this may obstruct the development of RE technology. Enterprises or people do not have enough incentive to install high cost RE technology. Therefore, regulating the electricity generation cost structure must be given top priority.
5. Feed in Tariffs should be reduced, and this would also encourage stakeholders to invest in the RE market.
6. New transmission and distribution (T&D) networks should be developed to increase efficiency and decrease T&D losses in the energy system.
7. There are many countries where RE technology accounts for more than 50% of electricity generation, such as Norway (96%), Austria (68%), Colombia (70%), Denmark (57%), Brazil (85%), Sweden (55%), and Iceland (100%) [101]. Thus, the GoP can refer to successful foreign cases

to identify feasible implementation plans and accelerate the promotion and development of RE technology.

Further, for future research, more stakeholders, the inclusion of different experts, and the application of other MCDM methods such as VIKOR, DEMATEL, ELECTRE, ANP, and PROMETHEE can be utilized to refine results and can be explored to compare the results in search of any changes. In addition, more criteria and RE resources (such as offshore wind and tidal power) could be considered. We believe that further research can shed much more light on this subject.

Author Contributions: Conceptualization, Y.A.S. and Q.T.; Data curation, Y.A.S. and N.H.M.; Formal analysis, Y.A.S. and G.D.W.; Funding acquisition, Q.T.; Investigation, Y.A.S., N.H.M. and M.W.A.K.; Methodology, Y.A.S., N.H.M. and G.D.W.; Software, Y.A.S.; Supervision, N.H.M. and M.I.; Validation, M.I.; Writing—original draft, Y.A.S.; Writing—review & editing, Y.A.S. and N.H.M.

Funding: This paper was supported by the Social Science Foundation of China (15BGL029) and the Social Science Fund Major Project of Jiangsu Province (16ZD008).

Conflicts of Interest: The authors declare no conflict of interest.

Abbreviations

MCDM	multi-criteria decision making
AHP	analytical hierarchy process
F-TOPSIS	fuzzy technique for order of preference by similarity to ideal solution techniques
CR	consistency ratio
CI	consistency index
RI	random index
CV	coefficient of variation
CVR	content validity ratio
RE	renewable energy
GoP	government of Pakistan
PMD	Pakistan meteorological department
AEDB	alternative energy development board
PCRET	Pakistan council of renewable energy technologies

Appendix A

Table A1. Demographic information of experts.

Designation	Qualification	Age	Organization
Stakeholder	Graduate	55	REAP, Lahore
Stakeholder	Graduate	56	Resource Future, Islamabad
Professor	PhD	43	Mehran U.E.T, Jamshoro
Professor	PhD	38	University of Sindh, Jamshoro
Manager	PhD	52	HESCO, Hyderabad
Director	Graduate	50	NTDC, Islamabad
Secretary	PhD	48	MoPW, Islamabad
Director	Graduate	45	AEDB, Islamabad
Senior Manager	PhD	48	PAEC, Islamabad
Deputy Director	Graduate	40	PCRET, Islamabad

Note: Names of the experts not revealed at their request. Questionnaire survey on the development of criteria for the selection of renewable energy resources.

Table A2. Coefficient of variation (CV) and content validity ratio (CVR) results of final evaluation criteria.

Criteria	CV (Less than 0.50)	CVR (Greater than 0.29)
1. Economic Aspect (ECA)		
1. Initial cost (ECA_1)	0.29	0.56
2. O&M cost (ECA_2)	0.27	0.48
3. Resource Potential (ECA3)	0.17	0.75
4. Energy Generation Cost (ECA_4)	0.23	0.80
5. Expected life of RE plant (ECA_5)	0.25	0.65
2. Environmental Aspect (ENA)		
1. CO_2 Emission reduction (ENA_1)	0.32	0.80
2. Impact on Environment (ENA_2)	0.15	0.60
3. Land Requirement (ENA_3)	0.18	0.56
4. Noise (ENA_4)	0.31	0.65
3. Technical Aspect (TA)		
1. Technology maturity (TA_1)	0.34	0.60
2. Efficiency (TA_2)	0.21	0.45
3. Capacity factor (TA_3)	0.19	0.50
4. Human Resource Expertise (TA_4)	0.26	0.60
5. Climate conditions (TA_5)	0.14	0.40
6. Reliability/Feasibility (TA_6)	0.16	0.40
4. Socio-Political Aspect (SPA)		
1. Public acceptance (SPA_1)	0.32	0.45
2. Job creation (SPA_2)	0.11	0.75
3. Energy security (SPA_3)	0.12	0.80
4. Institutional arrangement (SPA_4)	0.37	0.80
5. Regulatory mechanism (SPA_5)	0.22	0.60

Appendix B

Table A3. Pair-wise comparison matrix of criteria with respect to goals along with the priority weight.

	ECA	ENA	TA	SPA	Priority Weight	Rank
ECA	1	2.6673	1.7826	1.3195	0.3695	1st
ENA	0.3749	1	1	0.4251	0.1487	4th
TA	0.5610	1	1	0.6988	0.1859	3rd
SPA	0.7579	2.3522	1.4310	1	0.2959	2nd
CR = 0.0102 < 0.10 and λmax = 4.1080						

Table A4. Pair-wise comparison matrix of economic sub-factor.

	(ECA_1)	(ECA_2)	(ECA_3)	(ECA_4)	(ECA_5)	Priority Weight	Rank
(ECA_1)	1	1.3195	0.4152	0.6084	1.9332	0.1688	4th
(ECA_2)	0.7579	1	0.4503	1.1487	1.7826	0.1725	3rd
(ECA_3)	2.4082	2.2206	1	1.4310	3.3227	0.3475	1st
(ECA_4)	1.6438	0.8706	0.6988	1	2.2206	0.2164	2nd
(ECA_5)	0.5173	0.5610	0.3010	0.4503	1	0.0949	5th
CR = 0.0157 < 0.10 and λmax = 5.2937							

Table A5. Pair-wise comparison matrix of environmental sub-factor.

	(ENA$_1$)	(ENA$_2$)	(ENA$_3$)	(ENA$_4$)	Priority Weight	Rank
(ENA$_1$)	1	1.8882	1.1487	3.5195	0.3695	1st
(ENA$_2$)	0.5296	1	0.5173	18882	0.1886	3rd
(ENA$_3$)	0.8706	1.9332	1	3.1777	0.3382	2nd
(ENA$_4$)	0.2841	0.5296	0.3147	1	0.1037	4th

CR = 0.0011 < 0.10 and λmax = 4.0545

Table A6. Pair-wise comparison matrix of technical sub-factor.

	(TA$_1$)	(TA$_2$)	(TA$_3$)	(TA$_4$)	(TA$_5$)	(TA$_6$)	Priority Weight	Rank
(TA$_1$)	1	1.1487	2.4082	4.2823	1.8882	1.8882	0.2804	1st
(TA$_2$)	0.8706	1	1.8882	3.8981	1.3195	1.1487	0.2203	2nd
(TA$_3$)	0.4152	0.5296	1	2.0477	1.1487	1	0.1354	5th
(TA$_4$)	0.2335	0.2565	0.4884	1	0.4066	0.3749	0.0608	6th
(TA$_5$)	0.5296	0.7579	0.8706	2.4595	1	0.8027	0.1415	4th
(TA$_6$)	0.5296	0.8706	1	2.6673	1.2457	1	0.1617	3rd

CR = 0.0057 < 0.10 and λmax = 6.2112

Table A7. Pair-wise comparison matrix of socio-political sub-factor.

	(SPA$_1$)	(SPA$_2$)	(SPA$_3$)	(SPA$_4$)	(SPA$_5$)	Priority Weight	Rank
(SPA$_1$)	1	0.4152	0.2453	0.4884	0.3701	0.0771	5th
(SPA$_2$)	2.4082	1	0.3615	1.1487	0.6598	0.1563	3rd
(SPA$_3$)	4.0760	2.7663	1	3.3935	2.5508	0.4275	1st
(SPA$_4$)	2.0477	0.8706	0.2947	1	0.6988	0.1385	4th
(SPA$_5$)	2.7019	1.5157	0.3920	1.4310	1	0.2006	2nd

CR = 0.0095 < 0.10 and λmax = 5.1250

Appendix C

Table A8. Decision matrix.

	ECA_1	ECA_2	ECA_3	ECA_4	ECA_5	ENA_1	ENA_2	ENA_3	ENA_4	TA_1	TA_2	TA_3	TA_4	TA_5	TA_6	SPA_1	SPA_2	SPA_3	SPA_4	SPA_5
SE	2.6, 4.6, 6.4	4.2, 6.2, 8	7, 8.6, 9.4	6.6, 8.4, 9.4	3.8, 5.8, 7.6	5, 7, 8.6	2.6, 4.6, 6.6	5, 6.8, 8.2	2.6, 4.6, 6.6	6.6, 8.6, 9.8	5.4, 7.4, 9	4.6, 6.6, 8.6	4.2, 6.2, 8	4.2, 6.2, 7.8	4.6, 6.6, 8.4	2.2, 4.2, 6.2	5.4, 7.4, 9	6.6, 8.4, 9.4	5.8, 7.8, 9.2	7, 8.6, 9.4
HE	3.8, 5.8, 7.8	4.2, 6.2, 8.2	6.2, 8.2, 9.6	4.6, 6.6, 8.6	3.4, 5.4, 7.4	4.2, 6.2, 8.2	3.4, 5.4, 7.2	3, 5, 6.8	2.2, 4.2, 6.2	4.2, 6.2, 8	4.2, 6.2, 8	4.2, 6.2, 8.2	3.8, 5.8, 7.8	5.4, 7.4, 9.2	6.2, 8, 9.2	2.6, 4.6, 6.6	4.2, 6.2, 8.2	6.2, 8.2, 9.6	5.4, 7.4, 9	5.8, 7.8, 9.4
BE	3, 5, 7	4.2, 6.2, 8.2	7.4, 9, 9.8	7, 8.8, 9.8	4.2, 6.2, 8	4.2, 6.2, 8.2	2.6, 4.6, 6.6	4.2, 6.2, 8	1.8, 3.8, 5.8	5.8, 7.8, 9.4	5.8, 7.6, 9	4.6, 6.6, 8.4	5.8, 7.8, 9.2	5.8, 7.8, 9.2	5, 7, 8.8	4.6, 6.4, 8	5.4, 7.4, 9	6.6, 8.4, 9.4	6.2, 8.2, 9.6	7.4, 9.2, 10
WE	2.6, 4.6, 6.6	3, 5, 7	7.8, 9.4, 10	4.6, 6.6, 8.6	3.4, 5.4, 7.4	4.2, 6.2, 8.2	3.4, 5.4, 7.2	3, 5, 6.8	2.2, 4.2, 6.2	4.2, 6.2, 8	4.2, 6.2, 8	4.2, 6.2, 8.2	3.8, 5.8, 7.8	5.4, 7.4, 9.2	6.2, 8, 9.2	2.6, 4.6, 6.6	6.2, 8, 9.4	7, 8.8, 9.8	5.4, 7.4, 9	5.8, 7.8, 9.4
GE	3, 5, 7	4.2, 6.2, 8.2	7.4, 9, 9.8	7, 8.8, 9.8	4.2, 6.2, 8	4.2, 6.2, 8.2	2.6, 4.6, 6.6	4.2, 6.2, 8	1.8, 3.8, 5.8	5.8, 7.8, 9.4	5.8, 7.6, 9	5, 7, 8.8	5.8, 7.8, 9.2	5.8, 7.8, 9.2	5, 7, 8.8	4.6, 6.4, 8	5.4, 7.4, 9	6.6, 8.4, 9.4	6.2, 8.2, 9.6	7.4, 9.2, 10

Table A9. Normalized decision matrix.

	ECA_1	ECA_2	ECA_3	ECA_4	ECA_5	ENA_1	ENA_2	ENA_3	ENA_4	TA_1	TA_2	TA_3	TA_4	TA_5	TA_6	SPA_1	SPA_2	SPA_3	SPA_4	SPA_5
SE	0.406, 0.565, 1	0.375, 0.484, 0.714	0.7, 0.86, 0.94	0.489, 0.548, 0.697	0.447, 0.586, 0.895	0.349, 0.429, 0.6	0.333, 0.478, 0.846	0.317, 0.382, 0.52	0.273, 0.391, 0.692	0.429, 0.488, 0.636	0.467, 0.568, 0.778	0.442, 0.576, 0.826	0.475, 0.613, 0.905	0.538, 0.677, 1	0.548, 0.697, 1	0.355, 0.524, 1	0.574, 0.787, 0.957	0.673, 0.857, 0.959	0.587, 0.692, 0.931	0.617, 0.674, 0.829
HE	0.333, 0.448, 0.684	0.366, 0.484, 0.714	0.62, 0.82, 0.96	0.535, 0.697, 1	0.459, 0.63, 1	0.366, 0.484, 0.714	0.306, 0.407, 0.647	0.382, 0.52, 0.867	0.29, 0.429, 0.818	0.525, 0.677, 1	0.525, 0.677, 1	0.463, 0.613, 0.905	0.487, 0.655, 1	0.457, 0.568, 0.778	0.5, 0.575, 0.742	0.333, 0.478, 0.846	0.447, 0.66, 0.872	0.633, 0.837, 0.98	0.6, 0.73, 1	0.617, 0.744, 1
BE	0.371, 0.52, 0.867	0.366, 0.484, 0.714	0.74, 0.9, 0.98	0.469, 0.523, 0.657	0.425, 0.548, 0.81	0.366, 0.484, 0.714	0.333, 0.478, 0.846	0.325, 0.419, 0.619	0.31, 0.474, 1	0.447, 0.538, 0.724	0.467, 0.553, 0.724	0.452, 0.576, 0.826	0.413, 0.487, 0.655	0.457, 0.538, 0.724	0.523, 0.657, 0.92	0.275, 0.344, 0.478	0.574, 0.787, 0.957	0.673, 0.857, 0.959	0.563, 0.659, 0.871	0.58, 0.63, 0.784
WE	0.394, 0.565, 1	0.429, 0.6, 1	0.78, 0.94, 1	0.535, 0.697, 1	0.459, 0.63, 1	0.366, 0.484, 0.714	0.306, 0.407, 0.647	0.382, 0.52, 0.867	0.29, 0.429, 0.818	0.525, 0.677, 1	0.525, 0.677, 1	0.463, 0.613, 0.905	0.487, 0.655, 1	0.457, 0.568, 0.778	0.5, 0.575, 0.742	0.333, 0.478, 0.846	0.66, 0.851, 1	0.714, 0.898, 1	0.6, 0.73, 1	0.617, 0.744, 1
GE	0.371, 0.52, 0.867	0.366, 0.484, 0.71	0.74, 0.9, 0.9	0.469, 0.523, 0.65	0.4, 0.54, 0.8	0.366, 0.484, 0.71	0.333, 0.478, 0.84	0.325, 0.419, 0.61	0.31, 0.474, 1	0.447, 0.538, 0.72	0.467, 0.553, 0.72	0.432, 0.543, 0.76	0.413, 0.487, 0.65	0.457, 0.538, 0.72	0.523, 0.657, 0.92	0.275, 0.344, 0.47	0.574, 0.787, 0.95	0.673, 0.857, 0.95	0.563, 0.659, 0.87	0.58, 0.6, 0.7

Table A10. Weighted normalized decision matrix.

	ECA_1	ECA_2	ECA_3	ECA_4	ECA_5	ENA_1	ENA_2	ENA_3	ENA_4	TA_1	TA_2	TA_3	TA_4	TA_5	TA_6	SPA_1	SPA_2	SPA_3	SPA_4	SPA_5
SE	0.02, 0.028, 0.05	0.019, 0.024, 0.036	0.035, 0.043, 0.047	0.024, 0.027, 0.035	0.022, 0.029, 0.045	0.017, 0.021, 0.03	0.017, 0.024, 0.042	0.016, 0.019, 0.026	0.014, 0.02, 0.035	0.021, 0.024, 0.032	0.023, 0.028, 0.039	0.022, 0.029, 0.041	0.024, 0.031, 0.045	0.027, 0.034, 0.05	0.027, 0.035, 0.05	0.018, 0.026, 0.05	0.029, 0.039, 0.048	0.034, 0.043, 0.048	0.029, 0.035, 0.047	0.031, 0.034, 0.041
HE	0.017, 0.022, 0.034	0.018, 0.024, 0.036	0.031, 0.041, 0.048	0.027, 0.035, 0.05	0.023, 0.031, 0.05	0.018, 0.024, 0.036	0.015, 0.02, 0.032	0.019, 0.026, 0.043	0.015, 0.021, 0.041	0.026, 0.034, 0.05	0.026, 0.034, 0.05	0.023, 0.031, 0.045	0.024, 0.033, 0.05	0.023, 0.028, 0.039	0.025, 0.029, 0.037	0.017, 0.024, 0.042	0.022, 0.033, 0.044	0.032, 0.042, 0.049	0.03, 0.036, 0.05	0.031, 0.037, 0.05
BE	0.019, 0.026, 0.043	0.018, 0.024, 0.036	0.037, 0.045, 0.049	0.023, 0.026, 0.033	0.021, 0.027, 0.04	0.018, 0.024, 0.036	0.017, 0.024, 0.042	0.016, 0.021, 0.031	0.016, 0.024, 0.05	0.022, 0.027, 0.036	0.023, 0.028, 0.036	0.023, 0.029, 0.041	0.021, 0.024, 0.033	0.023, 0.027, 0.036	0.026, 0.033, 0.046	0.014, 0.017, 0.024	0.029, 0.039, 0.048	0.034, 0.043, 0.048	0.028, 0.033, 0.044	0.029, 0.032, 0.039
WE	0.02, 0.028, 0.05	0.021, 0.03, 0.05	0.039, 0.047, 0.05	0.027, 0.035, 0.05	0.023, 0.031, 0.05	0.018, 0.024, 0.036	0.015, 0.02, 0.032	0.019, 0.026, 0.043	0.015, 0.021, 0.041	0.026, 0.034, 0.05	0.026, 0.034, 0.05	0.023, 0.031, 0.045	0.024, 0.033, 0.05	0.023, 0.028, 0.039	0.025, 0.029, 0.037	0.017, 0.024, 0.042	0.033, 0.043, 0.05	0.036, 0.045, 0.05	.03, 0.036, 0.05	0.031, 0.037, 0.05
GE	0.019, 0.026, 0.043	0.018, 0.024, 0.036	0.037, 0.045, 0.049	0.023, 0.026, 0.033	0.021, 0.027, 0.04	0.018, 0.024, 0.036	0.017, 0.024, 0.042	0.016, 0.021, 0.031	0.016, 0.024, 0.05	0.022, 0.027, 0.036	0.023, 0.028, 0.036	0.022, 0.027, 0.038	0.021, 0.024, 0.033	0.023, 0.027, 0.036	0.026, 0.033, 0.046	0.014, 0.017, 0.024	0.029, 0.039, 0.048	0.034, 0.043, 0.048	0.028, 0.033, 0.044	0.029, 0.032, 0.039

References

1. Arto, I.; Capellán-Pérez, I.; Lago, R.; Bueno, G.; Bermejo, R. The Energy Requirements of a Developed World. *Energy Sustain. Dev.* **2016**, *3*, 1–13. [CrossRef]
2. Burke, M.J.; Stephens, J.C. Political Power and Renewable Energy Futures: A Critical Review. *Energy Res. Soc. Sci.* **2018**, *35*, 78–93. [CrossRef]
3. Aized, T.; Shahid, M.; Bhatti, A.A.; Saleem, M.; Anandarajah, G. Energy Security and Renewable Energy Policy Analysis of Pakistan. *Renew. Sustain. Energy Rev.* **2018**, *84*, 155–169. [CrossRef]
4. Kiani, K. Energy Sector Circular Debt Touches Record Rs922bn. Available online: https://www.dawn.com/news/1392681 (accessed on 7 November 2018).
5. Rafique, M.M.; Rehman, S. National Energy Scenario of Pakistan—Current Status, Future Alternatives, and Institutional Infrastructure: An Overview. *Renew. Sustain. Energy Rev.* **2017**, *69*, 156–167. [CrossRef]
6. Farooq, M.K.; Kumar, S. An Assessment of Renewable Energy Potential for Electricity Generation in Pakistan. *Renew. Sustain. Energy Rev.* **2013**, *20*, 240–254. [CrossRef]
7. Kamran, M. Current Status and Future Success of Renewable Energy in Pakistan. *Renew. Sustain. Energy Rev.* **2018**, *82*, 609–617. [CrossRef]
8. *Pakistan Energy Yearbook*; Hydrocarbon Development Institute of Pakistan: Islamabad, Pakistan, 2017.
9. Rauf, O.; Wang, S.; Yuan, P.; Tan, J. An Overview of Energy Status and Development in Pakistan. *Renew. Sustain. Energy Rev.* **2015**, *48*, 892–931. [CrossRef]
10. Awan, U.; Kraslawski, A.; Huiskonen, J. Understanding Influential Factors on Implementing Social Sustainability Practices in Manufacturing Firms: An Interpretive Structural Modelling (ISM) Analysis. *Procedia Manuf.* **2018**, *17*, 1039–1048. [CrossRef]
11. Lee, G.; Jun, K.S.; Chung, E.S. Integrated Multi-Criteria Flood Vulnerability Approach Using Fuzzy TOPSIS and Delphi Technique. *Nat. Hazards Earth Syst. Sci.* **2013**, *13*, 1293–1312. [CrossRef]
12. Marinakis, V.; Doukas, H.; Xidonas, P.; Zopounidis, C. Multicriteria Decision Support in Local Energy Planning: An Evaluation of Alternative Scenarios for the Sustainable Energy Action Plan. *Omega* **2017**, *69*, 1–16. [CrossRef]
13. Stewart, T.J.; French, S.; Rios, J. Integrating Multicriteria Decision Analysis and Scenario Planning-Review and Extension. *Omega (United Kingdom)* **2013**, *41*, 679–688. [CrossRef]
14. Marttunen, M.; Lienert, J.; Belton, V. Structuring Problems for Multi-Criteria Decision Analysis in Practice: A Literature Review of Method Combinations. *Eur. J. Oper. Res.* **2017**, *263*, 1–17. [CrossRef]
15. Wang, J.J.; Jing, Y.Y.; Zhang, C.F.; Zhao, J.H. Review on Multi-Criteria Decision Analysis Aid in Sustainable Energy Decision-Making. *Renew. Sustain. Energy Rev.* **2009**, *13*, 2263–2278. [CrossRef]
16. Amer, M.; Daim, T.U. Selection of Renewable Energy Technologies for a Developing County: A Case of Pakistan. *Energy Sustain. Dev.* **2011**, *15*, 420–435. [CrossRef]
17. Zuberi, M.J.S.; Hasany, S.Z.; Tariq, M.A.; Fahrioglu, M. Assessment of Biomass Energy Resources Potential in Pakistan for Power Generation. In Proceedings of the 4th International Conference on Power Engineering, Energy and Electrical Drives, Istanbul, Turkey, 13–17 May 2013; pp. 1301–1306.
18. Mousavi, S.M.; Tavakkoli-Moghaddam, R.; Heydar, M.; Ebrahimnejad, S. Multi-Criteria Decision Making for Plant Location Selection: An Integrated Delphi-AHP-PROMETHEE Methodology. *Arab. J. Sci. Eng.* **2013**, *38*, 1255–1268. [CrossRef]
19. Ho, L.W.; Lie, T.T.; Leong, P.T.; Clear, T. Developing Offshore Wind Farm Siting Criteria by Using an International Delphi Method. *Energy Policy* **2018**, *113*, 53–67. [CrossRef]
20. Shen, Y.C.; Chou, C.J.; Lin, G.T.R. The Portfolio of Renewable Energy Sources for Achieving the Three E Policy Goals. *Energy* **2011**, *36*, 2589–2598. [CrossRef]
21. Ghimire, L.P.; Kim, Y. An Analysis on Barriers to Renewable Energy Development in the Context of Nepal Using AHP. *Renew. Energy* **2018**, *129*, 446–456. [CrossRef]
22. Azizkhani, M.; Vakili, A.; Noorollahi, Y.; Naseri, F. Potential Survey of Photovoltaic Power Plants Using Analytical Hierarchy Process (AHP) Method in Iran. *Renew. Sustain. Energy Rev.* **2017**, *75*, 1198–1206. [CrossRef]
23. Wang, C.-N.; Huang, Y.F.; Cheng, I.F.; Nguyen, V.T. A Multi-Criteria Decision-Making (MCDM) Approach Using Hybrid SCOR Metrics, AHP, and TOPSIS for Supplier Evaluation and Selection in the Gas and Oil Industry. *Processes* **2018**, *6*, 252. [CrossRef]

24. Solangi, Y.A.; Tan, Q.; Khan, M.W.A.; Mirjat, N.H.; Ahmed, I. The Selection of Wind Power Project Location in the Southeastern Corridor of Pakistan: A Factor Analysis, AHP, and Fuzzy-TOPSIS Application. *Energies* **2018**, *11*, 1940. [CrossRef]

25. Jozaghi, A.; Alizadeh, B.; Hatami, M.; Flood, I.; Khorrami, M.; Khodaei, N.; Ghasemi Tousi, E. A Comparative Study of the AHP and TOPSIS Techniques for Dam Site Selection Using GIS: A Case Study of Sistan and Baluchestan Province, Iran. *Geosciences* **2018**, *8*, 494. [CrossRef]

26. Karatas, M.; Sulukan, E.; Karacan, I. Assessment of Turkey's Energy Management Performance via a Hybrid Multi-Criteria Decision-Making Methodology. *Energy* **2018**. [CrossRef]

27. Ervural, B.C.; Evren, R.; Delen, D. A Multi-Objective Decision-Making Approach for Sustainable Energy Investment Planning. *Renew. Energy* **2018**, *153*, 890–912.

28. Kim, M.-S.; Lee, E.-B.; Jung, I.-H.; Alleman, D. Risk Assessment and Mitigation Model for Overseas Steel-Plant Project Investment with Analytic Hierarchy Process—Fuzzy Inference System. *Sustainability* **2018**, *10*, 4780. [CrossRef]

29. Alamanos, A.; Mylopoulos, N.; Loukas, A.; Gaitanaros, D. An Integrated Multicriteria Analysis Tool for Evaluating Water Resource Management Strategies. *Water* **2018**, *10*, 1795. [CrossRef]

30. Hui, J.; Lim, S. An Analytic Hierarchy Process (AHP) Approach for Sustainable Assessment of Economy-Based and Community-Based Urban Regeneration: The Case of South Korea. *Sustainability* **2018**, *10*, 4456. [CrossRef]

31. Papapostolou, A.; Karakosta, C.; Doukas, H. Analysis of Policy Scenarios for Achieving Renewable Energy Sources Targets: A Fuzzy TOPSIS Approach. *Energy Environ.* **2017**, *28*, 88–109. [CrossRef]

32. Ligus, M.; Peternek, P. Determination of Most Suitable Low-Emission Energy Technologies Development in Poland Using Integrated Fuzzy AHP-TOPSIS Method. *Energy Procedia* **2018**, *153*, 101–106. [CrossRef]

33. Cayir Ervural, B.; Zaim, S.; Demirel, O.F.; Aydin, Z.; Delen, D. An ANP and Fuzzy TOPSIS-Based SWOT Analysis for Turkey's Energy Planning. *Renew. Sustain. Energy Rev.* **2018**, *82*, 1538–1550. [CrossRef]

34. Büyüközkan, G.; Karabulut, Y.; Güler, M. Strategic Renewable Energy Source Selection for Turkey with Hesitant Fuzzy MCDM Method. *Stud. Syst. Decis. Control* **2018**, *149*, 229–250.

35. Öztayşi, B.; Kahraman, C. Evaluation of Renewable Energy Alternatives Using Hesitant Fuzzy TOPSIS and Interval Type-2 Fuzzy AHP. In *Renewable and Alternative Energy: Concepts, Methodologies, Tools, and Applications*; IGI Global: Hershey, PA, USA, 2017; pp. 1378–1412.

36. Zhuang, Z.-Y.; Lin, C.-C.; Chen, C.-Y.; Su, C.-R. Rank-Based Comparative Research Flow Benchmarking the Effectiveness of AHP–GTMA on Aiding Decisions of Shredder Selection by Reference to AHP–TOPSIS. *Appl. Sci.* **2018**, *8*, 1974. [CrossRef]

37. Zare, K.; Mehri-Tekmeh, J.; Karimi, S. A SWOT Framework for Analyzing the Electricity Supply Chain Using an Integrated AHP Methodology Combined with Fuzzy-TOPSIS. *Int. Strateg. Manag. Rev.* **2015**, *3*, 66–80. [CrossRef]

38. Kaya, T.; Kahraman, C. Multicriteria Decision Making in Energy Planning Using a Modified Fuzzy TOPSIS Methodology. *Expert Syst. Appl.* **2011**, *38*, 6577–6585. [CrossRef]

39. Mirza, U.K.; Ahmad, N.; Harijan, K.; Majeed, T. Identifying and Addressing Barriers to Renewable Energy Development in Pakistan. *Renew. Sustain. Energy Rev.* **2009**, *13*, 927–931. [CrossRef]

40. Awan, U.; Imran, N.; Munir, G. Sustainable Development through Energy Management: Issues and Priorities in Energy Savings. *Res. J. Appl. Sci. Eng. Technol.* **2014**, *7*. [CrossRef]

41. Farooqui, S.Z. Prospects of Renewables Penetration in the Energy Mix of Pakistan. *Renew. Sustain. Energy Rev.* **2014**, *29*, 693–700. [CrossRef]

42. Shaikh, F.; Ji, Q.; Fan, Y. The Diagnosis of an Electricity Crisis and Alternative Energy Development in Pakistan. *Renew. Sustain. Energy Rev.* **2015**, *52*, 1172–1185. [CrossRef]

43. Zafar, U.; Ur Rashid, T.; Khosa, A.A.; Khalil, M.S.; Rahid, M. An Overview of Implemented Renewable Energy Policy of Pakistan. *Renew. Sustain. Energy Rev.* **2018**, *82*, 654–665. [CrossRef]

44. Mustafa, K. Pak Geothermal Energy Resources Have Potential to Generate 100,000MW Power: Research. Available online: https://www.thenews.com.pk/print/107384-Pak-geothermal-energy-resources-have-potential-to-generate-100000MW-power-Research (accessed on 20 January 2019).

45. Wakeel, M.; Chen, B.; Jahangir, S. Overview of Energy Portfolio in Pakistan. *Energy Procedia* **2016**, *88*, 71–75. [CrossRef]

46. NREL. Pakistan Resource Maps and Toolkit—NREL. Available online: https://www.nrel.gov/international/ra_pakistan.html (accessed on 1 January 2017).

47. Abdullah; Zhou, D.; Shah, T.; Jebran, K.; Ali, S.; Ali, A.; Ali, A. Acceptance and Willingness to Pay for Solar Home System: Survey Evidence from Northern Area of Pakistan. *Energy Rep.* **2017**, *3*, 54–60. [CrossRef]

48. Das Valasai, G.; Uqaili, M.A.; Memon, H.U.R.; Samoo, S.R.; Mirjat, N.H.; Harijan, K. Overcoming Electricity Crisis in Pakistan: A Review of Sustainable Electricity Options. *Renew. Sustain. Energy Rev.* **2017**, *72*, 734–745. [CrossRef]

49. Shakeel, S.R.; Takala, J.; Shakeel, W. Renewable Energy Sources in Power Generation in Pakistan. *Renew. Sustain. Energy Rev.* **2016**, *64*, 421–434. [CrossRef]

50. Siddiqi, A.; Wescoat, J.L.; Humair, S.; Afridi, K. An Empirical Analysis of the Hydropower Portfolio in Pakistan. *Energy Policy* **2012**, *50*, 228–241. [CrossRef]

51. Ministry of Water and Power. Hydro Power Resources of Pakistan. Available online: http://www.ppib.gov.pk/HYDRO.pdf (accessed on 22 November 2017).

52. Awan, A.B.; Khan, Z.A. Recent Progress in Renewable Energy—Remedy of Energy Crisis in Pakistan. *Renew. Sustain. Energy Rev.* **2014**, *33*, 236–253. [CrossRef]

53. Ullah, K. Electricity Infrastructure in Pakistan: An Overview. *Int. J. Energy Inf. Commun.* **2013**, *4*, 11–26.

54. *World Wind Energy Association-Market Report*; World Wind Energy Association: Bonn, Germany, 2016.

55. Lauha, F. *Global Wind Statistics 2017*; GWEC: Brussels, Belgium, 2017.

56. Sheikh, M.A. Energy and Renewable Energy Scenario of Pakistan. *Renew. Sustain. Energy Rev.* **2010**, *14*, 354–363. [CrossRef]

57. PMD WIND ENERGY PROJECT. Available online: http://www.pmd.gov.pk/wind/Wind_Project_files/Page767.html (accessed on 22 November 2017).

58. Younas, U.; Khan, B.; Ali, S.M.; Arshad, C.M.; Farid, U.; Zeb, K.; Rehman, F.; Mehmood, Y.; Vaccaro, A. Pakistan Geothermal Renewable Energy Potential for Electric Power Generation: A Survey. *Renew. Sustain. Energy Rev.* **2016**, *63*, 398–413. [CrossRef]

59. Dalkey, N.; Helmer, O. An Experimental Application of the DELPHI Method to the Use of Experts. *Manag. Sci.* **1963**, *9*, 351–515. [CrossRef]

60. Rowe, G.; Wright, G. The Delphi Technique as a Forecasting Tool: Issues and Analysis. *Int. J. Forecast.* **1999**, *15*, 353–375. [CrossRef]

61. Vidal, L.A.; Marle, F.; Bocquet, J.C. Using a Delphi Process and the Analytic Hierarchy Process (AHP) to Evaluate the Complexity of Projects. *Expert Syst. Appl.* **2011**, *38*, 5388–5405. [CrossRef]

62. Skulmoski, G.; Hartman, F.; Krahn, J. The Delphi Method for Graduate Research. *J. Inf. Technol. Educ. Res.* **2007**, *6*, 1–21. [CrossRef]

63. Dajani, J.S.; Sincoff, M.Z.; Talley, W.K. Stability and Agreement Criteria for the Termination of Delphi Studies. *Technol. Forecast. Soc. Chang.* **1979**, *13*, 83–90. [CrossRef]

64. Lawshe, C.H. A Quantitative Approach to Content Validity. *Pers. Psychol.* **1975**, *28*, 563–575. [CrossRef]

65. Wilson, F.R.; Pan, W.; Schumsky, D.A. Recalculation of the Critical Values for Lawshe's Content Validity Ratio. *Meas. Eval. Couns. Dev.* **2012**, *45*, 197–210. [CrossRef]

66. Kumar, A.; Sah, B.; Singh, A.R.; Deng, Y.; He, X.; Kumar, P.; Bansal, R.C. A Review of Multi Criteria Decision Making (MCDM) towards Sustainable Renewable Energy Development. *Renew. Sustain. Energy Rev.* **2017**, *69*, 596–609. [CrossRef]

67. Mirjat, N.H.; Uqaili, M.A.; Harijan, K.; Wazir, M.; Mustafa, M. Multi-Criteria Analysis of Electricity Generation Scenarios for Sustainable Energy Planning in Pakistan. *Energies* **2018**, *11*, 757. [CrossRef]

68. Saaty, T.L. *The Analytic Hierarchy Process*; McGraw-Hill: New York, NY, USA, 1980; Volume 48, pp. 109–121.

69. Saaty, T.L. Decision Making with the Analytic Hierarchy Process. *Int. J. Serv. Sci.* **2008**, *1*, 83–98. [CrossRef]

70. Franek, J.; Kresta, A. Judgment Scales and Consistency Measure in AHP. *Procedia Econ. Financ.* **2014**, *12*, 164–173. [CrossRef]

71. Cavallo, B. Computing Random Consistency Indices and Assessing Priority Vectors Reliability. *Inf. Sci.* **2017**, *420*, 532–542. [CrossRef]

72. Boran, K. An Evaluation of Power Plants in Turkey: Fuzzy TOPSIS Method. *Energy Sour. Part B Econ. Plan. Policy* **2017**, *12*, 119–125. [CrossRef]

73. Wang, C.N.; Huang, Y.F.; Chai, Y.C.; Nguyen, V.N. A Multi-Criteria Decision Making (MCDM) for Renewable Energy Plants Location Selection in Vietnam under a Fuzzy Environment. *Appl. Sci.* **2018**, *8*, 2069. [CrossRef]

74. Zadeh, L.A. Fuzzy Sets. *Inf. Control* **1965**, *8*, 338–353. [CrossRef]
75. Walczak, D.; Rutkowska, A. Project Rankings for Participatory Budget Based on the Fuzzy TOPSIS Method. *Eur. J. Oper. Res.* **2017**, *260*, 706–714. [CrossRef]
76. Vafaeipour, M.; Hashemkhani Zolfani, S.; Morshed Varzandeh, M.H.; Derakhti, A.; Keshavarz Eshkalag, M. Assessment of Regions Priority for Implementation of Solar Projects in Iran: New Application of a Hybrid Multi-Criteria Decision Making Approach. *Energy Convers. Manag.* **2014**, *86*, 653–663. [CrossRef]
77. Ishfaq, S.; Ali, S.; Ali, Y. Selection of Optimum Renewable Energy Source for Energy Sector in Pakistan by Using MCDM Approach. *Process Integr. Optim. Sustain.* **2018**, *2*, 61–71. [CrossRef]
78. EIA. Annual Energy Outlook 2017. Available online: https://www.eia.gov (accessed on 21 January 2019).
79. Chatzimouratidis, A.I.; Pilavachi, P.A. Multicriteria Evaluation of Power Plants Impact on the Living Standard Using the Analytic Hierarchy Process. *Energy Policy* **2008**, *36*, 1074–1089. [CrossRef]
80. Ribeiro, F.; Ferreira, P.; Araújo, M.; Braga, A.C. Public Opinion on Renewable Energy Technologies in Portugal. *Energy* **2014**, *69*, 39–50. [CrossRef]
81. Kassem, A.; Al-Haddad, K.; Komljenovic, D.; Schiffauerova, A. A Value Tree for Identification of Evaluation Criteria for Solar Thermal Power Technologies in Developing Countries. *Sustain. Energy Technol. Assess.* **2016**, *16*, 18–32. [CrossRef]
82. Brand, B.; Missaoui, R. Multi-Criteria Analysis of Electricity Generation Mix Scenarios in Tunisia. *Renew. Sustain. Energy Rev.* **2014**, *39*, 251–261. [CrossRef]
83. Ahmad, S.; Tahar, R.M. Selection of Renewable Energy Sources for Sustainable Development of Electricity Generation System Using Analytic Hierarchy Process: A Case of Malaysia. *Renew. Energy* **2014**, *63*, 458–466. [CrossRef]
84. Shen, Y.C.; Lin, G.T.R.; Li, K.P.; Yuan, B.J.C. An Assessment of Exploiting Renewable Energy Sources with Concerns of Policy and Technology. *Energy Policy* **2010**, *38*, 4604–4616. [CrossRef]
85. Kang, D.; Lee, D.H. Energy and Environment Efficiency of Industry and Its Productivity Effect. *J. Clean. Prod.* **2016**, *135*, 184–193. [CrossRef]
86. Troldborg, M.; Heslop, S.; Hough, R.L. Assessing the Sustainability of Renewable Energy Technologies Using Multi-Criteria Analysis: Suitability of Approach for National-Scale Assessments and Associated Uncertainties. *Renew. Sustain. Energy Rev.* **2014**, *39*, 1173–1184. [CrossRef]
87. Daim, T.; Yates, D.; Peng, Y.; Jimenez, B. Technology Assessment for Clean Energy Technologies: The Case of the Pacific Northwest. *Technol. Soc.* **2009**, *31*, 232–243. [CrossRef]
88. Shakouri G., H.; Aliakbarisani, S. At What Valuation of Sustainability Can We Abandon Fossil Fuels? A Comprehensive Multistage Decision Support Model for Electricity Planning. *Energy* **2016**, *107*, 60–77. [CrossRef]
89. Tasri, A.; Susilawati, A. Selection among Renewable Energy Alternatives Based on a Fuzzy Analytic Hierarchy Process in Indonesia. *Sustain. Energy Technol. Assess.* **2014**, *7*, 34–44. [CrossRef]
90. Shmelev, S.E.; Van Den Bergh, J.C.J.M. Optimal Diversity of Renewable Energy Alternatives under Multiple Criteria: An Application to the UK. *Renew. Sustain. Energy Rev.* **2016**, *60*, 679–691. [CrossRef]
91. Santos, M.J.; Ferreira, P.; Araújo, M.; Portugal-Pereira, J.; Lucena, A.F.P.; Schaeffer, R. Scenarios for the Future Brazilian Power Sector Based on a Multi-Criteria Assessment. *J. Clean. Prod.* **2018**, *167*, 938–950. [CrossRef]
92. Şengül, Ü.; Eren, M.; Eslamian Shiraz, S.; Gezder, V.; Sengül, A.B. Fuzzy TOPSIS Method for Ranking Renewable Energy Supply Systems in Turkey. *Renew. Energy* **2015**, *75*, 617–625. [CrossRef]
93. Singh, R.P.; Nachtnebel, H.P. Analytical Hierarchy Process (AHP) Application for Reinforcement of Hydropower Strategy in Nepal. *Renew. Sustain. Energy Rev.* **2016**, *55*, 43–58. [CrossRef]
94. Kahraman, C.; Kaya, I.; Cebi, S. A Comparative Analysis for Multiattribute Selection among Renewable Energy Alternatives Using Fuzzy Axiomatic Design and Fuzzy Analytic Hierarchy Process. *Energy* **2009**, *34*, 1603–1616. [CrossRef]
95. Vidadili, N.; Suleymanov, E.; Bulut, C.; Mahmudlu, C. Transition to Renewable Energy and Sustainable Energy Development in Azerbaijan. *Renew. Sustain. Energy Rev.* **2017**, *80*, 1153–1161. [CrossRef]
96. Shen, P.; Lior, N. Vulnerability to Climate Change Impacts of Present Renewable Energy Systems Designed for Achieving Net-Zero Energy Buildings. *Energy* **2016**, *114*, 1288–1305. [CrossRef]
97. Chaiamarit, K.; Nuchprayoon, S. Modeling of Renewable Energy Resources for Generation Reliability Evaluation. *Renew. Sustain. Energy Rev.* **2013**, *26*, 34–41. [CrossRef]

98. Sahir, M.H.; Qureshi, A.H. Specific Concerns of Pakistan in the Context of Energy Security Issues and Geopolitics of the Region. *Energy Policy* **2007**, *35*, 2031–2037. [CrossRef]
99. Xin-Gang, Z.; Tian-Tian, F.; Lu, C.; Xia, F. The Barriers and Institutional Arrangements of the Implementation of Renewable Portfolio Standard: A Perspective of China. *Renew. Sustain. Energy Rev.* **2014**, *30*, 371–380. [CrossRef]
100. Shah, A.A.; Qureshi, S.M.; Bhutto, A.; Shah, A. Sustainable Development through Renewable Energy-The Fundamental Policy Dilemmas of Pakistan. *Renew. Sustain. Energy Rev.* **2011**, *15*, 861–865. [CrossRef]
101. Lee, H.C.; Chang, C. Ter. Comparative Analysis of MCDM Methods for Ranking Renewable Energy Sources in Taiwan. *Renew. Sustain. Energy Rev.* **2018**, *92*, 883–896. [CrossRef]

Article

Cogeneration Process Technical Viability for an Apartment Building: Case Study in Mexico

Hugo Valdés [1],* and Gabriel Leon [2]

[1] Centro de Innovación en Ingeniería Aplicada (CIIA), Departamento de Computación e Industrias, Universidad Catolica del Maule (UCM), 3460000 Talca, Chile

[2] Departamento de Sistemas Energéticos, Universidad Nacional Autonoma de Mexico (UNAM), Ciudad Universitaria, 04510 México, Mexico; tesgleon@yahoo.com

* Correspondence: hvaldes@ucm.cl; Tel.: +56-71-220-3541

Received: 29 December 2018; Accepted: 7 February 2019; Published: 13 February 2019

Abstract: The objective of this paper is to evaluate and to simulate the cogeneration process applied to an apartment building in the Polanco area (Mexico). Considering the building's electric, thermal demand and consumption data, the cogeneration process model was simulated using Thermoflow[©] software (Thermoflow Inc., Jacksonville, FL, USA), in order to cover 1.1 MW of electric demand and to supply the thermal needs of hot water, heating, air conditioning and heating pool. As a result of analyzing various schemes of cogeneration, the most efficient scheme consists of the use of a gas turbine (Siemens model SGT-100-1S), achieving a cycle with efficiency of 84.4% and a heat rate of 14,901 kJ/kWh. The economic results of this evaluation show that it is possible to implement the cogeneration in the building with a natural gas price below US$0.014/kWh. The use of financing schemes makes the economic results more attractive. Furthermore, the percentage of the turbine load effect on the turbine load net power, cogeneration efficiency, chimney flue gas temperature, CO_2 emission, net heat ratio, turbine fuel flow and after burner fuel flow was also studied.

Keywords: cogeneration; technical viability; apartment building

---

1. Introduction

Rapidly increasing world energy use has already raised concerns over supply difficulties, exhaustion of energy resources and heavy environmental impacts (ozone layer depletion, global warming, climate change, etc.). Final energy consumption is usually shown split into three main sectors: industry, transport and 'other', including in the latter, agriculture, service sector and residential. This makes it difficult to gather information about building energy consumption [1].

Buildings account for approximately 40% of global energy consumption and play an important role in the energy market. The energy demands of buildings are predicted to continue growing worldwide in the coming decades [2–4]. Some authors [4–6] report that the energy demands of buildings (including residential and commercial buildings) have grown by 1.8% per year for forty years (see Figure 1a). Coal and oil use in buildings has remained fairly constant since then, while natural gas use grew steadily by about 1% per year. Global use of electricity in buildings grew on average by 2.5% per year since 2010, and in non-OECD countries it increased by nearly 6% per year. Global buildings sector energy intensity (measured by final energy per square meter) fell by 1.3% per year between 2010 and 2014, thanks to the continued adoption and enforcement of building energy codes and efficiency standards. Yet, progress has not been fast enough to offset growth in floor area (3% per year globally) and the increasing demand for energy services in buildings. More telling is energy demand per capita, where global average building energy use per person has remained practically constant since 1990, at just below 5 MWh per person per year (see Figure 1b) [6]. The Secretary of Energy (SENER) [7] has reported that buildings in Mexico are responsible for: 20% of total energy consumption, 27.8% of total electricity consumption, 78% of total gas consumption and 20% of CO_2 emissions. Ali [8] explored various architectural and building technologies that are employed to achieve a low-energy built environment. He concluded that the designers of the next generation of buildings, whether residential, commercial, or institutional, should aim for "zero energy" buildings in which there will be no need to draw energy from a region's power grid. In this approach, the climate and environment are used advantageously, rather than being treated as adversaries, and buildings become sources of energy.

On the other hand, by 2015, global energy generation was distributed in the following way: 78.3% fossil fuels, 2.6% nuclear energy and 19.1% renewable energy (9% biomass, 10.1% geothermal-solar-hydro-wind-biofuels). In addition, 15% of the world's population did not have access to electricity. Of this group, 87% belonged to rural areas, 55% to sub-Saharan Africa and 34% to South Asia [9]. SENER [7] has reported that the percentage distribution of the energy sources for buildings (Mexico) is: 42.2% fossil fuel, 29.2% electricity, 27.8% wood and 0.8% solar. In addition, until June 2015, the generation of electric energy was distributed in the following way: Photovoltaic <1%, Wind 0.8%, Geothermal 2.3%, Nuclear 4.3%, Hydroelectric 13.5%, Carboelectric 13.5% and Thermoelectric (combined cycle, steam cycle, turbo gas and internal combustion) 65.6%, which is why the Mexican electricity market is based mainly on power cycles that must use cogeneration in order to optimize their process.

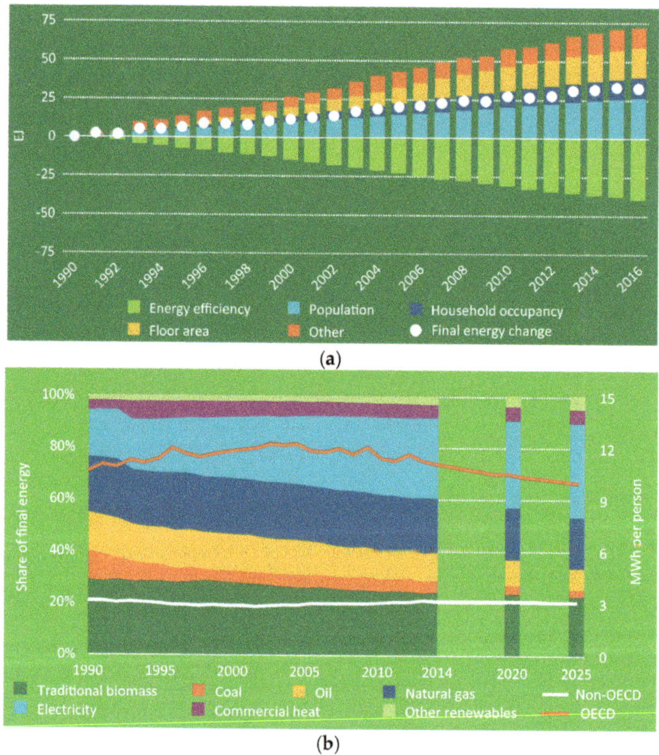

Figure 1. Recent trends in the buildings: (**a**) Decomposition of final energy demand; (**b**) Final energy use by fuel and per person [6].

There are other fuels that can be used in the cogeneration process or as part of a combined process, for example, Al-Aboosi and El-Halwagi [10] developed a design framework for integrating water and energy systems including multiple energy sources, the cogeneration process and desalination technologies in treating wastewater and fresh water for shale gas production. Solar energy was included to provide thermal power directly to a multi-effect distillation plant exclusively (to be more feasible economically) or indirect supply through a thermal energy storage system. On the other hand, if renewable energies are used, the cogeneration process can be more environmentally friendly; for example, Yan and Qin [11] designed an integrated heating system that incorporates geothermal energy into the framework of an integrated energy system of electricity, heating, and gas. An analysis of the environmental and economic benefits indicates that the system reduces pollutant emissions and decreases the cost of urban heating.

Cogeneration is defined as: "The production of more than one useful form of energy (such as process heat and electrical power) from the same energy source" [12]. This concept must be complemented by the use of waste-generated fuels in the same process as, for example, biogas. The cogeneration process has existed since 1882, when Thomas Edison designed and built the first commercial plant in the USA [5]. The basic elements of a cogeneration plant are: primary energy source, heat utilization systems, refrigeration systems, water treatment system, control system, electrical system and auxiliary systems. The different types of cogeneration can be distinguished by the equipment used in the production of energy, as, for example: cogeneration with steam turbine, cogeneration with gas turbine, cogeneration with an alternative engine, combined cycle cogeneration with gas turbine, combined cycle cogeneration with an alternative engine and tri-generation [13].

In conventional power plants, a large amount of heat is produced but not used. By using cogeneration, on the designed systems that can use heat, the efficiency of energy production can be increased starting from the current levels, ranging from 35% to 55%, to over 80% [14]. This increase in energy efficiency can be a result of spending less energy and reducing greenhouse gas emissions, when conventional methods of generating heat and electricity are compared separately [15,16]. Figure 2 shows the generation ranges and energy losses for the different cogeneration configurations. It is observed that in all four cases, a greater amount of heat, rather than electricity, is produced. Thus, the decision to choose one of the options is based on the technical-economic analysis of the process.

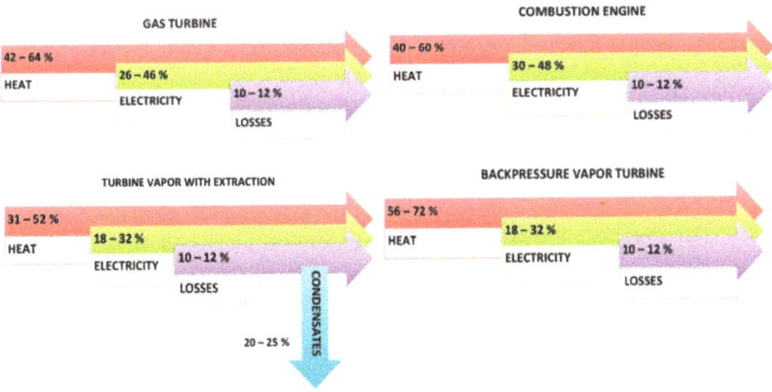

Figure 2. Heat and electricity production scheme for different cogeneration configurations.

Several authors have investigated cogeneration systems. Figure 3 shows the number of publications from 1978 to 2018. These graphs are for the word "cogeneration" in the title, keywords and abstract. Some authors on the subject of cogeneration are: Jana and De [17], who propose a biomass-based cogeneration plant with CO_2 capture. The thermodynamic modeling of the industrial plant was simulated by using ASPEN Plus; Ünal et al [18] have reported the techniques of optimization in the processes of trigeneration and poly-power generation; Shabbir and Mirzaeian [19] describe a feasibility study of the implementation of different cogeneration options to a paper mill to evaluate their energy saving potentials and economic benefits; Dincer and Zamfirescu [20] developed the concept of renewable-energy-based multigeneration options for producing a number of outputs, such as power, heat, hot water, cooling, hydrogen, fresh water, and so forth and discussed their benefits.

Such options obviously led to an improved system performance and reduced the environmental impacts; Buoro et al [21] identified the optimal energy production system and its optimal operation strategy required to satisfy the energy demand of a set of users in an industrial area. A distributed energy supply system is made up of a district heating network, a solar thermal plant with long term heat storage, a set of Combined Heat and Power units and conventional components also, such as boilers and compression chillers; and Yu et al [22] propose a general evaluation method to compare the performance of six different approaches for promoting wind power integration. In consideration of saving coal consumption, reducing CO_2 emissions, and increasing investment costs, the comprehensive benefits are defined as the evaluation index.

Current technology is making cogeneration cost-effective on increasingly smaller scales, which means that electricity and heat can be produced in neighborhoods, or even individual sites, in which cases the process is called micro-cogeneration [23]. Cogeneration applications in buildings include hospitals, institutional buildings, hotels, office and residential/housing buildings where several families live [24]. Thus, cogeneration systems for multifamily, commercial or institutional applications benefit from the thermal/electrical load diversity in the multiple loads required, which reduces the need for storage [25]. District energy systems reduce greenhouse gas emissions in two different ways: (i) In buildings, less efficient equipment is replaced by an efficient central power plant; and (ii) By producing electricity for the central grid which can replace, for example, coal and other sources of electricity that involve a large amount of greenhouse gas emission per each kWh [26]. Joelsson [27] determined district heating based on cogeneration of heat and electricity and bedrock heat pumps were found to be energy-efficient systems. The net emission of CO_2 is dependent on the fuel, and the CO_2 emissions from these systems are comparable to those from a wood pellet boiler, if biomass-based supply chains are used.

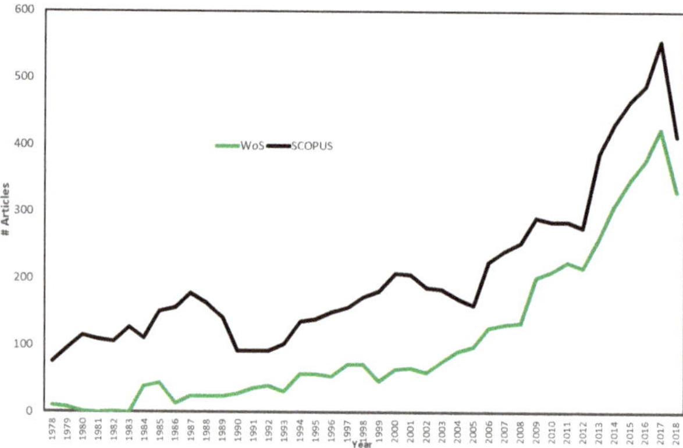

Figure 3. Representation of the number of articles published in WoS and Scopus from 1978 to 2018.

There are several publications about the cogeneration process in residential systems [25,28–34] (see Figure 3). Figure 4 shows the bibliographic mapping that connects the authors with the cited references. This was done with the search "Cogeneration AND Building". The circumferences indicate the volume of publications made by the authors. The proximity between these circumferences and lines account for the research network on the subject. The 9 authors with the highest number of citations are highlighted. E.g., Tchanche et al [32] presents existing applications and analyzes their maturity. Binary geothermal and binary biomass combined heat and power (CHP) are already mature. Provided the interest in recovering waste heat rejected by thermal devices and industrial processes continues to grow, and favorable legislative conditions are adopted, waste heat recovery organic Rankine cycle systems will experience rapid growth in the near future. Esen and Yuksel [33] experimentally investigated greenhouse heating by biogas, solar and ground energy in the climate conditions in Elazig, Turkey.

The greenhouse was constructed, and then the required heating load of the greenhouse was determined. For this purpose, biogas, solar and a ground source heat pump greenhouse heating system with a horizontal slinky ground heat exchanger was designed and set up. Chua et al [34] present a review of recent innovative cooling technology and strategies that could potentially lower the kW/R ton of cooling systems—from the existing mean of 0.9 kW/R ton towards 0.6 kW/R ton or lower.

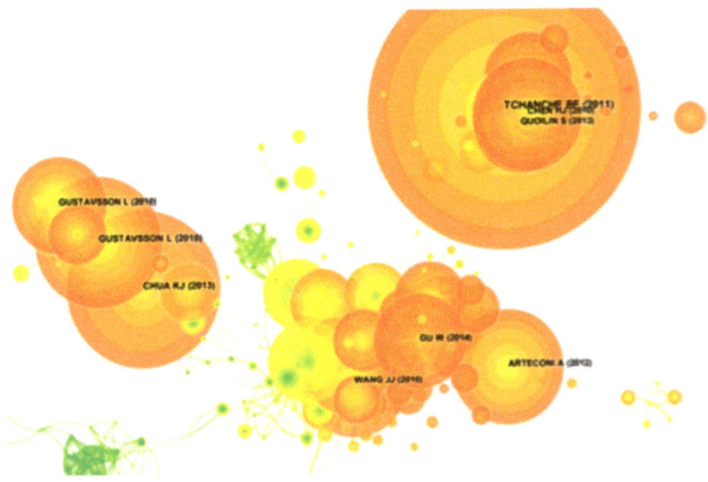

Figure 4. Authors network of research on cogeneration applied to buildings. Determined with the Citespace software.

Studies on the cogeneration process in Mexico have been carried out mainly in industrial areas [35–39]. Currently, despite the fact that the cogeneration process is considered an alternative to efficient and non-polluting energy generation, there is little scientific literature on cogeneration studies in buildings in Mexico, i.e., Fuentes-Cortes et al [40] presents a multi-objective optimization method for designing cogeneration systems in residential complexes and accounting for the involved uncertainty. The model accounts for satisfying the hot water and electric energy demands in a residential complex while minimizing the total annual cost and the associated greenhouse gas emissions. A housing complex in central Mexico is presented as a case study. Weber et al [41] compares the energy efficiency of two processes covering the thermal energy demand of a swimming pool: a CHP unit on the one hand, and a heat pump with internal combustion engine on the other. The energy losses for the CHP unit on-site are equivalent to half the losses caused by extraction and distribution of natural gas under current circumstances in Mexico.

Problem Statement

The problem is to determine the optimal configuration of a cogeneration system for a building. This configuration must meet the energy demands (electric and thermal), considering factors such as: operation scheme, size and type co-generator (turbine or combustion engine), capacity of an auxiliary thermal system (boiler) and energy purchase-sale scheme with the local company's grid, along with minimizing the total annual energy cost and greenhouse gas (GHG) emissions associated with the consumption of fuels in the process. In this research, the thermal load required by the building's facilities is much higher than the electric load, so to be able to comply with this, the electric energy produced will be divided into only what is absolutely necessary for the building, as well as for the sale to the electricity distribution company or sale to a private supplier under the current framework of the law in Mexico (since January 2015).

This research arises from recognition of the importance of the cogeneration process in Mexico. Therefore, it seeks to collect information from the analysis of the energy cogeneration system design applied to a housing building, using the computer tool Thermoflex 25 by Thermoflow© (Thermoflow Inc., Jacksonville, FL, USA).

The study contributes to the development of cogeneration in residential applications in warm and not so extreme climates. The study shows how to evaluate a cogeneration plant in a poorly-developed application in Mexico, because in the world, mainly in inhabited areas with colder and more extreme climates, cogeneration applications in apartment buildings are evaluated from a technical and economic point of view, and are very attractive for implementation. But in warmer regions such as Mexico City, this is not the case, since the need for hot and cold air are not as extensive as in cities on more northern latitudes. Hence, it is crucial to determine which cogeneration applications can be economically attractive, determining the values of economic variables and their prices-costs, in apartment buildings.

Knowing the technical, economic, and technological elements that help make cogeneration attractive for investment helps local regulators and governments design strategies for regulations, norms, and supports its development, accurately locating where and what type of support is required, given that these studies show where technical, economic, and even environmental variables are most sensitive, to give attractive results to investors or apartment owners to build and operate the plants.

This in turn favors the technological development of the equipment used by diversifying its applications and generating more market for the commercialization of equipment and services, which, in the long run, lowers costs and helps to increase the benefits of the projects, and also contributes to helping the emission reduction goals to be achieved on the regional, nation, and global scales by having more efficient energy sectors, supported by the contribution that cogeneration can make in applications for housing in countries and regions with warmer climates and not so extreme. And in particular for Mexico, it can also help accelerate the results of government programs for the development of cogeneration and exploitation of cogeneration potential, which has been sought after and promoted in Mexico since 1994, and has not yielded the desired results. This type of study and its results can contribute to providing certainty that cogeneration projects can be successfully evaluated and carried out.

2. Materials and Methods

In the first stage, the current legal framework for cogeneration in Mexico is thoroughly reviewed. This is quite important, as some conceptual design considerations of the cogeneration plant will come from this framework. Some of these considerations are the marketing and sale of the electric surplus, the availability and fees that pertain to the type of fuel as well as tax incentives and aid in fee payments, along with everything related to permits and procedures to follow for the implementation of the project, which will be reflected, firstly, as costs and economic savings in the initial stage of the project. Moreover, the available technical information on the property is collected in order to characterize the demand and consumption of the building's known average electrical and thermal energy; later, the thermal energy requirements are determined by estimating the demand and consumption of the building, based on the type of apartments, the number of people who could live in them, geographical conditions of the site and usage and customary habits of the potential tenants, keeping in mind the recommendations made by American Society of Heating, Refrigerating and Air-Conditioning Engineers (ASHRAE), equipment manufacturers and developers of this type of project. In this study, the Thermoflow© software was used, which allows the estimation of the operating conditions based on the building's energy requirements. Therefore, in order to carry out the Thermoflow© modeling of the building's cogeneration process, it is necessary to know the physical characteristics of the apartments and buildings, the number of users, energy consumption habits, environmental conditions, type of fuel available and current energy regulations. Then, using Thermoflex 25 software by Thermoflow©, two processes are proposed that comply with the building's thermal and electrical requirements. Finally, a feasibility and viability analysis of both proposed processes is carried out.

For the economic analysis, the value of money over time is considered, and it is necessary to define the discount rate. It is common to improperly use the interest rate paid for the debt as a discount rate, instead of a higher value that considers the opportunity cost for the investor; this is called the 'Minimum Attractive Rate of Return' (TREMA) or 'cost of capital'. This rate can be calculated as the rate that would be earned in an investment without risk (for example, Libor or CETES in Mexico), plus a premium that defines the risk level of the project. This methodology is typical for assessing projects feasibility [19,35].

2.1. Cogeneration Processes Simulation

Processes simulation in engineering is the digital representation of a set of unit operations, which allows determining process variables (flow, temperature, pressure, energy and power) through the use of computational tools. It also makes it possible to study existing processes in a faster, more economic and thorough way than in a real plant. The simulation of the cogeneration process was carried out in this research using the Thermoflex 25$^©$ software from Thermoflow$^©$, applied to a residential building. This program has been mainly used in cogeneration processes in industry [42–46]. In the Thermoflex simulation of this work, the planned cogeneration plant will simultaneous produce of electrical and thermal energy. Its production capacity will be designed to cover the entire thermal demand. Thus, this will make the electrical generation greater than the actual demand of the apartment building; therefore, the level of electric surplus and its possible economic income, benefiting the project due to its sale, must be determined. The plant will run uninterrupted 24 h a day, 365 days a year at 100% production capacity on-site and with the plant factor calculated for the proposed installation based on the equipment and the best operating practices. Moreover, in order to cover the scheduled and possible non-scheduled stoppages, the current fee structure provides for contracting an electrical backup service in the public network, with its corresponding cost which will be integrated into the operational cost of the cogeneration plant and into the economic evaluation of the project.

The simulations allow us to determine the performance of the cycle with different sizes of gas turbine and combustion engines. Both technologies are the most used in the cogeneration process. The gas turbine and the combustion engine chosen for the final analysis of the cycle differed mainly in the parameters, i.e., gross power, CHP efficiency, air stream, fuel stream, emissions of CO_2 and fuel consumption. These parameters have been evaluated using Thermoflex 25$^©$ software, as given by the equations described above.

The theoretical foundations of these simulations are based on thermodynamics and heat transfer. Mainly, energy and mass balances are considered in system flow. For the calculation of thermal energy fluxes in the non-phase-changing fluid, the sensible heat equation is considered [47]:

$$E = FC \cdot \Delta T, \tag{1}$$

where F is the mass flow (kg/s), C is the specific heat (kJ/kg·°C), ΔT is the temperature change of the fluid (°C) and E is the energy flow that absorbs or dissipates the fluid (kW). The determination of the energy flows is made based on the maximum requirement possible by all of the users, whereas for the calculation of the thermal energy flows in which the fluid changes phase, the latent heat, along with the sensible heat equation [47], are used:

$$E_{CF} = F \cdot C \cdot \Delta T + \Delta H, \tag{2}$$

where ΔH is the latent heat of the fluids (kW) and E_{CF} is the energy flow that absorbs or dissipates the fluid considering the phase change (kW). The determination of the energy flows is made based on the maximum requirement possible by all of the users. Furthermore, to determine the energy content of a flow, the following equation is considered valid:

$$H = F \cdot h, \tag{3}$$

where h is the specific enthalpy (kJ/kg) and H is the energy content (kW) of the flow. On the other hand, the following relationship is used to determine the power of a unit present in the cogeneration process:

$$\dot{W} = F \cdot \Delta h, \tag{4}$$

where Δh is the enthalpy change of the fluid (kW) and is the power of the equipment (kW). Furthermore, the equation for determining pump power ($\dot{W}$) is derived from the mechanical energy balance:

$$\dot{W} = \frac{F \cdot \Delta P}{\rho}, \tag{5}$$

where ΔP is the pressure change (Pa) and ρ is the density (kg/m^3). Equation (6) represents the energy balance of heat exchangers:

$$\dot{Q}_{HE} = F_H \cdot \Delta h_H = -F_C \cdot \Delta h_C, \tag{6}$$

where Δh_H is the specific enthalpy change of hot fluid (kJ/kg), Δh_C is the specific enthalpy change of cold fluid (kJ/kg), F_H is the mass flow of hot fluid (kg/s), F_C is the mass flow of cold fluid (kg/s) and $\dot{Q}_{HE}$ is the heat flow (kJ/s) transferred from the hot fluid to the cold fluid.

Other important energetic variables in the cogeneration cycles are energy intensity (kW/kg), which relates the amount of energy required to produce a mass unit of product, the Q/E ratio that specifies the ratio of the thermal and electrical energy needed to cover the requirements of the building and the efficiency of the cycle that indicates the relation between the energy demanded and the energy required by the cogeneration system.

The cogeneration is a process that is generates two products (thermal energy and electricity). The typical parameters used to determine the performance of a cogeneration plant are [48–50]: electrical efficiency (Equation (7)), thermal efficiency (Equation (8)) and total efficiency (Equation (9)).

$$\eta_e = \frac{Q_e}{Q_f} \tag{7}$$

$$\eta_h = \frac{Q_h}{Q_f} \tag{8}$$

$$\eta_{tot} = \frac{Q_h + Q_e}{Q_f} \tag{9}$$

where Q_e (kWh) is the gross generation of electricity, Q_h (kWh) is the net generation of heat Q_f (kWh) is the fuel used.

In the Thermoflow software, the plant model is built from the inside out. The users construct the subsystems from their basic elements. Then, the overall scheme emerges from the interconnected subsystems. Finally, the lowest level decisions are made, such as the fine details within the various subsystems. The structural approach automatically considers all interactions between subsystems. It also allows many lower lever inputs to be logically generated by the program, depending upon the user's higher level selection. At any level, however, the user is free to alter any or all of the program's automatic selections.

2.2. Case Study

In this research, the design of a cogeneration system for a building was analyzed. The building has electrical and thermal energy requirements (hot water and steam) in its apartments, pool and event hall. The building is located in a high-income area (Polanco) in Mexico City, so the occupants of the property have a high socioeconomic status. They are "L" shaped towers (see Figure 5); towers A and B have 19 and 14 floors, respectively. Their total height is approximately 52 m. There are 5 basement levels with a total of 528 parking lots, 233 cellars and an engine room. On the ground floor there are administrative offices, rooms with electric and gas meters, a gymnasium, entertainment rooms, an indoor pool and green areas.

Figure 5. Photographs of the building under study.

Medina et al. [51] describe the detail of the building's infrastructure. The building's apartments were still under construction at the time of the elaboration of this study; therefore, the electric consumption, the thermal demand and its consumption were estimated according to the average demand and consumption practices in apartments. The building has 233 apartments which are distinguished by the surface, number of people and number of showers (see Table 1).

Table 1. Number of apartments according to surface (m^2), number of people and showers [51].

Items	Amount	Amount	Amount
Surface (m^2)	<75	105–120	130–182
Apartments (n°)	60	88	85
People (n°)	2	4	5
Apartments (n°)	60	155	18
Showers (n°)	1	2	3
Apartments (n°)	60	93	80

By studying the energy conditions of the building, it was determined that it is necessary to supply it with thermal energy by way of steam and hot water, electric energy and cold and hot air thermal energy; therefore, throughout the day, users will have, at their disposal: Hot water service in the showers and swimming pool, and hot and cold air conditioning, all of which improve their comfort inside the property. In this case, the cogeneration plant will be able to sell surplus electrical energy, with additional benefits such as those provided by green certificates. The thermal demand is estimated according to the frequency of use given to the facilities (see Figure 6). It was considered that users would take showers mainly in the morning and a small fraction of them at night. This is the reason for the maximum values detected. It is important to highlight that a fraction of those who bathe in the morning during the summer will do so at night during the winter. On the other hand, the pool will be active 24 h a day, the kitchen sink will be used in the middle of the day and a rarely at night, and the washing machine will be used in the morning and evening. Finally, the trend in air conditioning usage has been estimated for both summer and winter.

Table 2 shows the thermal energy demand annual. Based on the lifestyles of the building users, a sharp decrease is shown during the following periods: the last week of March and the first week of April, as well as the three first weeks of August. This represents the holiday period, which is in line with the owners' standard of living, since they leave their apartments during these periods. Guelpa et al [52] examined the effects on the total load that can be obtained by adopting management strategies such as variation in the thermal request profile of the buildings or installation of local storage systems. Results show that even in the case only small changes being applied, reductions in annual primary energy consumption up to 0.4% can be obtained without any additional investment cost.

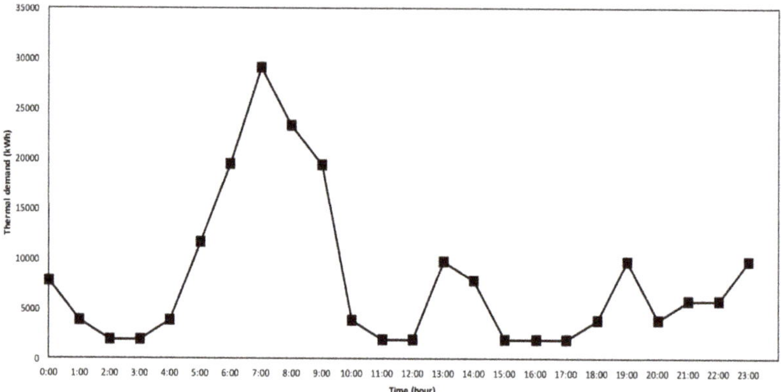

Figure 6. Thermal demand of users for winter and summer.

Table 2. Percentage of thermal demand of building.

Places of Energy Consumption	Winter (%)	Summer (%)
Shower	5.33	7.67
Bathroom sink	0.33	0.05
Kitchen sink	0.21	0.30
Washing machine	0.44	0.82
Swimming pools	6.21	7.61
Hot air	86.39	2.11
Cold air	1.40	81.45

Figure 7 shows the building's thermal hot and cold-water requirements, as well as cold air. This simulation was carried out with the Thermoflow© software which solves the mass and energy flows in detail. A summary of the most important energy requirements for the process is shown in Table 3.

Table 3. Energy requirements of the system.

Requirement	Value	Unit
Hot water flow (swimming pool)	3.16	kg/s
Energy for the swimming pool	52.06	kW
Hot water flow (Showers, washing machine, kitchen and bathroom sink)	34.40	kg/s
Energy for showers, washing machine, kitchen and bathroom sink	5033	kW
Hot water flow	32.33	kg/s
Hot air energy	426	kW
Cold air flow	28.40	kg/s
Cold air energy	1116	kW

The installed load of the building, based on the information reported by Medina et al. [51], is: contacts 35%, pumps 25%, lift force 14%, air force 14%, lighting common services 6%, parking lighting 3% and others 3%. It is observed that electricity is consumed mainly in plugs, pumps and elevators. This coincides with reported by Ali [8]; described in the introduction.

In an average day, there are two periods of greater electricity consumption (7:00–9:00 h and 19:00–21:00 h), which do not come close to the installed load or the maximum possible consumption of electricity, i.e., in the case that the building's entire electrical installation is working. On the other hand, Figure 8 shows that the electrical demand is between 19 and 62% of the installed load. The estimation of the annual electrical requirements was calculated based on the users' behavior with regards to their permanency in the building during the year.

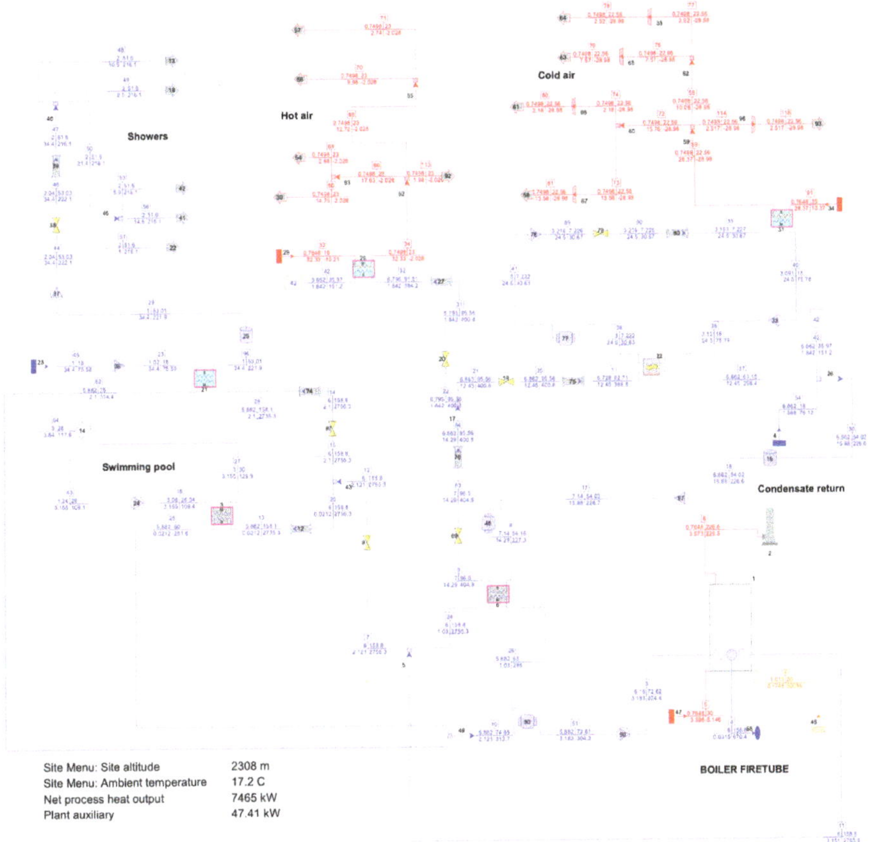

Figure 7. Thermoflow diagram of the thermal energy requirements for the building without cogeneration. Units: $\frac{bar}{kg/s}\left|\frac{^\circ C}{kJ/kg}\right.$.

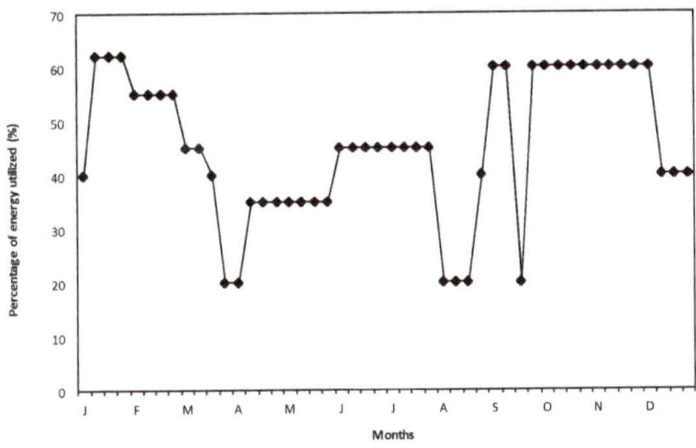

Figure 8. Percentage of annual electrical requirement for building users.

3. Results and Discussions

Energy efficiency should be understood as the intelligent use of energy, and not only as a decrease in energy consumption. That is to say, in the case of a building, residents should perform their activities using energy only at the time that it is absolutely necessary, without affecting their comfort and well-being.

In a cogeneration project, it is necessary to know the industrial context in which the process will be installed. Other authors [53–56] have also considered this prior information, i.e., to know in advance whether it is a manufacturing industry or a commercial building or a public building and establish the type of fuel to be used. Then, it is necessary to have the electrical requirements clearly quantified in order to define the electric demand profile of the process. On the other hand, it is necessary to establish the thermal requirements, i.e., hot water for showers, bathroom and kitchen sinks, or any other industrial use, as cold and/or hot water for air conditioning, water vapor for higher thermal requirements, such as large swimming pools. Once the above is defined, the electricity and natural gas fees for the central region of the country must be considered in the project, which in this case means using the current rates in Mexico City. Finally, it is also important to be informed of the environmental norms and regulations, techniques and the legal framework that will affect the decisions made during the project.

3.1. Choices of Cogeneration Systems

In this research two design process proposals were made: alternative A, which is based on combustion engines (EC process), and alternative B, which uses a gas turbine (GT process). Both comply with the basic conceptual model of a cogeneration plant, which is detailed in Figure 9.

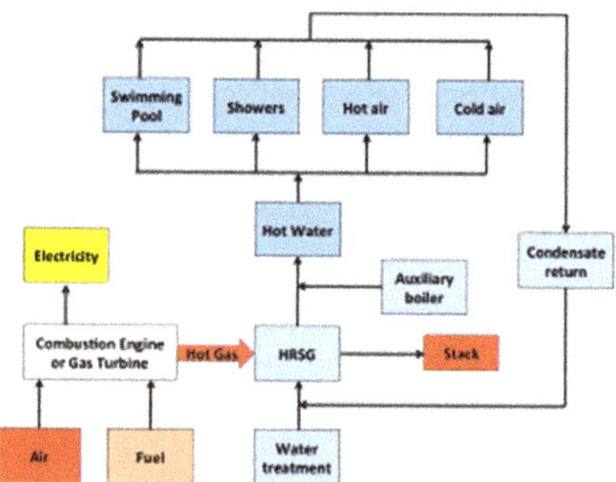

Figure 9. General scheme of the cogeneration system for the building under study.

Figure 9 shows that an internal combustion engine or a gas turbine will produce electricity and feed hot gas/air to the heat recovery steam generator (HRSG) in order to generate steam and hot water necessary for the pool, showers, toilets, washing machines, sinks and air conditioning systems. The generation of electricity will be distributed by a board to a system against fire, air conditioning, electrical system, auxiliary and substations, among others. The HRSG system will provide water at 133 and 99 °C. For the generation of cold air, a chiller of at least 300 tons of refrigeration is used. For both hot air and cold air, fan and coils are used as an air conditioning system. Once the water flow has passed through the corresponding stage, condensate recovery is performed.

The system will be installed in the basement to avoid interfering with the facilities at the top of the building. This will make the distribution of the thermal resource via variable speed pumps or hydropneumatics systems, to avoid the tanks in the highest part of the building. Regarding the electrical part, the system will be coupled to the current electrical connection, in order to connect the electric backup service in the public network. By means of a general board, the electrical load will be distributed to the different devices and substations that feed the building.

In both processes studied, compliance with 100% of the thermal and electrical requirements are achieved. The scheme (Figure 10) is divided into five sub-processes: I. Combustion engines or gas turbine and recovery boiler, II. Swimming pools, III. Showers IV. Hot air and V. cold air.

First, in the "Combustion Engine" stage of CE process where engines WAR 20V34SG (Wärtsilä North America Inc., Houston, TX, USA) and CAT G16CM34 (Caterpillar company, Deerfield, IL, USA)

(CH$_4$ fuel) were used to produce the necessary hot air sent to the HRSG. A cooling sub-process called "Refrigeration of engine system" was required to lower the heat of the engine due to the burning of fuel, without transforming into mechanical energy. Thus, this keeps the engine parts below their design temperature and avoids their deformation and destruction. In addition, part of the cooling heat is used to obtain a fraction of the necessary heat for the showers. The WAR 20V34SG engine has an electric efficiency of 45% and the output air flow temperature is 361 °C. The CAT G16CM34 engine has an electrical efficiency of 42.3%, and the output air flow temperature is 368 °C. Moreover, for proper operation at 100% capacity, 0.694 kg/s of fuel is required, thus generating 26.69 kg/s of air at 363.7 °C, which is necessary for the recovery boiler.

On the other hand, in GT process, the SIEMENS SGT-100-1S (Siemens Aktiengesellschaft, Munich, Germany) (CH$_4$ fuel) turbine was used to produce the necessary hot air which is sent to the recovery boiler (HRSG). Here, a post-combustion sub-process was required to increase the temperature of the exhaust gases, instead of using a higher capacity turbine, making the process more expensive.

For alternative A and B, in the "HRSG" and "Condensate return" steps, the sub-process consists of an evaporator (6; lower area Figure 10) and two economizers (7 and 8; lower-right area Figure 10), where the evaporator allows the saturated steam to be obtained and feed the thermal energy to the "swimming pool" and the "showers", while the economizer (7) generates the hot water for the stages of "hot air and "cold air". The exhaust gases from the economizer (8) must be greater than 100 °C to avoid the condensation of the water present in chimney flue gases. Finally, there is the "condensate return" stage which allows the mixing of the water-cooling flows in the different stages of the process, incorporating this flow to the economizer 7. The "swimming pool" stage is observed, in which a heat exchanger is used to contact, countercurrent wise, the saturated steam coming from the evaporator, with a flow of 3.64 kg/s of recirculated water from the pool (Process block with return 14). In the "showers" stage, hot water is used in the apartment's showers, washing machines, as well as bathroom and kitchen sinks. A heat exchanger (21; center-left area Figure 10) is used to obtain the hot water which is fed with saturated steam from the evaporator of the recovery boiler to achieve the hot water flow required for the building at 60 °C. Each process shaft (13, 18, 22, 41, 42) represents the hot water requirement for a section of the building, according to the number of people living in an apartment. The "hot air" stage is shown, which meets the needs of hot air by using a heat exchanger, where the hot water flow from the recovery boiler comes into contact, countercurrent wise, with the air that heats up from room temperature conditions on to 23 °C. Flows are separated per building and thereafter, the processes (30, 54, 56, 57 and 92) are distinguished according to living/dining room volume of <80 m^3 and between 80 and 144 m^3. Moreover, the hot air requirements of the building's entrance hall are included. In the "cold air" stage, the cold air is generated with the intervention of an absorption chiller (32; center-right area Figure 10), where hot water from the recovery boiler enters. This allows a flow of cooling water to be obtained and which comes in contact with the hot ambient air, cooling it down to 23 °C. The cold air is separated by buildings and also according to the cooling tons required by the apartment, in the living-dining room area. It is important to emphasize that fans appear in each flow representing the electric consumption by the air conditioning system, either for hot or cold air. Moreover, tag 42, which indicates the cooled water flow in exchanger 28 ("Hot air" stage; center area Figure 10), is returned to be mixed with the chiller outlet flow.

The main results of each process are reported in Table 4, where it can be observed that the GT process exceeds, in at least 13 aspects, the CE Process. For example, it has a CHP efficiency of 24.3%, consumes 40.9% less fuel, which translates into the same percentage of lower CO$_2$ emissions, the heat ratio for each kWh of electricity is 44.4% higher, the auxiliary energy consumption is 33% of the EC process and the water consumption is 8% lower. Therefore, the installation of the GT Process is proposed in order to comply with the energy requirements of the building. Onovwiona and Ugursal [25] reported that gas turbines offer a number of advantages when compared to reciprocating internal combustion based cogeneration systems. These include compact size, low weight, small number of moving parts and lower noise. In addition, gas turbine-based cogeneration systems have high-grade waste heat, low maintenance requirements (but require skilled personnel), low vibration and short delivery time. However, in the lower power ranges, reciprocating internal combustion engines have higher efficiency.

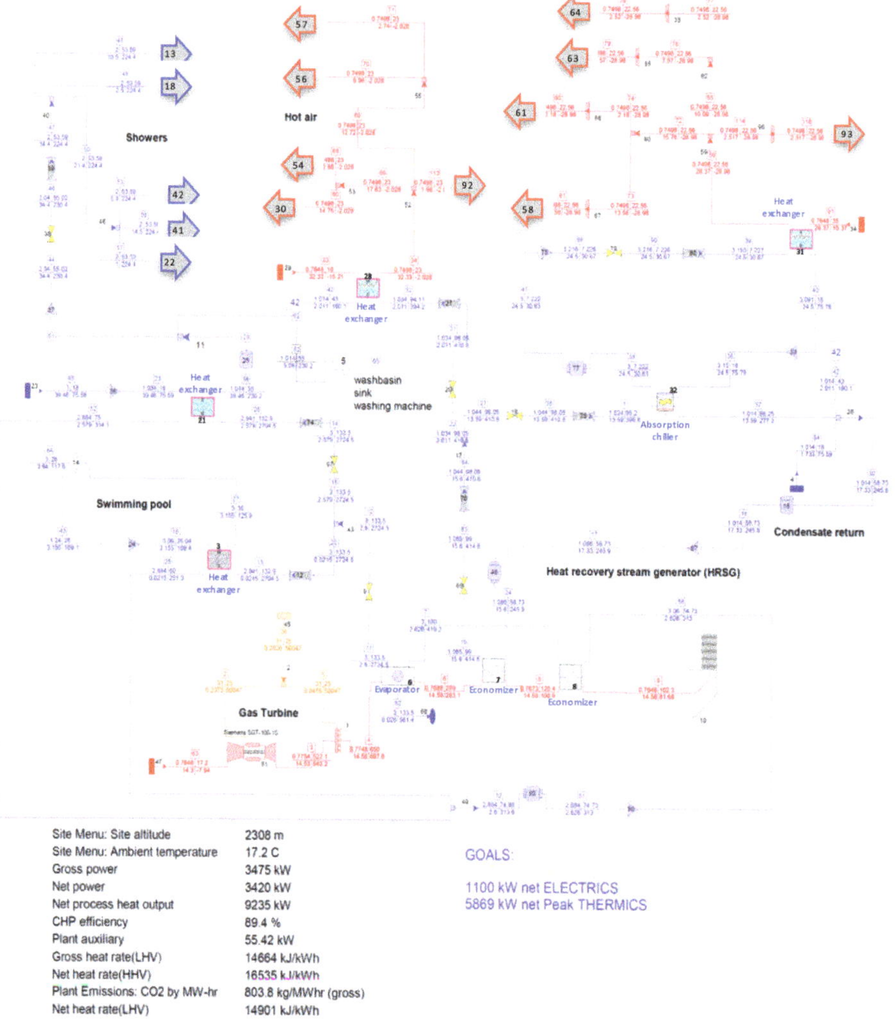

Site Menu: Site altitude	2308 m	
Site Menu: Ambient temperature	17.2 C	
Gross power	3475 kW	GOALS:
Net power	3420 kW	
Net process heat output	9235 kW	1100 kW net ELECTRICS
CHP efficiency	89.4 %	5869 kW net Peak THERMICS
Plant auxiliary	55.42 kW	
Gross heat rate(LHV)	14664 kJ/kWh	
Net heat rate(HHV)	16535 kJ/kWh	
Plant Emissions: CO2 by MW-hr	803.8 kg/MWhr (gross)	
Net heat rate(LHV)	14901 kJ/kWh	

Figure 10. Scheme of the simulated cogeneration process with the Thermoflow©, considering a gas turbine as an energy source. Carso II CHP/Gas Turbine with HRSG and natural gas to supply 100% of thermic demand.

3.2. GT Process

Figure 10 shows the simulation model of the cogeneration process using a gas turbine as the energy matrix. With this system, it is possible to comply with 100% of the thermal and electrical requirements. The results of the proposed GT process for the building are analyzed in detail below. First, the "Gas Turbine" section is explained. The gas turbine is a Siemens SGT-100-1S model with 30.6% efficiency, 4907 kWe (at sea level) electric power generation and gas production at 514 °C, which is sent to a post-combustion sub-process to raise its temperature to 650 °C. The air fed to the turbine is at an ambient temperature pressure of 0.765 bar (altitude 2308 m) and at an average temperature of the sector of 17.2 °C. Thus, due to these air conditions, the gross electric power generated is 3475 kW.

Table 4. Main results of the GT and EC Process.

Property	Unit	GT	CE
Gross power	kW	3475	15258
Net power	kW	3420	15090
Total auxiliares	kW	55.41	167.6
Net process heat output	kW	8533	8517
CHP efficiency	%	84.44	67.92
Net heat rate (LHV)	kJ/kWh	14901	8291
Net electric efficiency (LHV)	%	24.16	43.42
Air stream	N°	29	31
Water stream	N°	53	68
Fuel stream	N°	3	2
Cycle heat imbalance	%	0.0002	0.0011
Cycle mass imbalance	%	0	0
Water consumption	kg/s	3.66	3.97
Water discharge	kg/s	1.73	1.65
Emissions of CO_2	tonne/year	22626	55554
Fuel consumption	kg/s	0.283	0.694

3.2.1. Ambient Temperature Effect

Figure 11 shows the ambient temperature effect on the generation of electrical power and on the efficiency of cogeneration, where the minimum and maximum temperatures of the sector under study were considered. It was observed that in this range, the generation of power varies by 16.8%, while the cogeneration efficiency varies by 7.8%. The latter would indicate that the process is minimal affected by temperature changes in the sector. Basrawi et al [57] researched the effect of the inlet air temperature on the performance of a micro gas turbine (MGT) with cogeneration system (CGS) arrangement. The results showed that when ambient temperature increased, electrical efficiency of the MGT decreased but exhaust heat recovery increased.

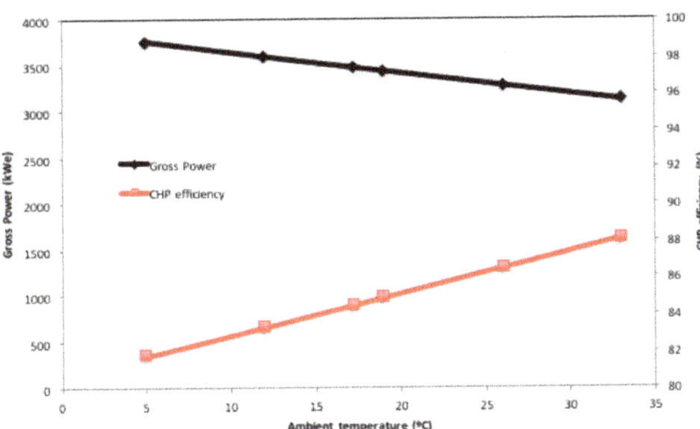

Figure 11. Variation of the gross generation of electrical power and the cogeneration efficiency according to the ambient temperature.

The sensitivity for the variation of the heat rate of the system in the function of the variation of the ambient temperature was analyzed (see Figure 12) as the base value for the analysis of 14901 kJ/kWh electrics. Therefore, the expected result (heat rate) for different environmental conditions of year is observed in Figure 12.

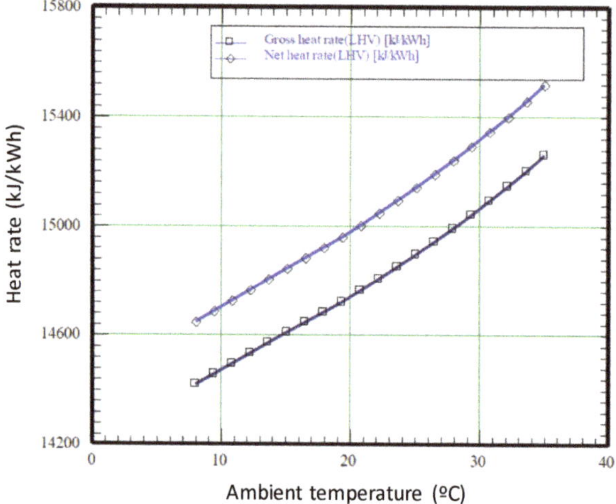

Figure 12. Heat rate of the system in the function of the variation of the ambient temperature. Image obtained from Thermoflow.

3.2.2. Fuel Effect

The Siemens SGT-100-1S turbine uses 0.24 kg/s of natural gas at 25 °C and 31 bar, obtaining the results reported in this work, but, moreover, the analysis was performed for different types of fuels. The composition of these gases is reported in Table 5, where the variability of the compounds present in the fuels is observed, so the contrast between the cogeneration efficiency and the gross generation of the electrical power was carried out (see Figure 13). It was then observed that Syngas allows a greater generation of gross electrical power, but with a low cogeneration efficiency compared to that obtained with the other fuels. This behavior of the GT process with Syngas is due to the high concentration of CO and CO_2, which gives this fuel a greater heat reaction. In addition, the installation requires a gasifier; thus, the proposal is intended to serve as an example. Figure 13 shows a linear trend with a negative slope (R^2: 0.981) between the cogeneration efficiency and the gross power generated. Hence, the higher the cogeneration efficiency, the lower the power generated for the different fuels. With Syngas as fuel, more electricity is generated, so it should be chosen according to an economic criterion. However, 60166 ton/year of CO_2 is emitted by using it, which goes against the environmental principles of the cogeneration process.

Table 5. Composition of gases tested in the GT process.

Combustible Molecule	Metane	Natural Gas (with H_2S)	Natural Gas (no H_2S)	Coke Oven Gas	Digester gas	Erdgas	Landfill Gas	Syngas
CH_4	100	87.00	87.00	33.9	62	97.65	63.5	5
H_2	0	0.36	0.36	47.9	0	0.00	2.5	30
O_2	0	0.07	0.07	0.6	0	0.00	0.0	0
N_2	0	3.61	3.65	3.7	2	0.86	0.0	5
CO	0	0.09	0.09	6.1	0	0.00	0.0	35
CO_2	0	0.34	0.34	2.6	36	0.08	33.0	25
C_2H_6	0	8.46	8.46	0.0	0	0.97	0.0	0
C_2H_4	0	0.03	0.03	5.2	0	0.00	0.0	0
C_3H_8	0	0.00	0.00	0.0	0	0.03	0.0	0
C_4H_{10}	0	0.00	0.00	0.0	0	0.11	0.0	0
C_5H_{12}	0	0.00	0.00	0.0	0	0.02	0.0	0
C_6H_{14}	0	0.00	0.00	0.0	0	0.01	0.0	0
H_2S	0	0.04	0.00	0.0	0	0.00	0.0	0

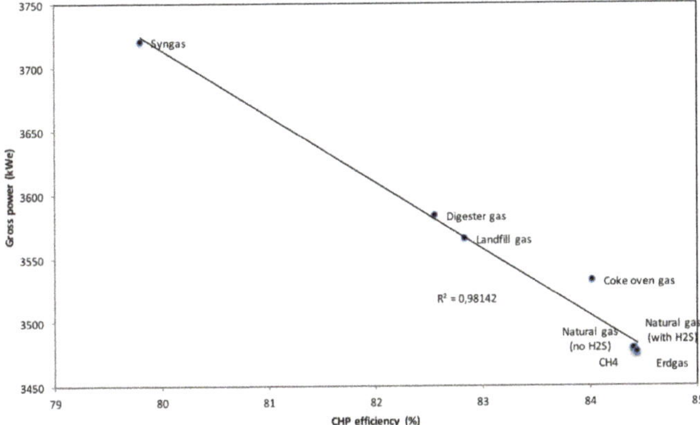

Figure 13. Contrast between cogeneration efficiency and gross electric power.

Figure 14 shows that the CO_2 emission is higher for lower cogeneration efficiency with regards to the different types of fuels. It should be highlighted that the CO_2 fuel emissions of renewable sources are considered neutral. This coincides with the report by Joelsson [27] which was described in the introduction.

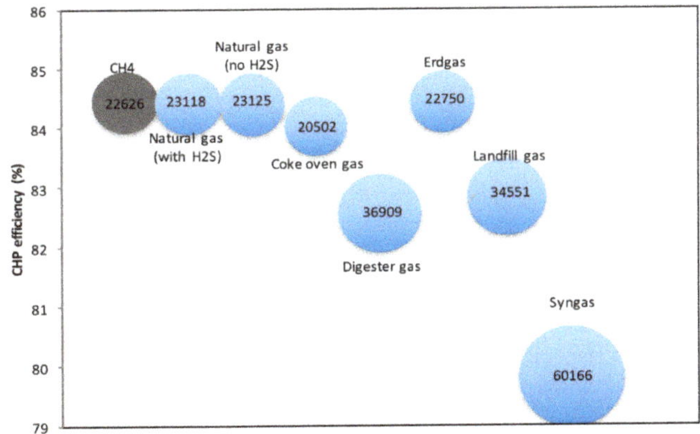

Figure 14. Emission of ton/year of CO_2 of each fuel according to the cogeneration efficiency.

The decrease in the cogeneration efficiency is due to the loss of useful energy that is reflected in the temperature of the gas outlet that leads to the chimney. As shown in Figure 15, there is a linear correlation (R^2: 0.99997) between the cogeneration efficiency and the temperature of the HRSG gas outlet. The efficiency of the cogeneration cycle is 84.4%, which is an expected value for this type of system, and therefore, is a value superior to the efficiency of a steam power cycle. In this case, the efficiency is greater in at least 44 percentage points. The energy used in auxiliary devices for the operation of the cycle corresponds to 1.6% of the gross energy generated, which can be considered a very acceptable value. On the other hand, the error percentages of the heat and mass balances show a very good simulation of the calculations obtained. Finally, the electric power generated is 3.48 MW, which greatly surpasses what is required by the building, i.e., 1.12 MW. This implies that the 2.36 MW surpluses can be commercialized according to the country's legal framework (active as of January 2015). However, the easiest option would be to sell them to the electricity company in Mexico City.

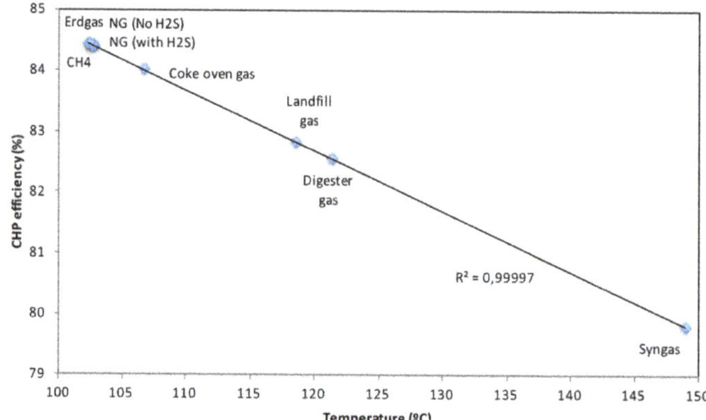

Figure 15. Linear trend between the cogeneration efficiency and the temperature of the HRSG gas outlet.

3.2.3. Load Effect

The analysis of the system was performed in case of a variation in the turbine load, decreasing from 100 to 30%, which complies with the building's thermal requirements, as summarized in Figure 16. Here, it can be observed that by lowering the turbine load, the fuel flow towards it decreases to 50%, while up to 141% more fuel is required in the afterburning, although the total fuel flow decreases by 19%, following the same trend as the CO_2 emissions. On the other hand, the temperature of the chimney flue gases decreases by 6.6%, reaching 95.6 °C, which implies that the water vapor present in them does not condense because the atmospheric pressure in the area is 0.7648 bar. Furthermore, the cogeneration efficiency decreases to 1.88%. Finally, the net power decreased to 1.01 MW, and this value is 9.8% less than the total load of the building. However, it must be considered that it is unlikely that 100% of the load will be used at a given time. In other words, with the proposed process, the thermal and electrical limit requirements can be met.

For cogeneration applications, the heat to power ratio of the engine is critical. Onovwiona and Ugursal [25] reported that the percentage of fuel energy input used in producing mechanical work, which results in electrical generation, remains fairly constant until 75% of full load, and thereafter starts decreasing. This means that more fuel is required per kWh of electricity produced at lower partial loadings, thereby leading to decreased efficiency.

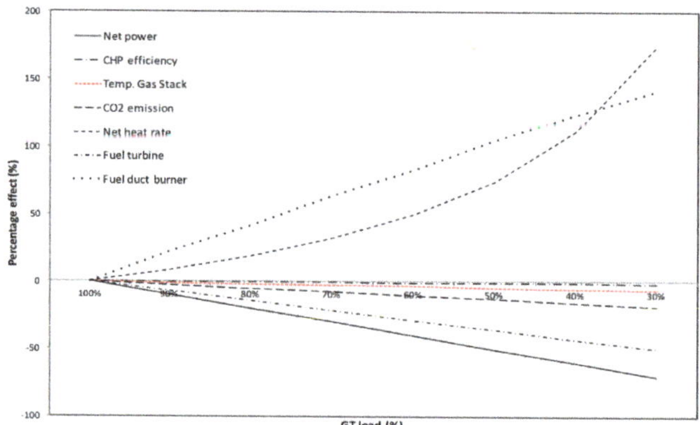

Figure 16. Percentage of the turbine load effect on the turbine load net power, cogeneration efficiency, chimney flue gas temperature, CO_2 emission, net heat ratio, turbine fuel flow and afterburner fuel flow.

3.2.4. Economic Analysis

Based on the building's energy study, the economic expenditure on electric energy, by the building's users, was determined (see Figure 17). It is important to highlight that the Mexican electric fee to be used for the calculation is the HM Central zone, where the cost of electricity depends on the time of use, seasonality and geographical area. Figure 17 shows that the highest economic expenditure on electricity consumption occurs in January because the winter rate increases with respect to the summer rate. The annual expenditure on electricity consumption then is US$ 29527.

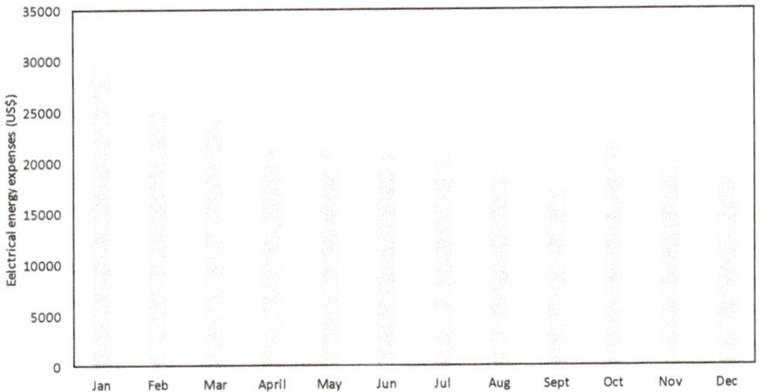

Figure 17. Monthly electricity expenses (US$) for the building.

On the other hand, the mean economic expenditure on apartment thermal energy was determined to be US$9471, annually, where the winter months showed greater spending on thermal energy (Figure 18). Moreover, there is the initial investment of the project which mainly considers the purchase of equipment (see Table 6). The largest investment in this cogeneration process is the acquisition of the Gas Turbine, which makes up 87.2% of equipment expenses.

Then, for the economic analysis of the process, the following annual costs are considered (Table 7): Energy of the conventional process and cogeneration, operation and maintenance of the conventional process and cogeneration, and investment in the cogeneration process. The values were compiled from quotations or estimates and information from similar projects. Considering the values in Table 7, the capital or investment recovery will be based on the savings in the consumption of electric and thermal energy, as well as the sale of electric power to the central electricity network, under the prices determined by the market.

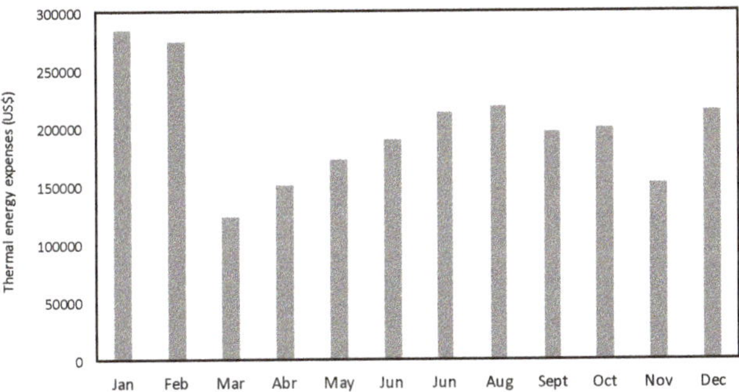

Figure 18. Monthly expenses (US$) on the building's thermal energy.

Table 6. Summary of the costs associated to the initial investment of the cogeneration process.

Equipment	Cost (US$)
Turbine Gas Model: Siemens SGT-100-1S	3,100,000
7 Centrifugal pumps	17,000
Absorption chiller	292,000
7 Heat exchangers	105,000
Auxiliary equipment	9000
Equipment import charges	20,000
Hand labor	7500
Equipment maintenance	10,000
Process administrator	7200
Miscellaneous	5000
Total	3,555,500

Table 7. Costs considered in the economic analysis.

Ítem	Cost (US$)
O&M Fixed	40,800
Conventional Process	
Energy/year	2,784,948
O&M/year	68,282
Cogeneration Process	
* Energy/year	2,000,781
O&M/year	66,973
Investment	3,555,500

* The term Energy consider fuel expenses and electric energy sale.

A positive NPV is then produced in the seventh year of operation (see Figure 19) of the cogeneration process, indicating that from that year on the cogeneration process will bring profits to the building being studied. The project is attractive, and an investment recovery period of 7 years is expected, with a natural gas price that does not exceed 0.014 dollars per kWh. However, it no longer is considered attractive if this price is set at around 5 dollars or more. As the gas price refers to that of South Texas (Henry Hub) or Ciudad Pemex to the south of Veracruz, Mexico, the final price is greatly increased by the cost of transportation to Mexico City.

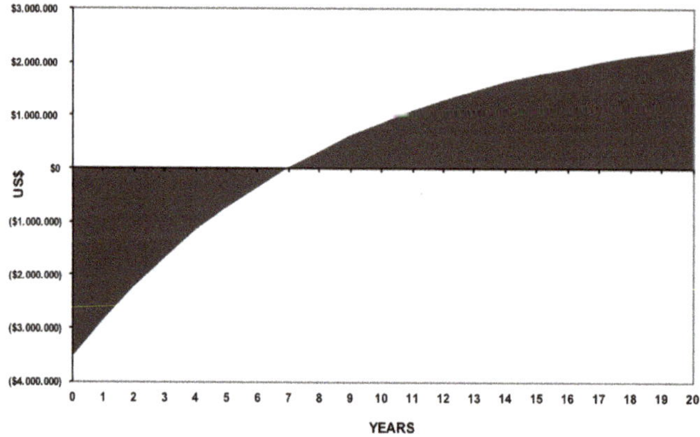

Figure 19. Recovery period of investment (MARR 12.1%).

4. Conclusions

The building's highest energy consumption is in the generation of hot water because it is mainly used in the swimming pool, showers and for the generation of hot and cold air (absorption chiller).

The cogeneration process based on a gas turbine gives better results in the main energy parameters than a process running with combustion engines, i.e. higher cycle efficiency, a lower amount of flows, lower water consumption, lower fuel consumption and therefore a lower amount of CO_2 emissions.

It was observed that the ambient temperature (5 to 33 °C) affects with an increasing linear tendency towards the efficiency of the cycle, increasing by 7%, while the generated gross power decreases linearly by 13%. This is because the greater the ambient temperature, the greater the greater enthalpy of the air necessary for the combustion inside the equipment. Such an effect can be visualized in the energy balance system. It is important to highlight that the temperature at the outlet of the chimney should be higher than 100 °C in order to avoid water condensation, along with corrosion of the ducts.

The type of fuel affects the efficiency of the cycle and the power obtained. It is observed that the use of syngas allows a lower cycle efficiency but, in turn, a higher gross power than for the rest of the fuels, while methane shows the opposite trend. These results are due to the calorific power of each compound, with that of Syngas being higher than the rest of the fuels. On the other hand, Syngas is the one that emits the largest amount of CO_2, while methane is the one that emits the least. Therefore, the latter is the most environmentally friendly fuel.

By making a variation in the turbine load by decreasing it from 100 to 30%, it was observed that when lowering the load on the turbine, the fuel flow towards the turbine decreases by 50%, while up to 141% more fuel is required in post-combustion, and the total fuel flow decreases by 19%, following the same trend of CO_2 emissions.

The results of the technical simulation and the economic and financial evaluation demonstrate that it is possible to take the industrial cogeneration process to the real estate industry; in this specific case, the project is attractive and an investment recovery period of 7 years is expected, with a natural gas price that does not exceed 0.014 dollars per kWh. However, it is not considered attractive any longer if this price is set at around 5 dollars or more. As the gas price refers to that of South Texas (Henry Hub) or Ciudad Pemex to the south of Veracruz, Mexico, the final price is greatly increased by the cost of transportation to Mexico City.

In conclusion, it was shown that the cogeneration process of an apartment building in Polanco area is technically feasible.

Author Contributions: Software, H.V.; Validation, G.L.; Investigation, H.V.; Writing-Original Draft Preparation, H.V.; Supervision, G.L.

Funding: This research was funded by The Pacific Alliance for the Alianza Pacifico Scholarship granted to Hugo Valdés of the Universidad Catolica del Maule, who agreed to finance the collaboration on this study together with Gabriel Leon de los Santos at the UNAM.

Acknowledgments: We wish to thank the Pacific Alliance for its Pacific Alliance Grant awarded to Hugo Valdés from the Catolica del Maule University (Chile), which funded the execution of this study, together with Gabriel León de los Santos from the Nacional Autonoma de Mexico University (Mexico).

Conflicts of Interest: The authors declare no conflict of interest. The funders had no role in the design of the study; in the collection, analyses, or interpretation of data; in the writing of the manuscript, and in the decision to publish the results.

References

1. Pérez-Lombard, L.; Ortiz, J.; Pout, C. A review on buildings energy consumption information. *Energy Build.* **2008**, *40*, 394–398. [CrossRef]
2. Xing, Y.; Hewitt, N.; Griffiths, P. Zero carbon buildings refurbishment—A hierarchical pathway. *Renew. Sustain. Energy Rev.* **2011**, *15*, 3229–3236. [CrossRef]

3. Ibn-Mohammed, T.; Greenough, R.; Taylor, S.; Ozawa-Meida, L.; Acquaye, A. Operational vs. embodied emissions in buildings—A review of current trends. *Energy Build.* **2013**, *66*, 232–245. [CrossRef]

4. Nejat, P.; Jomehzadeh, F.; Taheri, M.M.; Gohari, M.; Majid, M.Z.A. A global review of energy consumption, CO_2 emissions and policy in the residential sector (with an overview of the top ten CO_2 emitting countries). *Renew. Sustain. Energy Rev.* **2015**, *43*, 843–862. [CrossRef]

5. IEA. *Transition to Sustainable Buildings: Strategies and Opportunities to 2050*; International Energy Agency (IEA): Paris, France, 2013.

6. International Energy Agency. *Tracking Clean Energy Progress 2017*; International Energy Agency: Paris, France, 2017.

7. Secretaría de Energía, México. Balance nacional de energía. 2015. Available online: https://www.gob.mx/sener/documentos/balance-nacional-de-energia (accessed on 28 October 2017).

8. Ali, M.M. Energy efficient architecture and building systems to address global warming. *Leadersh. Manag. Eng.* **2008**, *8*, 113–123. [CrossRef]

9. REN21. *Renewables 2015 Global Status Report*; Renewable Energy Policy Network for 21st Century (REN21 Secretariat): París, France, 2015; ISBN 978-3-9815934-6-4.

10. Al-Aboosi, F.Y.; El-Halwagi, M.M. An Integrated Approach to Water-Energy Nexus in Shale-Gas Production. *Processes* **2018**, *6*, 52. [CrossRef]

11. Yan, Q.; Qin, C. Environmental and Economic Benefit Analysis of an Integrated Heating System with Geothermal Energy—A Case Study in Xi'an China. *Energies* **2017**, *10*, 2090. [CrossRef]

12. Boles, M.; Cengel, Y. *An Engineering Approach*, 8th ed.; McGraw-Hill Education: New York, NY, USA, 2014; ISBN 9814595292.

13. Rezaie, B.; Rosen, M.A. District heating and cooling: Review of technology and potential enhancements. *Appl. Energy* **2012**, *93*, 2–10. [CrossRef]

14. ASPEN Systems Corporation. *Combined Heat and Power: A Federal Manager's Resource Guide: Final Report*; Federal Energy Management Program; US Department of Energy: Washington, DC, USA, 2000.

15. Havelský, V. Energetic efficiency of cogeneration systems for combined heat, cold and power production. *Int. J. Refrig.* **1999**, *22*, 479–485. [CrossRef]

16. Onovwiona, H.; Ugursal, V. Modeling of internal combustion engine based cogeneration systems for residential applications. *Appl. Therm. Eng.* **2006**, *27*, 848–861. [CrossRef]

17. Jana, K.; De, S. Biomass integrated gasification combined cogeneration with or without CO_2 capture—A comparative thermodynamic study. *Renew. Energy* **2014**, *72*, 243–252. [CrossRef]

18. Ünal, A.; Ercan, S.; Kayakutlu, G. Optimisation studies on tri-generation: A review. *Int. J. Energy Res.* **2015**, *39*, 1311–1334. [CrossRef]

19. Shabbir, I.; Mirzaeian, M. Feasibility analysis of different cogeneration systems for a paper mill to improve its energy efficiency. *Int. J. Hydrog. Energy* **2016**, *41*, 16535–16548. [CrossRef]

20. Dincer, I.; Zamfirescu, C. Renewable-energy-based multigeneration systems. *Int. J. Energy Res.* **2012**, *36*, 1403–1415. [CrossRef]

21. Buoro, D.; Pinamonti, P.; Reini, M. Optimization of a distributed cogeneration system with solar district heating. *Appl. Energy* **2014**, *124*, 298–308. [CrossRef]

22. Yu, Y.; Chen, H.; Chen, L. Comparative Study of Electric Energy Storages and Thermal Energy Auxiliaries for Improving Wind Power Integration in the Cogeneration System. *Energies* **2018**, *11*, 263. [CrossRef]

23. Pereira, J.S.; Ribeiro, J.B.; Mendes, R.; Vaz, G.C.; André, J.C. ORC based micro-cogeneration systems for residential application—A state of the art review and current challenges. *Renew. Sustain. Energy Rev.* **2018**, *92*, 728–743. [CrossRef]

24. U.S. DOE Combined Heat and Power Installation Database. Available online: https://doe.icfwebservices.com/chpdb/ (accessed on 17 March 2018).

25. Onovwiona, H.; Ugursal, V. Residential cogeneration systems: Review of the current technology. *Renew. Sustain. Energy Rev.* **2006**, *10*, 389–431. [CrossRef]

26. Dennis, K. Environmentally beneficial electrification: Electricity as the end-use option. *Electr. J.* **2015**, *28*, 100–112. [CrossRef]

27. Joelsson, A. Primary Energy Efficiency and CO_2 Mitigation in Residential Buildings. Ph.D. Thesis, Mid Sweden University, Östersund, Sweden, 2008.

28. Fong, K.F.; Lee, C.K. System analysis and appraisal of SOFC-primed micro cogeneration for residential application in subtropical region. *Energy Build.* **2016**, *128*, 819–826. [CrossRef]
29. Mohamed, A.; Hasan, A.; Siren, K. Cost optimal and net zero energy office buildings solutions using small scale biomass-based cogeneration technologies. In Proceedings of the 2nd IBPSA-Italy Conference, Building Simulation Applications BSA 2015, Bozen-Bolzano, Italy, 4–6 February 2015; Available online: http://porto.polito.it/id/eprint/2626933 (accessed on 30 May 2016).
30. McDonald, E.; Kegel, M.; Tamasauskas, J. Building Energy Efficiency Assessment of Renewable and Cogeneration Energy Efficiency Technologies for the Canadian High Arctic. In Proceedings of the International High-Performance Buildings Conference, West Lafayette, Indiana, 11–14 July 2016; p. 175. Available online: http://docs.lib.purdue.edu/ihpbc/175 (accessed on 28 November 2016).
31. Somcharoenwattana, W.; Menke, C.; Kamolpus, D.; Gvozdenac, D. Study of operational parameters improvement of natural-gas cogeneration plant in public buildings in Thailand. *Energy Build.* **2011**, *43*, 925–934. [CrossRef]
32. Tchanche, B.F.; Lambrinos, G.; Frangoudakis, A.; Papadakis, G. Low-grade heat conversion into power using organic Rankine cycles–A review of various applications. *Renew. Sustain. Energy Rev.* **2011**, *15*, 3963–3979. [CrossRef]
33. Esen, M.; Yuksel, T. Experimental evaluation of using various renewable energy sources for heating a greenhouse. *Energy Build.* **2013**, *65*, 340–351. [CrossRef]
34. Chua, K.J.; Chou, S.K.; Yang, W.M.; Yan, J. Achieving better energy-efficient air conditioning—A review of technologies and strategies. *Appl. Energy* **2013**, *104*, 87–104. [CrossRef]
35. Rincón, L.; Becerra, L.; Moncada, J.; Cardona, C. Techno-economic analysis of the use of fired cogeneration systems based on sugar cane bagasse in south eastern and mid-western regions of Mexico. *Waste Biomass Valorization* **2014**, *5*, 189–198. [CrossRef]
36. Marin-Sanchez, J.E.; Rodriguez-Toral, M.A. An estimation of cogeneration potential by using refinery residuals in Mexico. *Energy Policy* **2007**, *35*, 5876–5891. [CrossRef]
37. Sheinbaum, C.; Jauregui, I.; Rodríguez, V.L. Carbon dioxide emission reduction scenarios in Mexico for year 2005: Industrial cogeneration and efficient lighting. *Mitig. Adapt. Strateg. Glob. Chang.* **1997**, *2*, 359–372. [CrossRef]
38. Huicochea, A.; Romero, R.J.; Rivera, W.; Gutierrez-Urueta, G.; Siqueiros, J.; Pilatowsky, I. A novel cogeneration system: A proton exchange membrane fuel cell coupled to a heat transformer. *Appl. Therm. Eng.* **2013**, *50*, 1530–1535. [CrossRef]
39. García, C.A.; García-Treviño, E.S.; Aguilar-Rivera, N.; Armendáriz, C. Carbon footprint of sugar production in Mexico. *J. Clean. Prod.* **2016**, *112*, 2632–2641. [CrossRef]
40. Fuentes-Cortés, L.F.; Santibañez-Aguilar, J.E.; Ponce-Ortega, J.M. Optimal design of residential cogeneration systems under uncertainty. *Comput. Chem. Eng.* **2016**, *88*, 86–102. [CrossRef]
41. Weber, B.; Cerro, E.; Martínez, I.G.; Rincón, E.; Duran, M.D. Efficient Heat Generation for Resorts. *Energy Procedia* **2014**, *57*, 2666–2675. [CrossRef]
42. Mokheimer, E.M.; Dabwan, Y.N.; Habib, M.A. Optimal integration of solar energy with fossil fuel gas turbine cogeneration plants using three different CSP technologies in Saudi Arabia. *Appl. Energy* **2017**, *185*, 1268–1280. [CrossRef]
43. Barigozzi, G.; Perdichizzi, A.; Ravelli, S. Performance prediction and optimization of a waste-to-energy cogeneration plant with combined wet and dry cooling system. *Appl. Energy* **2014**, *115*, 65–74. [CrossRef]
44. Bimüller, J.D.; Nord, L.O. Process Simulation and Plant Layout of a Combined Cycle Gas Turbine for Oshore Oil and Gas Installations. *J. Power Technol.* **2015**, *95*, 40.
45. Frunzulica, R.; Damian, A.; Baciu, R.; Barbu, C. Analysis of a CHP Plant Operation for Residential Consumers. In *Sustainable Energy in the Built Environment-Steps Towards nZEB*; Springer International Publishing: New York, NY, USA, 2014; pp. 77–86.
46. Murugan, S.; Horák, B. A review of micro combined heat and power systems for residential applications. *Renew. Sustain. Energy Rev.* **2016**, *64*, 144–162. [CrossRef]
47. Caramés, T. Diseño conceptual de una instalación de cogeneración con turbina de gas. Bachelor's Thesis, Escuela Técnica Superior de Náutica, Universidad de Cantabria, Cantabria, Spain, 2012.
48. Sartor, K.; Quoilin, S.; Dewallef, P. Simulation and optimization of a CHP biomass plant and district heating network. *Appl. Energy* **2014**, *130*, 474–483. [CrossRef]

49. Di Marcoberardino, G.; Roses, L.; Manzolini, G. Technical assessment of a micro-cogeneration system based on polymer electrolyte membrane fuel cell and fluidized bed autothermal reformer. *Appl. Energy* **2016**, *162*, 231–244. [CrossRef]

50. Mohsenzadeh, M.; Shafii, M.B. A novel concentrating photovoltaic/thermal solar system combined with thermoelectric module in an integrated design. *Renew. Energy* **2017**, *113*, 822–834. [CrossRef]

51. Medina, E.; Olmos, J.; Sánchez, A.; Valdez, E. Propuesta y análisis de un sistema fotovoltaico de autoabastecimiento eléctrico en un edificio de departamentos en Polanco. Bachelor's Thesis, Universidad Nacional Autónoma de México, México, Mexico, 2014.

52. Guelpa, E.; Barbero, G.; Sciacovelli, A.; Verda, V. Peak-shaving in district heating systems through optimal management of the thermal request of buildings. *Energy* **2017**, *137*, 706–714. [CrossRef]

53. Quispe, C. Análisis energético de un sistema de cogeneración con ciclo combinado y gasificación para la industria azucarera. Bachelor's Thesis, Universidad de Piura, Piura, Perú, 2010.

54. Ahmadi, G.; Toghraie, D.; Akbari, O.A. Efficiency improvement of a steam power plant through solar repowering. *Int. J. Exergy* **2017**, *22*, 158–182. [CrossRef]

55. Keynia, F. An optimal design to provide combined cooling, heating, and power of residential buildings. *Int. J. Model. Simul.* **2018**, 1–16. [CrossRef]

56. Ahmadi, G.; Toghraie, D.; Akbari, O.A. Technical and environmental analysis of repowering the existing CHP system in a petrochemical plant: A case study. *Energy* **2018**, *159*, 937–949. [CrossRef]

57. Basrawi, F.; Yamada, T.; Nakanishi, K.; Naing, S. Effect of ambient temperature on the performance of micro gas turbine with cogeneration system in cold region. *Appl. Therm. Eng.* **2011**, *31*, 1058–1067. [CrossRef]

Article

Energy-Efficient Train Driving Strategy with Considering the Steep Downhill Segment

Wentao Liu [1,2], **Tao Tang** [1,2], **Shuai Su** [1,3,*], **Jiateng Yin** [1], **Yuan Cao** [2,3] **and Cheng Wang** [4]

[1] State Key Laboratory of Rail Traffic Control and Safety, Beijing Jiaotong University, Beijing 100044, China; 17120250@bjtu.edu.cn (W.L.); ttang@bjtu.edu.cn (T.T.); jtyin@bjtu.edu.cn (J.Y.)

[2] School of Electronic and Information Engineering, Beijing Jiaotong University, Beijing 100044, China; ycao@bjtu.edu.cn

[3] National Engineering Research Center of Rail Transportation Operation and Control System, Beijing Jiaotong University, Beijing 100044, China

[4] School of IoT Engineering, Jiangnan University, Wuxi 214122, China; wangc@jiangnan.edu.cn

* Correspondence: shuaisu@bjtu.edu.cn

Received: 13 January 2019; Accepted: 31 January 2019; Published: 3 February 2019

Abstract: Implementation of energy-efficient train driving strategy is an effective method to save train traction energy consumption, which has attracted much attention from both researchers and practitioners in recent years. Reducing the unnecessary braking during the journey and increasing the coasting distance are efficient to save energy in urban rail transit systems. In the steep downhill segment, the train speed will continue to increase without applying traction due to the ramp force. A high initial speed before stepping into the steep downhill segment will bring partial braking to prevent trains from overspeeding. Optimization of the driving strategy of urban rail trains can avoid the partial braking such that the potential energy is efficiently used and the traction energy is reduced. This paper presents an energy-efficient driving strategy optimization model for the segment with the steep downhill slopes. A numerical method is proposed to calculate the corresponding energy-efficient driving strategy of trains. Specifically, the steep downhill segment in the line is identified firstly for a given line and the solution space with different scenarios is analyzed. With the given cruising speed, a primary driving strategy is obtained, based on which the local driving strategy in the steep slope segment is optimized by replacing the cruising regime with coasting regime. Then, the adaptive gradient descent method is adopted to solve the optimal cruising speed corresponding to the minimum traction energy consumption of the train. Some case studies were conducted and the effectiveness of the algorithm was verified by comparing the energy-saving performance with the classical energy-efficient driving strategy of "Maximum traction–Cruising–Coasting–Maximum braking".

Keywords: rail transit; train control; energy-efficient driving strategy; steep downhill segment; local optimization

---

1. Introduction

Owing to the advantages in safety, high capacity and efficiency, urban rail transit has rapidly developed worldwide in recent years. However, with the massive construction and short headway, the energy consumption of urban rail systems has increased dramatically. Consequently, how to reduce the total energy consumption has become an important and urgent concern for a sustainable development of rail transit systems. The traction energy consumption of trains accounts for about 53% of the total energy consumption in urban rail transit system [1]. Thus, the total energy consumption of the system can be effectively decreased if the train traction energy consumption is reduced, which would also contribute to the reduction of operational cost and carbon emission [2].

Implementation of energy-efficient train driving strategy in the automatic train operation (ATO) system contributes greatly to reducing the traction energy consumption. In recent years, many scholars have conducted a lot of research on the energy-efficient train control problem, which is mainly divided into continuous control models and discrete control models.

In 1980, Milroy [3] developed an approach to optimize train speed trajectory based on the continuous train control model, which established the theoretical foundation of the optimal train control problem. Afterwards, the problem with constant gradients was analyzed and the Pontryagin maximum principle was applied to obtain the optimal speed trajectory by Asnis [4]. Howlett [5,6] formulated a finite dimensional constrained optimization model and used the maximum principle to solve the energy-efficient train driving regimes and the corresponding switching points. Considering varying gradients and speed limits, Khmelnitsky [7] built a continuous train driving model, in which the kinetic energy was considered as the state variable. Liu [8] proposed an analytic algorithm to solve the optimal switching points among different regimes by applying the Pontryagin maximum principle. Taking variable traction efficiency into consideration, Su [9,10] developed a numerical algorithm based on an energy consumption allocation method, in which the energy-efficient driving strategy among multi-stations was calculated by optimizing the multi-station running time distribution. Except for the analytical and numerical algorithms, many scholars also used intelligent algorithms based on modern heuristic search methods to study the energy-efficient driving strategy of trains. To reduce traction energy consumption by making full use of coasting, Chang and Sim [11] applied the genetic algorithm to optimize the position of the coast starting points in the coast control table. Ma [12] used real-coded genetic algorithm to automatically calculate the optimal coasting points in the energy-efficient driving strategy of subway trains. Jin [13] used the neural network technology and genetic algorithm to study the energy-efficient driving strategy of trains on the undulating track, which could adapt to different line conditions and meet the requirements of real-time optimization. Ke et al. [14,15] presented a method of designing block-layout between successive stations, in which the Max-Min ant colony algorithm was used to optimize the train speed curve for a significant improvement in computational efficiency.

In practice, the control output of the diesel electric locomotives used for the heavy freight is discrete. Each different handle position corresponds to a fixed fuel supply rate and output power, thereby determining that the traction cannot change continuously. Therefore, many scholars have developed discrete train control models for this feature [16]. In the 1990s, Cheng and Howlett [17,18] studied the energy-efficient driving problem with a constant gradient and speed limit, and proved that the energy-efficient driving regimes of discrete control model trains included "maximum traction, coasting, and maximum braking". Then, Pudney and Howlett [19] took varying speed limits into consideration, and proved that the train must run at limits on the segments where the speed limits were lower than the expected cruising speed. Besides, Howlett and Cheng [20] considered the problem of continuously changing gradients. They solved critical equations of switching points in different operating regimes by using Lagrange Function and Kahn–Tucker Conditions to find an optimal type of driving strategy. Importantly, Howlett [21] proved that an arbitrary continuous energy-efficient operating sequence can be approximated by discrete "traction–coasting" pairs, establishing a connection between continuous control and discrete control models. In addition, considering non-constant gradients, curve and speed limits, Han [22] used genetic algorithms to optimize the driving strategy of the train ATO system. Ding [23] also designed a genetic algorithm to find the optimal solution of the energy-efficient train operation problem. In 2014, based on a real ATO system, Dominguez [24] introduced a Multi-Objective Particle Swarm Optimization algorithm to obtain the consumption/time Pareto front, which solved the optimization problem more efficiently than the previous algorithms.

It is concluded from the above studies that the classical energy-efficient driving strategy consists of maximum traction, cruising, coasting, and maximum braking as well as their corresponding switching points [2]. However, if there are steep slope segments in the route, the position of the switching points will be affected. For an individual steep uphill segment, Howlett [25] introduced an analytical method to obtain the optimal switching points. Furthermore, Albrecht [26–28] proved that the optimal driving strategy always existed and was unique via a perturbation analysis. Considering the regenerative braking in the train model, Ko [29,30] adopted dynamic programming to optimize the train driving strategy with the confined state space and irregular lattice for trains running on the route containing a steep downhill. For a steep downhill segment, if entering the ramp at a relatively high speed, the train would probably brake on the segment to avoid exceeding the speed limit, which increases the global energy consumption. If the entering speed is relatively low, the trip time would be longer and the operational efficiency and service quality would consequently be reduced. Hence, to make full use of the potential energy of the steep downhill segment, this paper proposes an approach to solving the energy-efficient control problem for trains running in the steep downhill segment based on the classical energy-efficient driving strategy.

The main contributions of this paper are stated as follows:

1. Considering the route with a steep downhill, the solution space for a given cruising speed is analyzed to obtain the classical energy-efficient driving strategy for a given trip time.
2. A local optimization approach is developed to reduce the traction energy consumption for trains running in the steep downhill segment by applying the dichotomization algorithm.
3. A global optimization is achieved with the utilization of the adaptive gradient descent method to calculate the optimal cruising speed, which corresponds to the minimum traction energy consumption.

The remainder of the paper is organized as follows. Section 2 describes the optimized problem including the formulation of an energy-efficient train control model as well as the definition of the steep downhill. In Section 3, a numerical method is designed to calculate the optimal driving strategy for trains running in a steep downhill segment. In Section 4, simulations with actual data of Beijing Yizhuang line is presented to illustrate the effectiveness of the proposed approach, followed by the conclusion in Section 5.

2. Problem Description

2.1. Assumptions

To simplify the train model, we make the following assumptions:

- The train is considered as a mass point when running on the track, and its mass is fixed.
- The slope of the centroid of the train represents the gradient of the entire vehicle.
- The traction efficiency is assumed to be constant, and the mechanical energy is used as the traction energy during the trip.
- The regenerative braking is not considered in the train model, because we only consider optimizing the driving strategy of a single train in this work.

2.2. Model Formulation

The train driving strategy, which is intuitively reflected with a train running speed curve, is the combination of an operating control sequence and the corresponding switching points among different control regimes. With the given planned trip time, line conditions, vehicle performances, etc., a set of train driving strategies between two successive stations satisfy the operation constraints [31] (see Figure 1).

Although the above driving strategies can make the train arrive at the target station on a punctual basis, the accelerating distances and positions are not the same between different driving strategies, resulting in different traction energy consumption for the interstation. The purpose of energy-efficient driving is to find a driving strategy with minimum traction energy consumption among these feasible strategies.

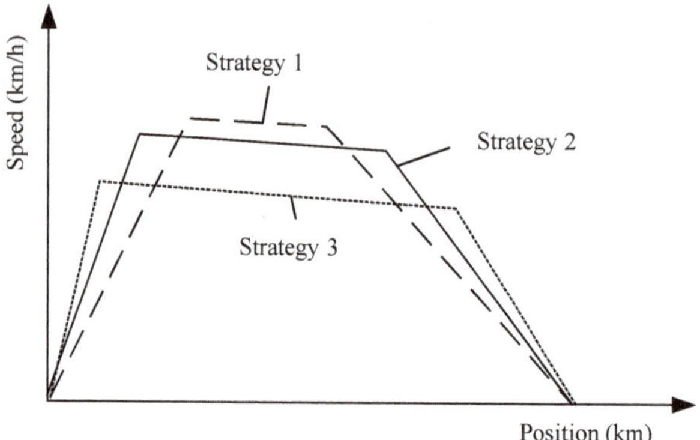

Figure 1. Different driving strategies of trains between successive stations.

The objective function of the energy-efficient train control problem can be generally written as [8]

$$\min E = \int_0^S u_f \cdot F(v)dx. \tag{1}$$

where E represents the traction energy consumption; u_f is the relative traction force; F is the maximum traction force; v is the train speed; and x and S are the train position and the trip distance, respectively.

For a mass-point train, the equation of motion can be described as [8]

$$\begin{cases} \dfrac{dv}{dt} = \dfrac{u_f F(v) - u_b B(v) - R_b(v) - R_g(x) - R_c(x)}{m\rho}, \\ \dfrac{dx}{dt} = v, \end{cases} \tag{2}$$

where m is the train mass; ρ is the rotating mass factor; u_b and B denote the relative braking force and maximum braking force, respectively; $R_b(v)$ is the basic resistance including rolling resistance and air resistance; and $R_g(x)$ and $R_c(x)$ represent the grade resistance and curve resistance, respectively.

The traction force and the braking force should be bounded by the maximum traction and braking force. Thus, we have

$$u_f \in [0,1], u_b \in [0,1]. \tag{3}$$

The train speed should satisfy the maximum train speed and trip time constraints:

$$0 \le v(x) \le V_{max}(x). \tag{4}$$

where V_{max} is the maximum allowable train speed with respect to x.

To arrive at the next station on time and stop precisely, the boundary conditions of the train movement are described as follows.

$$\begin{aligned} v(0) &= v_{start} = 0, x(0) = x_{start} = 0, \\ v(T) &= v_{end} = 0, x(T) = x_{end} = S. \end{aligned} \tag{5}$$

where T is the planned trip time given by the timetable.

Based on Equations (1)–(5), the optimization model is formulated to minimize the energy consumption during the trip.

2.3. Definition of Steep Downhill

The steep downhill is a piece of route where the train speed will increase without applying traction force (see Figure 2). In the steep downhill segment, the gradient force is larger than the sum of the curve resistance and line resistance.

$$R_g(x) - (R_c(x) + R_b(v_c)) > 0, \tag{6}$$

In Equation (6), $R_b(v_c)$ can be described as follows [32]

$$R_b(v_c) = m(a_r v_c^2 + b_r v_c + c_r), \tag{7}$$

$R_g(x)$ is calculated by

$$R_g(x) = mg \sin \alpha(x), \tag{8}$$

and the curve resistance can be described by empirical formulas [33]

$$R_c(x) = f_c(r(x)) = \begin{cases} m\dfrac{6.3}{r(x) - 55}, & r(x) \geq 300m \\ m\dfrac{4.91}{r(x) - 30}, & r(x) < 300m \end{cases}, \tag{9}$$

where a_r, b_r, and c_r are non-negative constants, which can be identified from the historical data. g, $\alpha(x)$ and $r(x)$ are the gravitational acceleration, the slope and the radius of curvature, respectively. With the estimate train speed for a fixed position, the basic and curve resistances can be calculated. By comparing with the gradient force, the steep downhill slope can be easily identified.

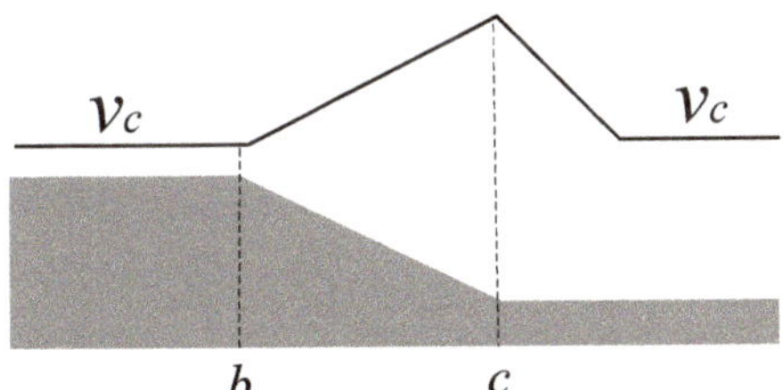

Figure 2. Train speed curve in a steep downhill segment.

3. Solution Approach

Given a specific planned trip time, the proposed solution algorithm starts from calculating the classical energy-efficient driving strategy with an initial cruising speed. With the obtained driving strategy, a local optimization approach is then developed to optimize the driving strategy in the steep downhill segment. Furthermore, an adaptive gradient method is applied to adjust the cruising speed such that the traction energy consumption during the trip is minimized.

The algorithm for solving the energy-efficient driving strategies of trains with considering the steep downhill segment is mainly divided into the following four parts and the overall framework of the proposed algorithm is shown in Figure 3.

- Initialization: Load line data and vehicle data and set a planned trip time T.
- Solution to classical energy-efficient driving strategy: Initialize a cruising speed v_c, identify steep segment $[b, c]$ and solve classical energy-efficient driving strategy.
- Local optimization on $[b, c]$: Judge the driving regime of the train in the steep downhill; if the train is cruising in this segment, then optimize the local driving strategy of the segment.

- Optimal strategy search with adaptive gradient method: Calculate energy consumption and use the gradient method to solve the cruising speed corresponding to the minimum traction energy consumption, and then obtain the optimal energy-efficient driving strategy.

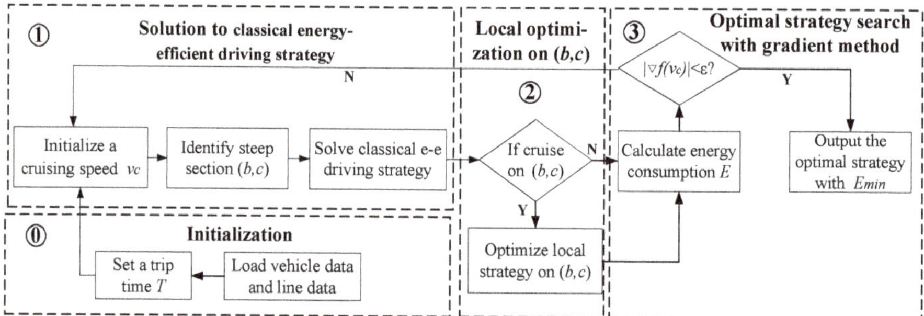

Figure 3. The structure of the proposed optimization approach.

3.1. Calculation of Classical Energy-Efficient Driving Strategy

3.1.1. Analysis of Energy-Efficient Driving Regimes

In this subsection, the energy-efficient driving regimes are analyzed by applying the Pontryagin maximum principle, according to which the following Hamiltonian function should be maximized with optimal control sequences [8,10]

$$H = \frac{\eta_1(x)}{m\rho v} \times (u_f F - u_b B - R_b - R_g - R_c) + \eta_2(x)v - u_f F. \tag{10}$$

where η_1 and η_2 and the complementary slackness condition $M(x)$ should satisfy the following differential equations:

$$\frac{dv}{dt} = \frac{dH}{d\eta_1}, \quad \frac{dx}{dt} = \frac{dH}{d\eta_2},$$
$$\frac{dM}{dx}(v - V_{max}) = \frac{dH}{d\eta_1}, \quad \frac{dM}{dx} \geq 0 \tag{11}$$

Equation (10) can be rewritten as

$$H = \left(\frac{\eta_1(x)}{m\rho v} - 1\right)u_f F - \frac{\eta_1(x)}{m\rho v}(u_b B + R_b + R_g + R_c) + \eta_2(x)v. \tag{12}$$

Thus, the four energy-efficient driving regimes are derived by maximizing Equation (12) in the following cases with respect to the control variables u_f and u_b [10]:

- $\eta_1 > m\rho v$: u_f should be 1 and u_b is 0, which implies the Maximum traction regime.
- $\eta_1 = m\rho v$: u_b should be 0 and u_f may vary in (0,1), which indicates the Partial traction phase; $\eta_1 = 0$: u_f should be 0, and u_b may vary in (0,1), which suggests the Partial braking phase. These two phases only exist for the Cruising regime by analyzing Equation (11).
- $0 < \eta_1 < m\rho v$: u_b should be 1 and u_f is 0, which implies the Maximum braking regime.
- $\eta_1 < 0$: Both u_f and u_b should be 0, which suggests the Coasting regime.

3.1.2. Solution Space Analysis

For an individual given cruising speed, the steep downhill segment $[b, c]$ in the line is identified by Equation (6). Then, four critical states of the driving strategy, as shown in Figure 4, and the corresponding running time t_1, t_2, t_3 and t_4 are calculated. t_i is explained as follows:

- t_1 denotes the trip time of the driving strategy that the train begins to coast as soon as it reaches the cruising speed at the position s_1.
- t_2 denotes the trip time of the driving strategy that the train begins to coast at the initial position s_2 of the steep slope.
- t_3 denotes the trip time of the driving strategy that the train begins to coast at the final position s_3 of the steep slope.
- t_4 denotes the trip time of the driving strategy that the train begins to coast at the position s_4 when the train speed reaches the braking profile.

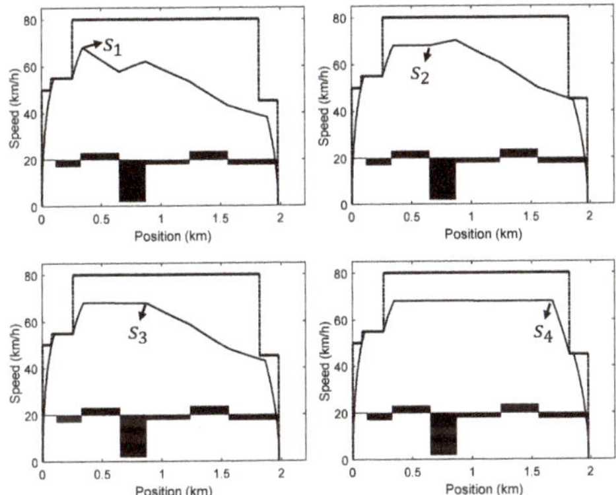

Figure 4. Driving strategies for the train, respectively, coasting from s_1, s_2, and s_3,s_4.

By the analysis, the length and position of the steep slope differ in the different steep slope lines, which leads to five typical scenarios of the running time t_1, t_2, t_3 and t_4 (see Figure 5).

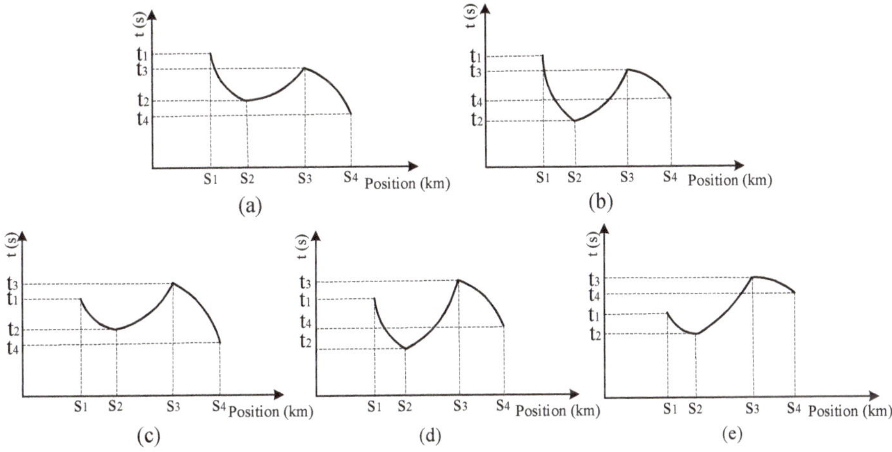

Figure 5. Five typical scenarios of the running time t_1, t_2, t_3 and t_4. (**a**) $t_1 > t_3 > t_2 > t_4$; (**b**) $t_1 > t_3 > t_4 > t_2$; (**c**): $t_3 > t_1 > t_2 > t_4$; (**d**) $t_3 > t_1 > t_4 > t_2$; (**e**) $t_3 > t_4 > t_1 > t_2$.

- Scenario 1: $t_1 > t_3 > t_2 > t_4$ (Figure 5a). This situation always happens for the scenario that the steep downhill is relatively short and exists in the middle of the line. For the segment $[s_1, s_2]$, when the train begins to coast later, the trip time with a higher average coasting speed will be shorter, i.e., $t_1 > t_2$. Coasting from the position s_2 makes the train speed increase due to the steep slope such that the speed in the steep slope and the following segment is higher, which indicates $t_3 > t_2$. In addition, a small steep slope has a small effect on the average train speed. Keeping the cruising speed on the segment $[s_1, s_3]$ can achieve a short trip time, i.e., $t_1 > t_3$. The train speed will decrease without traction and steep slopes after s_3. Hence, we have $t_3 > t_4$. In conclusion, t_1, t_2, t_3 and t_4 will satisfy $t_1 > t_3 > t_2 > t_4$.
- Scenario 2: $t_1 > t_3 > t_4 > t_2$ (Figure 5b). This situation generally happens for the scenario that the steep downhill exists in the second half of the line. Differing from the first situation, due to the steep slope exists in the second half, coasting from the position s_2 makes the average train speed higher than the cruising speed in the segment $[s_2, s_4]$, i.e., $t_4 > t_2$. As a result, t_1, t_2, t_3 and t_4 will satisfy $t_1 > t_3 > t_4 > t_2$.
- Scenario 3: $t_3 > t_1 > t_2 > t_4$ (Figure 5c). This situation always happens when the steep downhill is relatively short and exists in the first half of the line. After accelerating to the cruising speed, the train immediately enters the steep downhill segment, and the speed rises to exceed the cruising speed such that the average running speed of the train in the segment $[s_1, s_3]$ is greater than the cruising speed, i.e., $t_3 > t_1$. Thus, t_1, t_2, t_3 and t_4 will satisfy $t_3 > t_1 > t_2 > t_4$.
- Scenario 4: $t_3 > t_1 > t_4 > t_2$ (Figure 5d). This situation exists when the steep downhill is relatively long and exists in the first half of the line. Because a longer steep slope has a greater effect on the average train speed, coasting from the position s_2 makes the average train speed higher than the cruising speed for the segment $[s_2, s_4]$, i.e., $t_4 > t_2$. Therefore, t_1, t_2, t_3 and t_4 will satisfy $t_3 > t_1 > t_4 > t_2$.
- Scenario 5: $t_3 > t_4 > t_1 > t_2$ (Figure 5e). This situation happens when the relatively long steep downhill exists in the first half of the line, and there is a downward slope, but not a steep downhill, in the segment $[s_1, s_2]$. Compared with Scenario 4, a downward slope in the segment $[s_1, s_2]$ brings a higher initial speed before stepping into the steep downhill, hence, the average train speed will be higher than the cruising speed in the segment $[s_1, s_4]$, that is, the trip time will be shorter ($t_4 > t_1$). Consequently, t_1, t_2, t_3 and t_4 will satisfy $t_3 > t_4 > t_1 > t_2$.

3.1.3. Coasting Position Calculation

As shown in Figure 5, the running time t in each section is monotonously increasing or decreasing. According to the planned trip time T and the relationships among t_1, t_2, t_3, t_4, the distribution of the coasting position is firstly determined in Table 1.

Table 1. Five scenarios of t_1, t_2, t_3, and t_4 and distribution of the coasting point s.

$t_1 > t_3 > t_2 > t_4$		$t_1 > t_3 > t_4 > t_2$		$t_3 > t_1 > t_2 > t_4$		$t_3 > t_1 > t_4 > t_2$		$t_3 > t_4 > t_1 > t_2$	
$t_2 > T > t_4$	$[s_3, s_4]$	$t_4 > T > t_2$	$[s_1, s_2]$	$t_2 > T > t_4$	$[s_3, s_4]$	$t_4 > T > t_2$	$[s_1, s_2]$	$t_1 > T > t_2$	$[s_1, s_2]$
$t_3 > T > t_2$	$[s_1, s_2]$	$t_3 > T > t_4$	$[s_1, s_2]$	$t_1 > T > t_2$	$[s_1, s_2]$	$t_1 > T > t_4$	$[s_1, s_2]$	$t_4 > T > t_1$	-
$t_1 > T > t_3$	$[s_1, s_2]$	$t_1 > T > t_3$	$[s_1, s_2]$	$t_3 > T > t_1$	-	$t_3 > T > t_1$	-	$t_3 > T > t_4$	-

Taking the first scenario as an example, the detailed analysis process is described as the following:

- $t_2 > T > t_4$ or $t_1 > T > t_3$: In Figure 5a, the horizontal line from the ordinate T intersects the curve in the figure at a unique point $s \in [s_3, s_4]$ or $s \in [s_1, s_2]$, which indicates the optimal coasting position with the trip time T exists in the segment $[s_3, s_4]$ or $[s_1, s_2]$, and the solution is unique in these two situations.

- $t_3 > T > t_2$: There will be three solutions, which belong to $[s_1, s_2]$, $[s_2, s_3]$ and $[s_3, s_4]$, respectively. Due to the principle that, when the running times are the same, a longer coasting distance contributes to a smaller traction energy consumption, the optimal coasting position should be found in $[s_1, s_2]$.

It is noted that cases such as $t_3 > T > t_1$ in Scenario 3 should be treated differently. As shown in Figure 5c, if the planned trip time T satisfies $t_3 > T > t_1$, there will be two different solutions that belong to $[s_2, s_3]$ and $[s_3, s_4]$, respectively. By applying the principle that a longer coasting distance leads to a smaller traction energy consumption, the solution belonging to $[s_3, s_4]$ is undesirable. Additionally, if the train switches from cruising to coasting in position s_c that belongs to $[s_2, s_3]$, the potential energy in the first half of the segment will be wasted (see Figure 6). Compared to the strategy that trains begin to coast at the position s, the trip time is the same but the latter strategy will cost less energy consumption, i.e., $E_{s_c} > E_s$. Thus, the solution s_c is also undesirable. In addition, though the strategy with coasting from s is dropped as the speed of s is less than the current given cruising speed, it will be found in another circumstance when the other cruising speed is given.

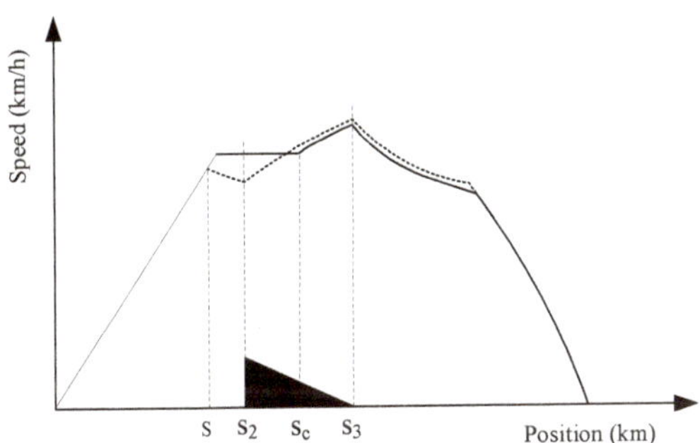

Figure 6. Driving strategies with coasting from s and s_c, respectively.

Summarizing Sections 3.1.2 and 3.1.3, the starting point s of the coasting regime is found according to the time of four critical states and the given trip time, and then a classical energy-efficient driving strategy with a fixed cruising speed is obtained.

3.2. Local Optimization on Steep Downhill Segment

As shown in Figure 7, $[b, c]$ is the steep downhill segment. If the train uses the cruising regime at the segment $[a, d]$ in the classical energy-efficient driving strategy, the speed profile is shown as the horizontal dotted line and the running time in $[a, d]$ is T_{hold}. Specifically, the train may apply partial traction at the segment $[a, b]$ to maintain the cruising speed v_c. The traction energy consumption at such segment is $E_{(a,b)}$. The train applies partial braking at the segment $[b, c]$ to maintain the train drive at the cruising speed. The traction energy consumption at this segment is $E_{(b,c)} = 0$. The regime at segment $[c, d]$ is the same as that at the segment $[a, b]$, and the traction energy consumption is $E_{(c,d)}$. Thus, before optimizing the driving strategy of steep downhill segment, the actual traction energy consumption at the segment $[a, d]$ can be expressed as

$$E_{(a,d)} = E_{(a,b)} + E_{(c,d)}. \tag{13}$$

Meeting the trip time constraint, the proposed local optimal driving strategy in the steep downhill is to switch the regime from cruising to coasting at point $x = a$ before the steep segment and from coasting to cruising at point $x = d$ after passing by the steep segment. The train coasts during the segment $[a, d]$ and the coasting time is T_{coast}. In this way, the potential energy of the steep downhill segment can be fully used. Furthermore, substituting the cruising with the coasting consumes no traction energy, i.e., $E'_{(a,d)} = 0$. Hence, the traction energy consumed by the optimized energy-efficient driving strategy is obviously reduced compared to the classical driving strategy [2], i.e., we have the following inequality:

$$E_{(a,d)} > E'_{(a,d)}. \tag{14}$$

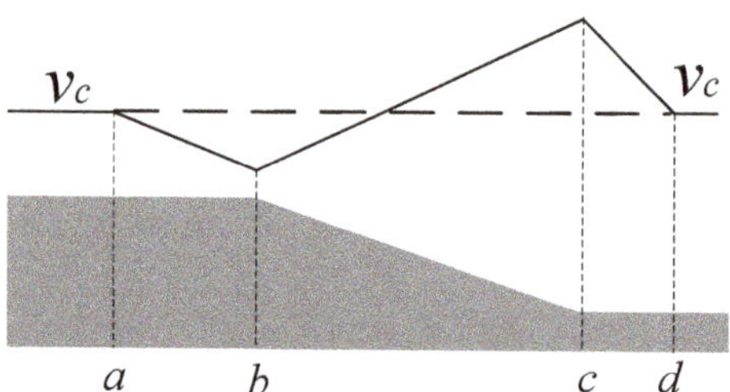

Figure 7. Local optimization at the steep downhill segment.

The trip time for the segment $[a, d]$ is continuous with respect to the coasting position s. The "$s*$" must exist, which satisfies

$$T_{hold} = T_{coast}. \tag{15}$$

Equations (14) and (15) indicate that the effect of local optimization in the steep slope segment is realized with the same trip time. To obtain the optimal coasting position s^*, the dichotomization method is used in this paper.

Specifically, we define a time function

$$T(s_a) = T_{hold}(s_a) - T_{coast}(s_a), \tag{16}$$

then the purpose of the dichotomization algorithm for calculating the coasting position is to find the position s^* that satisfies

$$s^* = \arg_{s_a} \left(T(s_a) = 0 \right). \tag{17}$$

The specific steps of the dichotomization algorithm can be described as Algorithm 1.

Algorithm 1: The dichotomization algorithm for calculating the optimal coasting position.

Step 1. Give a search segment $[x_s, x_e]$ (initially setting $x_s = s_1$ and $x_e = s_2$) and an error ξ, and verify
$\quad T(x_s) \cdot T(x_e) < 0$;

Step 2. Calculate the midpoint x_m of the segment $[x_s, x_e]$ and its time function $T(x_m)$;

Step 3. If $T(x_s) \cdot T(x_m) < 0$, then turn to step (4), otherwise, turn to step (5);

Step 4. If $x_m - x_s < \xi$, then output the coasting point $s^* = x_m$; otherwise, return to step (2) with setting
$\quad x_e = x_m$;

Step 5. If $x_e - x_m < \xi$, then output the coasting point $s^* = x_m$; otherwise, return to step (2) with setting
$\quad x_s = x_m$.

3.3. Calculation of the Optimal Cruising Speed With Adaptive Gradient Descent Method

The above subsections have described how to obtain an energy-efficient driving strategy for a fixed cruising speed. In this subsection, the adaptive gradient descent (AGD) method is applied to solve the optimal cruising speed corresponding to the minimum traction energy consumption of the trip. By dynamically incorporating knowledge of the geometry of the gradient in earlier iterations, AGD is able to perform informative gradient-based search, which ensures more robust performance [34].

The objective of gradient search is to find the optimal cruising speed

$$v_c^* = \arg\min_{v_c} E(v_c^k). \tag{18}$$

The specific solution to the strategy search problem is described in Algorithm 2.

Algorithm 2: The ADG algorithm for calculating the optimal cruising speed.

Step 1. Give a cruising speed $v_c^0 \in [v_c^{min}, v_c^{max}]$, set the allowable error ε, the number of iterations $k = 0$ and initialize a step size parameter λ_0;

Step 2. Calculate the traction energy consumption $E_k = E(v_c^k)$ and the gradient $g_k = \nabla E(v_c^k)$ with v_c^k;

Step 3. Take the negative gradient direction as the search direction $d_k = -g_k$, and determine the step size λ_k by $\lambda_k = \dfrac{\lambda}{\sqrt{\sum\limits_{i=0}^{k} g_i^2}} * g_k$.

Step 4. Search the next cruising speed $v_c^{k+1} = v_c^k + \lambda_k d_k$, and calculate $E_{k+1} = E(v_c^{k+1})$ and $g_{k+1} = \nabla E(v_c^{k+1})$.

Step 5. Determine whether the termination condition $|E_{k+1} - E_k| \leq \varepsilon$ is satisfied. If yes, then output the optimal strategy with v_c^*. Otherwise, set $k = k + 1$, and return to step (2).

---

4. Simulation Results

In this Section, we present some simulations to illustrate the effectiveness of the proposed optimization approach. In Case 1, a case study based on the practical data of Yizhuang line was conducted and a comparison of energy efficiency was derived between the two algorithms, i.e., the proposed and the classical algorithms. In Case 2, the route was gently modified to verify the availability of the proposed approach in the different situations shown in Figure 5.

4.1. Case 1

In this case, we chose the interval from Jinghai to Ciqu as an example. The distance between Jinghai and Ciqu stations is 2086 m and the speed limit of the whole segment is 80 km/h. The gradient information is shown in Table 2. The vehicle data of Beijing Yizhuang rail transit line were used for simulations. Characteristics of the traction force $F(v)$, the braking force $B(v)$ and the basic resistance $R_b(v)$ can be found in [10].

Table 2. Gradient information between Jinghai and Ciqu.

Segments (m)	Gradients (‰)
0–19	−1.4006
20–339	0
340–689	−15.6250
690–1389	−24.3900
1390–1629	3.0030
1630–1979	−10.1010
1980–2086	−2

For the cruising speed ranging 65–78 km/h, the segment from 690 m to 1389 m was identified as the steep downhill segment according to Equation (6).

When the planned trip time T was set to be 160 s, the energy-efficient driving strategies calculated with the classical algorithm and the proposed approach are shown in Figure 8. As shown in Figure 8a, the cruising speed of the classical energy-efficient driving strategy is 69 km/h and the traction energy consumption is 16.56 kW·h.

Additionally, Figure 8b shows that the optimal cruising speed solved by the proposed approach is 72 km/h and the traction energy consumption is reduced to 14.79 kW·h. It is worth noting that there is no unnecessary braking on the steep slope segment in the optimized energy-efficient driving strategy, which consists of maximum traction regime in $[x'_1, x'_2]$, cruising regime in $[x'_2, x'_3]$, coasting regime in $[x'_3, x'_4]$ and maximum braking in $[x'_4, x'_5]$.

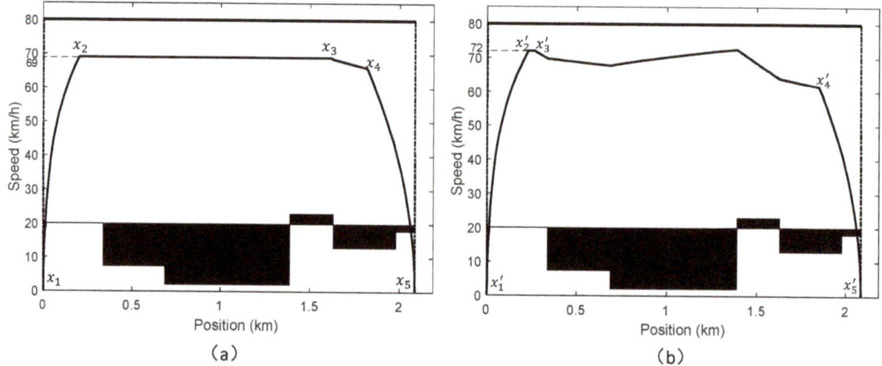

Figure 8. Energy-efficient driving strategies calculated with: (**a**) the classical algorithm; and (**b**) the proposed approach.

The energy-efficient driving strategies with the proposed approach and the classical energy-efficient algorithm were solved for other trip times. Then, the energy consumption was compared, as shown in Table 3.

Table 3. Energy efficiency comparison between the proposed and classical algorithm with different trip times.

E (kW·h) Methods Trip Times	Proposed	Classical	Proportion
155 s	17.52	19.27	9.08%
157 s	16.35	18.48	11.52%
160 s	14.79	16.56	10.69%
162 s	13.84	15.38	10.01%
164 s	13.00	15.03	13.50%
Average	-	-	10.96%

It is shown from the simulation results that the energy consumption is reduced by 10.96% in average. In addition, the energy-efficient performance is low when the trip time is short. To reduce the trip time, the train has to apply partial braking to keep a high speed, although the steep downhill segment exists. As a result, the potential energy cannot be used efficiently and the energy reduction is less in this situation.

The simulation was conducted on a personal desktop PC with processor speed of 3.6 GHz and memory size of 16 GB. The average computation time of the proposed approach was about 0.12 s, which can satisfy the requirement of real-time control.

4.2. Case 2

As stated in Section 3.1, there are five kinds of lines, in which the length and position of the steep downhill segment are different. In this case, the availability of the proposed method was tested with the different line data (see Table 4).

Table 4. Gradient information of different lines.

Lines	Segments (m)	Gradients (‰)	Lines	Segments (m)	Gradients (‰)
	0–120	0		0–100	0
	121–330	−3		101–380	3
	331–650	3		381–660	−4
Line 1	651–870	−26	Line 2	661–1200	2.5
	871–1240	−2		1201–1620	−28
	1241–1560	3.5		1621–1880	3.5
	1561–1980	−2.5		1881–1980	2
	0–100	0		0–120	0
	101–320	3.5		121–420	2
	321–460	2		421–1400	−26
	461–780	−26	Line 4	1401–1650	3.5
Line 3	781–970	3		1651–1800	−2
	971–1320	−3		1801–2000	2
	1321–1580	4		2001–2200	1.5
	1581–1760	−2			
	1761–1980	2			
	0–120	0			
	121–280	2			
	281–420	−4			
Line 5	421–1400	−26			
	1401–1650	3.5			
	1651–1800	−2			
	1801–2000	2			

Taking Line 1 as an illustrative example, the planned trip time T was set to be 160 s. The optimal cruising speed calculated by the proposed approach is 75 km/h, and the corresponding energy consumption is 19.93 kW·h. The energy-efficient driving strategies solved by the classical (the black line) and the proposed (the red line) approaches are shown in Figure 9. Obviously, tn the proposed approach, the cruising regime is substituted with coasting on segment $[x_3, x_4]$ within the same trip time.

Under this circumstance, the critical states are $s_1 = 426$ m, $s_2 = 651$ m, $s_3 = 870$ m and $s_4 = 1626$ m; the running times are $t_1 = 167.61$ s, $t_2 = 160.06$ s, $t_3 = 160.98$ s and $t_4 = 151.91$ s; and the coasting position s, i.e., x_5 in Figure 9, is 914 m. It is clearly shown that, because T is set to be within the time interval $(t_4, t_2]$, the coasting point s is obtained from the interval $(s_3, s_4]$, which is in accordance with Scenario 1 described in Table 1, i.e., $s \in (s_3, s_4]$.

Furthermore, when different line conditions and planned trip times are given, the results solved by the proposed approach are shown in Tables 5 and 6. Table 5 shows the optimal cruising speed and corresponding energy consumption of each scenario with the given planned trip time. Moreover, the critical states and the corresponding running time of each scenario are given in Table 6. It can be observed that the coasting point s of each scenario is obtained according to the planned trip time, as analyzed in Section 3.1. For instance, the planned trip time T of Line 2 was set as 160 s, which belongs to the interval $(149.84, 168.02]$, i.e., $T \in (t_3, t_1]$. Thus, according to Scenario 2 illustrated in Table 1, the coasting point s found in $(333, 1201]$ is 622 m, i.e., $s \in (s_1, s_2]$.

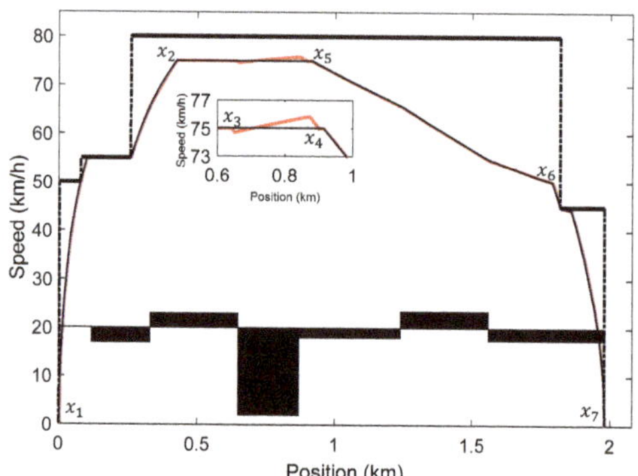

Figure 9. Energy-efficient driving strategy for the train running in Line 1 with $T = 160$ s.

Table 5. The optimal cruising speed v_c^* and energy consumption E of each line with a planned trip time T.

Lines	T (s)	v_c (km/h)	E (kW·h)
Line1	160	75	19.93
Line2	160	74	19.19
Line3	160	73	19.38
Line4	164	73	15.85
Line5	154	70	14.29

Table 6. Four critical states s_1–s_4, the corresponding running time t_1–t_4 and the coasting point s of different lines.

Lines	s_1 (m)	s_2 (m)	s_3 (m)	s_4 (m)	s (m)
Line1	426	651	870	1626	914
Line2	333	1201	1620	1689	622
Line3	430	461	780	1579	1056
Line4	319	421	1400	1906	343
Line5	287	421	1400	1723	310

Lines	t_1 (s)	t_2 (s)	t_3 (s)	t_4 (s)	T (s)
Line1	167.61	160.06	160.98	151.91	160
Line2	168.02	149.22	149.84	149.76	160
Line3	166.74	165.69	167.98	155.48	160
Line4	164.56	162.21	166.26	161.77	164
Line5	154.35	152.35	156.80	154.85	154

In addition, we present comparisons among the proposed and the classical strategies to illustrate the energy-saving performance, respectively, with the same planned trip time (see Table 7). It can be concluded that the greater the proportion of the steep slope in the line accounts for, the more obvious the energy-saving effect is (e.g., see Line 4 and Line 5).

Table 7. Energy-saving performance comparisons for trains running in different lines.

E (kW·h) \ Methods Lines	Proposed	Classical	Proportion
Line 1	19.00	19.48	2.46%
Line 2	21.65	23.37	7.36%
Line 3	18.37	19.27	4.67%
Line 4	16.16	19.23	15.9%
Line 5	15.25	18.07	15.6%
Average	-	-	9.20%

5. Conclusions

Based on the classical energy-efficient train control approach, this paper proposes an optimization approach focusing on solving the energy-efficient control problem for trains running on a line with a steep downhill segment. The solution space for a given cruising speed is firstly analyzed to obtain the classical energy-efficient driving strategy for a given trip time. With the same trip time, a local optimization approach is developed to replace the partial braking in the downhill segment by the coasting regime such that the local energy consumption is reduced. Further, the adaptive gradient descent method is utilized to obtain the optimal cruising speed with minimum traction energy consumption to achieve a global optimization. Some simulations based on practical data of the Yizhuang line showed that the proposed approach can averagely reduce the energy consumption by 10.96%, compared with the classical energy-efficient train control approach. Simulations with five typical lines were also conducted to indicate that the proposed method has a good availability for variable lines. The average computation time of this method was about 0.12 s, thus the proposed approach can be applied in the real-time control system. Future work could extend this approach to study the energy-efficient train control strategy with considering multi-slopes in the line.

Author Contributions: Conceptualization, T.T.; Methodology, S.S., W.L.; Validation, J.Y.; Formal Analysis, Y.C., C.W.; Writing-Original Draft Preparation, W.L.; Writing-Review and Editing, S.S., W.L.; Funding Acquisition, S.S.

Funding: This research was funded by Beijing Laboratory of Urban Rail Transit, Beijing Key Laboratory of Urban Rail Transit Automation and Control, by the National Natural Science Foundation of China (U1734210, 61803021), the Beijing Natural Science Foundation "The Joint Rail Transit" (L171007) and the Fundamental Research Funds for the Central Universities (2018YJS195).

Conflicts of Interest: The authors declare no conflict of interest.

References

1. Su, S.; Tang, T.; Li, X.; Gao, Z.Y. Optimization on multi-train operation in subway system. *IEEE Trans. Intell. Transp.* **2014**, *15*, 673–684.
2. Liu, W.T.; Tang, T.; Su, S.; Chang, Y.H.; Li, X.L. Study on the energy-efficient driving strategy for trains running in the steep downhill section. In Proceedings of the International Conference on Electronic, Control Automation and Mechanical Engineering, Sanya, China, 19–20 November 2017; pp. 534–539.
3. Milroy, I.P. Aspects of Automatic Train Control. Ph.D. Thesis, Loughborough University, Loughborough, UK, 1980.
4. Asnis, I.; Dmitruk, A.; Osmolovskii, N. Control of the motion of a train by the maximum principle. *Energy Optim. USSR Comput. Math. Math. Phys.* **1985**, *25*, 37–44. [CrossRef]
5. Howlett, P.G. An optimal strategy for the control of a train. *J. Aust. Math. Soc. B* **1990**, *31*, 454–471. [CrossRef]
6. Howlett, P.G.; Milroy, I.P.; Pudney, P.J. Energy-efficient train control. *Control Eng. Pract.* **1994**, *2*, 193–200. [CrossRef]

7.   Khmelnitsky, E. On an optimal control problem of train operation. *IEEE Trans. Autom. Control* **2000**, *45*, 1257–1266. [CrossRef]

8.   Liu, R.F.; Golovitcher, I. Energy-efficient operation of rail vehicles. *Transp. Res. A Policy Pract.* **2003**, *37*, 917–932. [CrossRef]

9.   Su, S.; Li, X.; Tang, T. Driving strategy optimization for trains in subway systems. *Proc. Inst. Mech. Eng. F J. Rail Rapid Transit* **2018**, *232*, 369–383. [CrossRef]

10.   Su, S.; Tang, T.; Chen, L.; Liu, B. Energy-efficient train control in urban rail transit systems. *Proc. Inst. Mech. Eng. F J. Rail Rapid Transit* **2015**, *229*, 446–454. [CrossRef]

11.   Chang, C.S.; Sim, S.S. Optimising train movements through coast control using genetic algorithms. *IEE Proc. Electr. Power Appl.* **1997**, *144*, 65–73. [CrossRef]

12.   Ma, C.Y.; Mao, B.H.; Liang, X. Coast control of urban train movement for energy efficiency. *J. Transp. Inf. Saf.* **2010**, *28*, 37–42.

13.   Jin, W.D.; Wang, Z.L.; Li, C.W. Study on optimal method of train operation for saving energy. *J. China Railw. Soc.* **1997**, *6*, 58–62.

14.   Ke, B.R.; Chen, M.C.; Lin, C.L. Block-layout design using max-min ant system for saving energy on mass rapid transit systems. *IEEE Trans. Intell. Transp.* **2009**, *10*, 226–235.

15.   Ke, B.R.; Lin, C.L.; Yang, C.C. Optimisation of train energy-efficient operation for mass rapid transit systems. *IET Intell. Transp. Syst.* **2012**, *6*, 58–66. [CrossRef]

16.   Yang, X.; Li, X.; Ning, B. A survey on energy-efficient train operation for urban rail transit. *IEEE Trans. Intell. Transp.* **2016**, *17*, 2–13. [CrossRef]

17.   Cheng, J.; Howlett, P.G. Application of critical velocities to the minimisation of fuel consumption in the control of trains. *Automatica* **1992**, *28*, 165–169.

18.   Cheng, J.; Howlett, P.G. A note on the calculation of optimal strategies for the minimization of fuel consumption in the control of trains. *IEEE Trans. Autom. Control* **1993**, *38*, 1730–1734. [CrossRef]

19.   Pudney, P.J.; Howlett, P.G. Optimal driving strategy for a train journey with speed limits. *J. Aust. Math. Soc. B* **1994**, *1*, 38–49. [CrossRef]

20.   Howlett, P.G.; Cheng, J. Optimal driving strategies for a train on a track with continuously varying gradient. *J. Aust. Math. Soc. B* **1995**, *3*, 388–410. [CrossRef]

21.   Howlett, P.G.; Milroy, I.P.; Pudney, P.J. Energy-efficient train control. In *Advances in Industrial Control*; Springer: Berlin, Germany, 1995.

22.   Han, S. An optimal automatic train operation (ATO) control using genetic algorithms (GA). In Proceedings of the IEEE Region 10 Conference, Cheju Island, Korea, 15–17 September 1999; pp. 360–362.

23.   Ding, Y.; Liu, H.; Bai, Y. A two-level optimization model and algorithm for energy-efficient urban train operation. *J. Transp. Syst. Eng. Inf. Technol.* **2011**, *11*, 96–101. [CrossRef]

24.   Domínguez, M.; Fernández, A.; Cucala, A.P. Multi-objective particle swarm optimization algorithm for the design of efficient ATO speed profiles in metro lines. *Eng. Appl. Artif. Intell.* **2014**, *29*, 43–53. [CrossRef]

25.   Howlett, P.G.; Pudney, P.J.; Vu, X. Local energy minimization in optimal train control. *Automatica* **2009**, *45*, 2692–2698. [CrossRef]

26.   Albrecht, A.R.; Howlett, P.G.; Pudney, P.J.; Vu, X. Energy-efficient train control: From local convexity to global optimization and uniqueness. *Automatica* **2013**, *19*, 3072–3078. [CrossRef]

27.   Albrecht, A.; Howlett, P.G.; Pudney, P.J. The key principles of optimal train control-Part 1: Formulation of the model, strategies of optimal type, evolutionary lines, location of optimal switching points. *Transp. Res. B Meth.* **2016**, *94*, 482–508. [CrossRef]

28.   Albrecht, A.; Howlett, P.G.; Pudney, P.J. The key principles of optimal train control-Part 2: Existence of an optimal strategy, the local energy minimization principle, uniqueness, computational techniques. *Transp. Res. B Meth.* **2016**, *94*, 509–538. [CrossRef]

29.   Ko, H.; Koseki, T.; Miyatake, M. Application of dynamic programming to the optimization of the running profile of a train. In *Computers in Railways IX*; WIT Press: Southampton, UK, 2004; pp. 103–112.

30.   Ko, H.; Koseki, T.; Miyatake, M. Numerical study on dynamic programming applied to optimization of running profile of a train. *IEEE Trans. Ind. Appl.* **2005**, *125*, 1084–1092. [CrossRef]

31.   Su, S.; Li, X.; Tang, T.; Gao, Z.Y. A subway train timetable optimization approach based on energy-efficient operation strategy. *IEEE Trans. Intell. Transp.* **2013**, *14*, 883–893. [CrossRef]

32. Rochard, B.P.; Schmid, F. A review of methods to measure and calculate train resistances. *Proc. Inst. Mech. Eng. F J. Rail Rapid Transit* **2000**, *214*, 185–199. [CrossRef]

33. Wang, Y.H.; Schutter, B.D.; Ton, J.J. Optimal trajectory planning for trains-a pseudospectral method and a mixed integer linear programming approach. *Transp. Res. C Emerg. Technol.* **2013**, *29*, 97–114. [CrossRef]

34. John, D.; Elad, H.; Yoram, S. Adaptive subgradient methods for online learning and stochastic optimization. *J. Mach. Learn. Res.* **2011**, *12*, 2121–2159.

 *processes*

Article

A Hybrid Energy Feature Extraction Approach for Ship-Radiated Noise Based on CEEMDAN Combined with Energy Difference and Energy Entropy

Yuxing Li [1,*], Xiao Chen [2] and Jing Yu [3]

[1] School of Information Technology and Equipment Engineering, Xi'an University of Technology, Xi'an 710048, China
[2] College of Electrical & Information Engineering, ShaanXi University of Science & Technology, Xi'an 710021, China; chenxiao@sust.edu.cn
[3] School of Marine Science and Technology, Northwestern Polytechnical University, Xi'an 710072, China; yujing@nwpu.edu.cn
* Correspondence: liyuxing@xaut.edu.cn

Received: 17 December 2018; Accepted: 27 January 2019; Published: 1 February 2019

Abstract: Influenced by the complexity of ocean environmental noise and the time-varying of underwater acoustic channels, feature extraction of underwater acoustic signals has always been a difficult challenge. To solve this dilemma, this paper introduces a hybrid energy feature extraction approach for ship-radiated noise (S-RN) based on complete ensemble empirical mode decomposition with adaptive noise (CEEMDAN) combined with energy difference (ED) and energy entropy (EE). This approach, named CEEMDAN-ED-EE, has two main advantages: (i) compared with empirical mode decomposition (EMD) and ensemble EMD (EEMD), CEEMDAN has better decomposition performance by overcoming mode mixing, and the intrinsic mode function (IMF) obtained by CEEMDAN is beneficial to feature extraction; (ii) the classification performance of the single energy feature has some limitations, nevertheless, the proposed hybrid energy feature extraction approach has a better classification performance. In this paper, we first decompose three types of S-RN into sub-signals, named intrinsic mode functions (IMFs). Then, we obtain the features of energy difference and energy entropy based on IMFs, named CEEMDAN-ED and CEEMDAN-EE, respectively. Finally, we compare the recognition rate for three sorts of S-RN by using the following three energy feature extraction approaches, which are CEEMDAN-ED, CEEMDAN-EE and CEEMDAN-ED-EE. The experimental results prove the effectivity and the high recognition rate of the proposed approach.

Keywords: complete ensemble empirical mode decomposition with adaptive noise (CEEMDAN); energy difference (ED); energy entropy (EE); hybrid energy feature extraction; ship-radiated noise (S-RN)

1. Introduction

Due to the complexity of ocean ambient noise and the time-varying of underwater acoustic channels, feature extraction of underwater acoustic signals has always been a difficult problem in the area of underwater acoustic signal processing [1,2]. In order to solve this problem, some feature extraction approaches for underwater acoustic signals have been proposed, including a time domain analysis, a spectral analysis, a time–frequency analysis, a high-order statistics analysis and a complexity analysis. In recent years, with the development of mode decomposition, feature extraction approaches have been proposed based on mode decomposition [3].

After Empirical mode decomposition (EMD) was first proposed as a classical mode decomposition approach, it has become widely used [4,5]. The research history and the current status of EMD mainly include two parts. On the one hand, it is the improvement of EMD, in particular for mode mixing. Two of the revised EMD approaches are generally accepted to be effective, they are ensemble EMD (EEMD) [6] and complete ensemble empirical mode decomposition with adaptive noise (CEEMDAN) [7]. In addition, CEEMDAN is an upgrade of EEMD, which can better suppress mode mixing than EEMD. On the other hand, their application areas are expanding and deepening gradually. EMD has been widely used in different fields, such as short-term wind speed forecasting combined with hybrid linear and nonlinear models [8], the detection and location of pipeline leakage [9], the detection of incipient damages for truss structures [10], denoising for grain flow signal [11], biomedical photoacoustic imaging optimization [12] and heart rate variability analysis [13]. Many scholars have also applied EEMD to their research fields, such as wind speed forecasting combined with the cuckoo search algorithm [14], machine feature extraction combined with a kernel-independent component [15], feature extraction for motor bearing combined with multi-scale fuzzy entropy [16], a bearing fault diagnosis combined with correlation coefficient analysis [17], a partial discharge feature extraction combined with sample entropy [18] and monthly streamflow forecasting combined with multi-scale predictors selection [19]. In addition, CEEMDAN is used in machinery, electricity and medicine, such as impact signal denoising [20], daily peak load forecasting [21], health degradation monitoring for rolling bearings combined with multi-scale sample entropy [22], planetary gear fault diagnosis combined with permutation entropy [23], denoising for gear transmission system [24], friction signal denoising combined with mutual information [25] and electrocardiogram signal denoising combined with wavelet threshold [26]. Generally, the three EMD approaches can solve practical problems in different fields. Some comparative studies have also proven that CEEMDAN has a better decomposition performance.

In the past three years, the mode decomposition approach has been applied to the underwater acoustic field. Two frequency feature extraction approaches were proposed, based on maximum energy intrinsic mode function (IMF) by using EEMD and variational mode decomposition (VMD), respectively [27,28]. In addition, two complexity feature extraction approaches were proposed based on the permutation entropy of maximum energy IMF by EMD [29] and the multi-scale permutation entropy of maximum energy IMF by VMD [30]. Energy feature extraction approaches for underwater acoustic signals were seldom proposed by scholars. In Reference [31], an energy feature extraction approach was put forward based on EEMD, which extracted the energy difference between the high-frequency and the low-frequency bands as a new energy feature. However, this energy feature extraction approach has limited recognition ability for different sorts of ship-radiated noise (S-RN) signals.

In this paper, we propose a new energy feature extraction approach to effectively extract the energy feature for underwater acoustic signals. The method we propose, named CEEMDAN-ED-EE, is based on CEEMDAN, energy difference (ED) and energy entropy (EE). We use CEEMDAN to decompose three sorts of S-RN signals into IMFs. According to the rule of ED and EE, we can obtain the features of ED and EE for three sorts of S-RN. Compared with CEEMDAN-ED and CEEMDAN-EE, the proposed CEEMDAN-ED-EE approach can extract energy features more effectively and has a relatively higher recognition rate.

The following section presents the theory related to CEEMDAN, ED and EE; the novel energy feature extraction approach for underwater acoustic signal is presented in Section 3; the proposed energy feature extraction approach is used to three sorts of S-RN signals in Section 4; finally, the concluding remarks are made in the last section.

2. Theory

2.1. CEEMDAN

In this study, we use the CEEMDAN approach for two main reasons: (i) CEEMDAN has a better anti-mode mixing performance than EMD and EEMD and (ii) thus far, an energy feature extraction approach using CEEMDAN has not been found for underwater acoustic signals. The main procedures of CEEMDAN can be summarized as follows:

CEEMDAN, as an improved algorithm of EMD and EEMD, can adaptively decompose complex signals into IMFs in order. The specific steps of CEEMDAN are summarized as follows [8]:

(1) Construct the noise signal $f_i(t)$ by combining the original signal $f(t)$ and white noise $n_i(t)$, N noisy signals $f_i(t)$ can be obtained:

$$f_i(t) = f(t) + n_i(t), \quad i = 1, 2, \cdots, N \tag{1}$$

(2) Decompose each $f_i(t)$ by using EMD in order to get the IMF1 $c_{i1}(t)$ and its residual item, $r_i(t)$:

$$\begin{bmatrix} f_1(t) \\ f_2(t) \\ \cdots \\ f_i(t) \\ \cdots \\ f_N(t) \end{bmatrix} \xrightarrow{\text{EMD}} \begin{bmatrix} c_{11}(t) & r_1(t) \\ c_{21}(t) & r_2(t) \\ \cdots & \cdots \\ c_{i1}(t) & r_i(t) \\ \cdots & \cdots \\ c_{N1}(t) & r_N(t) \end{bmatrix} \tag{2}$$

(3) Calculate the average value of $c_{i1}(t)$ to get the IMF1 $c_1(t)$ of CEEMDAN:

$$c_1(t) = \frac{1}{N} \sum_{i=1}^{N} c_{i1}(t) \tag{3}$$

(4) Subtract $c_1(t)$ from $f(t)$ to get the residual item $R_1(t)$:

$$R_1(t) = f(t) - c_1(t) \tag{4}$$

(5) White noise $n_i(t)$ participates in subsequent decompositions at different scales. Here we use EMD to decompose white noise as follows:

$$\begin{bmatrix} n_1(t) \\ n_2(t) \\ \cdots \\ n_i(t) \\ \cdots \\ n_N(t) \end{bmatrix} \xrightarrow{\text{EMD}} \begin{bmatrix} c_{n_1 1}(t) & c_{n_1 2}(t) & \cdots & c_{n_1 j}(t) & r_{n_1}(t) \\ c_{n_2 1}(t) & c_{n_2 2}(t) & \cdots & c_{n_2 j}(t) & r_{n_2}(t) \\ \cdots & \cdots & \cdots & \cdots & \cdots \\ c_{n_i 1}(t) & c_{n_i 2}(t) & \cdots & c_{n_i j}(t) & r_{n_i}(t) \\ \cdots & \cdots & \cdots & \cdots & \cdots \\ c_{n_N 1}(t) & c_{n_N 2}(t) & \cdots & c_{n_N j}(t) & r_{n_N}(t) \end{bmatrix} \tag{5}$$

where $c_{n_{ij}}(t)$ is the j-th IMF of the i-th white noise $n_i(t)$, and $r_{n_i}(t)$ is the residual item of $n_i(t)$. For convenience, we define $E_j(g_i(t))$ as a set, which consists of the j-th IMFs of $g_i(t)$. Therefore, $E_1(n_i(t))$ can be expressed as:

$$E_1(n_i(t)) = \begin{pmatrix} c_{n_1 1}(t) & c_{n_2 1}(t) & \cdots & c_{n_i 1}(t) & \cdots & c_{nN}(t) \end{pmatrix}^T \tag{6}$$

(6) Construct $fnew_1(t)$ by combining $R_1(t)$ and $E_1(n_i(t))$. We can decompose $fnew_1(t)$ as follows:

$$fnew_1(t) = R_1(t) + E_1(n_i(t)) \tag{7}$$

$$f new_1(t) = R_1(t) + \begin{bmatrix} c_{n_1 1}(t) \\ c_{n_2 1}(t) \\ \cdots \\ c_{n_i 1}(t) \\ \cdots \\ c_{n_N 1}(t) \end{bmatrix} \xrightarrow{EMD} \begin{bmatrix} c_{r_1 n_1 1}(t) \\ c_{r_1 n_2 1}(t) \\ \cdots \\ c_{r_1 n_i 1}(t) \\ \cdots \\ c_{r_1 n_N 1}(t) \end{bmatrix} \tag{8}$$

(7) Calculate the average value of $c_{r_1 n_i 1}(t)$ to get the IMF2 $c_2(t)$ of CEEMDAN, $c_2(t)$. Residual item $R_2(t)$ of CEEMDAN can be expressed as follows:

$$c_2(t) = \frac{1}{N}\sum_{i=1}^{N} c_{r_1 n_i 1}(t) \tag{9}$$

$$R_2(t) = R_1(t) - c_2(t) \tag{10}$$

(8) In order to get the rest of IMFs $c_j(t)$ and the residual item $R_j(t)$, we can construct $f new_{j-1}(t)$ and repeat step (6) and step (7). We can express $f new_{j-1}(t)$ $c_j(t)$ and $R_j(t)$ as follows:

$$f new_{j-1}(t) = R_{j-1}(t) + E_{j-1}(n_i(t)) \tag{11}$$

$$c_j(t) = \frac{1}{N}\sum_{i=1}^{N} c_{r_{j-1} n_i 1}(t) \tag{12}$$

$$R_j(t) = R_{j-1}(t) - c_j(t) \tag{13}$$

(9) If the new IMF cannot be extracted from $R_j(t)$, we make $R_j(t)$ equal to $R(t)$. We can express $f(t)$ as follows:

$$f(t) = \sum_{j=1}^{M} c_j(t) + R(t) \tag{14}$$

where M and $R(t)$ are the number of IMFs and the residual item of $f(t)$.

2.2. ED

According to the definition of energy difference in Reference [31], we define an instantaneous frequency that is equal or less than 1 kHz as the low-frequency band, and an instantaneous frequency that is more than 1 kHz as the high-frequency band. Therefore, ED is defined as the difference between the high-frequency band energy and the low-frequency band energy. In this paper, the specific calculation steps of ED for S-RN are as follows:

(1) Decompose S-RN into IMFs by CEEMDAN, and then process each IMF $c_i(t)$ through the Hilbert transform:

$$\hat{c}_i(t) = \frac{1}{\pi}\int_{-\infty}^{\infty} \frac{c_i(\tau)}{t - \tau}d\tau \tag{15}$$

(2) The analytic signal of each IMF is represented as:

$$z_i(t) = c_i(t) + j\hat{c}_i(t) = \lambda_i(t)e^{j\theta_i(t)} \tag{16}$$

(3) We can obtain the instantaneous amplitude $\lambda_i(t)$, instantaneous phase $\theta_i(t)$ and instantaneous frequency $f_i(t)$ as follows:

$$\lambda_i(t) = \sqrt{c_i^2(t) + \hat{c}_i^2(t)} \tag{17}$$

$$\theta_i(t) = \arctan(\frac{\hat{c}_i(t)}{c_i(t)}) \tag{18}$$

$$f_i(t) = \frac{1}{2\pi}\frac{d\theta_i(t)}{dt} \tag{19}$$

(4) Calculate the instantaneous energy intensity of each sampling point by using the instantaneous amplitude of each sampling point. For example, the q-th sampling point of p-th IMF is b_{pq}, the instantaneous energy intensity Q_{pq} of this sampling point can be represented as:

$$Q_{pq} = b_{pq}^2 \tag{20}$$

(5) Calculate the high-frequency band energy P_H and the low-frequency band energy P_L, respectively.

$$P_H = 10 \log \sum_{k=1}^{m} Q_{Hk} \tag{21}$$

$$P_L = 10 \log \sum_{k=1}^{n} Q_{Lk} \tag{22}$$

where Q_{Hk} and Q_{Lk} are the instantaneous energy intensity of the k-th sampling point in the high-frequency and the low-frequency bands, respectively.

(6) ED is represented as:

$$\Delta P = P_H - P_L \tag{23}$$

2.3. EE

The difference in time–frequency distribution can be expressed by the uncertainty of energy distribution in different time–frequency bands. Time–frequency bands can be provided by using IMFs. In this paper, we propose an energy feature extraction approach for S-RN, based on CEEMDAN and EE. The specific calculation steps of EE for S-RN are as follows:

(1) Decompose S-RN into M IMFs by CEEMDAN. The energy sum of each IMF equals the total energy of the S-RN signal without considering the residual item.

$$E = \sum_{i=1}^{M} E_i \tag{24}$$

where E and E_i are the energy of the S-RN signal and the energy of i-th IMF.

(2) Calculate the energy proportion of each IMF in the S-RN signal.

$$C_i = \frac{E_i}{E} \tag{25}$$

where C_i is the energy proportion of i-th IMF.

(3) According to the definition of information entropy, we can express EE of the S-RN signal as:

$$H(C) = -\sum_{i=1}^{M} C_i \ln C_i \tag{26}$$

3. Hybrid Energy Feature Extraction Approach for S-RN

This paper presents a hybrid energy feature extraction approach for S-RN, based on CEEMDAN, ED and EE. The proposed CEEMDAN-ED-EE approach, combining the advantages of CEEMDAN-ED and CEEMDAN-EE, can reflect the energy distribution of a S-RN signal at different scales. The flowchart of the hybrid energy feature extraction approach is shown in Figure 1. The specific steps of the hybrid energy feature extraction approach are as follows:

Step 1: The S-RN signal decomposition.
(1) Collect S-RN signals under sensor measurement;
(2) Decompose S-RN signals into N IMFs by CEEMDAN.

Step 2: hybrid energy feature extraction.
(1) Extract ED features for S-RN signals;
(2) Extract EE features for S-RN signals;
(3) Extract hybrid energy features for S-RN signals by combining ED and EE;

Step 3: classification and recognition.
(1) Input hybrid energy features of different S-RN signals into a support vector machine (SVM);
(2) Obtain the classification accuracy for S-RN signals.

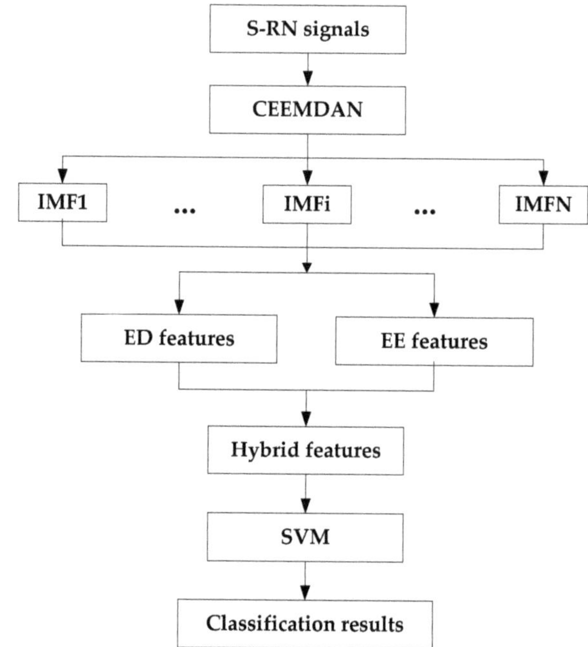

Figure 1. The flowchart of the hybrid energy feature extraction approach; S-RN: ship-radiated noise, CEEMDAN: complete ensemble empirical mode decomposition with adaptive noise; IMF: intrinsic mode function, ED: energy difference, EE: energy entropy, SVM: support vector machine.

4. Energy Feature Extraction for S-RN

4.1. Data Measurement

In this paper, three sorts of S-RN signals were measured in the South China Sea, called Ship-1, Ship-2 and Ship-3. In order to reduce the influence of artificial and ocean background noise, we obtained the data under a sea state of level 1. The depths of the measurement area and the hydrophones were about 4 km and 30 m, respectively. Each sample of S-RN had 5000 sampling points; the three types of S-RN signals, which were normalized are shown in Figure 2.

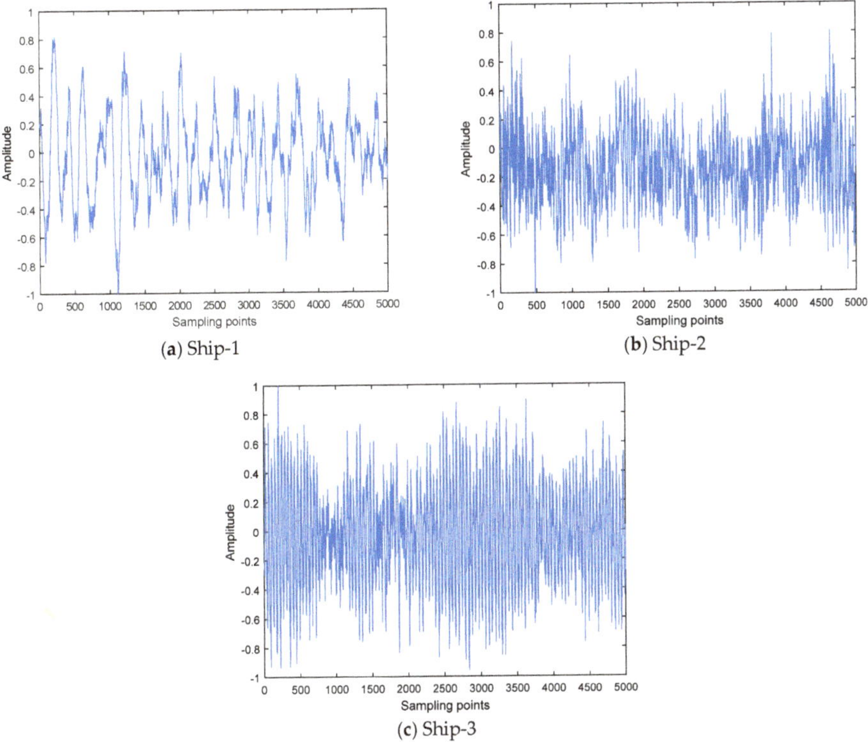

(a) Ship-1

(b) Ship-2

(c) Ship-3

Figure 2. Three sorts of S-RN signals.

4.2. CEEMDAN for S-RN

Traditional feature extraction approaches usually extract features from original target signals, which can only acquire limited features. In this study, three sorts of S-RN signals were decomposed from a high-frequency to a low-frequency by using CEEMDAN. The CEEMNAN results for S-RN signals are shown in Figure 3. By observing Figure 3, it can be seen that the amplitude and the number of IMFs were different for all three types of S-RN signals.

4.3. CEEMDAN-ED

The ED of the high-frequency and the low-frequency bands can reflect the energy distribution of S-RN signals on the macroscopic scale. The CEEMDAN-ED approach was first used to calculate the analytic signal of each IMF; we then obtained the energy of the high-frequency and the low-frequency bands, according to the instantaneous frequency and amplitude of each sampling point. The energy of the low-frequency and the high-frequency bands for S-RN signals are shown in Figure 4. Finally, the ED was obtained by a subtraction operation. The ED for S-RN signals are shown in Table 1. As can be seen from Figure 4 and Table 1, the ED of Ship-1 was distinctly different from the other two ships, while the EDs of Ship-2 and Ship-3 were very close.

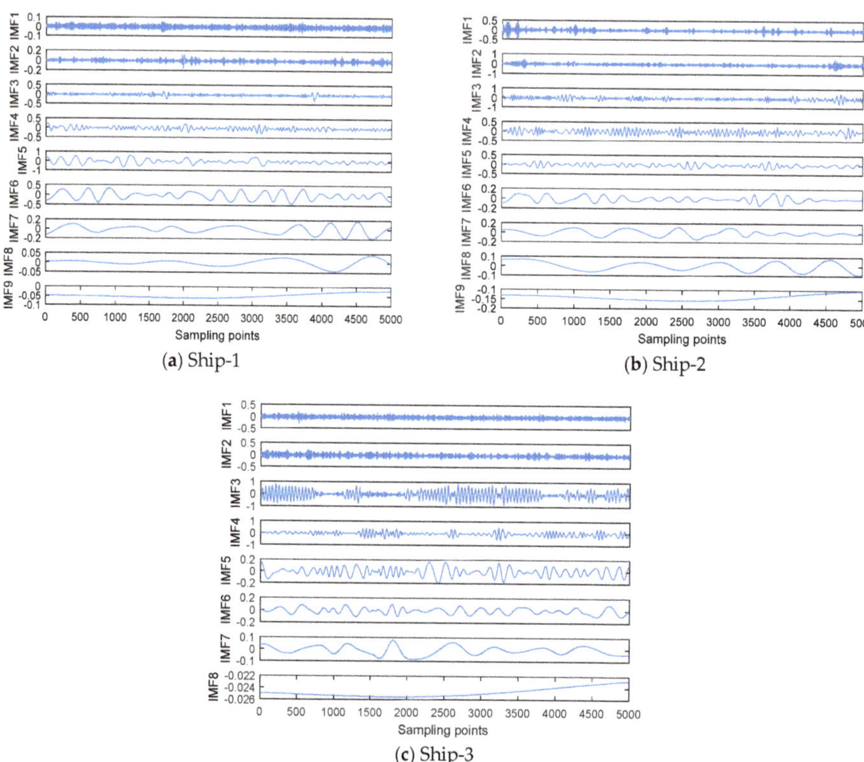

(**a**) Ship-1

(**b**) Ship-2

(**c**) Ship-3

Figure 3. CEEMNAN results for S-RN signals.

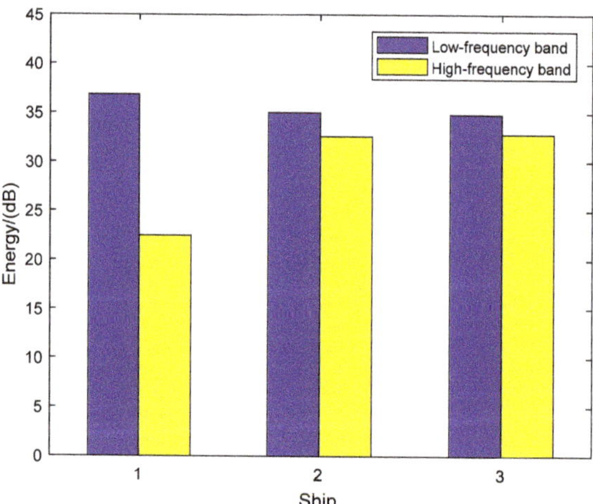

Figure 4. The energy of the low-frequency and the high-frequency band for S-RN signals.

Table 1. The ED for S-RN signals. S-RN: ship-radiated noise, ED: energy difference.

Ship-1	Ship-2	Ship-3
−14.3713 dB	−2.3841 dB	−2.0176 dB

The ED distribution for S-RN signals (20 samples for each ship) is shown in Figure 5. As can be seen from Figure 5, the ED of the same ship remained at the same level. We could easily distinguish Ship-1 by using ED, however, it was difficult to distinguish between Ship-2 and Ship-3, due to their similar EDs.

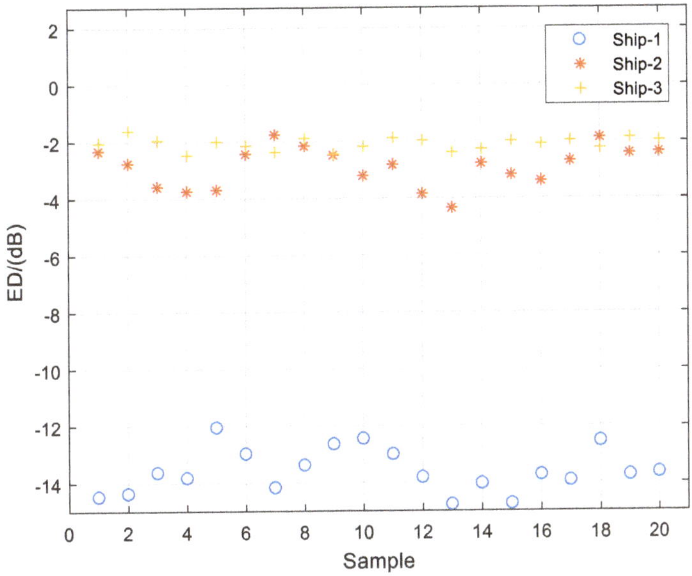

Figure 5. The ED distribution for S-RN signals.

4.4. CEEMDAN-EE

The energy of IMF by CEEMDAN can reflect the energy distribution of S-RN signals on the IMF scale. The CEEMDAN-EE approach was first used to calculate the energy of each IMF according to the instantaneous amplitude of each sampling point; we then obtained the energy proportion of each IMF. The energy proportion for S-RN signals is shown in Figure 6. Finally, the EE was calculated according to information entropy. The EE for S-RN signals is shown in Table 2. As can be seen from Figure 6 and Table 2, the EE of Ship-2 was distinctly different from the other two ships, while the EE difference between Ship-1 and Ship-3 was small.

Table 2. The EE for S-RN signals. EE: energy entropy.

Ship-1	Ship-2	Ship-3
1.2138	1.9125	0.9494

The EE distribution for S-RN signals (20 samples for each ship) is shown in Figure 7. As was the case in the measurement of the ED distribution, the EE of the same ship was also at the same level, as can be seen in Figure 7; we could distinguish Ship-2 by using EE, however, it was hard to distinguish between Ship-1 and Ship-3 because of their similar EEs.

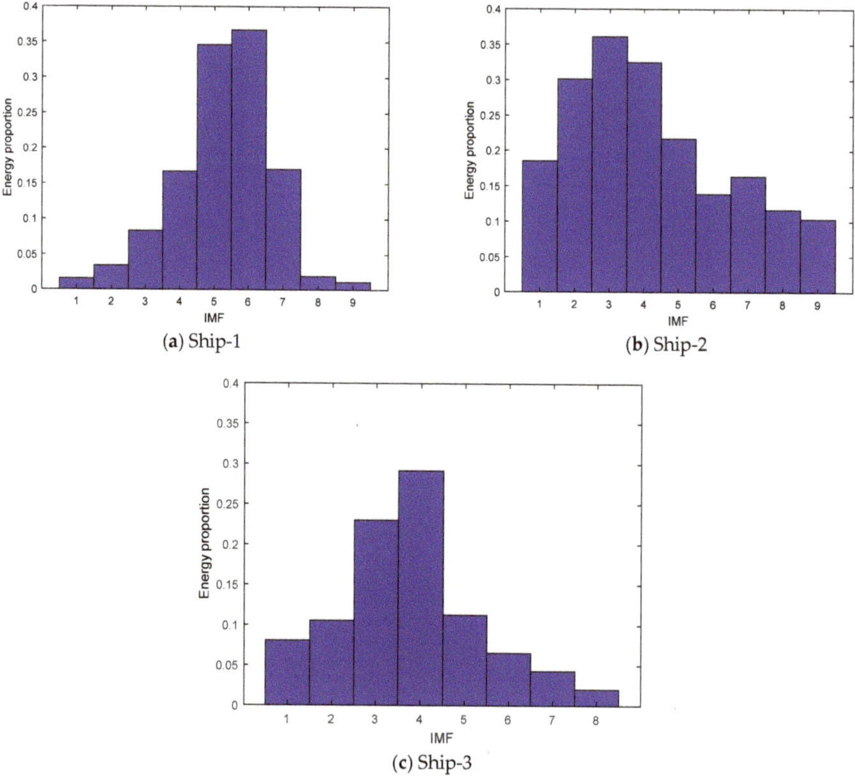

(**a**) Ship-1

(**b**) Ship-2

(**c**) Ship-3

Figure 6. The energy proportion for S-RN signals.

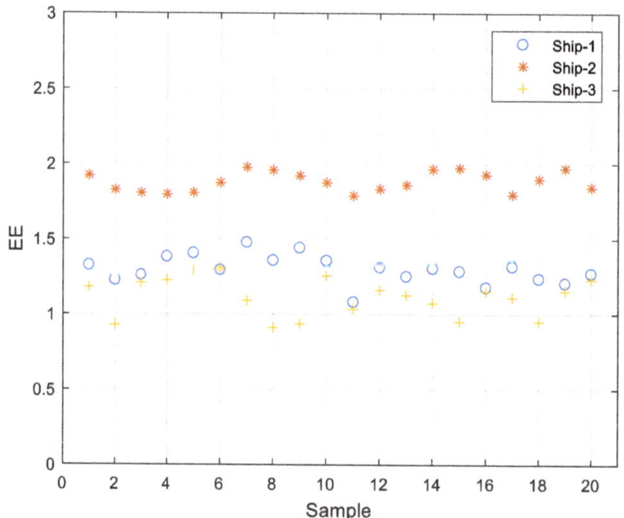

Figure 7. The EE distribution for S-RN signals.

4.5. CEEMDAN-ED-EE

CEEMDAN-ED and CEEMDAN-EE are all energy feature extraction approaches for S-RN signals. The two approaches extract energy features based on the macroscopic scale and the IMF scale, respectively. However, CEEMDAN-ED and CEEMDAN-EE have limited and different capabilities for S-RN signals that distinguish them. In this paper, CEEMDAN-ED-EE, as a hybrid energy feature extraction approach, was proposed because it combines the advantages of CEEMDAN-ED and CEEMDAN-EE. The hybrid feature distribution for S-RN signals (20 samples for each ship) is shown in Figure 8. As can be seen from the horizontal and vertical coordinates in Figure 8, representing ED and EE respectively, the hybrid features of the same ship were distributed in a limited region, and the hybrid features of the different ships were independent and non-overlapping. Therefore, we could easily distinguish the three sorts of S-RN.

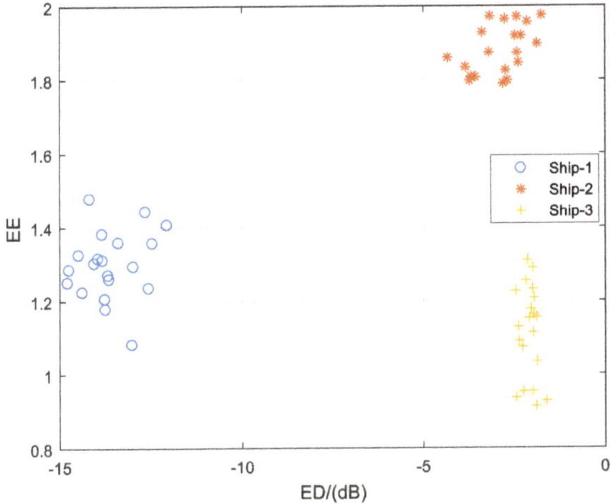

Figure 8. The hybrid feature distribution for S-RN signals.

In order to further prove the effectiveness of CEEMDAN-ED-EE, we used SVM for the classification of the three sorts of S-RN. The number of samples for each ship was 100, and the classification results for S-RN signals are listed in Table 3. As shown in the Table 3, CEEMDAN-ED and CEEMDAN-EE had a higher recognition rate than EMD-ED, EMD-EE, EEMD-ED and EEMD-EE; CEEMDAN-ED-EE also had a higher recognition rate than EMD-ED-EE and EEMD-ED-EE; in addition, the classification result of CEEMDAN-ED-EE was shown to be 100%, which was better than CEEMDAN-ED and CEEMDAN-EE.

Table 3. The classification results for S-RN signals. CEEMDAN: complete ensemble empirical mode decomposition with adaptive noise; EMD: empirical mode decomposition; EEMD: ensemble empirical mode decomposition.

(a)	EMD-ED	EMD-EE	EMD-ED-EE
	67.33%	66.33%	90.67%
(b)	EEMD-ED	EEMD-EE	EEMD-ED-EE
	70.67%	69.67%	96.33%
(c)	CEEMDAN-ED	CEEMDAN-EE	CEEMDAN-ED-EE
	79.67%	76.67%	100%

5. Conclusions

A hybrid energy feature extraction approach for S-RN was proposed in this paper based on CEEMDAN, ED and EE. The main contributions to this work are as follows:

(1) CEEMDAN was first used to extract the energy features of IMF for underwater acoustic signals in this paper.

(2) An energy feature extraction approach for S-RN was proposed in this paper based on IMFs by CEEMDAN and EE.

(3) CEEMDAN-ED-EE was successfully applied to extract the energy feature of S-RN signals. CEEMDAN-ED-EE can be more accurate and efficient in extracting the energy feature.

(4) Compared with CEEMDAN-ED and CEEMDAN-EE, CEEMDAN-ED-EE was shown to have a better performance, which proves its effectivity and its high recognition rate.

Author Contributions: Y.L. contributed the energy feature extraction method, all authors designed and analyzed the experiments.

Funding: Authors gratefully acknowledge the supported by National Natural Science Foundation of China (No. 51409214 and No. 51179157).

Conflicts of Interest: The authors declare no conflict of interest.

References

1. Li, Y.; Li, Y.; Chen, X.; Yu, J. Denoising and Feature Extraction Algorithms Using NPE Combined with VMD and Their Applications in Ship-Radiated Noise. *Symmetry* **2017**, *9*, 256. [CrossRef]
2. Li, Y.; Li, Y.; Chen, X.; Yu, J. Research on ship-radiated noise denoising using secondary variational mode decomposition and correlation coefficient. *Sensors* **2018**, *18*, 48.
3. Wang, S.; Zeng, X. Robust underwater noise targets classification using auditory inspired time-frequencyanalysis. *Appl. Acoust.* **2014**, *78*, 68–76. [CrossRef]
4. Huang, N.E.; Shen, Z.; Long, S.R.; Wu, M.C.; Shih, H.H.; Zheng, Q.; Yen, N.C.; Tung, C.C.; Liu, H.H. The empirical mode decomposition and the Hilbert spectrum for nonlinear and non-stationary time series analysis. *Proc. R. Soc. Lond. A* **1998**, *454*, 903–995. [CrossRef]
5. Ge, H.; Chen, G.; Yu, H.; Chen, H.; An, F. Theoretical Analysis of Empirical Mode Decomposition. *Symmetry* **2018**, *10*, 623. [CrossRef]
6. Wu, Z.; Huang, N.E. Ensemble empirical mode decomposition: A noise-assisted data analysis method. *Adv. Adapt. Data Anal.* **2009**, *1*, 1–41. [CrossRef]
7. Li, Y.; Li, Y.; Chen, X.; Yu, J.; Yang, H.; Wang, L. A New Underwater Acoustic Signal Denoising Technique Based on CEEMDAN, Mutual Information, Permutation Entropy, and Wavelet Threshold Denoising. *Entropy* **2018**, *20*, 563. [CrossRef]
8. Han, Q.; Wu, H.; Hu, T.; Chu, F. Short-Term Wind Speed Forecasting Based on Signal Decomposing Algorithm and Hybrid Linear/Nonlinear Models. *Energies* **2018**, *11*, 2976. [CrossRef]
9. Pan, S.; Xu, Z.; Li, D.; Lu, D. Research on Detection and Location of Fluid-Filled Pipeline Leakage Based on Acoustic Emission Technology. *Sensors* **2018**, *18*, 3628. [CrossRef] [PubMed]
10. Moreno-Gomez, A.; Amezquita-Sanchez, J.P.; Valtierra-Rodriguez, M.; Perez-Ramirez, C.A.; Dominguez-Gonzalez, A.; Chavez-Alegria, O. EMD-Shannon Entropy-Based Methodology to Detect Incipient Damages in a Truss Structure. *Appl. Sci.* **2018**, *8*, 2068. [CrossRef]
11. Wang, H.; Song, H. A Filtering Method for Grain Flow Signals Using EMD Thresholds Optimized by Artificial Bee Colony Algorithm. *Symmetry* **2018**, *10*, 575. [CrossRef]
12. Guo, C.; Chen, Y.; Yuan, J.; Zhu, Y.; Cheng, Q.; Wang, X. Biomedical Photoacoustic Imaging Optimization with Deconvolution and EMD Reconstruction. *Appl. Sci.* **2018**, *8*, 2113. [CrossRef]
13. Bin Queyam, A.; Kumar Pahuja, S.; Singh, D. Quantification of Feto-Maternal Heart Rate from Abdominal ECG Signal Using Empirical Mode Decomposition for Heart Rate Variability Analysis. *Technologies* **2017**, *5*, 68. [CrossRef]

14. Liu, T.; Liu, S.; Heng, J.; Gao, Y. A New Hybrid Approach for Wind Speed Forecasting Applying Support Vector Machine with Ensemble Empirical Mode Decomposition and Cuckoo Search Algorithm. *Appl. Sci.* **2018**, *8*, 1754. [CrossRef]

15. Fang, L.; Sun, H. Study on EEMD-Based KICA and Its Application in Fault-Feature Extraction of Rotating Machinery. *Appl. Sci.* **2018**, *8*, 1441. [CrossRef]

16. Zhao, H.; Sun, M.; Deng, W.; Yang, X. A New Feature Extraction Method Based on EEMD and Multi-Scale Fuzzy Entropy for Motor Bearing. *Entropy* **2017**, *19*, 14. [CrossRef]

17. Jiang, F.; Zhu, Z.; Li, W.; Ren, Y.; Zhou, G.; Chang, Y. A Fusion Feature Extraction Method Using EEMD and Correlation Coefficient Analysis for Bearing Fault Diagnosis. *Appl. Sci.* **2018**, *8*, 1621. [CrossRef]

18. Shang, H.; Lo, K.L.; Li, F. Partial Discharge Feature Extraction Based on Ensemble Empirical Mode Decomposition and Sample Entropy. *Entropy* **2017**, *19*, 439. [CrossRef]

19. Chu, H.; Wei, J.; Qiu, J. Monthly Streamflow Forecasting Using EEMD-Lasso-DBN Method Based on Multi-Scale Predictors Selection. *Water* **2018**, *10*, 1486. [CrossRef]

20. Zhan, L.; Li, C. A Comparative Study of Empirical Mode Decomposition-Based Filtering for Impact Signal. *Entropy* **2017**, *19*, 13. [CrossRef]

21. Dai, S.; Niu, D.; Li, Y. Daily Peak Load Forecasting Based on Complete Ensemble Empirical Mode Decomposition with Adaptive Noise and Support Vector Machine Optimized by Modified Grey Wolf Optimization Algorithm. *Energies* **2018**, *11*, 163. [CrossRef]

22. Lv, Y.; Yuan, R.; Wang, T.; Li, H.; Song, G. Health Degradation Monitoring and Early Fault Diagnosis of a Rolling Bearing Based on CEEMDAN and Improved MMSE. *Materials* **2018**, *11*, 1009. [CrossRef] [PubMed]

23. Kuai, M.; Cheng, G.; Pang, Y.; Li, Y. Research of Planetary Gear Fault Diagnosis Based on Permutation Entropy of CEEMDAN and ANFIS. *Sensors* **2018**, *18*, 782. [CrossRef] [PubMed]

24. Bai, L.; Han, Z.; Li, Y.; Ning, S. A Hybrid De-Noising Algorithm for the Gear Transmission System Based on CEEMDAN-PE-TFPF. *Entropy* **2018**, *20*, 361. [CrossRef]

25. Li, C.; Zhan, L.; Shen, L. Friction Signal Denoising Using Complete Ensemble EMD with Adaptive Noise and Mutual Information. *Entropy* **2015**, *17*, 5965–5979. [CrossRef]

26. Xu, Y.; Luo, M.; Li, T.; Song, G. ECG Signal De-noising and Baseline Wander Correction Based on CEEMDAN and Wavelet Threshold. *Sensors* **2017**, *17*, 2754. [CrossRef]

27. Li, Y.; Li, Y.; Chen, X. Ships'radiated noise feature extraction based on EEMD. *J. Vib. Shock* **2017**, *36*, 114.

28. Li, Y.; Li, Y.; Chen, X.; Yu, J. Feature extraction of ship-radiated noise based on VMD and center frequency. *J. Vib. Shock* **2018**, *37*, 213.

29. Li, Y.; Li, Y.; Chen, Z.; Chen, X. Feature Extraction of Ship-Radiated Noise Based on Permutation Entropy of the Intrinsic Mode Function with the Highest Energy. *Entropy* **2016**, *18*, 393. [CrossRef]

30. Li, Y.; Li, Y.; Chen, X.; Yu, J. A Novel Feature Extraction Method for Ship-Radiated Noise Based on Variational Mode Decomposition and Multi-Scale Permutation Entropy. *Entropy* **2017**, *19*, 342.

31. Yang, H.; Li, Y.; Li, G. Energy analysis of ship-radiated noise based on ensemble empirical mode decomposition. *J. Vib. Shock* **2015**, *34*, 55–59.

Article

Smart Community Energy Cost Optimization Taking User Comfort Level and Renewable Energy Consumption Rate into Consideration

Kun Shi [1], Dezhi Li [1], Taorong Gong [1], Mingyu Dong [1], Feixiang Gong [1,*] and Yajie Sun [2]

[1] Department of Power Consumption, China Electric Power Research Institute, Beijing 100192, China; shikun@epri.sgcc.com.cn (K.S.); lidezhi@epri.sgcc.com.cn (D.L.); gongtaorong@epri.sgcc.com.cn (T.G.); dongmingyu@epri.sgcc.com.cn (M.D.)

[2] Department of Electrical Engineering, Northeast Electric Power University, Jilin 132012, China; yjsun_neepu@foxmail.com

* Correspondence: gongfeixiang@epri.sgcc.com.cn; Tel.: +86-432-6308-3214

Received: 14 January 2019; Accepted: 24 January 2019; Published: 26 January 2019

Abstract: With the rapid development of smart community technologies, how to improve user comfort levels and make full use of renewable energy have become urgent problems. This paper proposes an optimization algorithm to minimize daily energy costs while considering user comfort level and renewable energy consumption rate. In this paper, the structure of a typical smart community and the output models of all components installed in the community are introduced first. Then, the characteristics of different types of loads are analyzed, followed by defining the coefficients of user comfort level. In this step, the influence of load-scheduling on user comfort level and the renewable energy consumption rate is emphasized. Finally, based on the time-of-use gas price, this paper optimizes the daily energy costs for an off-grid community under the constraints of the comfort level and renewable energy consumption rate. Results show that scheduling transferable loads and interruptible loads are not independent to each other, and improving user comfort level requires spending more money as compensation. Moreover, fully consuming renewable energy has side effects on energy bills and battery lifetime. It is more conducive to system economy and stability if the maximum renewable energy consumption rate is restricted to 95%.

Keywords: smart communities; user comfort levels; renewable energy consumption rate

---

1. Introduction

A smart community is a standard architecture group, which is planned and constructed by governments. It has complete supporting facilities, including energy supply, communication, transportation, and warehousing, which makes it possible to meet the needs of industrial production and specific scientific experiments. More importantly, an advanced smart community can provide convenient and personalized services in terms of human work and life [1]. Since smart communities have obvious advantages compared with traditional communities, the development of smart communities has prompted considerable awareness from all over the world, which thus accelerates the pace of planning and constructing of smart communities [2–4].

With the rapid development of renewable energy and internet of things technology, the proportion of distributed power generation increases significantly in smart communities. By increasing the amount of renewable generation, smart communities can not only reduce the exploitation of fossil energy, but also promote the sustainable development of society [5,6]. In addition, a smart community has the characteristic of local consumption of renewable energy, which is considered to be an effective means of dealing with energy efficiency [7,8]. Therefore, it plays an important role in improving energy efficiency and promoting the use of low-carbon energy [9,10]. However, at present, the operation and management of smart communities still lacks perfect user comfort evaluation methods and research on renewable energy consumption, which will seriously hinder the further development of smart communities [11]. Therefore, it is of great significance to study further the influence of user comfort level, renewable energy consumption, and other factors on the optimal operation of the smart community.

In the study of controllable load optimization in smart communities, user comfort level is an important evaluation index and an essential constraint condition. An optimal and automatic residential energy consumption scheduling framework, proposed in [12], aimed to find a desired trade-off between the optimal energy costs and the waiting time of equipment operation under the incentive of electricity price. It was mentioned in [13] that an approximate greedy iterative algorithm had been employed to adjust the use time of electric equipment to reduce the cost of electricity. Both above-mentioned articles used the same criteria to judge user comfort level for all the electrical appliances in a unified way. However, regarding controllable loads, the influences of shifting transferable loads and shedding interruptible loads on user comfort level are significantly different, and therefore the lack of classification of user comfort level would lead to a large error in optimization results. The coordination scheduling method of electric vehicles and home energy was described in [14] based on energy costs and comfort levels. The authors only considered the influence of a detailed controllable load (electric vehicle) optimization on user comfort level, while there are multiple controllable loads in smart communities. Therefore, it is necessary to classify controllable loads and propose a universal comfort evaluation method. In this paper, the controllable loads in smart communities were divided into transferable loads and interruptible loads, and the coefficients of user comfort levels relating to transferable and interruptible loads were proposed according to their respective characteristics, to optimize the load-scheduling of the smart community.

With the rapid development of clean energy technologies, how to improve the consumption rate of renewable energy has become a research hotspot of power system optimization. To enhance a network's peak load regulation capacity and improve system efficiency, an additional portable energy system and a heat storage tank were introduced in [15] to optimize the operation of a distributed network. This study significantly improved the stability of network operation, but it ignored renewable energy consumption rate. Based on the state-queuing model, a renewable energy output tracking control algorithm was put forward in [16] to consume renewable energy. To maximize the direct self-consumption of photovoltaic power, [17] proposed a new method to determine the power generation of photovoltaic generators based on the cost-competitiveness. A logarithmic mean divisia index method was proposed in [18], proving that urbanization had a positive effect on the growth of renewable energy consumption. All the aforementioned articles promote the consumption rate of renewable energy. However, there was no quantitative analysis of the impact of promoting renewable consumption rate on energy costs. This paper compared and analyzed the cost changes under different renewable energy consumption rates and revealed the relationship between the renewable energy consumption rate and the cost change.

To make full use of demand-side resources to participate in power system interaction programs, many experts have carried out studies on tariff policies such as time-of-use tariff. The authors of [19,20] put forward optimization methods of scheduling controllable loads according to the electricity price policy. Yu et al. proposed a new operation method of risk-averse to adjust hybrid power generation with the purpose of dealing with price fluctuations in the power market and saving electricity costs for industrial users [21]. Nojavan et al. developed a model relating to user demand response under the time-of-use tariff policy, and discussed the impact of time-of-use tariff on load-scheduling and electricity purchase [22]. The authors of [23] proposed an EV impact analysis approach and analyzed the impact of EV charging on distributed network under time-of-use pricing. The above articles, in terms of the influence of electricity price policy upon power grid and users, played an important guiding role in reducing electricity costs and improving system stability. It is worth noting that with the rapid development of power generation technologies in recent years, many fuel cells and diesel generators fueled by natural gas and diesel emerge on the demand side, causing price fluctuations of gas and diesel. However, there is a lack of research on power system optimization based on gas price or diesel price. Therefore, this paper proposed a fuel cell operation optimization strategy considering time-of-use gas price.

The main contributions of this paper can be summarized as follows:

(1) Controllable loads were classified as transferable loads and interruptible loads, and then the coefficients of user comfort level relating to transferable loads and interruptible loads were defined, respectively. Additionally, the influences of load-shifting and interruption on user comfort level were studied.

(2) Based on different renewable energy consumption rates, daily energy costs of the smart community were optimized, which revealed how the renewable energy consumption rate influenced the daily energy costs and user comfort level.

(3) Considering the influence of user comfort level and renewable energy consumption rate on system optimization, the optimal operation strategy of the smart community was proposed based on time-of-use gas price, and subsequently verified by the case study.

2. System Definition and Its Modelling

2.1. System Definition

Compared with a traditional community, a smart community integrates the most advanced technologies of information, communication, and renewable energy, and these technologies enable the smart community to interlace information with energy generation system, storage system, and loads. In addition, a clear classification of electrical loads and proper configuration of energy storage system capacity can help smart communities optimize power flow in multiple time scales.

In this paper, a distributed generation system with photovoltaic, wind turbine, and fuel cells installation is selected as an example to show the typical structure of the distributed generation system. Additionally, lead-acid batteries are used as an energy storage system. Figure 1 is the structure diagram of a smart community.

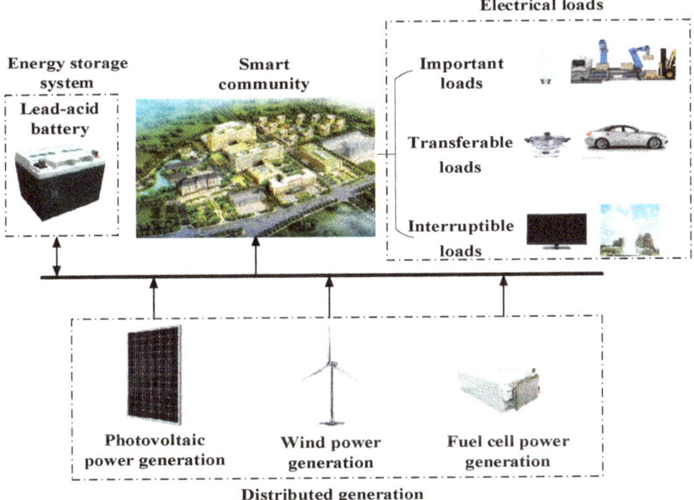

Figure 1. A typical structure of a smart community.

2.2. Modelling of Distributed Power Systems

To accurately balance power generation and consumption within smart communities, the mathematical output power models of the distributed power sources (including distributed photovoltaic, wind turbine power and fuel cell), distributed energy storage systems, and loads are developed in this part.

2.2.1. Modelling of Renewable Energy Generation Systems

As four primary types of renewable energy generation, solar generation, wind turbine power generation, geothermal generation, and tidal generation have their own remarkable advantages of less pollution and strong sustainability, and therefore are widely deployed in distributed smart communities to reduce carbon emissions in the process of power generation [24]. Considering the fact that solar energy generation and wind turbine power generation are easy to access and are complementary to each other, they are selected as examples to show the output power models of renewable energy generation systems.

- Modelling of Photovoltaic Generation

Photovoltaic generation is a kind of noise-free and pollution-free energy technology, which can be installed on roofs to reduce the footprint of power generation [25]. Since the output of photovoltaic generation is strongly affected by solar radiation intensity, ambient temperature, etc., its output power needs to be revised under the standard test condition (STC). The realistic output power of photovoltaic can be expressed as [26]:

$$P_{pv} = P_{STC}\frac{G_S}{G_{STC}}[1 + k(T_c - T_0)] \tag{1}$$

- Modelling of Wind Turbine Generation

Similarly to photovoltaic generation, wind turbine power generation also has the advantage of environmentally friendliness, and additionally it achieves much lower levelized cost of energy (LCOE). Because the output powers of photovoltaic generation and wind turbine power generation are complementary to each other, they can be installed as a package to compensate for the uncertainty of renewable generation.

Based on the operation characteristics, wind turbines can be divided into two categories, namely the fixed-speed generator and the variable-speed generator. Compared with the fixed-speed generator, the variable-speed generator is more cost efficient, stable, and lightweight. Therefore, it is preferable to install variable-speed generators in smart communities. The mathematical output power model of variable-speed generators can be expressed as follows [27]:

$$P_w = \begin{cases} 0, v < v_{ci}, v > v_{co} \\ av + b, v_{ci} \le v < v_r \\ P_0, v_r \le v \le v_{co} \end{cases} \tag{2}$$

where:

$$a = \frac{P_0}{v_r - v_{ci}} \tag{3}$$

$$b = -\frac{P_0 v_{ci}}{v_r - v_{ci}} \tag{4}$$

2.2.2. Modelling of Fossil Power Generation Systems

Fossil power generation systems have characteristics of high energy efficiency, flexible use, and low environmental dependency. They can be used to supply supplementary power for loads when renewable generation cannot meet loads. In this paper, the fuel cell is selected as an example to demonstrate the output power of a fossil power generation system [28].

$$P_{fc} = \eta_{fc} \cdot P_{fcin} \tag{5}$$

2.2.3. Modelling of Distributed Energy Storage Systems

Since renewable generation is the dominant form of power generation in most smart communities, the great uncertainty of renewable energy may break the balance of power generation and loads. To improve power quality and maintain system stability, an energy storage system is generally installed in smart communities.

The optimization of an energy storage system helps to boost system flexibility and reduce system loss of load probability (LOLP). As an important parameter that needs to be monitored, battery state of charge (SOC) represents the proportion of remaining battery capacity to battery installation capacity at the current time. Therefore, the mathematical expression of battery SOC can be represented by [29]:

$$SOC_{es}(t) = (1 - \delta) \cdot SOC_{es}(t-1) - \frac{P_{es}(t) \cdot \eta_{es}(t) \cdot \Delta t}{E_{es}} \tag{6}$$

2.2.4. Modelling of Loads

According to the importance of user loads, electrical demands can be classified into three types: important loads, transferable loads, and interruptible loads. Among them, important loads, which include basic lighting, production equipment, etc., must be satisfied in all operation conditions, and cannot be removed or shifted. However, transferable loads, which include residential electric cookers, electric vehicles etc., can be transferred from one time period to another without changing total power consumption. Finally, interruptible loads, including air conditioning and electrical heating systems, can be cut off for some time according to system scheduling plans. Since important loads have no effects on user comfort levels, this paper focuses on modelling the influences of transferable loads and interruptible loads on user comfort levels.

The length of operation time can be different for different types of transferable loads. However, for a specific transferable load, the amount of energy consumption cannot be changed within an optimization period. When scheduling transferable loads, it is not only necessary to consider current transferable loads, but also the number of loads shifted in/out to the current/next period. Therefore, the net transferable loads at time t can be expressed as:

$$H_{TLN}(t) = H_{TLC}(t) + H_{TLI}(t) - H_{TLO}(t) \tag{7}$$

where:

$$H_{TLC}(t) = \sum_{i=1}^{N} P_{TLCi}(t) \tag{8}$$

$$H_{TLI}(t) = \sum_{m=1}^{J} P_{TLI_m}(t) \tag{9}$$

$$H_{TLO}(t) = \sum_{n=1}^{K} P_{TLOn}(t) \tag{10}$$

When renewable power generation is less than the predicted value or system spinning reserve is insufficient to cover all electrical loads, some of interruptible loads should be cut off to ensure network safety operation. Total electrical loads that need to be cut off at time t can be expressed as:

$$H_{IL}(t) = \max\{\Delta P_{r\_d}(t) - f_{sr\_u}(t), 0\} \tag{11}$$

As demonstrated in the introduction, most previous studies focus on minimizing user energy costs without considering customer comfort level. A few studies may consider user comfort level when optimizing user energy costs, but in this process, they do not clearly specify the types of loads. In the following parts, two coefficients of customer comfort level relating to transferable loads and interruptible loads are defined separately.

For transferable loads, user comfort level is directly related to the time interval of load-shifting. If the time interval of load-shifting is shorter, less impact will be given to users and higher comfort can be experienced by users. Therefore, the coefficient of user comfort level relating to transferable loads (φ_{TL}) is defined as follows:

$$\varphi_{TL} = \frac{|t_s - t_\alpha|}{t_\beta - t_\alpha} \tag{12}$$

In contrast to transferable loads, interruptible loads can be directly cut off at peak demand time and do not need to make up for it at low demand time. For interruptible loads, user comfort level is mainly affected by the actual interruption power. For a given amount of interruptible energy, if it is removed in a short period (i.e., the interruption power is high), user comfort level can be significantly affected. By contrast, if it is removed for a relatively long period of time (i.e., the interruption power is lower), user comfort level be will less affected. In other words, it is preferable to cut off interruptible loads with lower power and longer period of time. In summary, the coefficient of user comfort level relating to interruptible loads (φ_{IL}) can be expressed as:

$$\varphi_{IL} = \begin{cases} 0 & U_{IL} = 0 \\ \sum_{i=1}^{T_{IL}} \left(\frac{P_{IL_i}\Delta t}{U_{IL}} \right)^2 & U_{IL} \neq 0 \end{cases} \tag{13}$$

3. System Optimization Strategy

To balance the calculation speed and accuracy of simulation results, this paper takes day-ahead planning to optimize power flow within the smart community. Day-ahead planning focuses on optimizing daily energy costs on the time scale of 24 h.

3.1. Objective Function

This paper selects daily energy costs as the objective to optimize power flow within the smart community. For a smart community, daily energy costs mainly include two parts, which are the costs of renewable generation and fuel cell. Therefore, the objective function can be expressed as:

$$\min C = \min \sum_{t=1}^{T} \sum_{i=1}^{M} C_i(P_i(t))\Delta T \tag{14}$$

The operation costs of a distributed fuel cell generally include fuel costs, maintenance costs and emission penalty costs, which can be listed as follows:

$$C_{fc} = C_{fuel} + C_{op} + \sum_{y=1}^{Y} C_{yen} \tag{15}$$

3.2. Constraints

- Power Conservation

To improve the reliability of energy supply and avoid system loss of load, energy generation must be greater than community electrical demands all the time. Moreover, since the distance of power transmission within the community is relatively short, electricity transmission loss is neglected in this paper. Equation 16 shows the constraint of power conservation [30]:

$$P_{pv}(t) + P_w(t) + P_{fc}(t) + P_{es}(t) = P_{loads}(t) \tag{16}$$

- Fuel Cell Operation Constraints

Fuel cell, as a typical fossil generator, is constrained by its rated maximum output power and minimum output power, which can be written as [31]:

$$P_{fcmin} < P_{fc}(t) < P_{fcmax} \tag{17}$$

In addition, the operation of a fuel cell is also limited by its ramp rate and can be described as [31]:

$$- \Delta P_{fcdmax} < P_{fc}(t+1) - P_{fc}(t) < \Delta P_{fcumax} \tag{18}$$

- Battery Operation Constraints

To maintain battery safety operation, its charging/discharging power needs to be limited within a certain range, which can be written as [29]:

$$- P_{escmax} < P_{es}(t) < P_{esdmax} \tag{19}$$

Moreover, the energy storage system SOC needs to be constrained within a certain range to extend the life of battery. Therefore, the inequality constraint of SOC can be expressed as [29]:

$$SOC_{esmin} < SOC_{es}(t) < SOC_{esmax} \tag{20}$$

- User Comfort Level Constraints

Considering the fact that the reduction of interruptible loads and the shift of transferable loads have direct influence on user comfort level, it is necessary to set the threshold of the coefficients of user comfort level relating to the interruptible and transferable loads.

$$0 < \varphi_{IL} \leq \varphi_{ILmax} \tag{21}$$

$$0 < \varphi_{TL} \leq \varphi_{TLmax} \tag{22}$$

- Renewable Energy Consumption Rate Constraints

To make full use of renewable energy generator installation capacity and reduce carbon emissions, renewable energy consumption rate needs to be designed.

$$\gamma_{min} < \gamma < \gamma_{max} \tag{23}$$

In this paper, the Particle Swam Optimization (PSO) is selected to optimize daily energy costs of the aforementioned smart community under the equality and inequality constraints proposed in this section. Figure 2 shows the flow diagram of the proposed PSO-based optimization model.

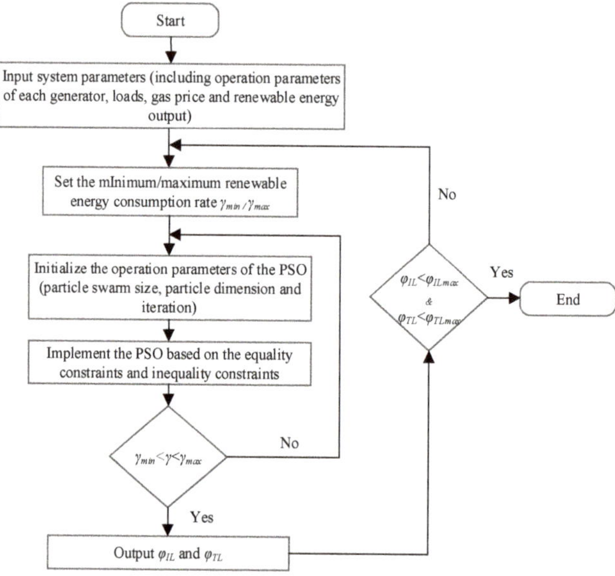

Figure 2. The flow diagram of the proposed operation model.

4. Case Study

In this paper, a smart community (shown in Figure 1) located in the north of China is selected as an example to verify the proposed operation strategy. In this community, the installation capacity of the distributed photovoltaic generation system is 5 MW, and the rated power of the distributed wind turbine generation system is 2.5 MW. To improve the power consumption rate of renewable generation and avoid system loss of loads, a 3 MWh lead-acid battery is installed as the energy storage system. Meanwhile, the smart community is equipped with a 3 MW fuel cell as a backup power generator and its energy conversion efficiency is 40%. Table 1 shows the rated parameters of the fuel cell and energy storage system. Figure 3 reveals the initial load curve and the output curves of photovoltaic and wind turbine power.

Table 1. Rated parameters of the fuel cell and energy storage system.

Distributed Generators	Rated Parameters	Value
Fuel cell	Ramp-up rate (MW/h)	1.5
	Maximum discharge power (MW)	2
Energy storage	Maximum charging power (MW)	1.5
	Overall efficiency	80%
	SOC	20~90%

Because of the irregular use of nature gas [32], this paper takes time-of-use gas price to optimize system operation cost. The time-of-use gas price is shown in Table 2. In addition, the costs of photovoltaic and wind turbine power generation are set as 10 and 14 US cents/kWh [33].

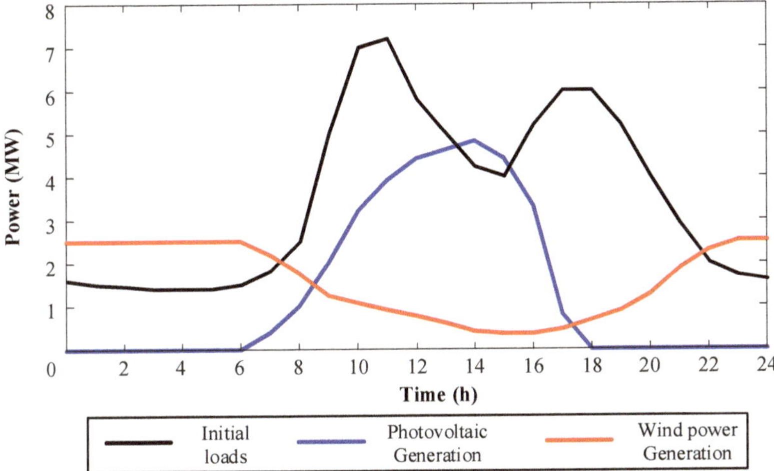

Figure 3. Daily load curve and the output curves of photovoltaic and wind turbine power.

Table 2. Daily time-of-use gas price.

Natural Gas Price ($/m^3)	Time Period
0.45	23:00–05:00; 13:00–15:00
0.6	05:00–09:00; 15:00–17:00; 20:00–23:00
0.75	09:00–13:00; 17:00–20:00

5. Results and Analysis

5.1. The Influence of Load-Scheduling on User Comfort Level

To show the influence of load-scheduling on user comfort levels, Figure 4 shows the optimal loads scheduling results under different comfort levels.

By ignoring user comfort level (shown in Figure 4, Case 1), the optimal daily energy cost reduces to $8894.90. In this case, three types of transferable loads are shifted to other time periods and the interruptible loads are largely cut off between 10:00–12:00 and 16:00–19:00. To be specific, the operation time periods of transferable loads are shifted to a high photovoltaic output period or a lower gas price period. For interruptible loads, they are mainly cut off in a period of higher gas price. By reducing interruptible loads and transferable loads during peak gas price time, Case 1 minimizes users' energy costs. However, this method neglects user comfort levels, which results in a relatively long load-shifting time period for transferable loads and high interruption power for interruptible loads. Simulation results show that φ_{TL} and φ_{IL} are 28.5 and 0.182, if user comfort levels are ignored.

Case 2 shows the optimization results of load-scheduling with the restriction of user comfort level. In this case, the coefficients of user comfort level relating to the transferable loads and interruptible loads are set as 8.5 and 0.150, respectively. It can be seen from Case 2 that the transferable load TLa shifted its demands from the period of 16:00–19:00 to the period of 13:00–16:00. Compared with Case 1, the length of the load-shifting period reduces from 16 h to 3 h. In addition, the length of the load-shifting periods for TLb and TLc are also shortened to some extent. For interruptible loads, the interruption power is much lower compared with Case 1, but the length of time that needs to cut off loads is extended. Simulation results show that the optimal daily energy cost is $9055.00 in this case, which gives a 1.8% of increase compared with Case 1. However, this optimization method significantly enhances user comfort level and is more practical in engineering applications.

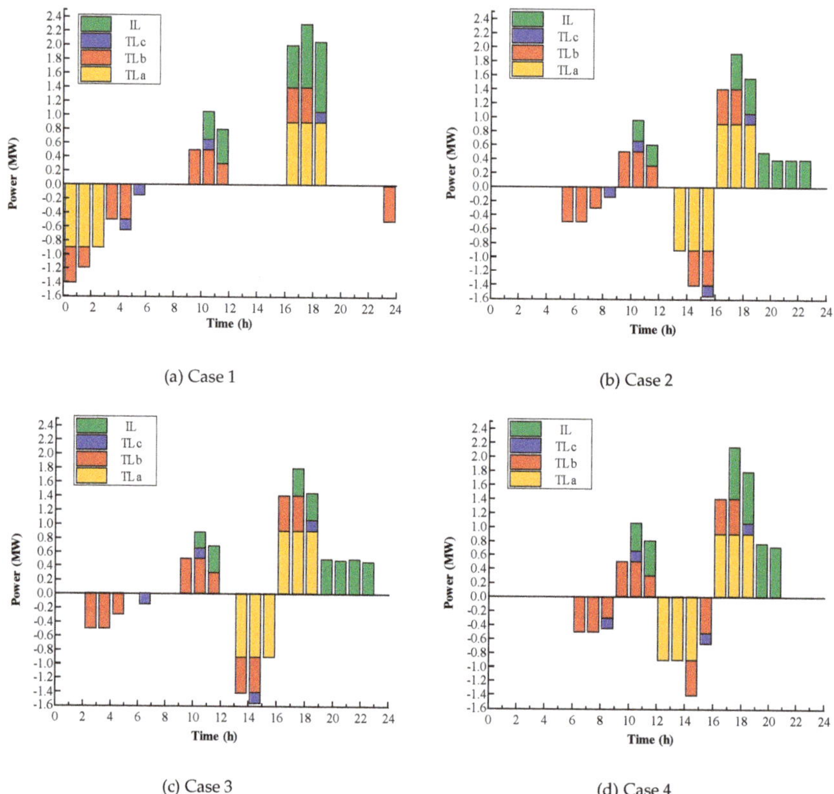

(a) Case 1

(b) Case 2

(c) Case 3

(d) Case 4

Figure 4. Optimal load-scheduling results under different comfort levels. (IL represents interruptible loads; TLa, TLb, and TLc represent three types of transferable loads; positive power represents the amount of active power shifted out/cut off and negative power represents the amount of active power shifted in).

The optimization results of Case 3 and Case 4 are obtained by increasing the coefficients of user comfort level relating to the transferable loads and interruptible loads based on Case 2 (the threshold of φ_{TL} in Case 3 are increased to 13 and the threshold of φ_{IL} in Case 4 are increased to 0.170). It can be concluded from Figure 4c,d, the length of the load-shifting period is extended in Case 3 compared with Case 2 and the average interruption power in Case 4 is increased compared with Case 2. The results show that the optimal daily energy costs in Case 3 and Case 4 are \$8976.70 and \$9018.40, which are 0.9% and 0.4% decreases compared with Case 2. Simulation results indicate that the increase of the coefficients of user comfort level can reduce users' energy costs, which further proves the inverse proportional relationship between user energy costs and the coefficients of comfort levels. It is worth noting that the optimal results of Case 2 and Case 3 are not same in terms of interruptible loads reduction, given that the coefficient of user comfort level relating to interruptible loads stays the same in Case 3. This is because scheduling transferable loads and scheduling interruptible loads are not independent of each other. Similarly, changing the coefficient of user comfort level relating to interruptible loads will also affect the optimization results of transferable loads.

5.2. The Influence of Increasing Renewable Energy Consumption Rate on User Comfort Level

Simulation results show that when only selecting the daily energy costs as the criterion, the renewable energy consumption rate in the aforementioned Case 1 is 82.5%, which is about 7.5% lower than the minimum requirement of the renewable energy consumption rate published by Chinese National Energy Administration. Therefore, to reveal the impact of increasing renewable energy consumption rate on user comfort level, this part presents further research.

Figure 5 shows the optimal daily load curves (including transferable loads, interruptible loads, and important loads) of the proposed community, given that the minimum renewable energy consumption rates are set as 82.5%, 85%, 90%, 95%, and 100%, respectively. It can be seen from Figure 5 that with the increase of the renewable energy consumption rate, the demands increase during the maximum PV output period (13:00–14:00). When the renewable energy consumption rate reaches 100%, the peak demand time period will shift from 18:00–19:00 to 13:00–14:00, exerting a significant impact on user comfort levels.

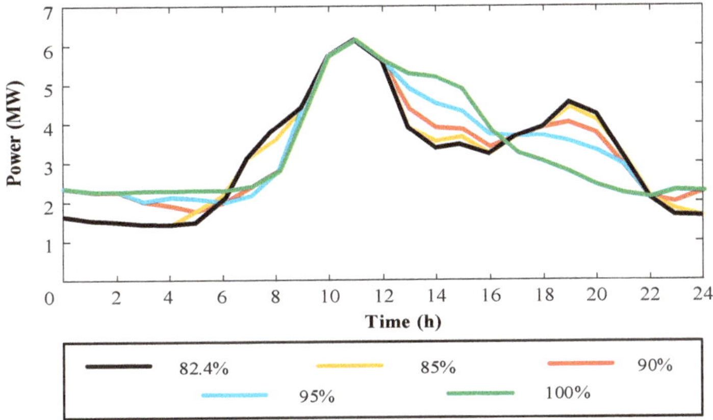

Figure 5. Optimal daily load curves of the proposed community under different renewable energy consumption rates.

Table 3 shows the optimization results of the system operating parameters under different renewable energy consumption rates. It can be concluded that when the renewable energy consumption rate increases from 82.5% to 85%, 90%, and 95%, daily energy costs can increase about 0.23%, 0.33%, and 0.74%, respectively. In these situations, user average daily energy costs almost remain the same and user comfort levels will not be strongly affected. However, when the renewable energy consumption rate reaches 100%, daily energy costs increase to $9295.20, which is about 3.16% of increase. In addition, compared with the case of 95% of renewable energy consumption, the coefficients of user comfort level relating to the transferable and interruptible loads increase significantly as well. Simulation results show that the coefficients of user comfort level relating to the transferable loads (φ_{TL}) and interruptible loads (φ_{IL}) increase from 31.5 and 0.193 to 36.5 and 0.213, respectively. The growth rates of these two parameters are more than 10%, which is mainly caused by the mismatch between renewable energy generation and community demands.

To completely consume renewable energy generated during the period of 13:00–14:00, an excessive number of transferable loads need to be shifted to this period.

Table 3. Optimization results of system operating parameters.

Renewable Energy Consumption Rate	Average Daily Energy Costs ($)	User Comfort Level		Energy Storage System	
		φ_{TL}	φ_{IL}	Range of SOC	Total Energy Interaction (MWh)
82.5%	8894.9	28.5	0.182	20%–88.7%	3.92
85%	8915.4	28.5	0.184	20%–88.1%	3.98
90%	8944.7	29	0.187	20%–87.3%	4.11
95%	9010.5	31.5	0.193	20%–85.9%	4.37
100%	9295.2	36.5	0.213	20%–80.2%	5.31

To improve the renewable energy consumption rate, one possible solution is to shift loads, or alternatively to store redundant electricity generated by the renewable energy sources in energy storage systems. However, since the loss of energy storage system is relatively high (20%), inappropriate use of energy storage can lead to the decrease of battery lifetime and the increase of daily energy costs. As can be seen from Table 3, the operation range of battery SOC in a day does not change significantly during the process of increasing the renewable energy consumption rate from 82.5% to 95%. In this process, the maximum SOC of battery decreases from 88.7% to 85.9%, which is 3.2 percent of reduction. However, when the renewable energy consumption rate increases to 100%, the operation range of battery SOC decreases to 20%–80.2%. In this situation, the maximum SOC of battery reduces by 9.6%. In addition, it can be concluded from Table 3 that when the renewable energy consumption rate is lower than 95%, total energy interaction between the energy storage system and the smart community is almost the same. While, when the renewable energy consumption rate increases from 95% to 100%, total daily energy interaction between the energy storage system and the smart community increases rapidly, which is about 21.5% increase. This indicates that when fully consume renewable energy, 0.19 MWh extra electricity will be generated to compensate the loss of energy storage system compared to the case of 95% of renewable energy consumption. Therefore, fully consuming renewable energy will lead to the decrease of battery lifetime and the increase of users' daily energy costs.

5.3. Optimal Load-Scheduling Results with Considering Comfort Level and Renewable Consummation Rate

To avoid the serious impacts of load-scheduling on user comfort levels and avoid the reduction of renewable energy consumption rate caused by large-scale installation of renewable energy sources, this part demonstrates system optimal operation results under the conditions of $\varphi_{TL} < 15$, $\varphi_{IL} < 0.17$ and $90\% < \gamma < 95\%$. Figure 6 shows the optimal scheduling results of transferable loads and interruptible loads under the aforementioned constraints.

It can be concluded from Figure 6 that compared with the optimization results of Figure 4a (only considering energy costs), the demands of transferable loads, shown in this part, increase about 2.93 MWh during the period of low renewable energy consumption rate (2:00–7:00 and 13:00–16:00). In addition, during the period of the maximum renewable generation (13:00–14:00), the demands of transferable loads are the highest, which are 1.54 MWh. In this situation, the coefficients of user comfort level relating to transferable loads and interruptible loads are 14.8 and 0.165, which is 48.1% and 9.3% reduction compared with results of Figure 4a. Figure 7 shows the output power of all generators and the operation states of the energy storage system taking user comfort levels and renewable energy accommodation rate into account.

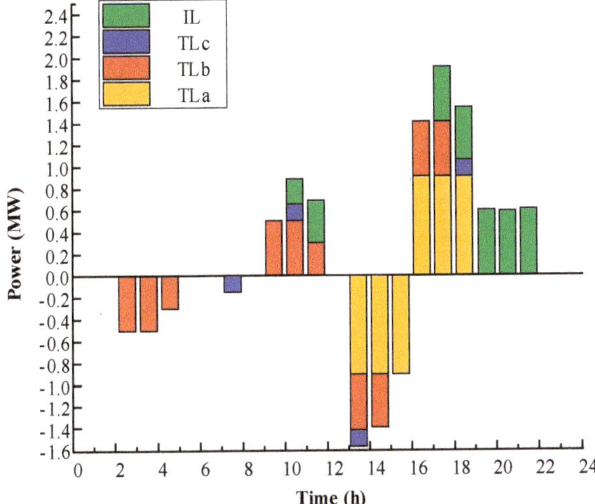

Figure 6. Optimal load-scheduling results of transferable and interruptible loads given $\varphi_{TL} < 15$, $\varphi_{IL} < 0.17$ and $90\% < \gamma < 95\%$.

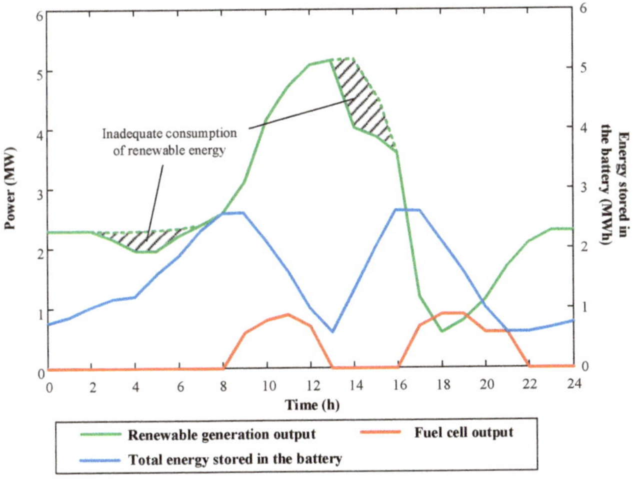

Figure 7. Optimization results of renewable generation, fuel cell generation, and operation state of battery.

As can be seen from Figure 7, the fuel cell, working as a backup power source, generates 6.79 MWh electricity during the periods of 8:00–13:00 and 16:00–22:00, which accounts for 8.2% of the total electrical demands. In addition, the energy storage system has two complete charge/discharge cycles in a day and the operation range of battery of SOC is between 20% and 86.8%. To be specific, the SOC of energy storage system starts at 25%, and reaches its maximum value of 86.8% from 16:00–17:00. Moreover, from 2:00–7:00 and 13:00–16:00 (shadows in Figure 7), renewable energy generation is not sufficiently consumed by the smart community. Optimization results shows that the optimal daily energy costs of the community are $9023.10 when the renewable energy consumption rate reaches 91%. Compared to Case 1 of Section 5.1, the renewable energy generation rate increases by 10.3%, with only a small cost of 1.4% of energy costs increase. It should be noted that the energy storage system is in charging state when the renewable energy consumption is insufficient, because the loads are less than the renewable power generation in these periods. To ensure that more than 90% of the renewable generation is consumed by community, it is necessary to make full use of the energy storage system to store redundant electricity.

6. Conclusions

To further improve the renewable energy consumption rate and save energy costs in smart communities, this paper proposes a novel load-scheduling method for optimizing a smart community's daily energy costs while taking user comfort level and the renewable energy consumption rate into consideration. Then, by implementing the PSO algorithm, the proposed strategy schedules the transferable loads and interruptible loads, optimizes the output of each renewable generator and controls the state of energy storage system. Finally, a case study is developed to validate the effects of improving user comfort levels and the renewable energy consumption rate on daily energy costs. Meanwhile, mathematical optimization results are analyzed in detail. Three main findings of this paper can be concluded as follows:

(1) To reduce daily energy costs in a smart community, it must increase the coefficients of user comfort level relating to transferable loads and interruptible loads, and this will lead to a decrease in user comfort. Therefore, it is important to consider user comfort level when scheduling loads.

(2) Compared with the case of 95% of renewable energy consumption, fully consuming renewable energy can increase daily energy costs, the coefficient of user comfort level relating to transferable loads, and the coefficient of user comfort level relating to interruptible loads by as much as 3.16%, 15.9%, and 10.4%, respectively. This proves that the excessive increase of renewable energy consumption rate will result in the increase of electricity cost and the decrease of user comfort.

(3) For the given constraints of renewable energy consumption rate and user comfort levels, a smart community can acquire minimum operation costs if the system renewable energy consumption rate, the coefficient of user comfort level relating to transferable loads, and the coefficient of user comfort level relating to interruptible loads are 91%, 14.8%, and 0.165%, respectively. In this case, user comfort level and renewable energy consumption rate increase significantly while the community's daily energy costs only increase by 1.4%.

Author Contributions: Conceptualization, Y.S.; Methodology, M.D.; Validation, F.G.; Formal Analysis, D.L.; Writing-Original Draft Preparation, Y.S.; Writing-Review & Editing, T.G.; Funding Acquisition, K.S.

Funding: This research was funded by the project of "research and validation on multi-user energy supply and demand interactive simulation model in smart community" of CEPRI, grant number "YD83-18-002".

Conflicts of Interest: The authors declare no conflict of interest.

Nomenclature

Nomenclature	Meaning	Nomenclature	Meaning
P_{pv}	Actual output power of PV generation	P_{STC}	Maximum output power under the standard test condition
G_s	Actual solar radiation intensity	G_{STC}	Solar radiation intensity under the standard test condition
k	A coefficient of temperature	T_c	Cell temperature
T_0	Reference ambient temperature	P_w	Actual output power of a wind turbine system
P_0	Rated power of a wind turbine system	v_{ci}	Cut-in speed of a wind turbine system
v_r	Rated wind speed of a wind turbine system	v_{co}	Cut-out speed of a wind turbine system
v	Actual wind speed of a wind turbine system	a, b	Power coefficients of a wind turbine system
P_{fc}	Output power of a fuel cell generator	P_{fcin}	Input power of a fuel cell generator
η_{fc}	Power generation efficiency of a fuel cell generator	$SOC_{es}(t)$	Energy storage system state of charge at time t
δ	Self-discharge efficiency of an energy storage system	$P_{es}(t)$	Charging/discharging power of an energy storage system at time t
$\eta_{es}(t)$	Energy storage system charging/discharging efficiency at time t	$\triangle t$	Time interval
E_{es}	Installation capacity of an energy storage system	$H_{TLN}(t)$	Net transferable loads at time t

$H_{TLC}(t)$	Original transferable loads at time period t	$H_{TLI}(t)$	Transferable loads shifted into current time period t
$H_{TLO}(t)$	Transferable loads shifted out to next period at time t	N	Total number of transferable loads
$P_{TLCi}(t)$	The ith transferable load's original electrical demands at time t	J	Total number of transferable loads shifted into current time t
$P_{TLIm}(t)$	The mth transferable load's power shifted into current time t	K	Total number of loads shifted out to next period at time t
$P_{TLOn}(t)$	The nth transferable load's power shifted out to next period at time t	$H_{IL}(t)$	Total electrical loads need to be cut off at time t
$\triangle P_{r\_d}(t)$	Difference between predicted and actual renewable power generation at time t	$f_{sr\_u}(t)$	System up spinning reserve capacity
φ_{TL}	Coefficient of user comfort level relating to the transferable loads	t_s	Actual start time of the transferable loads
t_α	Expected start time	t_β	Expected end time
φ_{IL}	Coefficient of user comfort level relating to the interruptible loads	T_{IL}	Number of time periods existing interruption loads
P_{ILi}	Actual interruption power in the ith time period	U_{IL}	Total amount of actual interruption energy
C	Daily energy costs of a smart community	T	Total number of time period
M	Number of distributed power generators	C_i	Energy costs of the ith renewable generation
$P_i(t)$	Output power of the ith renewable generator at time t	$\triangle T$	Scheduling time period
C_{fc}	Energy costs of a fuel cell generator	C_{fuel}	Fuel costs
C_{op}	Maintenance costs	Y	Total types of pollutant
C_{yen}	Penalty costs of the yth pollutant	$P_{pv}(t)$	Actual output power of PV generation at time t
$P_w(t)$	Actual output power of a wind turbine system at time t	$P_{fc}(t)$	Actual output power of a fuel cell generator at time t
$P_{loads}(t)$	Total amount of loads of a smart community at time t	P_{fcmin}	Minimum output power of a fuel cell generator
P_{fcmax}	Maximum output power of a fuel cell generator	$\triangle P_{fcdmax}$	Maximum fuel cell down ramp rate
$\triangle P_{fcumax}$	Maximum fuel cell up ramp rate	P_{escmax}	Maximum charging power of an energy storage system
P_{esdmax}	Maximum discharging power of an energy storage system	SOC_{esmin}	Lower limit of energy storage SOC
SOC_{esmax}	Upper limit of energy storage SOC	φ_{ILmax}	Maximum coefficient of user comfort level relating to the interruptible loads
φ_{TLmax}	Maximum coefficient of user comfort level relating to the transferable loads	γ	Renewable energy consumption rate
γ_{min}	Minimum renewable energy consumption rate	γ_{max}	Maximum renewable energy consumption rate

References

1. Li, X.; Lu, R.; Liang, X.; Shen, X.; Chen, J.; Lin, X. Smart community: An internet of things application. *IEEE Commun. Mag.* **2011**, *49*, 68–75. [CrossRef]
2. China Industrial Research Network. 2018 Development Investigation and Development Trend Analysis Report of China Smart Community. Available online: https://www.cir.cn/R_ITTongXun/25/ZhiHuiYuanQuDeXianZhuangHeFaZhanQuShi.html (accessed on 26 January 2019).
3. Dewit, A. Chapter 21—Japanese Smart Communities as Industrial Policy. In *Sustainable Cities & Communities Design Handbook*; Elsevier: Amsterdam, The Netherlands, 2018.
4. Smart and Connected Communities Framework 2015. Available online: https://www.nitrd.gov/sccc/materials/scccframework.pdf (accessed on 26 January 2019).
5. Liang, Y.; Yu, B.; Wang, L. Costs and benefits of renewable energy development in China's power industry. *Renew. Energy* **2019**, *131*, 700–712. [CrossRef]
6. Sullivan, J.L.; Lewis, G.M.; Keoleian, G.A. Effect of mass on multimodal fuel consumption in moving people and freight in the U.S. *Transp. Res. Part D Transp. Environ.* **2018**, *63*, 786–808. [CrossRef]
7. Khuong, P.M.; McKenna, R.; Fichtner, W. Analyzing drivers of renewable energy development in Southeast Asia countries with correlation and decomposition methods. *J. Clean. Prod.* **2019**, *213*, 710–722. [CrossRef]
8. Zhong, S.; Niu, S.; Wang, Y. Research on Potential Evaluation and Sustainable Development of Rural Biomass Energy in Gansu Province of China. *Sustainability* **2018**, *10*, 3800. [CrossRef]
9. Liu, Y.; Hu, S. Renewable Energy Pricing Driven Scheduling in Distributed Smart Community Systems. *IEEE Trans. Parallel Distrib. Syst.* **2017**, *28*, 1445–1456. [CrossRef]
10. Huang, Z. Evaluating intelligent residential communities using multi-strategic weighting method in China. *Energy Build.* **2014**, *69*, 144–153. [CrossRef]

11. Barone, G.; Brusco, G.; Burgio, A.; Menniti, D.; Pinnarelli, A.; Motta, M.; Sorrentino, N.; Vizza, P. A Real-Life Application of a Smart User Network. *Energies* **2018**, *11*, 3504. [CrossRef]
12. Mohsenian-Rad, A.H.; Leon-Garcia, A. Optimal Residential Load Control with Price Prediction in Real-Time Electricity Pricing Environments. *IEEE Trans. Smart Grid* **2010**, *1*, 120–133. [CrossRef]
13. Chavali, P.; Yang, P.; Nehorai, A. A Distributed Algorithm of Appliance Scheduling for Home Energy Management System. *IEEE Trans. Smart Grid* **2014**, *5*, 282–290. [CrossRef]
14. Nguyen, D.T.; Le, L.B. Joint Optimization of Electric Vehicle and Home Energy Scheduling Considering User Comfort Preference. *IEEE Trans. Smart Grid* **2014**, *5*, 188–199. [CrossRef]
15. Zhang, H.; Zhang, Q.; Gong, T.; Sun, H.; Su, X. Peak Load Regulation and Cost Optimization for Microgrids by Installing a Heat Storage Tank and a Portable Energy System. *Appl. Sci.* **2018**, *8*, 567. [CrossRef]
16. Wu, X.; Liang, K.; Han, X. Renewable Energy Output Tracking Control Algorithm Based on the Temperature Control Load State-Queuing Model. *Appl. Sci.* **2018**, *8*, 1099. [CrossRef]
17. Talavera, D.L.; Muñoz-Rodriguez, F.J.; Jimenez-Castillo, G.; Rus-Casas, C. A new approach to sizing the photovoltaic generator in self-consumption systems based on cost–competitiveness, maximizing direct self-consumption. *Renew. Energy* **2018**, *130*, 1021–1035. [CrossRef]
18. Yang, J.; Zhang, W.; Zhang, Z. Impacts of urbanization on renewable energy consumption in China. *J. Clean. Prod.* **2015**, *114*, 443–451. [CrossRef]
19. Materassi, D.; Bolognani, S.; Roozbehani, M.; Dahleh, M.A. Optimal Consumption Policies for Power-Constrained Flexible Loads Under Dynamic Pricing. *IEEE Trans. Smart Grid* **2015**, *6*, 1884–1892. [CrossRef]
20. Xu, Z.; Callaway, D.S.; Hu, Z.; Song, Y. Hierarchical Coordination of Heterogeneous Flexible Loads. *IEEE Trans. Power Syst.* **2016**, *31*, 4206–4216. [CrossRef]
21. Yu, D.; Liu, H.; Bresser, C. Peak load management based on hybrid power generation and demand response. *Energy* **2018**, *163*, 969–985. [CrossRef]
22. Nojavan, S.; Ghesmati, H.; Zare, K. Robust optimal offering strategy of large consumer using IGDT considering demand response programs. *Electr. Power Syst. Res.* **2016**, *130*, 46–58. [CrossRef]
23. Dubey, A.; Santoso, S. Electric Vehicle Charging on Residential Distribution Systems: Impacts and Mitigations. *IEEE Access* **2015**, *3*, 1871–1893. [CrossRef]
24. Wang, D.D.; Toshiyuki, S. Climate change mitigation targets set by global firms: Overview and implications for renewable energy. *Renew. Sustain. Energy Rev.* **2018**, *94*, 386–398. [CrossRef]
25. Abdelsamad, S.F.; Morsi, W.G.; Sidhu, T.S. Probabilistic Impact of Transportation Electrification on the Loss-of-Life of Distribution Transformers in the Presence of Rooftop Solar Photovoltaic. *IEEE Trans. Sustain. Energy* **2015**, *6*, 1565–1573. [CrossRef]
26. Mathew, D.; Rani, C.; Rajesh Kumar, M.; Wang, Y.; Binns, R.; Busawon, K. Wind-Driven Optimization Technique for Estimation of Solar Photovoltaic Parameters. *IEEE J. Photovolt.* **2018**, *8*, 248–256. [CrossRef]
27. Wang, G.; Tan, Z.; Tan, Q.; Yang, S.; Lin, H.; Ji, X.; Gejirifu, D.; Song, X. Multi-Objective Robust Scheduling Optimization Model of Wind, Photovoltaic Power, and BESS Based on the Pareto Principle. *Sustainability* **2019**, *11*, 305. [CrossRef]
28. Lindahl, P.A.; Shaw, S.R.; Leeb, S.B. Fuel Cell Stack Emulation for Cell and Hardware-in-the-Loop Testing. *IEEE Trans. Instrum. Meas.* **2018**, *67*, 214–2152. [CrossRef]
29. Ma, H.; Wang, B.; Gao, W.; Liu, D.; Sun, Y.; Liu, Z. Optimal Scheduling of an Regional Integrated Energy System with Energy Storage Systems for Service Regulation. *Energies* **2018**, *11*, 195. [CrossRef]
30. Mavrotas, G.; Karmellos, M. Multi-objective optimization and comparison framework for the design of Distributed Energy Systems. *Energy Convers. Manag.* **2019**, *180*, 473–495.
31. Vahidi, A.; Kolmanovsky, I.; Stefanopoulou, A. Constraint Handling in a Fuel Cell System: A Fast Reference Governor Approach. *IEEE Trans. Control Syst. Technol.* **2007**, *15*, 86–98. [CrossRef]

32. Li, K.; Yan, H.; He, G.; Zhu, C.; Liu, K.; Liu, Y. Seasonal Operation Strategy Optimization for Integrated Energy Systems with Considering System Cooling Loads Independently. *Processes* **2018**, *6*, 202. [CrossRef]
33. Liu, H.; Li, D.; Liu, Y.; Dong, M.; Liu, X.; Zhang, H. Sizing Hybrid Energy Storage Systems for Distributed Power Systems under Multi-Time Scales. *Appl. Sci.* **2018**, *8*, 1453. [CrossRef]

MDPI

St. Alban-Anlage 66

4052 Basel

Switzerland

Tel. +41 61 683 77 34

Fax +41 61 302 89 18

www.mdpi.com

Processes Editorial Office

E-mail: processes@mdpi.com

www.mdpi.com/journal/processes

9 783039 366828